Springer Proceedings in Earth and Environmental Sciences

The series Springer Proceedings in Earth and Environmental Sciences publishes proceedings from scholarly meetings and workshops on all topics related to Environmental and Earth Sciences and related sciences. This series constitutes a comprehensive up-to-date source of reference on a field or subfield of relevance in Earth and Environmental Sciences. In addition to an overall evaluation of the interest, scientific quality, and timeliness of each proposal at the hands of the publisher, individual contributions are all refereed to the high quality standards of leading journals in the field. Thus, this series provides the research community with well-edited, authoritative reports on developments in the most exciting areas of environmental sciences, earth sciences and related fields.

Tatiana Chaplina
Editor

Complex Investigation of the World Ocean (CIWO-2023)

Proceedings of the VII International Conference of Young Scientists

Springer

Editor
Tatiana Chaplina
Antalya, Türkiye

ISSN 2524-342X ISSN 2524-3438 (electronic)
Springer Proceedings in Earth and Environmental Sciences
ISBN 978-3-031-47850-5 ISBN 978-3-031-47851-2 (eBook)
https://doi.org/10.1007/978-3-031-47851-2

This Springer imprint is published by the registered company Springer Nature Switzerland AG
The registered company address is: Gewerbestrasse 11, 6330 Cham, Switzerland

Paper in this product is recyclable.

Introduction

The book presents the most relevant extended abstracts of the participants of the VII all-Russian Conference of Young Scientists “Complex Investigation of the World Ocean” (CIWO-2023), held at Saint Petersburg State University on May 15–19, 2023 (Saint Petersburg, Russia). The conference covers a wide range of fundamental and applied marine and limnology studies combined in eight sections: Ocean Physics, Ocean Biology, Ocean Chemistry, Marine Geology, Marine Geophysics, Marine Ecology and Environmental Management, Physical and Biological interactions (interdisciplinary section), Oceanological Technology and Instrumentation. The aim of this book is to show the relevance of the marine research due to the crucial role of the World Ocean in determining climate change on Earth, huge resources (fish resources, oil, gas and ore deposits, etc.) and intensive development of infrastructure in coastal and offshore zones. All these topics were marked within the framework of realization of the United Nations Decade of Ocean Science for Sustainable Development (2021–2030).

The studies presented in the book cover the wide spectrum of different and the most important marine and limnology issues: thermohaline structure of water body and interactions between ocean and atmosphere, dynamic of the ocean, marine ice in polar regions, biodiversity of the marine ecosystems, adaptation of marine life to climate changes, geological and geophysical investigations in oil and gas regions, sedimentation and biostratigraphy, hydrochemistry of estuary regions and carbon fluxes, microplastic pollution of the ocean, etc.

One of the areas of studies presented in the book is the Arctic region, and the changes taking place there are due to modern climate instability and growing anthropogenic impact. The Arctic theme is traditional for CIWO, and reports on this topic were presented in all eight sections of CIWO-2023. In total, young scientists presented about 100 reports on the Arctic topic, which accounted for more than 1/3 of all applications selected for participation in CIWO-2023. In this book, the papers are about different issues on the Arctic Basin including carbon fluxes and methane sources on the shelf, estimates of climate changes in seas, estimates of microplastic compounds in surface waters, quantity and quality assessments of hydrocarbons, inter-annual variability of sea ice extend, etc.

Contents

Ocean Physics

Climate assessment of water, heat and salt fluxes in the Pechora Sea 3
D. V. Deshova and S. M. Gordeeva

Assessment of Seasonal and Multi-year Variations in Ice Age Composition of Greenland Sea and Barents Sea 11
E. S. Egorova and Ye. U. Mironov

Laboratory Investigations of the Evolution of Baroclinic Eddies in a Two-Layer Rotating Fluid 19
D. N. Elkin and A. G. Zatsepin

Analysis of Hydrophysical and Energy Fields in the Northern Part of the Black Sea on the Basis of the Numerical Model With the Assimilation of the Hydrological Data in 2016 29
N. A. Evstigneeva and S. G. Demyshev

Water Structure and Dynamics in the Transform Fracture Zones of the Tropical Atlantic According to Instrumental Measurements 35
F. N. Gippius, A. N. Demidov, K. V. Artamonova, and S. B. Krasheninnikova

The Spatio-Temporal Variability of the Ocean-Atmosphere Heat Exchange and the Sea-Ice Cover in the Nordic Seas 42
A. S. Kaledina and I. L. Bashmachnikov

"1.5D" Vertical Turbulent Exchange Model Verification Based on Microstructural Probe Data on Cruise 122 of the R/V Professor Vodyanitsky 48
D. A. Kazakov, M. I. Pavlov, and A. M. Chukharev

Seasonality of Pacific Water Distribution in the Sea of Okhotsk Revealed by Lagrangian Approach 59
K. K. Kivva, M. V. Budyansky, M. Yu. Uleysky, and S. V. Prants

Turbulence Influence on the Thickness of the Mixed Layer in the Coastal Zone of the Black Sea 69
V. A. Korzhuev

Heat Fluxes in a Multilayer Atmosphere-Snow-Ice-Water Environment Using the Gulf of Finland as an Example 80
K. V. Kravtsova and N. A. Podrezova

Spatial Trend Analysis of Significant Wave Heights in the Caspian Sea 95
E. E. Kruglova and S. A. Myslenkov

An Influence of Meridional Oceanic and Atmospheric Heat Fluxes to the Atlantic Sector of the Arctic on Arctic Amplification 104
M. M. Latonin, I. L. Bashmachnikov, and L. P. Bobylev

Lagrangian Method for Studying Agulhas Leakage in the Water Mixing Area 111
A. A. Malysheva, T. V. Belonenko, and M. V. Budyansky

Wind Profile Evaluation by Using Result of Wave Forecast Model 118
D. D. Romashchenko and K. Yu. Bulgakov

Features of the Ice Extent Dynamics in the Western Sector of the Arctic on the Example of the Archipelagos of Svalbard and Franz Josef Land 126
N. O. Rudenko and B. V. Ivanov

Estimation of Ice Age Categories Composition in the East Siberian Sea at the End of Winter Period 135
M. V. Sharatunova and V. V. Ivanov

Surface Manifestations of Short-Period Internal Waves in the Sea of Okhotsk and the Kuril-Kamchatka Region of the Pacific Ocean 141
E. I. Svergun, A. V. Zimin, K. V. Kulik, and N. S. Frolova

Topographic Waves of the Aleutian Trench 150
S. P. Khudyakova, V. S. Travkin, and T. V. Belonenko

Probability Distributions for Finite Ensembles of Irregular Waves 160
A. S. Tregubov, E. G. Didenkulova, and A. V. Kokorina

Model of the Schooner "Saint Anna" Drift with the Usage of NOAA-CIRES-DOE 20CR and ERA-20C Reanalises 175
S. V. Tsedrik

The Spatial and Temporal Distribution Patterns of Icebergs in the Barents Sea Based on Satellite Radar Data 181
T. S. Ustinova

Small Eddies Observed in Ladoga and Onega Lakes Using SAR Data 191
A. V. Zimin, E. A. Blagodatskikh, and O. A. Atadzhanova

Antarctic Bottom Water in the Vema Channel ... 198
O. A. Zuev and A. M. Seliverstova

Ocean Biology

Trophic Ecology of Chaetognaths in the Barents Sea Revealed by Nitrogen Stable Isotope Analysis ($\delta^{15}N$) ... 207
G. M. Artemev, D. V. Zakharov, G. A. Batalin, B. I. Gareev, G. Z. Mingazov, R. M. Sabirov, and A. V. Golikov

Vertical Distribution of Viral Particles in the Deep Part of the Black Sea ... 217
M. A. Ufimtseva and V. S. Mukhanov

MtDNA Polymorphism of *Macrohectopus branickii* Dybowsky, 1974 (Amphipoda) – An Endemic Pelagic Key Species of Lake Baikal ... 223
I. Y. Zaidykov, E. Y. Naumova, and L. V. Sukhanova

Ocean Chemistry

Marine Carbonate System Parameters of the West Spitsbergen Fjords in Late Summer 2022 ... 233
N. K. Alekseeva, A. L. Nikulina, I. V. Ryzhov, A. E. Novikhin, R. V. Kornilova, N. A. Smirnov, and A. A. Fedorova

The Hydrochemical Composition of the Watercourses of the Bay Oga, Tzivolki, Sedova (The Coast of the Kara Sea, Novaya Zemlya) ... 239
G. V. Borisenko, A. A. Polukhin, and D. V. Sharmar

Characteristics of Bottom Sediments of the North-Eastern Part of the Black Sea Coastal Zone ... 245
Y. S. Gurova, K. I. Gurov, and N. A. Orekhova

Marine-Freshwater Ratio by Hydrochemical Proxies in the Surface Desalinated Layer of the Kara Sea ... 254
U. A. Kazakova and A. A. Polukhin

Evidence of Atlantic-Origin Water Shoaling in the Western Chukchi Sea ... 262
K. V. Kodryan and K. K. Kivva

Carbonate System and CO_2 Fluxes in the Partizanskaya River Estuary (Nakhodka Bay, Sea of Japan) ... 268
K. A. Kryzhova and P. Yu. Semkin

Results of the Carbonate System Features Study of the South-Eastern Part of the Baltic Sea in 2022 ... 280
A. A. Muratova, E. S. Bubnova, and A. A. Polukhin

Joint Analysis of the Variability of Dissolved Organic and Inorganic Carbon Concentrations as Markers of Biogeochemical Processes in Kara Sea ... 287
Y. V. Murzakova, N. A. Belyaev, A. V. Kostyleva, V. Y. Fedulov, and A. A. Polukhin

Distribution of Hydrochemical Parameters in the Surface Water Layer of the Zone of Mixing River and Sea Waters of the Amazon River Basin During the 52nd Voyage of the R/V "Academik Boris Petrov" ... 295
A. M. Seliverstova, O. A. Zuev, and A. L. Chultsova

Marine Geology

Features of the Behavior of Oil Hydrocarbons in the Barents Sea in 2019–2020 ... 303
E. V. Koltovskaya and I. A. Nemirovskaya

Features of the Distribution of Dispersed Sedimentary Matter Composition in the Nordic Seas ... 308
I. A. Migdisova, D. P. Starodymova, A. A. Klyuvitkin, A. N. Novigatsky, and A. V. Bulokhov

New Data on the Age of the Khvalynian Strait ... 319
D. V. Semikolennykh and A. V. Panin

A CSM-Based Python Bot for Estimating the Stability Zone of Gas Hydrates ... 326
Yu.Yu. Smirnov, T. V. Matveeva, N. A. Shchur, A. A. Shchur, and A. V. Bochkarev

Study of Background and Focused Hydrocarbon Discharge, NE Kara Sea ... 332
A. K. Yumasheva, E. D. Basova, E. N. Poludetkina, M. Y. Tokarev, A. Y. Rybalko, and A. O. Kil

Marine Geophysics

Features of Multiple Waves Suppression in the Seismoacoustic Data Processing Obtained During Near Surface Geophysical Surveys in the Northern Part of the Kara Sea ... 341
I. O. Paltsev, V. V. Polovkov, A. A. Kudinov, and A. E. Rybalko

Joint Processing Technology of MCS Data (600 m Streamer) and Wide-Angle Reflection/refraction Seismic Sonobuoy Soundings Acquired in Ice Conditions During Expedition "ARCTIKA-2014" 350
S. N. Tabyrtsa, A. S. Zholondz, and V. V. Polovkov

Marine Ecology and Environmental Management

Persistent Organic Pollutants in Rhinoceros Auklet (*Cerorhinca Monocerata*) from the Sea of Japan 363
M. A. Belanov, M. Yu. Shchelkanov, D. V. Pankratov, M. M. Donets, and V. Yu. Tsygankov

Ichthyofauna of the Karkinitsky Gulf (Black Sea) and Its Changes Under the Anthropogenic Impact 370
R. E. Belogurova and E. P. Karpova

Comparative Analysis of Contamination with Microplastic Particles of Ice in the Curonian Lagoon (Baltic Sea) and Novik Bay (Sea of Japan) 376
I. Yu. Bocherikova, O. I. Lobchuk, and I. P. Chubarenko

Chlorinated Organic Compounds in Mytilidae Mollusks from Nakhodka Bay (the Sea of Japan) 381
A. D. Borovkova and V. Yu. Tsygankov

Individual Components of Microbiological Community in Water Northeast Part of the Kara Sea 387
V. Yu. Budarova, E. A. Tikhonova, N. V. Burdiyan, and Yu. V. Doroshenko

Influence of External Factors on the Seaweeds Productivity in the Coastal Zone of Crimea 396
T. A. Filippova and E. F. Vasechkina

Microlitter Distribution on Sandy Beaches of the Neva Bay of the Gulf of Finland (Baltic Sea) in 2018–2022 404
E. A. Golubeva and A. A. Ershova

Data on the Microplastics Content in the Bottom Sediments of the Gulf of Finland and Southeastern Part of the Baltic Sea According to the Data of 39th Cruise of the R/v "Akademik Nikolai Strakhov" 412
P. N. Krivoshlyk and I. P. Chubarenko

^{137}Cs Latitudinal Distribution in the Surface Waters of the Atlantic Ocean: Results of Expeditionary Research in 2020 418
O. N. Miroshnichenko, I. G. Sidorov, and A. A. Paraskiv

Qualitative and Quantitative Assessment of Hydrocarbons Composition in the Water of the Kara Sea Northeastern Part 425
Y. S. Tkachenko and E. A. Tikhonova

Physical and Biological Interactions

Results of Hydrooptical Field Studies in the Barents and Kara Seas in September 2022 439
D. I. Glukhovets, E. A. Aglova, V. A. Artemiev, O. V. Glitko, V. A. Glukhov, D. N. Deryagin, S. K. Klimenko, M. A. Pavlova, and I. V. Sahling

The Impact of Mesoscale Eddies on the Vertical Distribution of Oxygen in the Black Sea 446
A. A. Kubryakov, S. V. Svishchev, P. N. Lishaev, and E. A. Kubryakova

Modeling Chlorophyll *a* Concentration for the European Arctic Corridor Based on Environmental Parameters 456
S. K. Kuzmina, P. V. Lobanova, and S. S. Chepikova

Optical and Microphysical Characteristics of Tropospheric Aerosol over Kara Sea using Measurements Onboard TU-134 "Optic" Aircraft Laboratory in 2014, 2020, and 2022 463
P. N. Zenkova, B. D. Belan, D. G. Chernov, D. M. Kabanov, and I. A. Kruglinsky

Oceanological Technology and Instrumentation

Application of Microwave Resonant Measuring Transducer for Express Assessment of Water Salinity 473
D. A. Poletaev, B. V. Sokolenko, and V. I. Voytitsky

Investigation of Winter Plankton Ecology of Lake Baikal Using Complex Instrumental Methods 483
T. P. Rzhepka and E. Y. Naumova

Author Index 491

Ocean Physics

Climate assessment of water, heat and salt fluxes in the Pechora Sea

D. V. Deshova(✉) and S. M. Gordeeva

Russian State Hydrometeorological University, 192007 Saint Petersburg, Russia
ya.deshova-diana@yandex.ru

Abstract. The Pechora Sea is a Barents Sea part located in the southeast. The natural conditions of the Pechora Sea are very different from the Barents Sea water area. There are various types of water masses and many dynamic processes. The purpose of the study is to assess the components of the water, heat and salt balances of the Pechora Sea. Based on the monthly average data of the CMEMS GLORYS12V1 and ERA5 reanalyses for the period 1993–2018, the water, heat and salt transport at the boundaries of the Pechora Sea (in longitude 50 E on the west, latitude 71 N on the north and in the Kara Strait) and the characteristics of the sea surface-atmosphere interaction are calculated. The average long-term resulting water transport on the Pechora Sea boundaries coming from the west (0.36 Sv) and from the north (0.04 Sv), leave the Pechora Sea through the Kara Gates (0.40 Sv). It is shown that in the region there is heat excess coming due to advection, which is compensated by sea-atmosphere interaction. All transports during the study period have a significant positive trend, which characterizes the growth of the transit of North Atlantic water through the Pechora Sea.

Keywords: Water transport · Water balance · Heat balance · Salt balance · Sea-atmosphere interaction · Pechora Sea · Kara Gates

1 Introduction

The Pechora Sea is a Barents Sea part located in the southeast. The natural conditions of the Pechora Sea are very different from the Barents Sea water area. There is a more severe climate and ice conditions. The influence of the Norwegian Current, which gets into the Barents Sea, is weakened in the southeastern part [1].

The Pechora Sea area is affected by many dynamic processes and many various types of water masses there are [2, 3]: transport of North Atlantic waters from the west, transport of waters from the White Sea and the Pechora River runoff, transport of waters from the Kara Sea through the Kara Gates Strait, as well as interaction with the atmosphere through the surface. The main water dynamic regime of the Pechora Sea is determined by the impact of the Pechora River runoff [4]. All of these processes are changing under current climate change.

Thus, the combination of climatic, hydrological, and oceanological factors creates an environment in the Pechora Sea that is different from the adjacent Barents Sea basin [5].

T. Chaplina (Ed.): CIWO 2023, SPEES, pp. 3–10, 2023.
https://doi.org/10.1007/978-3-031-47851-2_1

Many natural resources are concentrated on the shelf of the Pechora Sea [6], so it is the great strategic importance for Russia.

The purpose of the study is to estimate the components of the water, heat and salt balances of the volume of the Pechora Sea.

Water balance is a quantitative ratio of the inflow, flow and accumulation (reserve) of water for the object under study for a certain period of time. The water balance method makes it possible to compare and evaluate the degree of impact of individual sources on the general regime of a water body over different periods of time [7].

Thermal balance is a quantitative ratio of income, consumption and heat retention during various thermal processes for the object under study for a certain period of time. The heat balance method helps to identify patterns of changes in hydrological processes occurring under the influence of heat exchange between neighboring water bodies and the atmosphere [7].

Salt balance is the quantitative ratio of salt input and output during various processes for the object under study for a certain period of time [7].

2 Material and Methods

The GLORYS12V1 reanalysis of the Copernicus marine service, based on the existing CMEMS real-time global forecasting system, was taken as the initial data for hydrological parameters (water temperature, salinity, current velocity, sea level, ice concentration). The output files of the global ocean are displayed on a standard regular grid with a step of $1/12^0$ (about 8 km) and at 50 standard levels [8].

Atmospheric parameters (precipitation, evaporation, components of heat exchange with the atmosphere) are taken from the ERA5 reanalysis of the Copernicus climate variability service. ERA5 is a fifth generation ECMWF reanalysis of global climate and weather over the past 8 decades [9].

All of data was converted to a regular latitude-longitude grid of $0{,}25^0$.

As the western and northern boundaries of the Pechora Sea, the meridional and latitudinal sections from point 71^0N 50^0E to the south and east to the coast were considered; for the Kara Strait – along the 70^0N parallel at longitude from 57^05' to 58^045'E (from Novaya Zemlya in the west to the mainland in the east). The average monthly values of the parameters at the database horizons, based on the bottom topography from the surface to the bottom, were selected for the period from January 1993 to December 2018.

Estimation of water, heat and salt transport in the water column for each cell of the section was calculated separately in Eqs. (1), (2) and (3) [10], which were then summed over the section.

Water transport – the liquid volume flowing through the cross-section of the stream per unit time, can be defined as:

$$Q = V \cdot \Delta H \cdot \Delta R \cdot 10^{-6}, \quad (1)$$

where Q is the fluid flow, Sv; V is the flow velocity in the center of each layer, m/s; ΔH is the layer thickness, m; ΔR is the distance between stations, m.

Heat transport – the amount of heat transferred per unit of time through a unit area, can be defined as [10]:

$$Ft = Cp \cdot \rho \cdot (T - Tf) \cdot V \cdot \Delta H \cdot \Delta R, \quad (2)$$

where Ft is the heat flow, W; Cp is the specific heat capacity of water, equal to 4200 J/(kg ·s); ρ is the water density, equal to 1025 kg/m^3; T is the sea water temperature, ^{0}C; Tf is the minimum temperature of sea water at the point, ^{0}C.

Salt transport – the amount of salt transferred per unit of time through a unit area, can be defined as [10]:

$$Fs = \rho \cdot Sw \cdot V \cdot \Delta H \cdot \Delta R, \quad (3)$$

where Fs is the salt flow, g/s; Sw is the sea water salinity, g/kg.

In addition to the main flows, the water and heat balances for the volume of the sea within the indicated boundaries were closed in Eqs. (4) and (5), only the salt balance was determined by the salt flow Fs.

Water balance:

$$Q + W + H + (P - E) = 0, \quad (4)$$

where Q is the fluid flow, water horizontal advection; W is the river runoff; H is the change in the sea volume due to level fluctuations; P is the atmospheric precipitation, E is the evaporation.

Heat balance:

$$Ft + \Phi = 0, \quad (5)$$

where Ft is heat advection by currents; $\Phi = (S + Eef - LE + F)$ is exchange of heat with the atmosphere; S is the solar radiation coming to the surface (shortwave radiation); Eef is thermal radiation from the surface (long-wave or terrestrial radiation); LE is the latent turbulent heat flux; F is the net of sea-atmosphere sensible heat flux.

Residuals for water, heat and salt balances were calculated, which show whether water, heat, and salt are accumulating in the region. They also help to understand whether there are enough components used up to determine the balance.

3 Results and Discussion

The long-term averaged values of flows across the sections (advection) are (incoming flow on western and northern boundaries, outgoing on Kara Gates):

- water flow: coming – 0,40 Sv; leaving – 0,40 Sv;
- heat flow: coming – 5,92 TW; leaving – 4,09 TW;
- salt flow: coming – 13,98 kt/s; leaving – 13,75 kt/s.

In long-term averaging, the main water transport and associated thermal and salt transport are directed from the northwestern corner of the region (Kolguyevo-Pechora Current) in transit to the Kara Sea through the Kara Gate Strait.

On long-term average, the sum of all the main components of the balances is shown in Fig. 1 and is:

- water balance: coming – 0,52 Sv, leaving – 0,53 Sv;
- heat balance: coming – 7,08 TW, leaving – 7,12 TW;
- salt balance: coming – 18,05 kt/s, leaving – 17,82 kt/s.

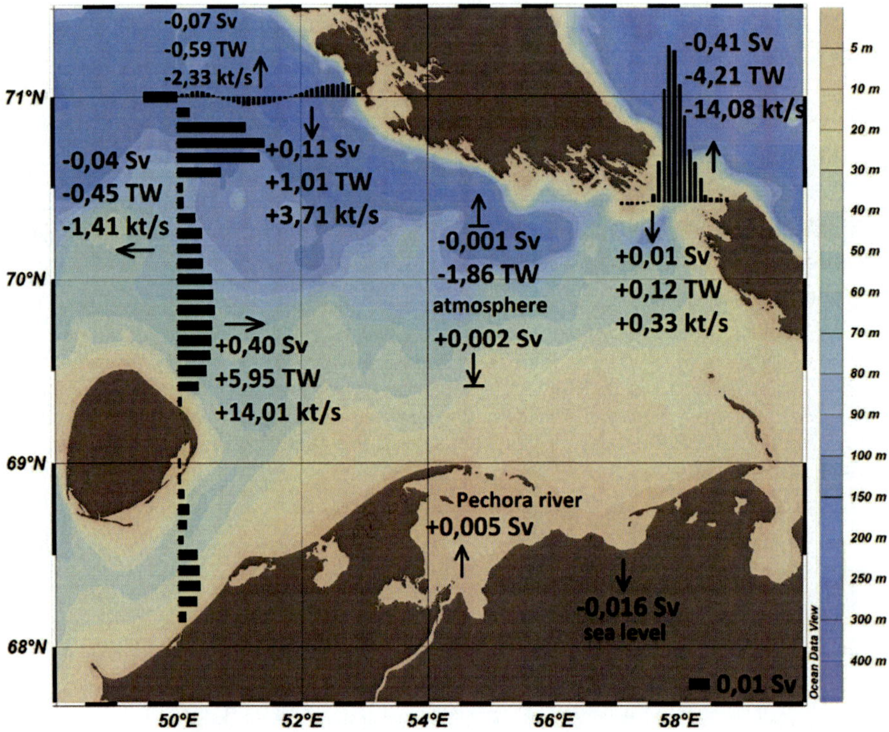

Fig. 1. Estimates of the water (Sv), heat (TW) and salt (kt/s) balances of the Pechora Sea (arrows show the flow directions)

In the composition of the components of the water, heat and salt balances of Table 1, the following can be noted.

On an average, all incoming through the borders with the Pechora river runoff and precipitation, the positive water balance components (0,52 Sv) is almost equal to negative water balance components: evaporation, escape through the boundaries of the water area and volume change due to sea level (0,53 Sv).

The heat excess entering the Pechora Sea water area with the transformed North Atlantic water due to water exchange with neighboring regions (7,08 TW) is almost completely compensated by the sea-atmosphere fluxes and outgoing flow (7,12 TW, where the escape through the boundaries is 5,25 TW and escape to the atmosphere 1,86 TW). The discrepancy in the heat balance can arise due to the neglect of heat coming with atmospheric precipitation and ice transformation. Also, an additional source of heat in the Pechora Sea is the Pechora river runoff, but that effect almost does not extend beyond the Pechora Bay.

In the salt balance there is a discrepancy in the inflow due to advection from neighboring water areas (18,05 kt/s) and outflow (17,82 kt/s) through the boundaries of the Pechora Sea. The discrepancy can be related to the continental freshwater runoff, atmospheric precipitation, evaporation and ice transformation processes, which were not considered.

Table 1. The main components of the balances on the long-term averaging for 1993–2018

Balance components	Coming	Leaving	Discrepancy	Share, %
Water balance, Sv				
Integral volume flow at the boundaries (advection)	0.515	0.513		
Pechora river runoff	0.005			
Precipitation	0.002			
Sea level change component		0.016		
Evaporation		0.001		
Water balance	**0.52**	**0.53**	**-0.01**	**1.63**
Heat balance, TW				
Integral heat flux at the boundaries (advection)	7.08	5.25		
Heat transfer to the atmosphere		1.86		
Heat balance	**7.08**	**7.12**	**-0.04**	**0.56**
Salt balance, kt/s				
Salt balance	**18.05**	**17.82**	**0.23**	**1.27**

In the water balance, according to the values of incoming water transport and outgoing water transport, due to advection, there is an accumulation of water in the region. As a result, the sea level rises.

In the heat balance, according to the values of advection by currents, heat accumulation in the Pechora Sea is observed, which is compensated by sea-atmosphere interaction.

The calculated discrepancy in the salt balance shows the accumulation of salt in the Pechora Sea, which is reflected as an excess of salt inflow over outflow. This can manifest itself in an increase in the salinity of the sea.

The discrepancies obtained in the course of balance calculations are small and amount to 1.63% for the water balance, 0.56% for the heat balance and 1.27% for the salt balance. The discrepancy may be due to the water volume from the neighboring water areas ice brought, its melting and formation was not considered. But this value is small, it can be concluded that the total volume of water from these parameters is insignificant.

In general, in the region all types of balances are carried out with sufficient accuracy. This may indicate that the data sources used for calculations (independent of each other) are quite well balanced among themselves even when assessing a small region and adequately reflects the physical processes occurring here.

According to the average annual estimates of the components of the water, heat and salt balances for the period 1993–2018, the linear trends of Table 2 were estimated.

During 1993–2018, the volume of warm and salty North Atlantic water entering the region through the northern part of the western border with a branch of the Kolguyevo-Pechora Current increased. Together with the revealed significant trends in heat and salt flows on the Kara Gate Strait [11], there is an increasing transit of North Atlantic water through the Pechora Sea basin into the Kara Sea can be concluded. It was also revealed that the Litke Current, which carries arctic waters from the Kara Sea along Novaya Zemlya, has intensified.

Table 2. Linear trends in average annual balances (statistically significant values are highlighted in red ($\alpha = 0.05$))

Balance components	Coming	Leaving
Linear trend of water balance, Sv/year		
Integral volume flow at the boundary of 50 E (advection)	0.003	-0.0002
Integral volume flow at the boundary of 71 N (advection)	−0.0003	0.002
Integral volume flow in the Kara Gates (advection)	0.00005	0.0057
Pechora river runoff	−0.00002	-
Precipitation	0.000004	-
Sea level change component	-	0.0001
Evaporation	-	0.00002
Water balance	**0.003**	**0.008**
Linear trend of heat balance, TW/year		
Integral volume flow at the boundary of 50 E (advection)	0.147	0.0007
Integral volume flow at the boundary of 71 N (advection)	0.0013	0.033
Integral volume flow in the Kara Gates (advection)	0.0029	0.156
Heat transfer to the atmosphere	-	0.067
Heat balance	**0.15**	**0.26**
Linear trend of salt balance, kt/s for the year		
Integral volume flow at the boundary of 50 E (advection)	0.114	-0.008
Integral volume flow at the boundary of 71 N (advection)	−0.009	0.058
Integral volume flow in the Kara Gates (advection)	0.002	0.201
Salt balance	**0.11**	**0.25**

4 Conclusion

Thus, the following conclusions were made.

Water advection at the boundaries of the water area, river runoff, sea level change, precipitation and evaporation were considered as components of the water balance. Heat flow by currents in the heat balance was compensated by sea-atmosphere interaction. The salt balance consisted only of advection from neighboring water areas.

On a long-term average, the sum of all the main components of the balance is:

- water balance: coming – 0,52 Sv, leaving – 0,53 Sv;
- heat balance: coming – 7,08 TW, leaving – 7,12 TW;
- salt balance: coming – 18,05 kt/s, leaving – 17,82 kt/s.

Advection across the boundaries of the water area plays a major role in the formation of the water balance. The main transport of waters is the Kolguyevo-Pechora Current, directed from the northwestern corner of the water area in transit through the Kara Gate Strait to the Kara Sea. A smaller flow enters the region through the Pomor Strait (White Sea Current). Sea level fluctuation is the second most important component. Pechora river runoff, precipitation and evaporation make the least contribution to the balance.

Heat advection by currents makes the main contribution to the heat balance, followed by heat transfer to the atmosphere. The salt balance in the water area is completely formed by water advection at the boundaries.

The discrepancies in the balance calculation are 1.63% for the water balance, 0.56% for the heat balance and 1.27% for the salt balance. The small value of residuals shows that the components used are sufficient to describe the water, heat, and salt balances in the Pechora Sea waters. All types of balances on the water area are performed with sufficient accuracy, which indicates the balance of independent data sources used for calculations.

The water, heat and salt transports of the Pechora Sea waters have significant positive trends for the period 1993–2018, which are formed to a greater extent by advective components. The trends of advection at the western boundary make the greatest contribution to the magnitude of the total inflow trends, and the flow into the Kara Sea to the magnitude of the outflow trends. The highest values of the trend coefficients in the current velocity, water temperature and salinity are observed in the surface layer in the northwest of the water area. In the Kara Gate Strait, a significant increase in temperature and velocity is noted in the central part in the 0–10 m layer, and salinity increases more strongly in the western part of the strait along the slope of Novaya Zemlya.

The increase in Atlantic water entering the region with the Kolguyevo-Pechora Current, together with positive trends in the heat and salt transport in the Kara Gates, indicate an increase in the transit of North Atlantic waters through the Pechora Sea basin into the Kara Sea. The Litke Current, which carries cold Arctic waters to the Barents Sea, has also intensified.

References:

1. Ozhigin, V.K., Ivshin, V.A., Trofimov, A.G., Karsakov, A.L., Antsiferov, M.Yu.: Waters of the Barents Sea: structure, circulation, variability. - Murmansk: PINRO, 2016, 260 p. (in Russian)
2. Pechora Sea. Ecological Atlas: Scientific and popular edition "Gazprom-Neft". - St. Petersburg: Print Gallery, 2018. – 144 p. (in Russian)
3. Pisarev, S.V.: Overview of hydrological conditions in the Barents Sea. Barents Sea System. Pod. editor acad. A.P. Lisitsyn. - M.: GEOS, 2021, pp. 153–166. (in Russian)
4. Boltunov, A.N., Dubinin, M.Yu., Ezhov, A.E., Larionov, M.V., Novoselov, A.P., Pukhova, M.A., Frolova, E.A.: Areas of limitation of anthropogenic activity: the Pechora Sea. Oil and gas complex. - Murmansk: World Wildlife Fund (WWF), 2014, 76 p. (in Russian)
5. Pavlidis, Yu.A., Nikiforov, S.L., Ogorodov, S.A., Tarasov, G.A.: Pechora Sea: past, present, future. Oceanology **47**(6), 865–876 (2007). https://doi.org/10.1134/S0001437007060112
6. Prishchepa, O.M., Nefedov, Yu.V., Airapetyan, M.G.: Hydrocarbon potential of the offshore Arctic sector of the north of the Timan-Pechora oil and gas province based on research results. Neftegazovaya geologiya. Theory and practice **15**(1), 7 (2020). (in Russian)
7. Chebotarev, A.I.: General hydrology (land waters) // ed. L.A. Chepelkina. - Leningrad: Gidrometeoizdat, 544 p. (1975). (in Russian)
8. E.U. Copernicus Marine Service Information, Global Ocean Physics Reanalysis. – URL: https://resources.marine.copernicus.eu/product-detail/GLOBAL_REANALYSIS_PHY _001_030
9. Copernicus Climate Change Service, Climate Data Store, (2023): ERA5 monthly averaged data on single levels from 1940 to present. Copernicus Climate Change Service (C3S) Climate Data Store (CDS). https://cds.climate.copernicus.eu/cdsapp#!/dataset/reanalysis-era5-single-levels-monthly-means
10. Doronin Yu.P. Physics of the ocean. - St. Petersburg: Ed. RSHU, 296 p. (2000). (in Russian)
11. Gordeeva, S.M., Deshova, D.V.: Transfer of water, heat and salt in the Kara Gate Strait. Abstracts of the All-Russian Scientific Conference "Russian Seas: Year of Science and Technology in the Russian Federation – UN Decade of Ocean Sciences". - Sevastopol, September 20–24, pp. 225–226 (2021). (in Russian)

Assessment of Seasonal and Multi-year Variations in Ice Age Composition of Greenland Sea and Barents Sea

E. S. Egorova(✉) and Ye. U. Mironov

State Scientific Center of the Russian Federation Arctic and Antarctic Research Institute, 199397 St. Petersburg, Russia
egorova@aari.ru

Abstract. We present the first winter estimates of the Greenland and Barents Seas ice age composition using the ice charts for the period from 1997 to 2022 compiled by the Arctic and Antarctic Research Institute (AARI). The interannual variations in the ice amount of different age categories in it relative fractions in the Greenland Sea and the Barents Sea do not show significant linear trends. When comparing the estimates, obtained in this study, with previous studies on this subject, we may state that quantitative changes in the Greenland and Barents Seas ice age composition began earlier than 1997. Only on the basis of data on the ice age composition over a 25-year long period of observations analyzed in this study, it is impossible to draw a clear conclusion about a decrease in its thickness in the Greenland and Barents Seas.

Keywords: Greenland Sea · Barents Sea · Ice cover · Ice age composition · Ice thickness · Winter period · Seasonal variations · Interannual variations

1 Introduction

The state of Arctic sea ice cover is a subject of much scientific debate and continuing research. It is a known fact that since the late 1980s the warming in the polar region has been much faster than in the rest of the world. Temperatures have increased about twice as fast in the Arctic as in the mid-latitudes [1], a phenomenon known as "Arctic amplification" [2]. There is a significant reduction in both ice area and thickness in the Arctic Ocean [3, 4].

According to [4, 5], the Northern Europe's sea basins, such as the Greenland Sea and the Barents Sea, hosts the most pronounced loss of Arctic winter sea ice. Sea ice coverage and thickness in the Greenland Sea and the Barents Sea are influenced by the inflow a warm, saline Atlantic Water as well as a cold fresh Arctic Water. The recent ice area and thickness reduction in the Greenland Sea and the Barents Sea linked to changes in the Atlantic Ocean heat transport (so-called "Atlantification") [6–9] and the ice outflow via the Fram Strait and some strait sites in northern part of the Barents Sea [10–12].

T. Chaplina (Ed.): CIWO 2023, SPEES, pp. 11–18, 2023.
https://doi.org/10.1007/978-3-031-47851-2_2

The ice thickness in the Arctic seas is a sensitive indicator of climate change. When estimating the ice thickness, if it cannot be measured by contact or non-contact methods, ice charts are used indicating the age categories of sea ice [13]. The study [14] presents the main features of the ice regime in the Greenland Sea and the Barents Sea, but the variability of sea ice age structure (hence, its average thickness) remains relatively underexplored.

Wittmann and Schule [15], for the first time, calculated from "Birds-eye" flight data in 1960 and 1962 the amount of young, first-year and old ice in Greenland Sea. For the following months: from January to May, from June to July, from August to October and from November to December. Mironov [14] used the ice charts of the Arctic and Antarctic Research Institute (AARI) for the 1988–1992 period to estimate the average monthly values of old ice areas of predominance in the Greenland Sea, as well as the ice areas of various age categories (young, first-year and old ice) for seasonal maximum (March-April) and seasonal minimum (September-October) of ice cover.

The age structure of the ice cover in the Barents Sea was estimated by Gudkovich et al. [16]. Based on the average monthly charts of surface atmospheric pressure, taking into account the average dates of the onset of ice formation, as well as the distribution of the boundaries of residual ice and fast ice at the end of May, the average values of the ice areas of the main age gradations, young, first-year and old ice were derived. Using a same method, Mironov [14] assesses the monthly mean values of ice areas of different age categories for the 1971–1976 period for three homogeneous sub-areas of the Barents Sea: the western, northeastern and southeastern ones.

This study presents the key results of the seasonal and interannual variations in Greenland Sea and the Barents Seas ice age composition during the winter season for seven standard age categories. To estimate the ice amount of different age categories, ice charts produced by the Arctic and Antarctic Research Institute (AARI) were used. The assessments are presented for the period 1997–2022 and are limited only to the winter season, from October to May, because reliable data on the ice age structure and its thickness are available only in these months. The results of this investigation expand the knowledge about the ice regime of the Greenland Sea and the Barents Sea in winter months [14].

2 Data and Methods

To obtain data on the age structure of ice, survey ice charts of the Greenland Sea and the Barents Sea were used, available in the electronic catalog of the World Sea Ice Data Center (available at http://wdc.aari.ru/datasets/) for the period 1997–2022. Charts are compiled every 10 days by ice experts from the "Sever" Center, which is part of the AARI, indicating the age composition only in the winter months, from the beginning of ice formation in October to its completion in May. Because of a layer of meltwater on the surface of the ice cover in the summer season significantly changes the ice cover picture on satellite images of all ranges, the differences between the ice age categories become almost invisible. A detailed methodology for compiling AARI ice charts using the satellite information is described in [13].

Work with vector ice charts was carried out using the geographic information system (GIS) QGIS, version 3.14 (available at https://qgis.org/ru/site/index.html). For the winter months of the period 1997–2022, the relative amount of drifting ice of various age categories was determined, expressed as a percentage of the total ice cover area in the Greenland Sea and the Barents Sea.

3 Results: Seasonal Variations

As is known, the old ice carried of the Arctic Basin is observed all year round in the Greenland Sea. During the winter period, old ice prevails in the ice cover and occupies at least 30% of the total ice area. There is a gradual decrease in the relative amount of old ice from 43% in October to 30% in February-May as the relative amount of ice of other age categories increases. The seasonal maximums of the old ice area *absolute* values are observed in December (155×10^3 km^2) and in April (143×10^3 km^2), which corresponds to two peaks in the seasonal ice outflow via the Fram Strait site [12]. As for the Barents Sea, although old ice is observed in the western and the northeastern sub-areas during the entire winter season, its amount does not exceed 4% of the total ice cover area, with the exception of October, when ice is just beginning to form. In the southeastern sub-area, where is the more favorable ice conditions among all the sub-areas of the Barents Sea, the presence of the old ice under average conditions is not typical.

The ice formation process in the Greenland Sea and the Barents Sea is observed during wintertime, so the categories of initial types and young ice are typical for the ice cover of all the months under consideration. The formation of first-year thin (medium, thick) ice in the Greenland Sea begins in November (December, January). As for the western sub-area of the Barents Sea, first-year thin (medium, thick) ice starts to form in November (December, March). In northeastern sub-area the formation of the first-year ice categories begins in November, December and February for the thin, medium and thick ice, respectively. Finally, the first-year thin (medium) ice in the Barents Sea southeastern sub-area is observed for the first time during wintertime in December (February).

In Greenland Sea, the total amount of initial types and young ice prevails varying from 46% to 27% (of the total ice area) from October to February. From March to May, most of the ice cover is occupied by first-year ice (up to 32%). The predominance of first-year thin ice category is observed from October to December (up to 17%), but in January-May, first-year medium ice already prevails in the ice cover of the Greenland Sea (up to 21%). First-year thick ice, the least prevalent among the first-year categories, makes up to 9% of the ice cover of the Greenland Sea. As for Barents Sea, the total amount of initial types and young ice prevails in the ice age structure in the western (northeastern) sub-area from October to March, varying from 81% to 51% (from 81% to 49%). In April and May, the ice age composition is determined to a greater extent by first-year ice of different thicknesses: from 59% to 72% in the western sub-area and from 50% to 65% in the northeastern one. In the Barents Sea southeastern sub-area, the gradation of first-year ice becomes predominant in the ice age composition only by the end of the winter season in May (55% of the total ice area). Under average conditions, first-year thick ice does not form here.

4 Results: Multi-year Variations

Let us consider the interannual variations in the ice age structure using the example of April, the month of annual maximum ice volume in the Greenland Sea and the Barents Sea. Relative to the total ice cover area in Greenland Sea the amount of initial types and young ice varies on average from 14% to 36%; for the first-year and old ice—from 20% to 40% and from 16% to 44% respectively. In particular, the first-year thin ice covers up to 10% of the Greenland Sea ice cover, medium—up to 28% and thick—from 4% to 30%. The amount of young ice in the western and northeastern sub-areas of the Barents Sea varies on average from 20% to 70%, in the southeastern one—from 20% to 85%. At this time, the proportion of first-year thin ice in the ice cover of the sea is from 10% to 55%, from 10% to 35% and from 5% to 50% in the western, northeastern and southeastern sub-areas, respectively. First-year ice of medium thickness occupies from 5% to 50% of the ice cover in the western, from 5% to 40% in the northeast, and finally from 0% to 50% in the southeast sub-area. The area of first-year thick ice, which is the least dominant among categories of first-year ice, varies on average in the western and northeastern sub-areas of the Barents Sea from 0% to 35%. Inside of natural variability, it is noted that both in the Greenland Sea and the Barents Sea with a relative increase in the amount of first-year thin ice in the ice age structure, the proportion of first-year medium and first-year thick ice decreases; true and vice versa. In April, among the first-year ice of the ice cover in Greenland Sea and the Barents Sea, the predominant age category is first-year medium ice, with a thickness of 70 to 120 cm.

The permanent presence of old ice in the Greenland Sea is a distinguishing feature of its ice regime. Figure 1 shows the average, minimum and maximum boundaries of old ice distribution (hereinafter, ice zones with a predominance of old ice were taken into account) in the Greenland Sea in April, obtained for the years 1997 to 2022 according to the AARI survey ice charts. The boundaries of old ice distribution in 2022 are taken as the average position, the minimum in 2003, and, finally, the maximum in 2019 (in the second decade of April); these years were chosen as the most representative.

The average boundary of the old ice distribution in 2022 generally coincides with the area of the East Greenland Current and reaches 70°N. Some old ice fields can be carried out to the marginal area due to the divergence of the ice cover, however, in winter, first-year and young ice prevails in the marginal zone. For the minimum old ice area in 2003, the boundary of its distribution barely reaches the northern boundary of the Greenland Sea and is concentrated between 81°N and 80°N; to the south, first-year and young ice also prevail. In 2019, with the maximum amount of old ice, its eastern boundary usually coincides with the boundary of the middle position (the deviation is no more than 2° in longitude at certain latitudes), however, its distribution zone goes beyond 68°N, i.e. south of the Greenland Sea.

Further, linear trends of the relative amount of ice of different age categories in the ice cover of the Greenland Sea were evaluated for statistical significance. Student's t-test was used as the main criterion for assessing the statistical significance of trends (at a significance level of 99%). For the period from 1997 to 2022, the trends turned out to be statistically insignificant. Thus, with the observed reduction in total ice area, the amount of ice of different age categories ranges within the limits of its own natural variability. Therefore, only on the basis of data on the ice age composition over a 25-year long

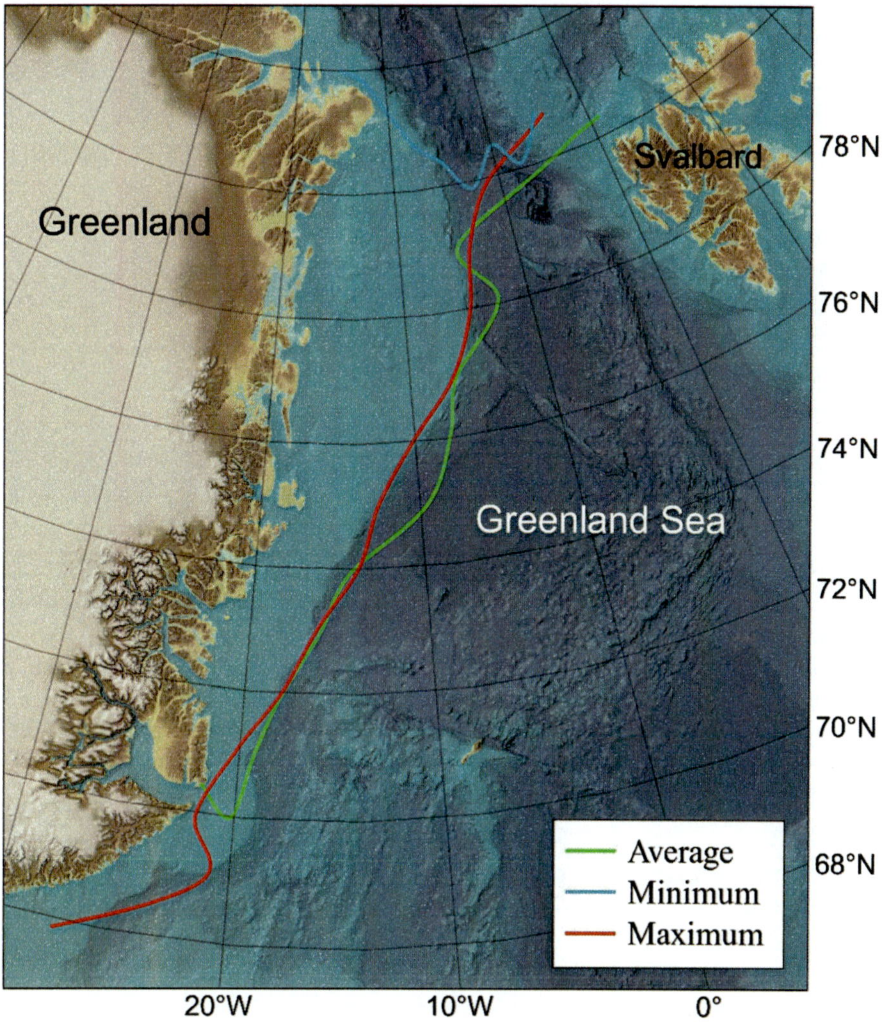

Fig. 1. Average, minimum and maximum boundaries of the old ice predominance in the Greenland Sea in April for the period from 1997 to 2022

period of observations analyzed in this study, it is impossible to draw a clear conclusion about a decrease in its thickness in the Greenland Sea and the Barents Sea.

Table 1 and Table 2 let us compare the results obtained in this study with other, earlier estimates of the ice age structure in the Greenland Sea and the Barents Sea, respectively. At the beginning of the 21st century, the ice age structure in the Greenland Sea and the Barents Sea underwent significant changes compared to the period 1970s-1990s. Consequently, taking into account the statistical insignificance of trends in interannual variations in the ice amount of different age categories, must be considered that quantitative changes in the age structure of the Greenland Sea and the Barents Sea ice cover began earlier than 1997.

According to assessment of Mironov [14] (1988–1992) and of the present study (1997–2022), there is a decrease in the average thickness of the Greenland Sea ice cover. In particular, a decrease in the amount of old ice is 15% (from 45% to 30%), as well as a decrease in the amount of initial types and young ice is 8% (from 33% to 25%); on the contrary, first-year ice in the sea ice cover increased by 8% (from 22% to 30%).

Table 1. Assessments of the ice age composition in the Greenland Sea in April (in % of the total ice area), according to different time periods

Ice age category/Assessment	Mironov [14] (1988–1992)	Present study (1997–2022)
Initial types and young	33	25
First-year	22	30
Old	45	30

Notes: The area of landfast ice is taken into account when calculating the ratios of ice of various age categories

The comparison the 1971–1976 estimates [14] and our estimates (1997–2022) show a decrease of the amount of first-year ice as well as an increase of amount of young ice in all the Barents Sea sub-areas. The difference in the amount of old ice in the ice cover of the Barents Sea western sub-area is interesting (24% according to [14] and only 1% according to our estimates). Most likely, the assessment [14] presents overestimated values, caused by an increase the ice outflow via the Fram Strait site. However, this issue requires more detailed consideration and is beyond the scope of this study.

Table 2. Assessments of the ice age composition in the Barents Sea in April (in % of the total ice area), according to different time periods

Ice age category / Barents Sea sub-areas	Mironov [14] (1971–1976)			Present study (1997–2022)		
	Western	Northeastern	Southeastern	Western	Northeastern	Southeastern
Initial types and young	12	13	20	38	41	45
First-year thin	0	0	64	27	18	26
First-year medium	10	17	16	28	24	16
First-year thick	54	69	0	5	8	0
Old	24	1	0	1	1	0

Notes: The area of landfast ice is taken into account when calculating the ratios of ice of various age categories

5 Conclusions

Here we have used, for the first time, survey ice charts compiling by AARI to quantify the amount of ice of various ice categories in the Greenland Sea and the Barents Sea. We performed a detailed analysis of the seasonal and interannual variations of ice age structure in the Greenland Sea and the Barents Sea for the years 1997 to 2022. Based on our analysis, the following conclusions can be drawn:

1. In the Greenland Sea the total amount of initial types and young ice prevails in the ice age structure of the Greenland Sea in October-February, and first-year ones—in March-May. In October-March, the age structure of the ice cover in the western and the northeastern sub-areas of the Barents Sea is dominated by initial types and young ice, in April-May—first-year ice. In the southeastern sub-area, the predominance of first-year ice is observed only in May, and from October to April.
2. Throughout the winter period, old ice prevails in the ice cover of the Greenland Sea, occupying at least 30% of the total ice area. As for the Barents Sea, although old ice is observed in the western and the northeastern sub-areas during the entire winter season, its amount does not exceed 4% of the total ice cover area, with the exception of October, when ice is just beginning to form. There is no old ice in the Barents Sea southeastern sub-area in the average conditions.
3. The interannual variations in the ice amount of different age categories in its relative fractions in the Greenland Sea and the Barents Sea do not show significant trends for the period from 1997 to 2022. Thus, with the observed reduction in total ice area, the amount of ice of different age categories ranges within the limits of its own natural variability. Therefore, only on the basis of data on the ice age composition over a 25-year long period of observations analyzed in this study, it is impossible to draw a clear conclusion about a decrease in its thickness in the Greenland Sea and the Barents Sea.
4. When comparing the estimates of the age structure of the ice cover in the Greenland Sea and the Barents Sea, obtained in this study, with previous studies on this subject, we may state that its average thickness at the beginning of the 21st century decreased, compared to the periods 1988–1992 and 1971–1976 respectively. Taking into account the statistical insignificance of trends in interannual variations in the ice amount of different age categories, must be considered that quantitative changes in the age structure of the Greenland Sea and the Barents Sea ice cover began earlier than 1997.

Acknowledgements. The work was supported by the Russian Science Foundation within the framework of the scientific project No. 22-27-00443.

References

1. IPCC Special Report on the Ocean and Cryosphere in a Changing Climate. – Cambridge, United Kingdom and New York, NY, USA: Cambridge University Press, (2019). In press
2. Previdi, M., Smith, K.L., Polvani, L.M.: Arctic amplification of climate change: a review of underlying mechanisms. Environ. Res. Lett. **16**, 093003 (2021)

3. Kwok, R.: Arctic sea ice thickness, volume and multiyear ice coverage: losses and coupled variability (1958–2018). Environ. Res. Lett. **13**(10), 105005 (2018)
4. Stroeve, J., Notz, D.: Changing state of Arctic sea ice across all seasons. Environ. Res. Lett. **13**(10), 103001 (2018)
5. Onarheim, I.H., Årthun, M.: Toward an ice-free barents sea. Geophys. Res. Lett. **44**, 8387–8395 (2017)
6. Aksenov, P.V., Ivanov, V.V.: «Atlantification» as a possible cause for reducing of the sea-ice cover in the Nansen Basin in winter. Arctic Antarctic Res. **64**(1), 42–54 (2018). (In Russian)
7. Smedsrud, L.H., et al.: The role of the Barents Sea in the Arctic climate system. Rev. Geophys. **51**, 415–449 (2013)
8. Tesi, T., et al.: Rapid Atlantification along the Fram Strait at the beginning of the 20th century. Sci. Adv. **7**(48), eabj2946 (2021)
9. Wang, Q., et al.: Intensification of the Atlantic water supply to the arctic ocean through fram strait induced by arctic sea Ice Decline. Geophys. Res. Lett. **47**(3), e86682 (2020)
10. Lind, S., Ingvaldsen, R.B., Furevik, T.: Arctic warming hotspot in the northern Barents Sea linked to declining sea-ice import. Nature Climate Change **8**, 634–639 (2018)
11. Wei, J., Zhang, X., Wang, Z.: Reexamination of Fram Strait sea ice export and its role in recently accelerated Arctic sea ice retreat. Climate Dyn. **53**, 1823–1841 (2019)
12. Egorova, E.S., Mironov, Ye.U.: Assessment of the seasonal and multiyear variability in the sea ice volume export via the Fram Strait. Int. J. Offshore Polar Eng. **33**(1), 18–26 (2023)
13. Afanasyeva, E.V., et al.: AARI methodology for sea ice charts composition. Russian Arctic **7**, 5–20 (2019)
14. Mironov, Ye.U.: Ice conditions in the Greenland Sea and the Barents Sea and its long-term forecast. Saint-Petersburg: AARI, 2004. – 319 p. [In Russian]
15. Wittmann, W.I., Schule, J.J., jr. Comments on the mass budget of Arctic pack ice. In: Proceedings of the symposium on the Arctic heat budget and atmospheric circulation. J. O. Fletcher (ed.) Santa Monica, Calif., Rand Corporation. – 1966, pp. 215–246
16. Gudkovich, Z.M., Kirillov, A.A., Kovalev, E.G., Smetannikova, A.V., Spichkin, V.A.: Fundamentals of the methodology of long-term ice forecasts for the Arctic seas. – Leningrad: Gidrometeoizdat, 348 p. (1972). (In Russian)
17. JCOMM Expert Team on Sea Ice. Sea-Ice Nomenclature: snapshot of the WMO Sea Ice Nomenclature No. 259. – Geneva, Switzerland: WMO-JCOMM, 121 p. (2014)

Laboratory Investigations of the Evolution of Baroclinic Eddies in a Two-Layer Rotating Fluid

D. N. Elkin(✉) and A. G. Zatsepin

Shirshov Institute of Oceanology of RAS, 117997 Moscow , Russia
dmelkin@mail.ru

Abstract. In this work, by means of laboratory modeling, the patterns of evolution of baroclinic eddies in a rotating two-layer fluid. Using the original application of the "cylinder method", single axisymmetric cyclonic and anticyclonic eddies were produced in the upper layer, and their evolution was traced over 100 or more platform rotation periods.

Keywords: Open ocean eddies · Laboratory modeling · Two-layer stratified rotating fluid · Baroclinic instability · Viscous degeneration viscous · Smooth · Rough · Sloping bottom

1 Introduction

The mesoscale baroclinic eddies are one of the important structural forms of ocean water movement organization [1]. The vortex formation occurs under the action of various mechanisms, which include the influence of topographic and orographic inhomogeneities on ocean currents, spatially non-uniform wind forcing, and baroclinic instability of density-stratified currents, which characterize by the presence of a vertical velocity gradient. The eddies which contain water with thermohaline characteristics different from the surrounding waters, are formed, as a result of the baroclinic instability of jet frontal currents (Gulf Stream, Kuroshio and others). This water is captured by them during the formation of a current from the meanders and is transferred from one side of the front to the other. The orbital flow velocity and vorticity of these eddies are formed due to the local rise or subsidence of isopycnal surfaces. In the first case, cyclonic eddies are formed, and in the second case, anticyclonic eddies. The first detection and instrumental investigation synoptic eddies in the open ocean was carried out by Soviet scientists as part of the natural experiment "Polygon-70" in the Atlantic in 1970. A more complete picture of this type of eddies was obtained during the Soviet-American experiment "POLYMODE" in the Sargasso Sea in 1977–1978. [2].

In this paper, the "cylinder" method for generating baroclinic eddies in a rotating two-layer water medium was tried. This method allows to simulate mesoscale open ocean eddies containing water of the same density, temperature, and salinity as the surrounding water in laboratory conditions [3–7]. The task of the evolution investigation

T. Chaplina (Ed.): CIWO 2023, SPEES, pp. 19–28, 2023.
https://doi.org/10.1007/978-3-031-47851-2_3

of a baroclinic vortex created in the upper layer of a two-layer rotating water medium depending on the dimensionless parameter – the initial Burger number $Bu_0 = (Rd/R_0)^2$, where $Rd = (g'(H^1H^2)^{1/2})^{1/2}/f$ – Rossby baroclinic deformation radius, R_0 – cylinder radius, $f = 2\Omega$ – Coriolis parameter, $g' = g\Delta\rho/\rho = g\beta\Delta S$, $\beta \approx 7*10–4$ – salinity compression factor, ΔS – drop salinity between the layers, and H^1, H^2 are the thicknesses of the upper and lower water layers. At a qualitative level, the influence of the roughness and slope of the basin bottom on the stability and viscous degeneration of the vortex is studied by conducting experiments with the same values of the governing parameters over a smooth, rough, and sloping bottom [8].

2 Experimental Setup and Methodology for Providing Experiments

The experiments were provided in a cylindrical tank with a diameter D = 60 cm and a height H = 28 cm, located on a rotating platform with a period T from 5 to 40 s. Initially, fresh water with a thickness of H_1 is poured into the tank. Then the platform rotates with a given period of rotation. For formation the two-layer stratification, when the fresh water layer (layer 1) is spun up to a state of solid rotation, slightly tinted water is poured from the reservoir (4) located on the platform post, and with salinity S_2, thickness H_2 the lower layer (layer 2) is formed. This reservoir is connected by a hose to a cylindrical tank, in the the bottom center of which there is an inlet, to which the specified hose is connected. The lower layer is separated from the upper layer by a sharp salinity boundary, the thickness of which does not exceed 1 cm. Then, it waits for some time until the formed two-layer-stratified by density (salinity) water medium reaches a state solid-state rotation (Fig. 1).

For formation a cyclonic eddy, a hollow cylinder is dived in a two-layer salinity-stratified water medium spun up to a solid-state rotation to a depth exceeding the thickness of the upper layer. The water upper layer in the cylinder is tinted with a dark blue dye. Then, a water given volume is pumped out from the upper layer inside the cylinder, which is poured a volumetric burette with a volume of 1200 cm^3 (7) fixed on the platform post, and the upper layer thickness in the cylinder decreases by the value h compared to its thickness H_1 in the surrounding aquatic environment of the upper layer (Fig. 2a). When the medium storm created by this procedure dies out, the cylinder is quickly removed vertically from the water and, together with the device for its movement, is removed from the rotating platform. This causes a convergent flow in the upper layer, which tends to fill the gap and thicken the upper layer in the area where the withdrawn cylinder is located (Fig. 2b). Then a cyclonic vortex motion occurs in the upper layer. The thickness decreases under the cylinder former location, an anticyclonic vortex motion arises from the conservation law of the potential vortex in the lower layer.

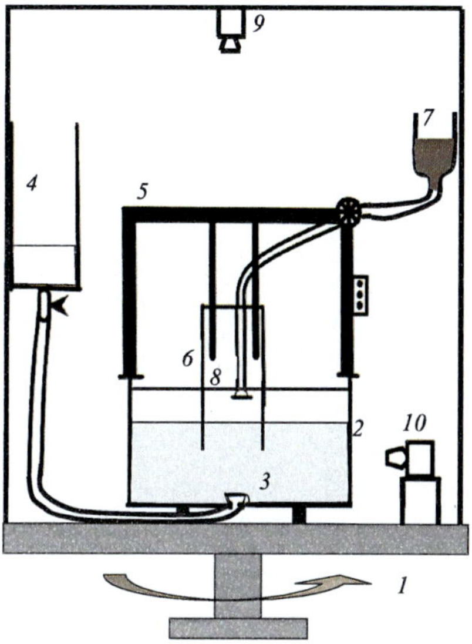

Fig. 1. Laboratory set-up: 1 – rotating platform; 2 – organic glass cylindrical; 3 – inlet connection at the tank bottom with a vertical velocity damping plate; 4 – reservoir to create a lower (more saline) water layer in the tank; 5 – device for automatic lowering (lifting) of a hollow cylinder - vortex generator; 6- hollow cylinder; 7 – volumetric burette with tap; 9 – video camera on the top; 10 – video camera on the side.

For formation an anticyclonic eddy, a given volume of tinted water with the density of the upper layer is poured into the cylinder from the burette from above. The thickness of this layer in the cylinder increases by h compared to its thickness H_1 outside the cylinder (Fig. 3c). After the of cylinder removal, an anticyclonic vortex is formed (Fig. 3d), and under it, due to the conservation law of the potential vortex, a cyclonic vortex arises.

Before lifting the cylinder, a video camera located above the platform is switched on, as well as a video camera located on the side near the outer surface of the annular wall of the tank. They record the process of evolution of the vortex motion of water visualized by dyes. To visualize the vortex motion, small paper particles are scattered at the surface of the water layer serving as flow tracers. With the program help, by calculating the movement of particles from figure to figure, their velocity is estimated at different points. To create a roughness of the bottom, a rug covered with beads with a characteristic size of beads $d = 0.4$ cm or a bath rug with an inhomogeneity size of about 0.5–1.0 cm is applied to the bottom. To create a bottom slope, an inscribed cone is installed on the flat bottom of the tank with the top down. The angle between the cone generatrix and the horizontal is 15°.

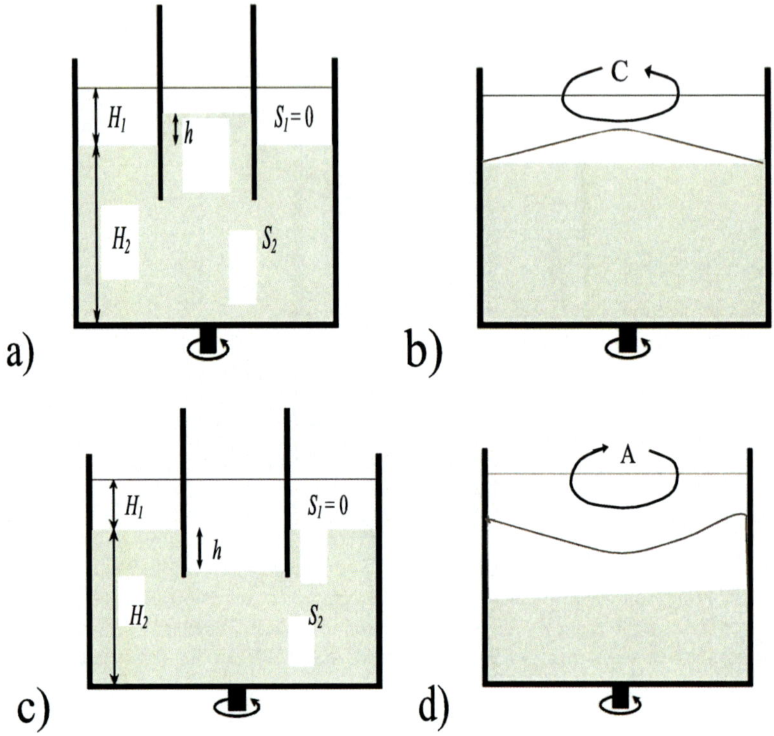

Fig. 2. Method for creating cyclonic eddies in the upper layer of a two-layer stratified aquatic environment in a basin on a rotating platform. Top layer - light, bottom layer - gray; a, b cyclonic eddies a) - before pulling out the cylinder; б) – after pulling out the cylinder. c, d antycyclonic eddies c) - before pulling out the cylinder; d) – after pulling out the cylinder.

3 Some Theoretical Aspects of the Dynamics of Baroclinic Vortices in a Rotating Fluid, their Instability and Viscous Degeneration

After it is pulled out, under the action of a radial pressure gradient, a vortex motion occurs in the upper layer, which, without taking into account non-stationarity and friction, is determined by the equation:

$$V^2/r + fV = g'\partial H^1/\partial r \quad (1)$$

V - the orbital velocity of water in the vortex, r - the radial coordinate measured from the vortex center.

From this equation, one can obtain a "geostrophic" estimate of the maximum orbital velocity of the cyclonic vortex, which has a positive sign for cyclone:

$$V_g = g'h/fR \quad (2)$$

And negative sign for anticyclone

$$V_g = -g'h/fR$$

An estimate should be made of the initial eddy radius after its adaptation.
In the case of an anticyclone:
In the case of a cyclone:

$$R_1 = \left(1 + h/H^1\right)^{1/2} \quad (3)$$

$$R_1 = \left(1 - h/H^1\right)^{1/2} R_0 \quad (4)$$

Now let's discuss the vorticity, which occurs in the upper and lower layers when the cylinder is removed.

In the lower layer under the anticyclone, it follows from the law of conservation of the potential vortex that a cyclone should appear in it:

$$(f + w_2)/H_2 = f/(H_2 - h);\ \omega_2 = (fh)/(H_2 - h) \quad (5)$$

When a cyclone occurs in the upper layer, an anticyclone occurs in the lower layer:

$$(f + w_2)/H_2 = f/(H_2 + h);\ \omega_2 = -(fh)/(H_2 + h) \quad (6)$$

It is follows from (5) and (6), that a very weak anticyclone generates in the lower layer under the cyclone, in comparison with the cyclone generating under the anticyclone.

Let us now consider the patterns of a barotropic eddy damping due to viscous friction on a smooth bottom. The vortex is determined by the core, in which the flow velocity V in the coordinate system associated with the rotating platform changes linearly with the radius: $V(r) = \omega r$, where - ω is its angular velocity. At the initial moment of time $\omega = \omega_0$.

The a barotropic vortex damping has an exponential character and is determined by the expression:

$$\omega = \omega_0 e^{-t/\tau} \quad (7)$$

$$\text{where } \tau = C_b H/(\nu\Omega))^{1/2} \quad (8)$$

Here determines τ, the vortex damping time constant [9]. In this case, C_b is some empirical coefficient that depends on the bottom roughness: for a smooth bottom, it should be close to one.

4 Description of Results of Experiments

Experiments were provided for cyclonic and anticyclonic eddies with the following constant values of the determining parameters: 1) the upper layer thickness - $H_1 = 4$ cm; bottom layer thickness - $H_2 = 10$ cm; inner radius of the cylinder - $R_0 = 10$ cm; change in the upper layer thickness inside the cylinder $h = 3$ cm. Accordingly, for an anticyclonic vortex, the upper layer thickness inside the cylinder is $H_1 + h = 7$ cm, and for a cyclonic vortex - H_1-h $= 1$ cm. In this case, such parameters change from experiment to experiment, as the platform rotation period Ω and salinity S_2 of the lower layer.

If an anticyclone is modeled, then after the removal of the cylinder, the water initially contained in the cylinder and observed with paint expands in the region of the upper layer. Its circle on the surface reaches its maximum size, then the process of compression occurs, and for some time radial oscillations continue with an amplitude of 10–20% of the radius of the tinted vortex core. If a cyclone is modeled, then, after removing the cylinder, the colored water spot first contracts and then expands, and intense radial oscillations are observed for some time. They decay with time.

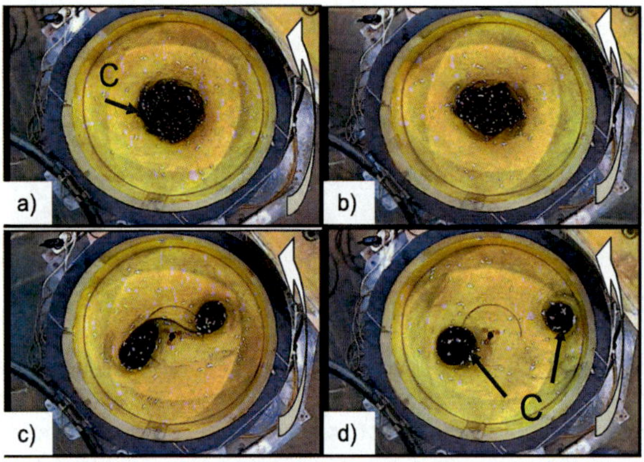

Fig. 3. Figure of strong instability and breakup of the cyclonic eddy for $Bu_0 = 0.28$ (T = 10 c; $\Delta S = 10$ psu): a) t/T = 1; b) t/T = 4; c) t/T = 9; d) t/T = 24. The white wide arrow on the right side of each figure shows the cyclonic platform rotation direction.

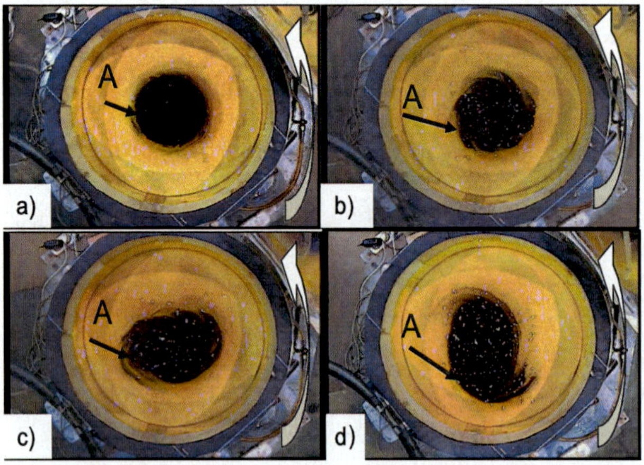

Fig. 4. Figure of weak instability of the anticyclonic eddy for $Bu_0 = 0.28$ (T = 10 c; $\Delta S = 10$ psu): a) t/T = 1; b) t/T = 4; c) t/T = 9; d) t/T = 24. The white wide arrow on the right side of each figure shows the cyclonic platform rotation direction.

The evolution of both cyclonic and anticyclonic eddies follows one of three scenarios. The first scenario: the eddy loses its axisymmetric shape, azimuthal waves run along its

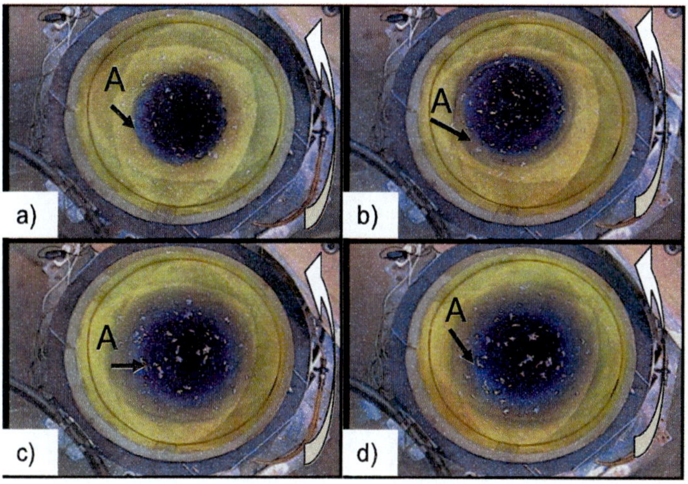

Fig. 5. Figure of stable evolution of the anticyclonic eddy for $Bu_0 = 1.13$ (T = 10 c; ΔS = 40 psu): a) t/T = 1; b) t/T = 4; c) t/T = 9; d) t/T = 24. The white wide arrow on the right side of each figure shows the cyclonic platform rotation direction.

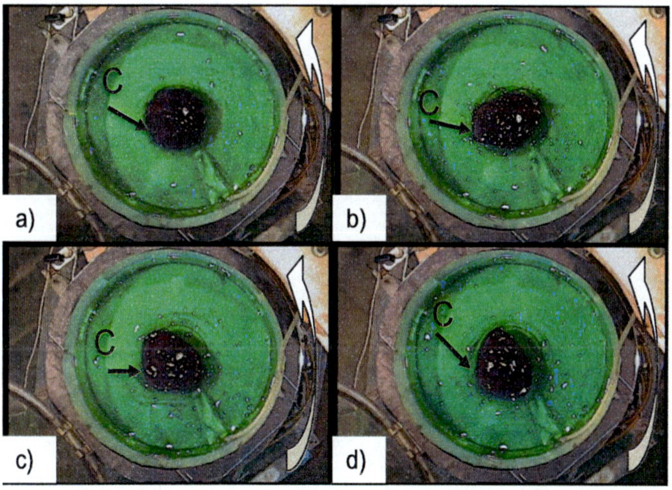

Fig. 6. Figure of weak instability of the cyclonic eddy in the case of an sloping bottom for $Bu_0 = 0.28$ (T = 10 c; ΔS = 10 psu): a) t/T = 1; b) t/T = 4; c) t/T = 9; d) t/T = 24. The white wide arrow on the right side of each figure shows the cyclonic platform rotation direction.

boundary it stretches and breaks into two or more secondary eddies of the same sign (Fig. 3). This is a strong instability.

The second scenario: the eddy loses its axisymmetry, but does not break, but exists in an "elongated" form, and vortices of the opposite sign join its "narrow" sides, which pull filaments from the original eddy, winding around the attached eddy (Fig. 4). This is a weak instability. The reason for the attached eddies generation may be the result of shear instability at the anticyclone periphery in the area of cyclonic velocity shear [10].

The third scenario: the eddy is stable, retains its axisymmetric shape until the end of the experiment, and its orbital velocity in the core slowly decreases with time due to viscous friction (Fig. 5).

The first scenario is implemented for small values of Bu_0. The second - at intermediate values of Bu_0. The third - at large values of Bu_0.

On Fig. 6 shows the video footage of a cyclone over a sloping bottom, the parameters of which are the same as for a cyclone over a smooth bottom, the video footage of which is shown in Fig. 4.

Experiments with the investigation of the stability/instability of cyclonic and anticyclonic eddies were provided at the following values of the initial Burger number: $Bu_0 = 2.27$; 1.12; 0.56; 0.28; 0.11. At $Bu_0 = 2.27$ - 0.56, cyclones and anticyclones are stable. For $Bu_0 = 0.28$ - 0.11 all eddies are unstable. The instability of cyclonic eddies at the same values of parameters is stronger than that of anticyclonic ones. At $Bu_0 = 0.28$, the cyclonic eddy experienced strong instability, while the anticyclonic eddy experienced weak instability (Fig. 5). The most probable reason for the stronger manifestation of the instability of cyclonic eddies is the influence of the ageostrophicity of cyclonic eddies.

The stability criterion should include not only the Burger number, but also the Froude number. On Fig. 7 shows the dependence of the parameter Bu/Fr on Bu_0. The parameter Bu/Fr, which takes into account ageostrophy, describes the transition from a stable to an unstable state of a eddy much better. In this case, the transition from the steady state of both cyclones and anticyclones occurs, apparently, at Bu/Fr = 1.0–1.5, if $H_1/H_2 = 0.4$, and $E \approx 10^{-4}$. This result needs additional experimental verification and refinement.

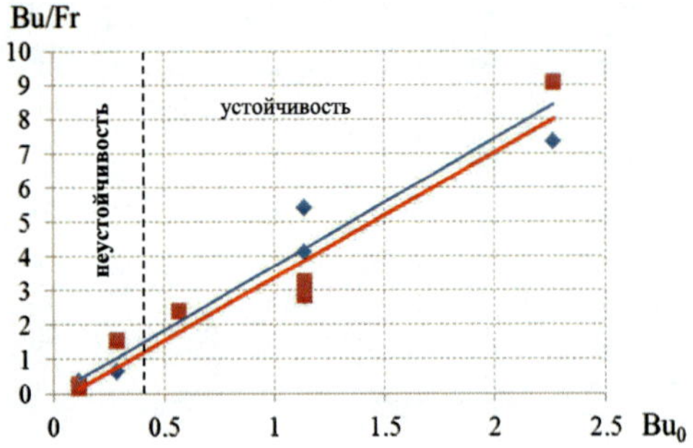

Fig. 7. a) Dependence of parameter Bu/Fr on Bu_0. Red squares are cyclones, blue diamonds are anticyclones. Lines – linear trends.

The attenuation of the maximum orbital velocity of the cyclone eddy with time occurs according to a close to exponential law. Plots of the dimensionless orbital velocity in a eddy (V/Vg) from dimensionless time (t/T) in Figs. 8 are presented in semi-logarithmic coordinates.

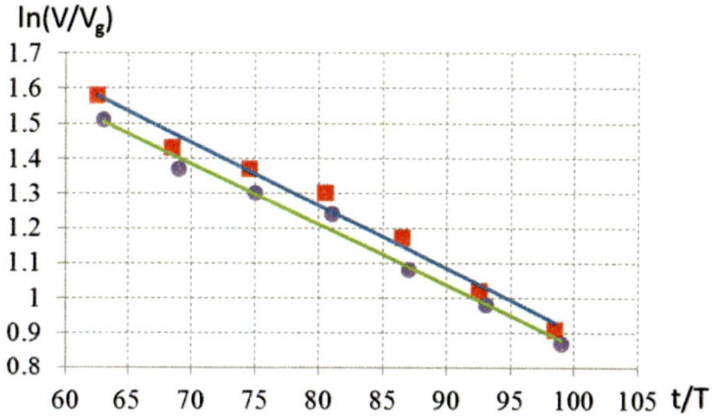

Fig. 8. Dependences of $\ln(V/V_g)$ on t/T for a stable eddy over a smooth and rough bottom. Cyclone, $\Delta S = 20$ psu, T = 10 s, $Bu_0 = 2.27$. Red dots are measured data for a smooth bottom, purple dots are measured data for a rough bottom, blue line is a linear trend for a smooth bottom, green line is a linear trend for a rough bottom. Roughness size - $r_0 = 0.2$ cm.

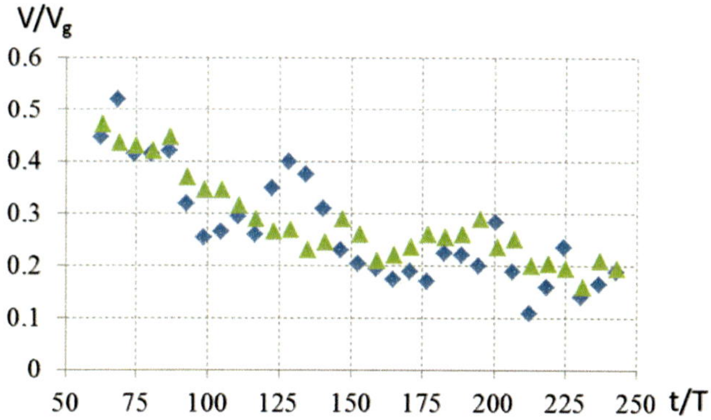

Fig. 9. Dependences of V/V_g on t/T. Anticyclone, $\Delta S = 40$ psu, T = 10 s, $Bu_0 = 1.13$. Blue dots are measured data for a smooth bottom, green dots are measured data for a rough bottom.

From Fig. 8 it follows that the bottom roughness does not affect the eddy damping rate. The eddy is damped due to friction effects in the upper layer. Considering the temporal evolution of stable anticyclonic eddies, we managed to reveal an interesting phenomenon. In several experiments, a non-monotonic decrease in the maximum orbital velocity with time was observed (Fig. 9) [11].

Acknowledgments. The work was supported in the frame of the government assignment theme No FMWE-2021-0002, and the financial support of the RSF, project No 21-77-10052.

References

1. Kamenkovich, V.M., Koshlyakov, M.N., Monin, A.S.: Mesoscale eddies in the ocean. Gidrometeoizdat, Leningrad, 264 p. (1982)

2. Koshlyakov, M.N., Belokopytov, V.N.: Mesoscale eddies in the open ocean: review of experimental investigations. Phys. Oceanography **27**(6), 559–572 (2020)
3. Saunders, P.M.: The Instability of a Baroclinic Vortex. J. Phys. Oceanogr. **3**(1), 61–65 (1973)
4. Griffiths, R.W., Hopfinger, E.J.: The Structure of Mesoscale Turbulence and Horizontal Spreading at Ocean Fronts. Deep Sea Research Part A. Oceanographic Research Papers **31**(3), 245–269 (1984). https://doi.org/10.1016/0198-0149(84)90104-3
5. Griffiths, R.W., Linden, P.F.: The stability of vorticies in a rotating stratified fluid. J. Fluid Mech. **105**, 283–316 (1981)
6. Kloosterziel, R.C., van Heijst, G.J.F.: An experimental study of unstable barotropic vorties in a rotating fluid. J. Fluid Mech. **223**, 1–24 (1991)
7. Zatsepin, A.G., Kostianoy, A.G.: Laboratory studies of the instability of baroclinic vortices and fronts. Coherent structures and self-organization of ocean motions. Moscow. Nauka, pp. 163–176 (1992)
8. Zatsepin, A.: Some experiments on rotating baroclinic vorvices.- Tech. Report. WHOI-83–41, 272–285 (1983)
9. Greenspan, H.P.: The Theory of Rotating Fluids. Gidrometeoizdat, Leningrad, 304 p. (1975)
10. Fedorov, K.N.: The Physical Nature and Structure of Oceanic Fronts. 333 p. Springer (1983)
11. Zatsepin, A.G., Elkin, D.N., Shvartsman, D.R.: Preliminary results of Laboratory investigatons of the evolution of nonfrontal eddies in a two-layer rotating fluid. J. Oceanological Res. **51**(1), 5–35 (2023)

Analysis of Hydrophysical and Energy Fields in the Northern Part of the Black Sea on the Basis of the Numerical Model With the Assimilation of the Hydrological Data in 2016

N. A. Evstigneeva(✉) and S. G. Demyshev

Marine Hydrophysical Institute RAS, 299011 Sevastopol, Russia
naevstigneeva@yandex.ru

Abstract. Three experiments were performed to reconstruct the fields of currents during the periods of hydrological cruises at the R/V "Professor Vodyanitsky" in 2016: June - July (summer season), September - October (autumn season) and November - December (autumn-winter season) on the basis of the z-coordinate hydrodynamic model of the Black sea and a procedure of assimilation of contact measurements of temperature and salinity. The components, that made the most significant contribution in the equations of dencity of kinetic and potential energy, were analyzed in order to study the possible mechanisms of formation of the reconstructed features of water dynamics.

Keywords: Numerical Modeling · Observational Data Assimilation · Fields of Currents · Mesoscale Features of Circulation

1 Introduction

The Marine Hydrophysical Institute (MHI) regularly monitors the state of the hydrology of the Black Sea, which allows replenishing sea database, identifying and specifying a number of peculiarities of hydrophysical fields at various scales [1–5]. Materials of expeditionary measurements expanded our understanding of the relationship between thermohaline and kinematic fields. In [1] on the basis of the materials of the 76th cruise in September 2013, it was shown that during the survey period the Rim Current (RC) meandered and, as it moved westward, turned in a southwestern direction. A strip of low-temperature waters, which was formed as a result of intensive coastal upwelling, was observed along the entire coast from the Belgorod–Dnestrovsky estuary to the Danube delta. In [2] the results of hydrological studies, performed in November 2015 near the Crimea and to the northwest of the Western cyclonic gyre of the Black Sea, were discussed. The observed differences in the thermohaline structure of waters from the climatic norms, obtained from hydrological data, were a manifestation of different types of variability (synoptic and interannual). In [3] hydrological studies of 2016 showed that a seasonal signal was traced in the distributions of all thermohaline characteristics according to the data of surveys of three seasons. In [4] according to the hydrological

T. Chaplina (Ed.): CIWO 2023, SPEES, pp. 29–34, 2023.
https://doi.org/10.1007/978-3-031-47851-2_4

measurements in November and December 2017, carried out during the 98th and 101st cruises, intense synoptic changes were found in the spatial structure of thermohaline fields due to the variability water dynamics. In [5] it was found that synoptic gyres in hydrological characteristics could manifest themselves in different ways. Most anticyclonic gyres in thermohaline fields were characterized by low temperatures and salinity. Some anticyclones were traced by thermohaline signs above the core of the cold intermediate layer by increasing temperature and decreasing salinity. Deepening of the core of that layer and the upper boundary of the hydrogen sulfide zone was observed in the zones of all anticyclones.

Several successive hydrological surveys were carried out by MHI in 2016. That fact made it possible to reconstruct continuous three-dimensional hydrophysical fields and study the obtained features of circulation in dependence of the season.

The purpose of the study was to reconstruct and analyze hydrophysical and energy fields with high spatial resolution near the coast of the Black Sea with the help of the assimilation of contact measurement data in a hydrodynamic model during hydrological surveys in the summer, autumn and autumn-winter seasons of 2016 and to investigate the formation of circulation features in dependence of the season. The relevance of the work lied in the possibility of more efficient use of the obtained expeditionary data, reconstructing the mesoscale structure of current fields in all layers and obtaining detailed characteristics of eddies and jet currents.

2 Material and Methods

All experiments were performed on the basis of methods of mathematical modeling, which included a nonlinear z-coordinate model of the MHI and the assimilation procedure, developed at the MHI. We used 27 vertical horizons and a horizontal resolution 1.64 km. To set the atmospheric effect close to the observed one, the Greek data of SKIRON (1/10°) [7] were considered. The processes of turbulent momentum exchange and turbulent vertical diffusion was parameterized using the Mellor–Yamada 2.5 theory [8]. The equations of changing the density of kinetic and potential energy [9] were used to calculate energy characteristics.

A four-dimensional analysis procedure [10], using the Kalman filter, was performed to assimilate the data of three hydrological surveys in 2016 [6]. Until the moments of receipt of the observational data, the fields of T and S were predicted, using the model equations. At the moments of assimilation, the hydrological fields were corrected by data interpolation, taking into account the correlation radius. Vertical profiles of temperature and salinity in 2016 were taken from the MHI data bank. Hydrological measurements (125, 128 and 107 stations, respectively) were made using the CTD SBE 911plus complex.

The currents and energy characteristics was calculated for three periods in 2016 – summer, autumn and autumn-winter seasons, including the days of expeditions). To verify obtained results, the results of geostrophic calculations and data of instrumental measurements of currents using the ADCP acoustic meter [3] were used.

3 Results and Discussion

Processing the results of numerical experiment in the summer period, we obtained some distinguishing features of circulation. Two anticyclonic eddies with a diameter of about 60 km were reconstructed to the west of the area under consideration and a cyclonic eddy with a diameter of about 80 km to the east. Along the Crimean coast the RC was distinctly observed (its velocity reached 40 cm/s at the depth of 3m). Eastern part of the polygon was characterized by the anticyclonic vorticity of currents in the coastal zone and the deep part of the sea – by cyclonic vorticity. An anticyclonic eddy with a diameter of about 50 km was periodically recorded in the vicinity of Yalta. Small–scale eddies were generated in the layer 1–10 m along the Crimean coast when flowing Rim Current around shoreline irregularities and the meandering of Rim Current.

Figure 1 demonstrated model fields of currents and energy characteristics at the depth of 5 m on 9th and 16th, July. Regions, in which the values of $\Pi \leftrightarrow E$ were negative (transition from kinetic energy to potential), increased over time and it meant that baroclinic instability could occured there.

The areas of mesoscale eddy formation along the shores were prevailingly characterized by negative values of the term determining the contribution to kinetic energy from the wind $\tau \rightarrow E$ (i.e., wind and currents on the surface were multidirectional).

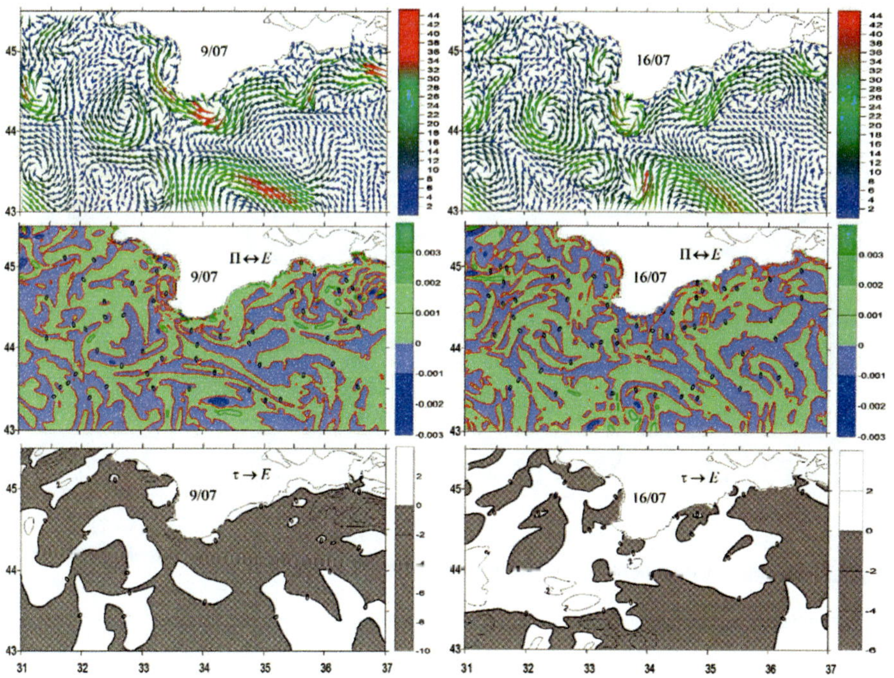

Fig. 1. Model fields of currents and energy characteristics in July, 2016

Processing the results of numerical experiment in the autumn period, we obtained the following features of circulation. There was a narrow jet, pressed to the Crimean coast, as well as anticyclonic eddies with a diameter of about 60 km near Sevastopol and eddies with a diameter of about 80 km of various directions of vorticity to west, center and east of the polygon. Small-scale eddies were generated in the layer 1–10 m along the eastern shore of Crimea when flowing Rim Current around coastline irregularities. When wind forcing reached to 15 m/s, the maximum velocity of the RC reached a value of 54 cm/s, all of the listed eddy formations were less distinguished.

Figure 2 demonstrated model fields of currents and energy characteristics at the depth of 5 m on 2nd and on 18th of October. The sections, where $\Pi \leftrightarrow E < 0$, alternated with the sections, where $\Pi \leftrightarrow E > 0$. Areas of negative values of $\Pi \leftrightarrow E$ prevailed along the eastern and western coasts of Crimea over the time.

From the analysis of the fields of $\tau \rightarrow E$, it was noted that wind forcing contributed significantly in energy inflow in almost the entire polygon. The values of $\tau \rightarrow E$ were prevailingly negative in the areas of mesoscale eddy formation along the coast.

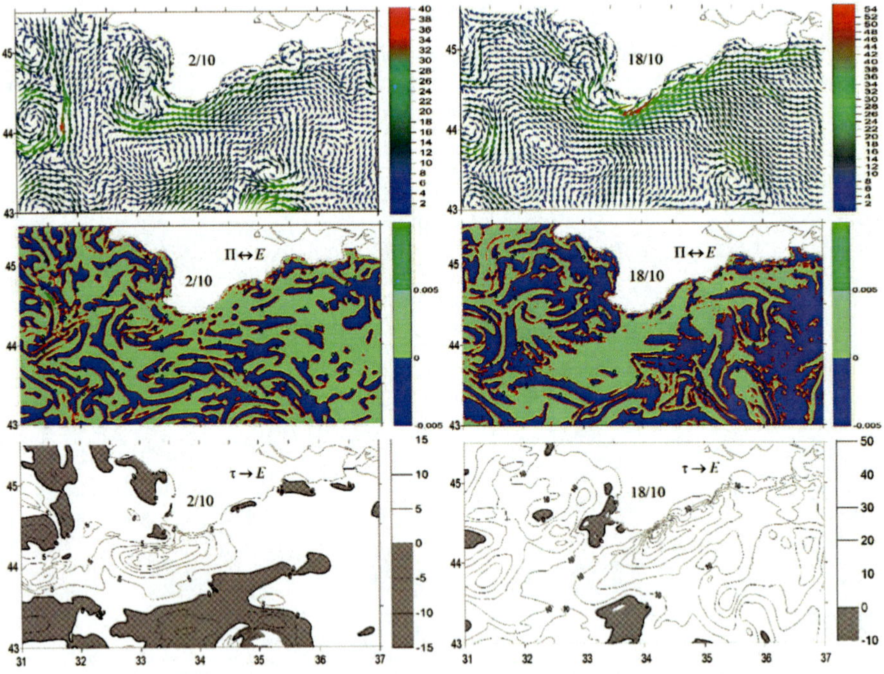

Fig. 2. Model fields of currents and energy characteristics in October, 2016

Processing the results of numerical experiment in the autumn-winter period, we obtained the following features of circulation [11]. The RC had a distinct jet character (with a maximum velocity of up to 52 cm/s with wind velocity up to 17 m/s). With some weakening of the wind forcing, eddies of different scales could be generated between the shore and the RC. The formation and development of cyclonic and anticyclonic eddies of different scales were noted in the western part. They could merge into larger one during the further calculation. It was also possible to note small-scale cyclonic and anticyclonic eddies, when flowing the Rim Current around shoreline irregularities.

Figure 3 demonstrated the model fields of currents and energy characteristics at the depth of in November and December, 2016. The values of energy term $\Pi \leftrightarrow E$ were prevailingly negative in the areas, where intense mesoscale variability observed. Baroclinic instability occured along the eastern and western Crimean coasts. Fields of $\tau \rightarrow E$ were characterized by high variability, negative values were observed in the areas of mesoscale eddy formation to the west.

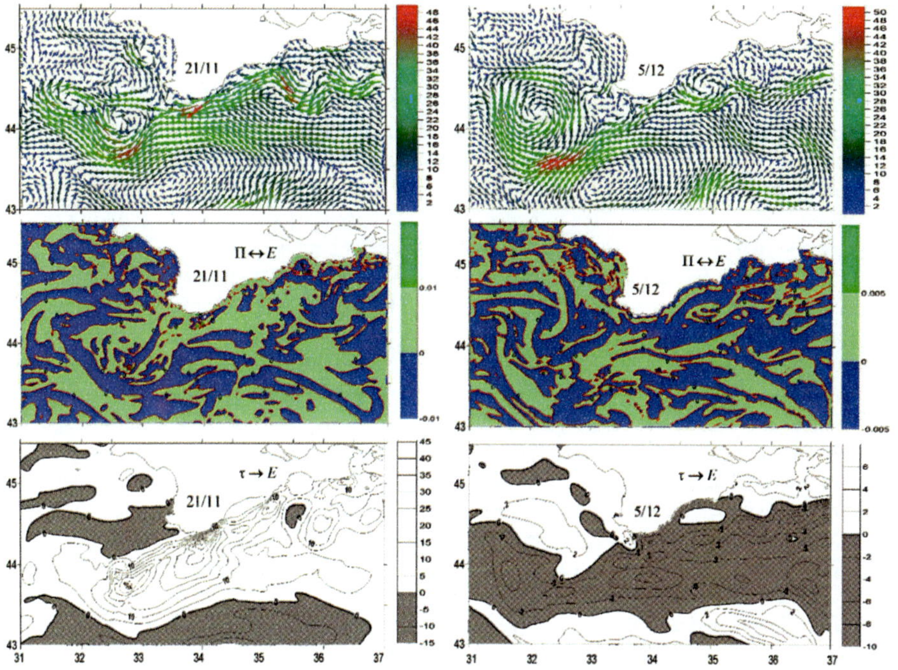

Fig. 3. Model fields of currents and energy characteristics in November and December, 2016

4 Conclusion

Hydrophysical and energy fields were reconstructed by z-coordinate model and contact data of three surveys of 2016 at the "Professor Vodyanitsky", in the coastal area of the Black Sea during periods of expeditions (summer, autumn and autumn-winter seasons).

The RC, uneven wind field and the influence on the current of the coastline and relief heterogeneities had a dominant influence on the water dynamics for all three seasons. Rim Current was clearly reproduced in autumn and autumn -winter, mesoscale variability was more intensive in summer. During all calculations, anticyclonic eddies were generated near Sevastopol. Cyclonic and anticyclonic eddies were also obtained on the west, center and east of the region. Small-scale eddies could be traced along the coast in the upper layer in case of low speed of wind. A shear of horizontal velocity and baroclinic instability of currents may cause the formation of mesoscale and submesoscale eddies.

Acknowledgments. Current fields were reconstructed by MHI model with hydrological data of 2016 within financial support of the state assignment on topic No. FNNN–2021–0003 (code "Operational Oceanology"). Energy characteristics were analysed with financial support of the state assignment on topic No. FNNN–2021–0004 (code "Oceanological processes").

References

1. Artamonov, Yu.V., Alekseev, D.V., Shutov, S.A., et al.: Dynamics and structure of the waters of the northwestern part of the Black Sea in September 2013. Ecological safety of the coastal and shelf zones and integrated use of shelf resources, vol. 1, 4–14 (2017). (in Russian)
2. Artamonov, Yu.V., Alexeev, D.V., Kondratyev, S.I., et al.: Hydrological Conditions in the Western Part of the Black Sea in November, 2015 (based on the Data Obtained in the 81st Cruise of R/V Professor Vodyanitsky). Phys. Oceanography **4**, 57–70 (2016)
3. Artamonov, Yu.V., Skripaleva, E.A., Alekseev, D.V., et al.: Hydrological research in the northern part of the black sea in 2016 (87th, 89th and 91st Cruises of R/V Professor Vodyanitsky). Phys. Oceanography **25**(3), 229–234 (2018)
4. Artamonov, Yu.V., Fedirko, A.V., Skripaleva, E.A., et al.: Seasonal and synoptic changes in the structure of waters southwest of the Crimean Peninsula in the autumn-winter period of 2017 (98th and 101st cruises of R/V "Professor Vodyanitsky"). Ecological safety of the coastal and shelf zones and integrated use of shelf resources, vol. 3, 4–18 (2019). (in Russian)
5. Artamonov, Yu.V., Skripaleva, E.A., Fedirko, A.V., et al.: Water circulation in the northern part of the Black Sea in summer-winter 2018. Ecological safety of coastal and shelf zones and integrated use of shelf resources **1**, 69–90 (2020). (in Russian)
6. Demyshev, S.G.: A numerical model of online forecasting Black Sea currents. Izv. Atmos. Ocean. Phys. **48**, 120–132 (2012)
7. NonHydrostatic SKIRON/Eta Modelling System. http://forecast.uoa.gr/forecastnew.php
8. Mellor, G., Yamada, T.: Development of a turbulence closure model for geophysical fluid problems. Rev. Geophys. Space Phys. **20**(4), 851–875 (1982)
9. Demyshev, S.G.: Energy of the Black Sea climatic circulation. Part I. Discrete equations for the rate of change of kinetic and potential energies. Meteorol. Hydrology **9**, 65–80 (2004). (in Russian)
10. Knysh, V.V., Moiseenko, V.A., Chernov, V.V.: Some results of four-dimensional analysis of hydrophysical fields in the Tropical Atlantic. Izv. Academy of Sciences of the USSR. Physics of the atmosphere and ocean **24**(7), 744–752 (1988). (in Russian)
11. Demyshev, S.G., Evstigneeva, N.A., Dymova, O.A.: Investigation of Features of Water Circulation in the Northern Part of the Black Sea on the Basis of the Assimilation of Observational Data in the Autumn–Winter Season of 2016. Springer Proceedings in Earth and Environmental Sciences, pp. 53–60 (2023)

Water Structure and Dynamics in the Transform Fracture Zones of the Tropical Atlantic According to Instrumental Measurements

F. N. Gippius[1(✉)], A. N. Demidov[1], K. V. Artamonova[1,2,3], and S. B. Krasheninnikova[4]

[1] Lomonosov Moscow State University, 119991 Moscow, Russia
gippiusfn@my.msu.ru

[2] Russian Federal Research Institute of Fisheries and Oceanography, 105187 Moscow, Russia

[3] Shirshov Institute of Oceanology of Russian Academy of Sciences, 117997 Moscow, Russia

[4] A.O. Kovalevsky Institute of Biology of the Southern Seas of Russian Academy of Sciences, 299011 Sevastopol, Russia

Abstract. The structure and dynamics of different water masses in the Tropical Atlantic is investigated using the results of instrumental measurements. The main attention is focused on the Vema fracture zone and the northern part of the WOCE A15 section. In the Vema Fracture zone an intradiurnal variability of hydrological parameters was observed. We also assess the long-term variability of hydrological parameters along the WOCE A15 section.

Keywords: Tropical Atlantic · Vema fracture zone · WOCE A15 section · Water masses · Water transport · Water structure

1 Introduction

The Atlantic Ocean holds a key position in the global thermohaline circulation since the formation of two deep and bottom waters occurs here – the North Atlantic Deep Water (NADW) and the Antarctic Bottom Water (AABW). The initial components of NADW are predominantly generated in Labrador, Greenland and Irminger Seas mainly due to deep convection, whereas the AABW is formed in the Weddell Sea. These waters propagate from their origin regions towards the equator generally via the abyssal regions of the western Atlantic. Their movement eastwards is limited by the Mid-Atlantic Ridge (MAR), which spans across the entire Atlantic Ocean from north to south. Several transform fracture zones (FZ) are present in the MAR. These FZ are relatively deep and narrow valleys. Thus, NADW and AABW can proceed from the Western to the Eastern Atlantic only via these FZ. According to previous studies, the most intense eastward water transport occurs in the Tropical Atlantic via the Vema, Romanche, Chain and Doldrums FZs [1]. However, despite numerous studies devoted to the deep and abyssal waters of the Tropical Atlantic and their circulation, which were carried out during the recent decades [1–3], a clear and consistent pattern of water distribution in this area is still

T. Chaplina (Ed.): CIWO 2023, SPEES, pp. 35–41, 2023.
https://doi.org/10.1007/978-3-031-47851-2_5

not constructed. Moreover, the transport of water through these FZs varies depending on the initial data and methods used for the assessments. Thus, the transport of AABW in the Vema FZ is estimated between 0.05 Sv and 2.24 Sv ([4] and [5], respectively).

The transport of deep and bottom waters, including NADW and AABW, was assessed in some recent oceanic reanalyses [6, 7]. The advantage of these products is the possibility to estimate long-term variations in different parameters of water masses, including their transport. However, such reanalyses usually have a very low vertical resolution in deep and bottom layers. Their accuracy at such depths is also debatable.

Therefore, instrumental *in situ* observations of principal hydrological and hydrochemical parameters in key regions of the World Ocean, including deep passages in the MAR, are of high value and interest, since they are the only mean to calibrate numerical ocean circulation models and assess the reliability of reanalyses.

The purpose of this paper is to give an overview of results, which are based on our own instrumental observations collected during research cruises in the Tropical Atlantic during the recent years.

2 Data and Methods

The data used in our study was collected during three research cruises – the 45th cruise of RV "Akademik Nikolaj Strakhov" and the 60th and 63rd cruises of RV "Akademik Ioffe". These cruises will be further noted as ANS-45, AI-60 and AI-63, respectively.

During the cruise AI-45 in October–November 2019 the research was focused on Vema, Doldrums and Vernadsky FZs [8]. In the cruise AI-60 in December 2021 – February 2022 hydrological and hydrochemical sampling was carried out in the Vema, Arkhangelsky and Vernadsky FZs [9]. In the cruise AI-63 in September–December 2022 we carried out studies in the Kane Gap, as well as in the Romanche and Chain FZs. Furthermore, some stations in this cruise were located along the WOCE A15 section.

The location of the study areas and stations carried out during the cruises ANS-45 and AI-60 can be found in [8] and [9], respectively, whereas the stations carried out during the cruise AI-63 are shown on Fig. 1.

During all cruises, hydrological parameters were determined at stations from the surface to the bottom. Temperature and salinity were measured with a SBE19plusV2 CTD profiler, which was equipped with additional sensors measuring the amount of dissolved oxygen (SBE43) and turbidity (Seapoint). The parameters of currents (direction and velocity) were measured with a Nortek Aquadopp 6000 current meter.

During the cruises AI-60 and AI-63 hydrochemical analyses of collected water samples were carried out in a shipborne laboratory. The samples were collected from various depths using a SBE 32 Carousel Water Sampler with Niskin bottles. All samples were processed immediately upon their collection. The amount of dissolved oxygen, phosphates, silicates and nitrates, as well as the pH value were determined during this analysis. The samples were processed according to the methods described in [10].

Data on all hydrological and hydrochemical parameters collected during the field works was stored and processed using the OceanDataView software [11].

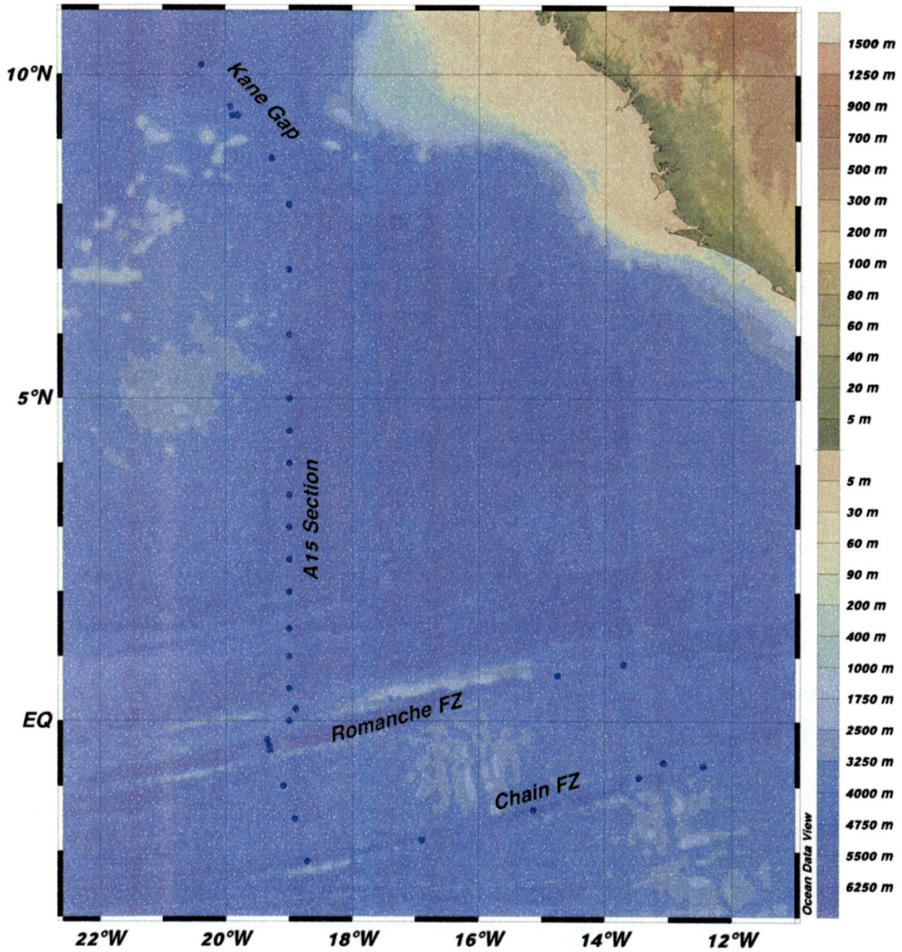

Fig. 1. Location of the stations during the AI-63 cruise

3 Results and Discussion

The biggest amount of data during our mentioned expeditions was collected at the Vema FZ during the cruise ANS-45. Moreover, the most interesting results correspond to a station located on the main sill of the Vema FZ (10.795° N, 41.02° W). At this station, quasi-diurnal series of CTD and currents measurements were carried out in the layer between the 3700 m depth and the bottom (4660 m). The experiment started at approx. 12:00 on November 13th, 2019 and ended at approx. 15:00 on the next day. During the experiment, the SBE 32 Carousel frame with the CTD profiler and the current meter moved constantly up and down between 3700 m depth and the bottom (its path can be seen as a thin black line on Fig. 2). Thus, we obtained about 20 subsequent profiles of CTD and current parameters. A technical interruption of the experiment was made around midnight of November 14th, during which data was downloaded from the equipment and fresh batteries were installed.

Figure 2 shows the vertical and temporal variability of potential temperature observed during the quasi-diurnal experiment. The most important issue here is the temporal variability of the isotherms' location, since the boundary of the lower NADW (LNADW) and the AABW is commonly attributed to one or the other isotherm. Initially, the isotherm $\theta = 2$ °C was considered as this boundary [12]. However, authors of [3] and [13] argue, that it is more correct to divide these waters at the isotherm $\theta = 1.7$ °C – in this case, layers of maximal dissolved oxygen and CFC concentrations correspond to the LNADW, which is more consistent from the hydrological point of view. As we can see from Fig. 2, both these isotherms experienced a quite strong variability of their vertical position during the experiment. Their vertical shift reached 100–150 m. Moreover, the amplitude of such oscillations increased with depth; for example, the isotherm $\theta = 1.45$ °C experienced a vertical shift of approx. 300 m. The potential temperature in the bottom layer varied between 1.36–1.41 °C during the experiment, which almost completely corresponds to the previously reported multiannual variability [14]. This observed quasi-diurnal variability can be a source of incorrect estimations of different water masses transports (first of all, AABW and lower components of NADW), because it is not reproduced by numerical models and reanalyses.

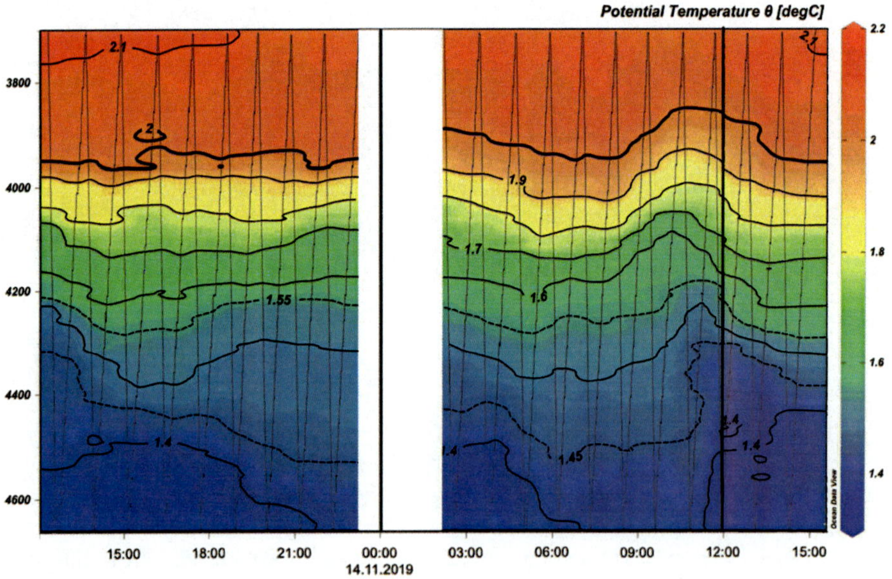

Fig. 2. Quasi-diurnal variability of potential temperature in the deep and bottom layers at the main sill of the Vema FZ. The vertical axis corresponds to depth (m), the horizontal axis represents the sampling time (UTC)

A reasonable question is the nature of the fluctuations, which were observed during this quasi-diurnal experiment at the main sill of the Vema FZ. The most likely reason, as we believe, is the effect of semi-diurnal tides, which affect the waters moving eastward along the Vema FZ. Evidence for this hypothesis can be the distribution of residual current velocities and directions. Figure 3 shows such a distribution in the bottom layer. Residual currents were calculated as the difference between the average current at this specific layer during the entire experiment and the current observed during each individual sampling. We found out that a variability of these residual currents was observed and that the period of this variability was close to 12 h. Moreover, the shapes that outline these residuals are close to typical tidal ellipses. Tidal effects in deep water passages were previously observed and described, e. g. measurements in the Kane Gap revealed tidal variations of currents with a pronounced M2 harmonic constituent [15].

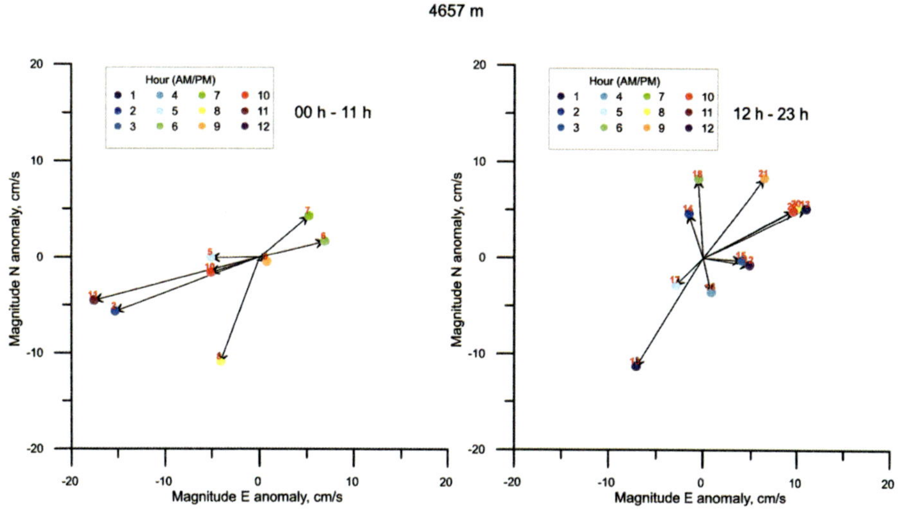

Fig. 3. Residual currents in the bottom layer at the main sill of the Vema FZ. The left panel corresponds to observations between 0 h and 11 h, the right panel – between 12 h and 23 h. The time of each individual observation is defined by the dot color and the number next to the dot

During the AI-63 cruise, we partially repeated the WOCE A15 Section, which follows the meridian 19° W. During the WOCE project, works on this section were carried out in 1994 onboard the RV Knorr [16]. Measurements and water sampling on this section were not repeated until the AI-63 cruise. One of our goals was to assess the multiannual variability on this section. Here, we will give some preliminary results of this assessment.

The highest variability of potential temperature was observed in the upper 1000 m layer – it is likely due to seasonal factors and should not be taken into account. We observed an increase in potential temperature in the layers between approx. 1000 and 3500 m. The layer between approx. 3500 and 4500 m experienced cooling by 0.1–0.2 °C. These depths correspond to lower layers of NADW. However, no cooling is observed in this layer approx. Between 3° and 5° N. Finally, warming waters were revealed in bottom layers of the Romance and Chain FZs at depths below 4500 m. These layers correspond to the AABW.

Regarding the salinity variations, we found an increase in salinity nearly everywhere below 1500 m. The values of this increase were of 0.01–0.02 PSU. An exception are the waters located at approx. 4000 m in the Romanche and Chain FZs and between them, which may indicate a freshening of the lower NADW.

4 Conclusion

Despite the advances in the development of numerical models and reanalyses, instrumental observations remain a source of valuable data on the structure and dynamics of oceanic waters. Using such data, we revealed some features of deep-water circulation and long-term variability in the Tropical Atlantic.

Thus, in the Vema FZ we described a quasi-diurnal variability in the location of isotherms as well as in current direction and velocities. The boundary between the lower NADW and AABW can experience a vertical displacement of about 100–150 m within a day.

We also described evidences of long-term variability along the WOCE A15 section during the last 28 years. Almost all layers corresponding to NADW and AABW experience an increase in both potential temperature and salinity. A decrease of potential temperature was observed in the layer corresponding to lower NADW.

Acknowledgments. Processing of data collected during ANS-45 and AI-60 cruises was supported by the Russian Science Foundation within the grant № 19-17-00110-P, processing of data collected during the AI-63 cruise was supported by the Russian Science Foundation within the grant № 23-17-00032.

The study was carried out within the state research task of the A.O. Kovalevsky Institute of Biology of the Southern Seas "Functional, metabolic and toxicological aspects of the existence of hydrobionts and their populations in biotopes with different physicochemical regimes" (№ 121041400077-1).

We are grateful to V. V. Mashura, A. S. Bich, N. A. Rykov, S. A. Babich, and E. A. Nizienko for their help in data acquisition during the cruises. We thank Prof. E. V. Ivanova and Dr. D. G. Borisov for their help in organizing the expeditions. We are also grateful to the crews of RV "Akademik Nikolaj Strakhov" and RV "Akademik Ioffe" for their assistance in the cruises.

References

1. Morozov, E.G., Demidov A.N., Tarakanov R.Y., Zenk W. Abyssal Channels in the Atlantic Ocean. – Dordrecht: Springer Netherlands, 266 p. (2010)
2. Rhein, M., Stramma, L., Send, U.: The Atlantic Deep Western Boundary Current: Water masses and transport near the equator. J. Geophys. Res. **100**(C2), 2441–2457 (1995)
3. Herrford, J., Brandt, P., Zenk, W.: Property changes of deep and bottom waters in the Western Tropical Atlantic. Deep. Res. Part I Oceanogr. Res. Pap. **124,** 103–125 (2017)
4. Vangriesheim, A.: Antarctic bottom water flow through the Vema fracture zone. Oceanologica Acta **3**(2), 199–207 (1980)
5. McCartney, M.S., Bennett, S.L., Woodgate-Jones, M.E.: Eastward flow through the Mid-Atlantic Ridge at 11 N and its influence on the abyss of the eastern basin. J. Phys. Oceanography **21**(8), 1089–1121 (1991)

6. Frey, D.I., Morozov, E.G., Fomin, V.V., Diansky, N.A.: Spatial structure of the Antarctic water flow in the Vema fracture zone of the Mid-Atlantic Ridge. Izvestiya, Atmospheric Oceanic Phys. **54**, 621–625 (2018)
7. Stephens, J.C., Marshall, D.P.: Dynamical pathways of Antarctic bottom water in the Atlantic. J. Phys. Oceanography **30**(3), 622–640 (2000)
8. Ivanova, E.V., et al.: Multidisciplinary investigation of the transform fault zones doldrums and vema during cruise 45 of the R/V "Akademik Nikolaj Strakhov". Oceanology **60**(3), 424–426 (2020)
9. Ivanova, E.V., et al.: Investigations of lateral sedimentation and water mass properties in the tropical atlantic during cruise 60 of the R/V Akademik Ioffe. Oceanology **62**(4), 581–583 (2022)
10. Sapozhnikov, V.V., et al.: Guidance on the chemical analysis of marine and fresh waters during ecological monitoring of fishery reservoirs and areas of the World Ocean promising for fishing. – M.: VNIRO Publishing House, 202 p. (in Russian) (2003)
11. Schlitzer, R.: Ocean Data View. https://odv.awi.de
12. Wüst, G.: 1936. Deutsche Atlantische Expedition "Meteor" 1925–1927. 6 (1). Berlin. 411
13. Rhein, M., Stramma, L., Krahmann, G.: The spreading of Antarctic bottom water in the tropical Atlantic. Deep Sea Research Part I: Oceanographic Research Papers **45**(4–5), 507–527 (1998)
14. Demidov, A. N., Ivanov, A. A., Gippius, F.N., Dobroliubov, S.A.: Transport of deep and bottom waters through the Mid-Atlantic Ridge in the Vema fracture zone. Doklady Earth Sci. **494,** 735–740 (2020)
15. van Haren, H., Morozov, E., Gostiaux, L., Tarakanov, R.: Convective and shear-induced turbulence in the deep Kane Gap. J. Geophys. Res. Oceans **118**(11), 5924–5930 (2013)
16. Goyet, C., Eischeid, G.: Partial pressure (or fugacity) of carbon dioxide, dissolved inorganic carbon, alkalinity, temperature, salinity and other variables collected from discrete sample, profile and underway - surface observations using CTD, Carbon dioxide (CO_2) gas analyzer and other instruments from the KNORR in the North Atlantic Ocean and South Atlantic Ocean from 1994-04-03 to 1994-05-21 (NCEI Accession 0115002). NOAA National Centers for Environmental Information. Dataset. https://doi.org/10.3334/cdiac/otg.woce_a15_1994

The Spatio-Temporal Variability of the Ocean-Atmosphere Heat Exchange and the Sea-Ice Cover in the Nordic Seas

A. S. Kaledina[1,2](✉) and I. L. Bashmachnikov[1,2]

[1] St Petersburg State University, 199034 St. Petersburg, Russia
a.kaledina@spbu.ru

[2] Nansen International Environmental and Remote Sensing Centre, 199034 St. Petersburg, Russia

Abstract. Turbulent heat flux in the Nordic Seas were decomposed into characteristic modes of variability using Empirical Orthogonal Functions (EOF). The first EOF describes the general warming trend in the Arctic Basin. The region of the highest fluctuations is consistent with the area of the sea-ice reduction by more than 20% between 1959–1980 and 2004–2021. In these areas the air temperature increases is a result of an increasing ocean heat release to the atmosphere. The ocean-atmosphere heat exchange over the ice-free ocean is a result of a change in atmospheric wind pattern.

Keywords: Nordic Seas · Heat flux · Sea-ice cover · Air temperature · Empirical Orthogonal Functions

1 Introduction

In the climate change research, the factors affecting an observed increase in the air temperature, as well as a reduction of the ice cover in the Arctic region, remain an important issue [1, 2]. The Barents Sea on-going warming is either associated with cloud feedback [3], or with an increasing transfer of the atmospheric heat and moisture, as well as of the oceanic heat, from the mid-latitudes [4–6]. In the Greenland Sea, the water temperature is linked to a change in the sea-ice area, which is a result of a variability of the sea-ice flux from the Arctic Basin and changes in atmospheric forcing primarily linked to NAO pattern. A heat flux from the ocean to atmosphere is also one of the reasons for the observed changes in the air temperature in the Arctic, in particular during the winter period [7]. The ice cover reduction increases the area of ocean-atmosphere contact and, thus, contributes to an increase of the atmospheric temperature [8, 9].

2 Data and Methods

In this study, we use ERA5 atmospheric reanalysis (1959 to the present), which is the fifth generation of global climate and weather products of the European Centre for Medium-Range Weather Forecasts (ECMWF). The reanalysis replaces the widely used

T. Chaplina (Ed.): CIWO 2023, SPEES, pp. 42–47, 2023.
https://doi.org/10.1007/978-3-031-47851-2_6

ECMWF ERA-Interim reanalysis and assimilates observations from the Earth surface to the upper atmosphere. We use monthly average data of mean surface latent and sensible heat fluxes of the ocean-atmosphere exchange, air temperature at 2 m above the sea surface, the eastward and northward components of the horizontal wind at 10 m, the mean sea level pressure and the sea-ice cover over the period 1959 – 2021 with a spatial resolution of 0.25° × 0.25°.

We decomposed the spatio-temporal variability of the ocean-atmosphere turbulent heat flux (a sum of the latent and sensible heat fluxes) into the natural variability modes using the method of empirical orthogonal functions (EOF). EOF analysis makes it possible to identify typical patterns of variability of geophysical parameters.

Variability of the sea-ice cover was estimated in the region 60°N–90°N, 45°W–105°E, which includes the Nordic Seas and the neighboring areas of the Arctic. We only consider the grid cell with the sea-ice concentrations over 15%. The change in the sea-ice concentration between two periods 1959–1980 and 2004–2021 was estimated. In further analysis we consider only the cells, where the change in the sea-ice concentration exceeded 20%.

3 Results

In practically the entire water area of the Nordic Seas, the average water temperature over the analysis period is higher than the air temperature and latent and sensible heat fluxes are directed from the ocean to atmosphere. The highest values of the total ocean-to-atmosphere heat flux are observed in and around the Norwegian Sea and in the western part of the Barents Sea, with the maximum values along the Norwegian and West Spitsbergen currents.

In the decomposition of the ocean-atmosphere heat exchange into EOF patterns, the first two EOF modes covered 63% of the total variance (Fig. 1). To explain the mechanisms of formation of the spatial and temporal variability in the first EOFs, the PCs of these EOFs were correlated with various atmospheric and oceanic indices. In addition, for high and low PC values, composite maps of the fields of surface wind speed and atmospheric pressure were computed. The "high" or "low" PC values were designated, when the data-point of the PC time series deviated by more than one standard deviation above or below the time mean value, respectively.

The first mode of the EOF (Fig. 1a,c) describes 39% of the dispersion of ocean-atmosphere heat exchange. The growing PC describes an increasing heat loss by the ocean since the 1970s over most of the study area of the North Atlantic and Arctic. This correlates with the global warming trends of the air temperature. The highest values of the ocean heat loss in EOF1 are observed in areas of a significant (by more than 20%) decrease in the sea-ice concentration over the last 50 years in the northern and eastern Barents Sea, northwest Kara Sea, areas around Spitsbergen and along the East Greenland Current (Fig. 4a). In ERA5 reanalysis the same areas correspond to the areas of the highest increase in the air temperature over 1959–2021 (Fig. 4b).

For the ice-free regions, the amplitude of EOF1 variability is much smaller than in those periodically covered with ice. Here we relate the growth of ocean heat release to changes in the atmospheric circulation patterns. Thus, for the Labrador and Irminger

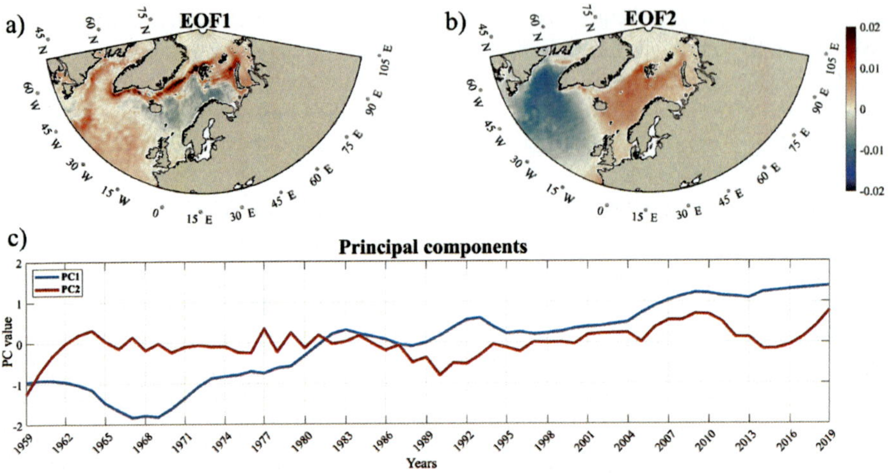

Fig. 1. Spatial distribution of the amplitude of variability of turbulent heat fluxes from the ocean to the atmosphere and interannual variability for 1959–2019. (a) EOF1 (39% variance), (b) EOF2 (24% variance), (c) normalized PCs of first two EOF modes.

Seas and the adjacent area of the North Atlantic, positive values of EOF1 are associated with an increase in cold westerly and northwesterly winds from the Canadian archipelago and a decrease in the ocean heat release over the Norwegian Sea, both associated with a decrease in frequency of the northerly winds from the Arctic (Fig. 2). The difference in the atmospheric pressure fields during periods of high minus low PC1 values suggests that the deepening of the Icelandic minimum over the Irminger Sea with a simultaneous air pressure increase over Scandinavia contribute to the wind field anomalies characteristic of the positive phase of the East Atlantic Pattern (EA). EA forms the second mode of the air pressure variability over the North Atlantic [10]. In fact, the PC1 of the ocean-atmosphere heat exchange significantly correlates with EA. Thus, similar to PC1, the EA index has been increasing since the 1970s from predominantly negative values in the 1950s-1970s to predominantly positive values in the 2000s-2010s (Fig. 1c). Thus, EOF1 has a mixed origin, combining the global warming trends manifested in both, the sea-ice retreat and the changes in the atmospheric circulation over the North Atlantic and Nordic Seas.

This atmospheric circulation variability is supposed also to form the EOF2 pattern (Fig. 3). This second EOF describes 24% of the variance of the ocean-atmosphere heat exchange. For high PC2 values the EOF2 pattern suggests an enhanced ocean heat release in the western subpolar North Atlantic, where the northwesterly winds intensify, and a decrease of the ocean heat release in the eastern subpolar North Atlantic, characterized by a simultaneous attenuation of the northern winds over the Norwegian-Greenland seas (Figs. 1b,c, 3a–c). The variability of PC2 shows moderately high and significant correlations (0.58) with the NAO index at zero lag. Pressure field anomalies (in years of high minus low PC2 values) only partly correspond to the structure of the NAO-related anomalies during its positive phase (Fig. 3d). Similar to the positive phase of the NAO,

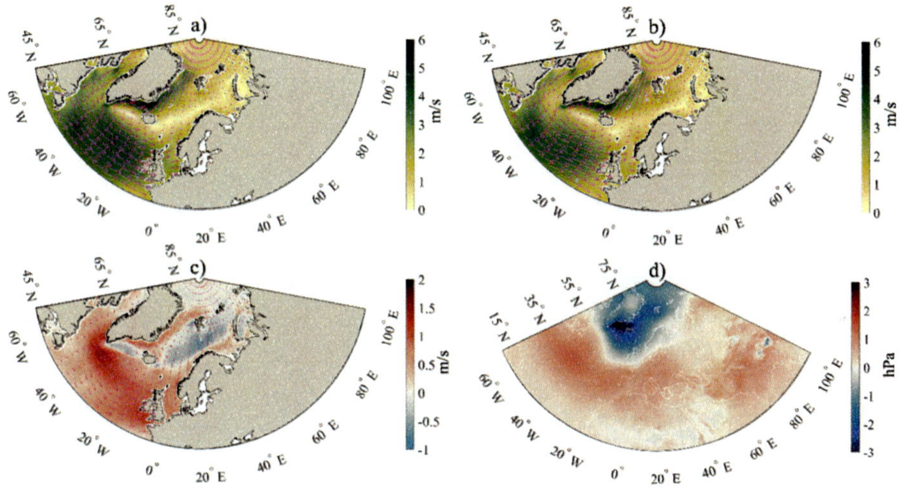

Fig. 2. Composite maps of the spatial distribution of average surface wind speed (m/s): (a) for the years with high values of PC1, (b) for the years with low values of PC1; (c) wind speed anomalies (a–b). (d) Anomalies of the sea level pressure (hPa) for the years with the high values of PC1 relative to the years with the low values of PC1.

the air pressure decreases over practically the entire subpolar North Atlantic (including Greenland) rather than being concentrated over the Iceland Sea as for EOF1.

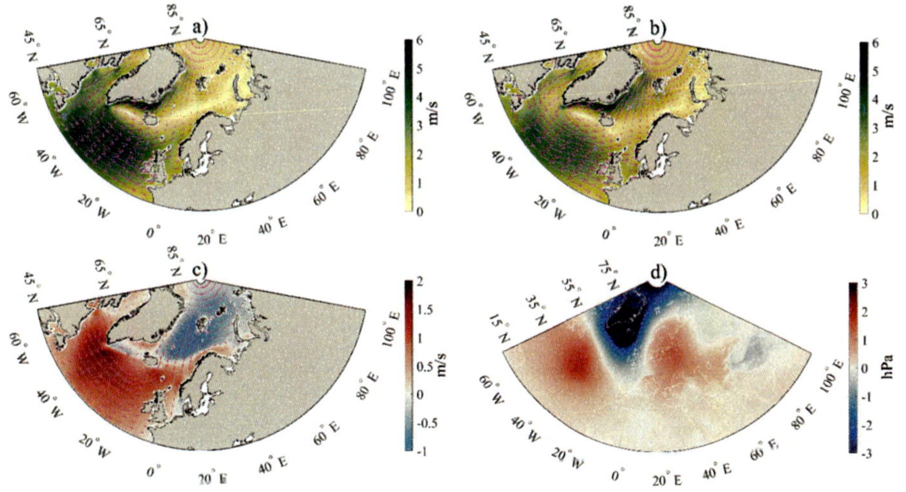

Fig. 3. Composite maps of the spatial distribution of average surface wind speed (m/s): (a) for the years with high values of PC2, (b) for the years with the low values of PC2; (c) wind speed anomalies (a–b). (d) Anomalies of the sea level pressure (hPa) for the years with the high values of PC2 relative to the years with the low values of PC2.

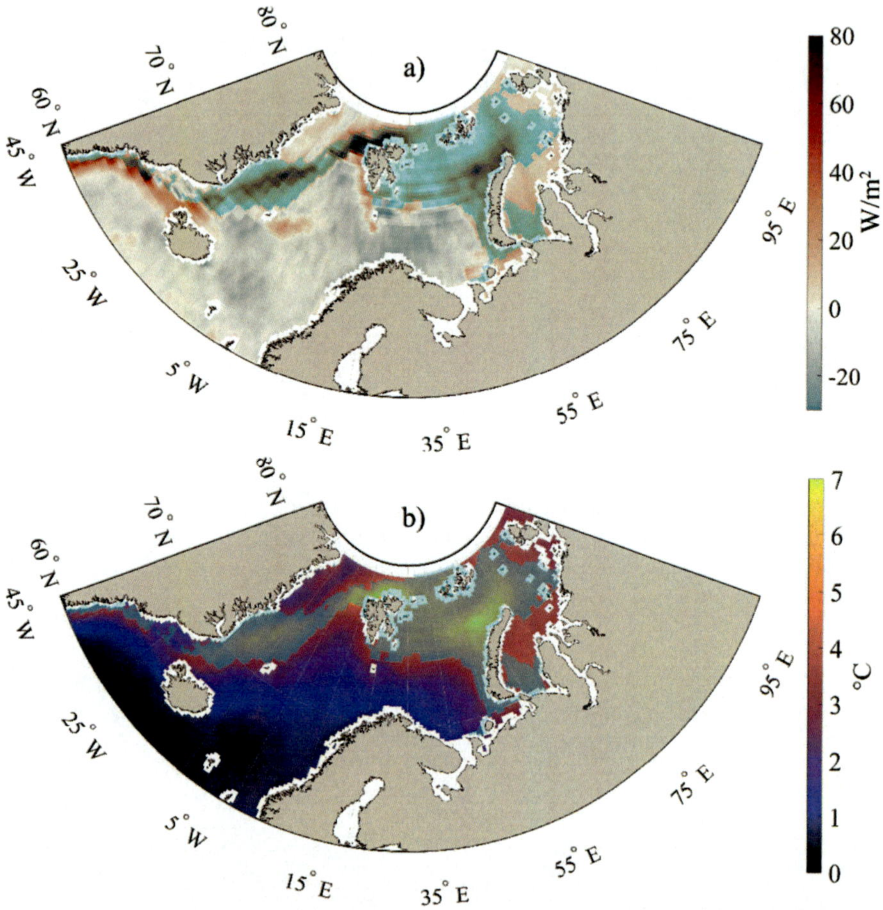

Fig. 4. a) The anomalies of the ocean-atmosphere heat exchange for 2004–2021 relative to 1959–1980, b) the anomalies of the air temperatures for 2004–2021 relative to 1959–1980. In blue color the sea-ice mask of the ice concentration change over 20% over the two periods in overlayed.

The region of the most rapid increase of the ocean-to-atmosphere heat release and of the air temperature growth (2004–2021 minus 1959–1980) (Fig. 4) is in perfect agreement with the mask of a decrease in the sea-ice concentration by more than 20% over the same periods (blue mask in Fig. 4). This allows to assume that, the sea-ice concentration decrease by more than 20% is the driving force for a rapid intensification of the ocean heat release in the northern Nordic Seas, which, in turn, leads to an increase in the air temperature. Within the area of the sea-ice reduction by more than 20% (masked in blue in Fig. 4), the integral increase in the ocean heat flux to the atmosphere reaches 150 TW, and the amount of the additional heat released over 2004–2021 relative to 1959–1980 is 290 kJ.

4 Conclusion

A change in the sea-ice concentration by more than 20% was the most efficient mechanism for a rising air temperature over the Nordic Seas during the latest decades. Outside of these areas the change in the ocean-atmosphere heat exchange, as well as the rise of the air temperature is much less drastic. These latter changes are linked to changes in atmospheric circulation pattern over the North Atlantic, which explains the observed spatial patterns of both, EOF1 and EOF2 of the ocean-atmosphere heat exchange in the ice-free regions. The EOF1 pattern is thus of a mixed origin, combining the warming trends, manifested both in the retreat of ice and in the change in the atmospheric circulation over the North Atlantic and the North European Basin. PC1 of the EOF1 correlates with variations of the EA index, while PC2 (EOF2) correlates with that of NAO index.

Acknowledgements. The study was funded by the St Petersburg University grant No. 94033410.

References

1. Semenov, V.A.: Modern arctic climate research: progress, change of concepts, and urgent problems. Izv. Atmos. Ocean. Phys. **57**, 18–28 (2021). https://doi.org/10.1134/S0001433821010114
2. Polyakov, I.V., et al.: One more step toward a warmer Arctic. Geophys. Res. Lett. **32**(L17605), 1–4 (2005). https://doi.org/10.1029/2005GL023740
3. Goosse, H., et al.: Quantifying climate feedbacks in polar regions. Nature Commun. **9**(1), 1919 (2018). https://doi.org/10.1038/s41467-018-04173-0
4. Henderson, G.R., Barrett, B.S., Wachowicz, L.J., Mattingly, K.S., Preece, J.R., Mote, T.L.: Local and remote atmospheric circulation drivers of Arctic change: a review. Front. Earth Sci. **9**, 709896 (2021). https://doi.org/10.3389/feart.2021.709896
5. Sang, X., Yang, X.-Q., Tao, L., Fang, J., Sun, X.: Decadal changes of wintertime poleward heat and moisture transport associated with the amplified Arctic warming. Climate Dyn. **58**(1–2), 137–159 (2022). https://doi.org/10.1007/s00382-021-05894-7
6. Cai, Z., et al.: Amplified wintertime Barents Sea warming linked to intensified Barents oscillation. Environ. Res. Lett. **17**(4), 04468 (2022). https://doi.org/10.1088/1748-9326/ac5bb3
7. Alexeev, V.A., Walsh, J.E., Ivanov, V.V., Semenov, V.A., Smirnov, A.V.: Warming in the Nordic Seas, North Atlantic storms and thinning Arctic sea ice. Environ. Res. Lett. **12**(8), 084011 (2017). https://doi.org/10.1088/1748-9326/aa7a1d
8. Årthun, M., Eldevik, T., Smedsrud, L.H.: The role of atlantic heat transport in future arctic winter sea ice loss. J. Clim. **32**(11), 3327–3341 (2019). https://doi.org/10.1175/JCLI-D-18-0750.1
9. Matveeva, T.A., Semenov, V.A., Astafyeva, E.S.: Arctic sea-ice coverage and its relation to the surface air temperature in the Northern Hemisphere. Ice Snow **60**(1), 134–148 (2020). https://doi.org/10.31857/S2076673420010029
10. Woollings, T., Hannachi, A., Hoskins, B.: Variability of the North Atlantic eddy-driven jet stream: variability of the North Atlantic Jet Stream. Q. J. Roy. Meteorol. Soc. **136**(649), 856–868 (2010). https://doi.org/10.1002/qj.625

"1.5D" Vertical Turbulent Exchange Model Verification Based on Microstructural Probe Data on Cruise 122 of the R/V Professor Vodyanitsky

D. A. Kazakov(✉), M. I. Pavlov, and A. M. Chukharev

Marine Hydrophysical Institute of RAS, 299011 Sevastopol, Russia
dk@mhi-ras.ru

Abstract. From June 7 to July 2, 2022, for the first time, information on the fluctuation characteristics of the flow in the stratified layers of the Black Sea was obtained with data on the vertical distribution of oxygen, chlorophyll A and turbidity at depths up to more than 900 m on cruise 122 of the R/V Professor Vodyanitsky. Studies in this region included obtaining the dependences of the vertical turbulent diffusion coefficient K on the buoyancy frequency N and on the depth z according to the pulsation characteristics of the flow. Calculation results were compared with the results of a 1.5D vertical turbulent exchange model, as well as the hydrophysical features general analysis of the region under study. Data analysis showed the presence of several layers with different characteristic features of mixing in them. The obtained power-law approximations of the dependences $K \cong A \cdot N^{\alpha}$ in these layers showed good convergence with the model ones in the power part.

Keywords: Black Sea · Energy dissipation · Stratified layer · Vertical turbulent exchange · Buoyancy frequency · Measuring complex · Field measurements · Turbulence models · 1.5D model

1 Introduction

In recent decades, the relevance of solving scientific and practical problems in the field of forecasting the state of ecosystems of oceans and seas has increased. One of the key approaches to identifying patterns of changes in the state of marine systems is the development of turbulent exchange models based on modern high-precision measurements. Accounting for a small number of aspects that affect mixing processes leads to the fact that turbulence is not adequately described in models of mesoscale and submesoscale circulations, which leads to insufficient correspondence to measurements. Therefore, the staff of the turbulence department of MHI RAS proposed the option of establishing dependence on local stratification and other hydrological conditions [1]. Such an approach to modeling is actually a simplification of the processes that actually occur in the basin under study, however, it offers a fairly reliable and physically substantiated version of the model of vertical turbulent exchange in the stratified layers of the Black Sea. This model can further be used in predictive models of climate, evolution of hydrological structure, biological systems, etc.

T. Chaplina (Ed.): CIWO 2023, SPEES, pp. 48–58, 2023.
https://doi.org/10.1007/978-3-031-47851-2_7

The purpose of this study is an attempt to identify patterns that determine the vertical structure of hydrophysical fields and the intensity of mixing, based on known semi-empirical approaches to determining the turbulent regime in the stratified layers of the Black Sea.

2 Methods and Data Processing

The data acquisition was carried out with an MSS-90L microstructure probe presented in Fig. 1. It is equipped with high-resolution turbulence (1 temperature and 2 shear) sensors, standard CTD sensors, and sensors for oxygen dissolved in water, turbidity, and chlorophyll content. It is intended for hydrophysical and hydrochemical processes studies, including measurements of average hydrological parameters and turbulence in sea and lake waters up to 1000 m deep.

Fig. 1. Microstructural Probe MSS90L

The cruise was conducted on the basis of the Center for Collective Use "R/V Professor Vodyanitsky" of the Federal State Budgetary Institution of Science of the Federal Research Center "Institute of Biology of the Southern Seas RAS", head of the expedition Burdiyan N.V. A total of 75 stations were made, the location of which you can see in Fig. 2, in the depth range from 40 to 970 m. Of these, 70 were of the greatest interest. Due to the political situation in the country, the measurements were carried out in the territorial waters of the Russian Federation within the 12-mile coastal zones. The data set was a table with the values of the following parameters of the environment and the probe: pressure (dbar), conductivity (mS/cm), temperature (microstructural and fine, °C), velocity shear in two axes, accelerometer readings (m/s^2), probe tilt angles in two axes (°), turbidity (FTU), chlorophyll-A content (mg/l), internal temperature (°C), oxygen content (mV, mg/l, ml/l, %), salinity (PSU), density (kg/m^3), date, time. We would like to pinpoint that the measurement data obtained by the probe fall mainly on the shelf and continental slope zone, while the 1.5D model was described and tested before mainly on the deep water data of the Black Sea basin.

A processing and analysis program was written to process data from microstructural sensors. Due to the design features of the microstructural sensor [2], the average sounding velocity vector of the device is aligned with the axis of the axisymmetric airfoil of the sensor. Thus, the sensor is not sensitive to axial forces, and the turbulent velocity transverse component creates a deflecting force on the airfoil, so the vertical component is not taken into account in the calculation. A vertical gradient of horizontal fluctuations is used: first, spikes and noises caused by oscillating movements and vibration of the device are removed, then a bandpass filter is used, which limits the frequencies of the measured

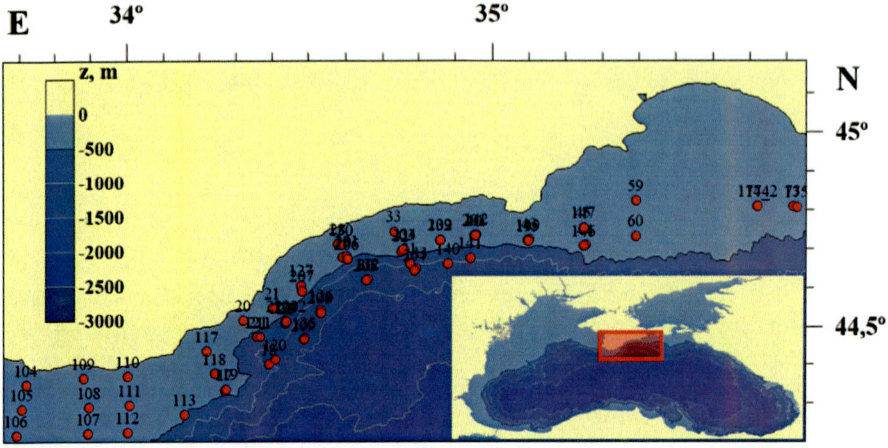

Fig. 2. Sounding stations location scheme on cruise 122 of the R/V Professor Vodyanitsky

fluctuations by the turbulence spectrum inertial subrange from the low-frequency side (k_l) and the Kolmogorov wave number $k_c = (1/2\pi)(\varepsilon/\nu^3)^{1/4}$ from high-frequency side, where ν is the kinematic viscosity, ε is the turbulence energy dissipation rate.

The dissipation rate is calculated by the iterative method: first, the boundary wave numbers k_l and k_{max} are determined, and then the *du*/*dz* value spectrum is calculated using the Welch's method [3]. This method allows us to estimate ε with a good vertical resolution; here we limited ourselves to a resolution of 2 m. In this case, the records are divided into overlapping segments, which are multiplied by the Hann time window, then the Fourier transform is performed, followed by averaging the spectral functions over all segments. The quality of the obtained pulsation spectra was estimated from the average absolute deviation from the Nasmyth model spectrum [4]. The dispersion $\left[\overline{(du/dz)^2}\right]$ is determined by integrating the spectrum values in the selected range of wave numbers. The dissipation rate of turbulent energy is determined by relation

$$\varepsilon = \frac{15}{2}\nu\left[\overline{(du/dz)^2}\right]. \tag{1}$$

This value of ε is used to calculate the Kolmogorov (non-dimensional) wavenumber. If the stop conditions are not met, the cycle is repeated, starting with the calculation of the spectrum. The stopping criteria are a small change in the value of k_c (less than the step in the spectrum Δk) and the excess of k_c over k_{max}. The final value of ε is obtained by averaging the values from both sensors installed in probe. If their values differ by more than an order of magnitude, the minimum value is used.

The coefficient of vertical turbulent diffusion is calculated according to the formula proposed by Osborne:

$$K \cong \frac{R_f}{1 - R_f}\frac{\varepsilon}{N^2}, \tag{2}$$

where R_f is the dynamic Richardson number (the ratio of the potential energy increase rate in the system to the rate of energy input there, spent on mixing). In papers [5–7] its constancy was established for the considered phenomena using various approaches. $N = \sqrt{\frac{g}{\rho}\frac{\partial \rho}{\partial z}}$ is the buoyancy frequency, or Brent-Väisälä, calculated through the potential density and its gradient, and accordingly implicitly including the temperature and salinity components. The R_f value in this study was taken equal to 1/4, in accordance with the approximate result from [6]. The coefficient profiles thus obtained are subjected to a 2-stage filtering. The Savitsky-Golay filter was used to highlight the main trend in K values with depth for subsequent comparison with the model distribution.

3 1.5D Model

The first similar vertical exchange model for the Black Sea, including vertical advection and vertical turbulent diffusion, was constructed in [8]. It was based on the conditions of conservation of salt mass, water mass and heat balance - the inverse problem was solved, when the known (measured) average vertical distributions of thermohaline characteristics $T(z)$, $S(z)$ were used to determine the vertical distributions of sources and absorbers of heat, salt and water. It was assumed that the processes under consideration took place in a basin with vertical walls of a certain height and a horizontal bottom of a fixed area. The upper boundary of the studied layer corresponded to the upper boundary of the Marble Sea jet flowing in from the Bosphorus and approximately corresponded to the level, starting from which the intensity of intra-annual temperature fluctuations sharply decreases. Two areas were distinguished: the area of transformed marble sea waters, or plumes, and the sea basin proper. The influence of the plume is realized as boundary conditions on the side wall. At the horizon z_*, the volume flow q_B of the Lower Bosphorus current was set, as well as its average thermohaline characteristics T_B, S_B. Under the action of buoyancy forces, the stratified Lower Bosphorus water descends and transforms when interacting with the surrounding fluid. The Lower Bosphorus current is the only source of salt in the basin. At the lower boundary $z = H$, the absence of liquid and salt flows and the known bottom heat flux were set. Thus, the "plume-pool" system under study was divided vertically into three layers (Fig. 3).

Subsequently, the models were refined in accordance with the depth-averaged area of the basin under study, and the average temperature and salinity profiles in the upper stratified layer were assimilated to obtain new dependences for the coefficients of vertical turbulent diffusion of temperature and salts [9]. Comparing to the works mentioned earlier, with this approach, the desired functions are calculated in the depth range of 50–1750 m.

Experimental data collected over the past decade in the Black Sea made it possible to develop ideas about the patterns of vertical turbulent exchange in the main pycnocline on the basis of theoretical and semi-empirical models [10–13]. New patterns closest to natural data were obtained for the energy dissipation rate, mainly depending on the vertical structure of the local buoyancy frequency. In this case, five conditional stratified layers with their characteristic features of vertical turbulent mixing have already been determined. The results of the model dependence $K(N)$ are shown in Fig. 4.

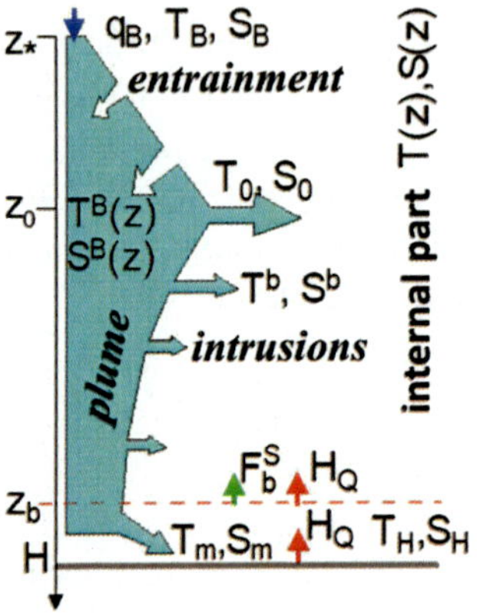

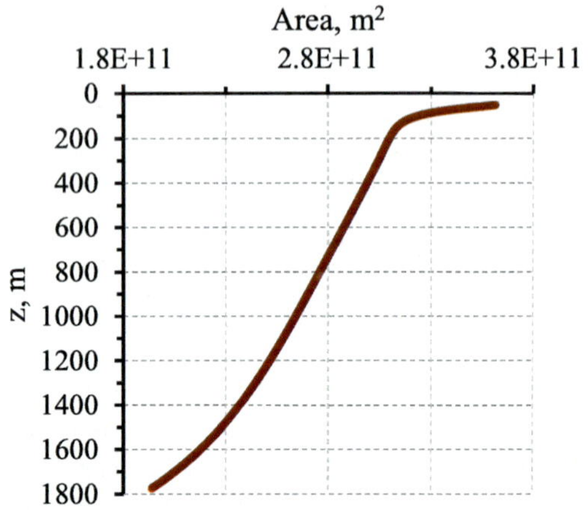

Fig. 3. Transformation scheme of the lower Bosphorus jet in the Black Sea [8] (left) and Dependence of the horizontal section area of the Black Sea on depth [9] (right)

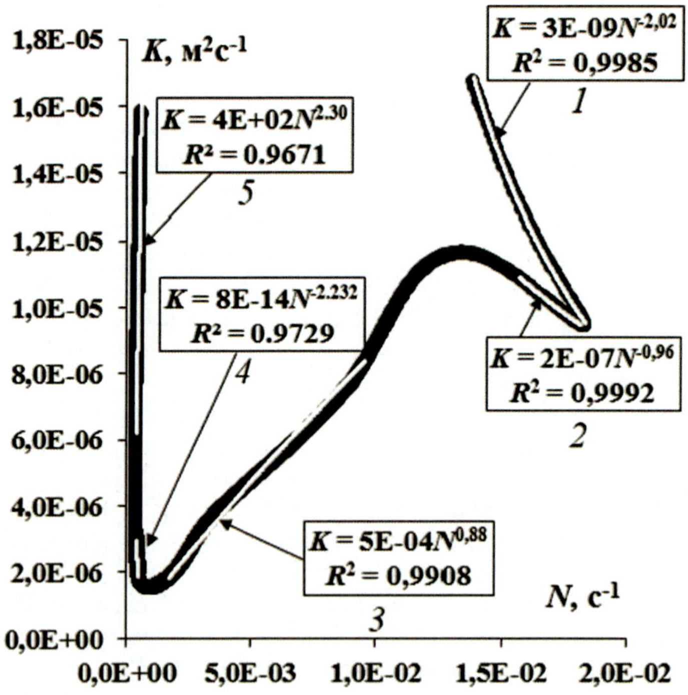

Fig. 4. 1.5D model dependence $K(N)$ for depths of 50–1750 m in the Black Sea [1]

4 Results and Discussion

Based on the calculated data, maps of the maximum vertical turbulent diffusion coefficient values distribution, the minimum buoyancy frequency values and their depths were constructed (see Fig. 5). The distribution of values in space was performed by the kriging method. The highest values of the coefficients fall on the areas of the shelf crest near Simeiz and a sharp drop in depths above the continental slope southeast of Alushta, which indicates the intensification of exchange processes in these areas, due to the effect of quasi-inertial internal waves packets re-reflection from the inclined bottom with an increase in potential energy in them, as it was described in [10].

Also, as part of the state task, an assessment was made of mixing in the suboxygen zone of the study region: maps of the depths of the upper boundaries of the suboxygen zone (with concentration of oxygen < 10 μmol/l) and the conditional beginning of the hydrogen sulfide zone, where oxygen is absent according to the sensor readings, were obtained, as well as the averages calculated between them K and N. Measurements have shown, the thickness of the suboxygen zone varies between 2 and 10 m. As can be seen from the Fig. 6, the areas of uplift of the suboxygen zone are due to the low coefficient of turbulent exchange in them. That is, due to weak mixing in the upper stratified layers of the identified zones, there was a weak transfer of oxygen from the upper quasi-homogeneous layer to a greater depth.

In order to improve the model of vertical turbulent exchange developed in our department, a comparative analysis of data obtained at deep-sea stations and simulation results was carried out. The model (Figs. 7b, 7d) showed good convergence with calculations (Figs. 7a, 7c) in the regions of the seasonal and main thermoclines, up to convergence in the power part of the approximation dependence $K \cong A \cdot N^{\alpha}$. For layer I $\alpha \approx -2$, for layer II $\alpha \approx -1$. The experimental dependences in five layers identified on the model distribution (Figs. 7b, 7d) correspond to the approximations from Fig. 4. However, new data showed the presence of at least three conditional layers, where the values of the coefficient differ from the model ones. Here we would like to emphasize that the calculated values of K exceed the 1.5D model ones by two orders of magnitude. These layers on 550–950 m depth differ in the intensity of turbulence generation, which is determined mainly by shear instability, and depend on various hydrophysical factors. In our opinion, the greatest influence on the intensity of exchange processes here is exerted by intrusions containing near-bottom water propagating along isopycna into the inner stratified region of the study area from the sloping bottom in addition to the internal waves packets re-reflection from the same bottom. A more detailed description of the modeling approach to bottom water rise along the sloping bottom and its distribution have been given in paper [14].

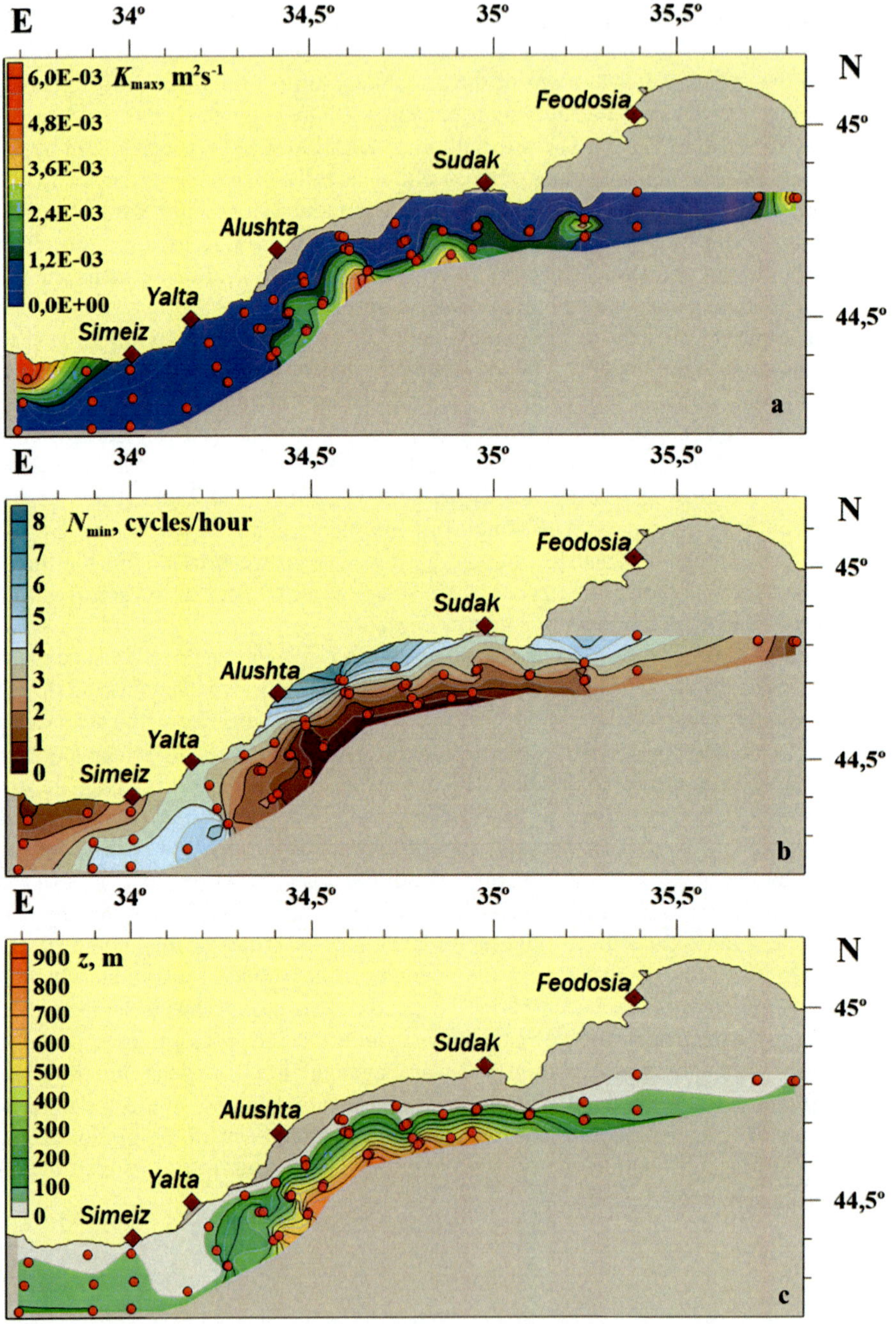

Fig. 5. Distribution maps of the vertical turbulent diffusion coefficient maximum value K_{max} (a), the buoyancy frequency minimum value N_{min} (b) and their depths z (c) according to sounding data

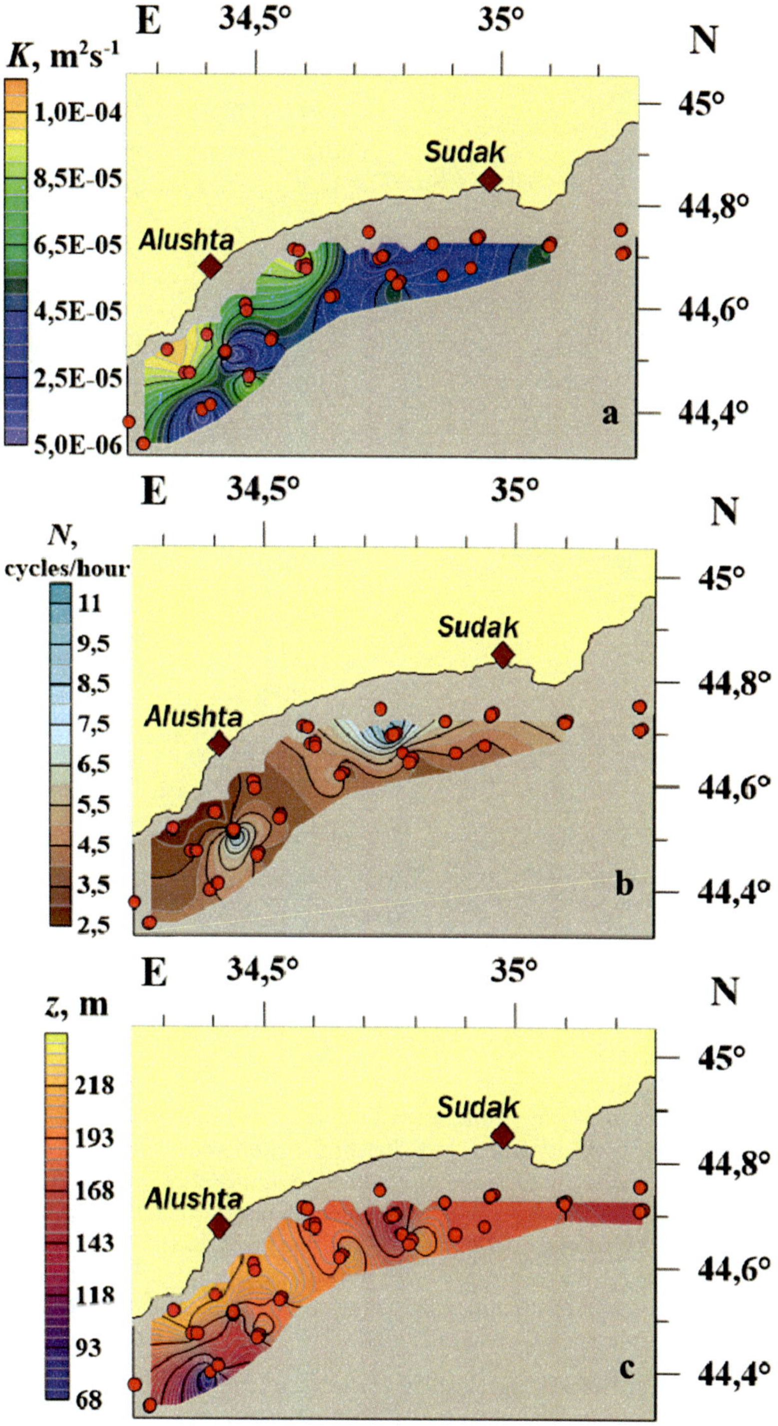

Fig. 6. Mixing in the suboxygen zone: distribution maps of the vertical turbulent diffusion coefficient value K (a), the buoyancy frequency value N (b) and the upper boundaries of the suboxygen zone depths z (c) according to sounding data

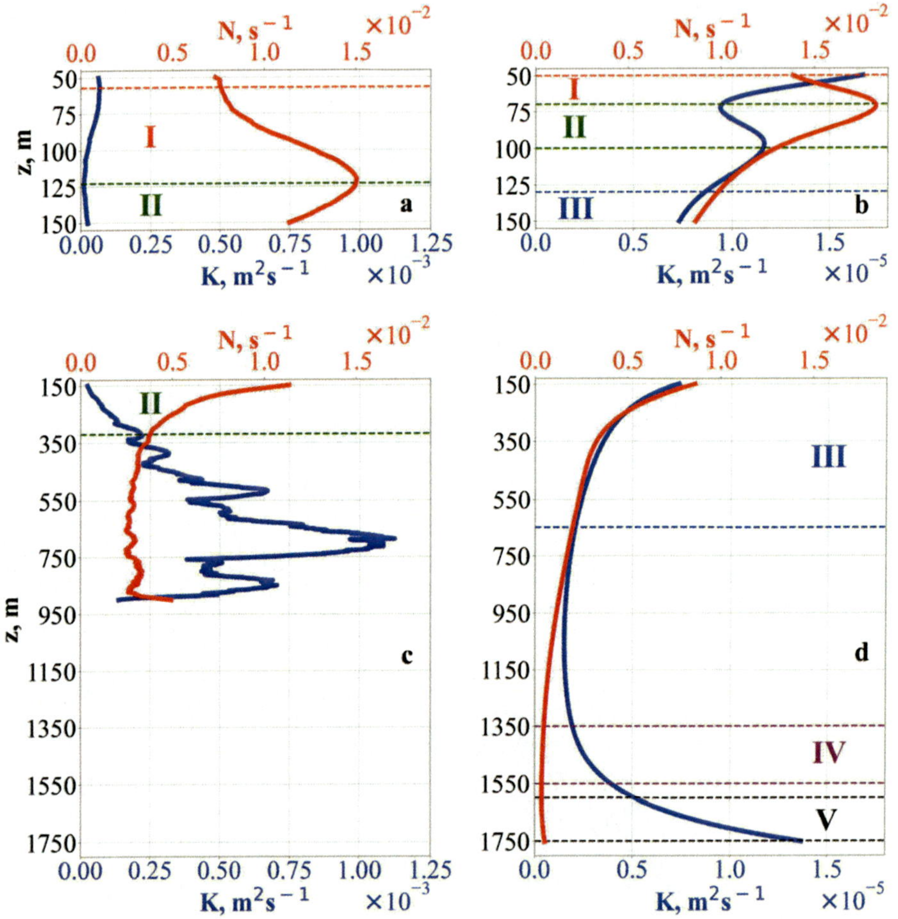

Fig. 7. Averaged profiles $K(z)$ and $N(z)$ based on the results of measurements of 122 cruises of the R/V Professor Vodyanitsky (a, c) and 1.5D model (b, d)

5 Conclusion

Summarizing the results of the work, we can say that 1.5D vertical turbulent exchange model, despite its simplicity and orientation to the deep water region of The Black Sea, must undergo reparametrization for application in models of mesoscale and submesoscale circulations in near a sloping bottom conditions. It is connected with the quasi-inertial internal waves packets re-reflection from the inclined bottom effect. Additional study must be made to confirm the presence of mixed near-bottom water intrusions in stratified layers lower then main thermocline with respect to local hydrodynamics. Nevertheless, following points can be distinguished:

- For the first time, data were obtained on the microstructure of hydrophysical fields in the territorial waters of the Black Sea off the Crimean coast to depths of 970 m, and the characteristics of turbulent mixing were calculated from them.

- The nature of the dependences of the vertical exchange coefficient on the stratification parameters in different layers generally corresponds to semi-empirical models for the upper stratified layers in the Black Sea.
- New results make it possible to significantly expand the area of research to greater depths, to clarify the form of semi-empirical dependencies and the values of numerical coefficients included in the calculation formulas.

Acknowledgements. The study was carried out within the framework of the state task FNNN-2021-0004 (code "Oceanological processes") and with the grant support of the Russian Science Foundation project No. 22-17-00150 (code "Turbulent structure").

References

1. Samodurov, A.S., Chukharev, A.M., Kazakov, D.A.: Basic regularities of vertical turbulent exchange in the mixed and stratified layers of the black sea. Phys. Oceanogr. [e-journal] **28**(4), 405–422 (2021). https://doi.org/10.22449/1573-160X-2021-4-376-391
2. Prandke, H., Stips, A.: Test measurements with an operational microstructure-turbulence profiler: detection limit of dissipation rates. Aquat. Sci. **60**, 191–209 (1998). https://doi.org/10.1007/s000270050036
3. Welch, P.D.: The use of fast Fourier transform for the estimation of power spectra: a method based on time averaging over short, modified periodograms (PDF). IEEE Trans. Audio Electroacoust. AU-**15**(2), 70–73 (1967). https://doi.org/10.1109/TAU.1967.1161901
4. Oakey, N.S.: Determination of the rate of dissipation of turbulent energy from simultaneous temperature and velocity shear microstructure measurements. J. Phys. Oceanogr. **12**, 256–271 (1982)
5. Thorpe, S.A.: Experiments of instability and turbulence in a stratified shear flow. J. Fluid Mech. **61**(4), 731–751 (1973). https://doi.org/10.1017/S0022112073000911
6. Mcewan, A.D.: The kinematics of stratified mixing through internal wavebreaking. J. Fluid Mech. **128**, 47–57 (1983). https://doi.org/10.1017/S0022112083000373
7. Mcewan, A.D.: Internal mixing in stratified fluids. J. Fluid Mech. **128**, 59–80 (1983). https://doi.org/10.1017/S0022112083000385
8. Samodurov, A.S., Ivanov, L.I.: Balance model for calculating average vertical fluxes of fluid, heat, salt and dissolved chemical substances in the black sea thermohaline. Morskoy Gidrofizicheskiy Zhurnal [Marine Hydrophysical J.] **1**, 7–24 (2002). (in Russian)
9. Samodurov, A.S., Kulsha, O.E., Belokopytov, V.N.: Vertical exchange stationary model in the black sea for real basin geometry. Ekologicheskaya bezopasnost' pribrezhnoy i shel'fovoy zon i kompleksnoye ispol'zovanie resursov shel'fa [Ecolog. Saf. Coastal Shelf Zones Compr. Use Shelf Resources] **14**, 517–523 (2006). (in Russian)
10. Samodurov, A.S., Chukharev, A.M., Nosova, A.V., Globina, L.V.: Internal waves intensification in the shelf break region as a factor of the vertical exchange intensification. Fundamentalnaya i Prikladnaya Gidrofizika [Fundamental Appl. Hydrophysics] **6**(2), 12–24 (2013). (in Russian)
11. Samodurov, A.S., Chukharev, A.M., Zubov, A.G., Pavlenko, O.I.: StructureFormation and vertical turbulent exchange in the coastal area of the Sevastopol region. Phys. Oceanogr. **6**, 3–14 (2015). https://doi.org/10.22449/1573-160X-2015-6-3-14
12. Samodurov, A.S., Chukharev, A.M.: Vertical turbulent exchange features in the layer of seasonal pycnocline in the northwestern part of the black sea. J. Phys. Conf. Ser. **1128**, 012148 (2018). https://doi.org/10.1088/1742-6596/1128/1/012148

13. Samodurov, A.S., Chukharev, A.M.: The vertical turbulent exchange features in the black sea active layer. In: Karev, V., Klimov, D., Pokazeev, K. (eds.) Physical and Mathematical Modeling of Earth and Environment Processes, pp. 148–156. Springer, Cham (2018). https://doi.org/10.1007/978-3-319-77788-7_16
14. Samodurov, A.S.: Intrusive lamination and vertical exchange in the black sea due to geothermal heat flow on a sloping bottom. In: MHI 2003, Ekologicheskaya Bezopasnost' Pribrezhnykh i Shel'fovykh Zon i Kompleksnoe Ispol'zovanie Resursov Shel'fa [Ecological Safety of Coastal and Shelf Zones and Comprehensive Use of Shelf Resources]. Sevastopol: ECOSI-Gidrofizika, (8), pp. 152–156 (2003). (in Russian)

Seasonality of Pacific Water Distribution in the Sea of Okhotsk Revealed by Lagrangian Approach

K. K. Kivva[1(✉)], M. V. Budyansky[2], M. Yu. Uleysky[2], and S. V. Prants[2]

[1] Russian Federal Research Institute of Fisheries and Oceanography, Moscow, Russia
kirill.kivva@gmail.com

[2] Pacific Oceanological Institute of the Russian Academy of Sciences, Vladivostok, Russia

Abstract. The eastern Sea of Okhotsk is commercially important ecosystem with substantial documented changes in physical setup and fish stocks. The productivity of this area is at least partly related to the water of Pacific origin which is rich in nutrients. However, the variability of Pacific-origin water distribution within this area is still purely studied. Thus the main objective of this study is to illustrate temporal variability of Pacific-origin water distribution. Another specific objective is to evaluate the degree of seasonal variability of this water distribution. Lagrangian approach related to the analysis of life history of water particles was used as the main instrument. The results revealed substantial inter-annual variability of Pacific-origin water distribution. The area of Pacific-origin water of particular residence time was suggested as an index of Pacific influence on the studied region. Area of Pacific-origin water with residence time from 61–180 days was chosen for further analysis. Seasonal component of this area index was distinguished with complete ensemble empirical mode decomposition with adaptive noise (CEEMDAN) method. The parameters of variability of this component were investigated. The analysis revealed substantial inter-annual variability of amplitude and period of seasonality.

Keywords: Pacific Ocean · Sea of Okhotsk · Pacific-origin water · current · water transport · intrusion · distribution · variability of seasonality · CEEMDAN

1 Introduction

The eastern Sea of Okhotsk is a productive ecosystem supporting considerable commercial fisheries. In the past ten years (2013–2022), the total catch in this area was between 753 and 1163 thousand metric tons. This region contributed from 17–23% to annual total catch of Russia [1]. The high biological productivity of the region is related to the inflow of Pacific water which passes through the northern straits of the Kuril islands. Intensive tidal mixing along the Kuril Islands [2–4] facilitates vertical flux of nutrients and enhances biological productivity downstream [5]. The main inflow of Pacific-origin water into the Sea of Okhotsk occurs through the Fourth Kuril Strait and Krusenstern Strait. The process of water transport through straits is known to increase during the cold season [6–8].

T. Chaplina (Ed.): CIWO 2023, SPEES, pp. 59–68, 2023.
https://doi.org/10.1007/978-3-031-47851-2_8

Despite commercial importance of the eastern Sea of Okhotsk and documented changes in its ecosystem [9, 10], the information on circulation in this region is still scarce and data on variability of Pacific-origin water inflow is virtually absent. Only few recent publications investigated variability of currents there based on high-resolution altimetry data or numerical modelling [1, 8, 11, 12]. At the same time, the variability of Pacific-origin water inflow is shown to be important for distribution of adult walleye pollock *Gadus chalcogrammus* [12], the dominating nekton species in the region, and may potentially influence the distribution and biology of other organisms including lower trophic levels. Two types of environmental indices reflecting the strength of Pacific-origin water inflow and spatial variability of such water distribution are suggested and discussed earlier [1]. Both of them are based on Lagrangian approach realized by water particle tracking back in time using data on surface current. One type of such indices is based on calculation of centroid of the region occupied by Pacific-origin water. The only information required for this index is residence time and initial location of each Pacific water particle within analyzed portion of the sea. The second index is more complex as it additionally requires information on place of entrance into the Sea of Okhotsk for each particle and implies cluster analysis of particles in terms of residence time and place of entrance in order to find bounded geographic area occupied by particles of same inflow event.

This paper aims to introduce another index of Pacific-origin water distribution in the eastern Sea of Okhotsk based on Lagrangian approach and area measure. Additional novelty of this paper is investigation of seasonal component of Pacific-origin water area variability.

2 Material and Methods

This paper is based on investigation of currents with Lagrangian approach. In general, this means analysis of water particles trajectories and history of transport of every particle. For every date in the interval between January 31, 1997, and April 17, 2022, the following procedure was carried out. The water particles were distributed in region of 42–56.5°N, 141–157°E in regular latitude-longitude grid of size 300 × 380 (spatial resolution approximately 0,038°N × 0,054°E). Each particle was tracked back in time for 1095 days. Gridded data product based on satellite altimetry distributed by the Copernicus Marine Environment Monitoring Service was used for calculations [13]. The spatial resolution of this product is 0.25° × 0.25°; temporal resolution is 1 day. The main advantage of this product is that it is based solely on satellite measurements with no model data involved. Spatial bi-cubic interpolation of current data was performed for every particle at every step of calculation. The calculations were carried out on computer cluster of POI FEB RAS. Each calculated trajectory was checked for intersection with conditional transect along the Kuril Islands, the straight line between 43.0°N, 144.5°E and 51.75°N, 157.5°E in cylindrical equidistant projection which is assumed to be a margin between the Pacific Ocean and the Sea of Okhotsk (red line in Fig. 1). On a given date within mentioned time interval, all particles which crossed this transect during previous 1095 days were considered water of Pacific origin. Residence time after transect crossing and point of intersection was recorded for such particles. All other particles including those whose trajectories crossed the coast mask were excluded from the analysis. Ocean current data, particle tracking methods, and water origination condition are same to published earlier [1, 12]. This paper is particularly focused on the region of 50–56.5°N, 150–157°E, where the main path of Pacific-origin water is located [1] (dashed line in Fig. 1).

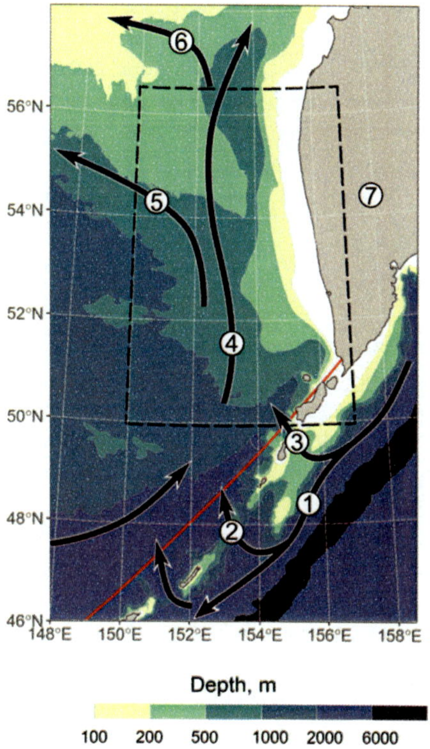

Fig. 1. Schematic chart representing bathymetry and main currents in the area of study [7, 8, 14]. Black dashed line represents the region of analysis. Red solid line denotes assumed margin between the Pacific Ocean and the Sea of Okhotsk (conditional Kuril transect). Major currents and geographic locations are marked with numbers. 1, Kuril current; 2, inflow of Pacific-origin water through Krusenstern Strait; 3, inflow of Pacific-origin water through the Fourth Kuril Strait; 4, West Kamchatka current; 5, Middle Okhotsk current; 6, slope branch of North Okhotsk current; 7, Kamchatka peninsula.

As a result, set of 9208 files describing Pacific-origin particles distribution and residence time after conditional transect crossing during 9208 days of analysis is obtained. The data was analyzed in R [15]. An indices representing area of Pacific-origin water with residence time between 1–30, 31–60, 61–180, 181–365, and 366–1095 days in a region 50–56.5°N, 150–157°E was calculated for each date. Area index for particles with residence time from 61–180 days was chosen for further analysis of seasonality. The infra-annual, annual, quasi-biannual, and multi-annual (low-frequency) modes of this index were separated using complete ensemble empirical mode decomposition with adaptive noise (CEEMDAN) method [16] by means of libeemd program package for

R [17]. This method returns so-called Intrinsic Mode Functions (IMFs) and residual (trend) component of the initial time-series data (Eq. 1).

$$S(t) = \sum_{i=1}^{n} IMF_i(t) + R(t), \tag{1}$$

where S is initial time-series, IMF_i is i-th Intrinsic Mode Function, R is residual component.

The main characteristic of those IMFs is mutual orthogonality. Each IMF should be symmetric regarding to zero line and should explain certain, usually periodic, mode of variability of the time-series being analyzed. The main advantage of the method is adaptivity and ability to distinguish modulated IMFs of different period.

The data manipulations were held with dplyr package for R [18].

3 Results and Discussion

Examples of Pacific-origin water particles distribution in the area of study are presented in Fig. 2. The color scale in Fig. 2 represents intervals of residence time of particles (1–30, 31–60, 61–180, 181–365, and 366–1095 days) which were chosen a priori based on assumption of water mass differences. Namely, as the main Pacific water inflow to the Sea of Okhotsk occurs during cold season, the new water of Pacific origin should be warmer at that time comparing to the water residing in the Sea of Okhotsk for several months and therefore exposed to stronger heat loss or even to ice formation. This assumption is partly supported by previously published data [12]. It is assumed here that Pacific-origin water keeps its properties for at least half a year.

Inflow of Pacific water into the Sea of Okhotsk takes place in all seasons. However the new Pacific-origin water with residence time lower than 180 days has wider spatial distribution in late winter through early summer (March and June in Fig. 2) comparing to season of late summer to early winter (September and December). This allows to conclude that seasonal component in Pacific-origin water spatial distribution is usually strong. This is an anticipated result as Pacific water inflow to the sea is known to strongly depend on season. However, considerable inter-annual variability of such water distribution is also obvious from Fig. 2.

Water of Pacific origin usually propagates along the Kamchatka Peninsula northward. The spatial structure of such water always includes many meanders, eddies, and relatively thin filaments. Such water with residence time larger than 180 days is usually represented by mosaic patches while fresher Pacific-origin water form distinct polygons (Fig. 2).

Portion of area of region 50–56.5°N, 150–157°E occupied by Pacific-origin water with different residence time was considered as an index of such water distribution. The time-series of such indices for water of different residence time intervals are present in Fig. 3.

Visual examination of those time-series allow to conclude that cycles with period of about 1 year (seasonal variability) is less pronounced for water with residence time from 1–30 and 31–60 days than for water with residence time from 61–180 and 181–365 days. Time-series of area of Pacific-origin water with residence time greater than 365 days has almost no seasonality and thus may be controlled mostly by low-frequency variability.

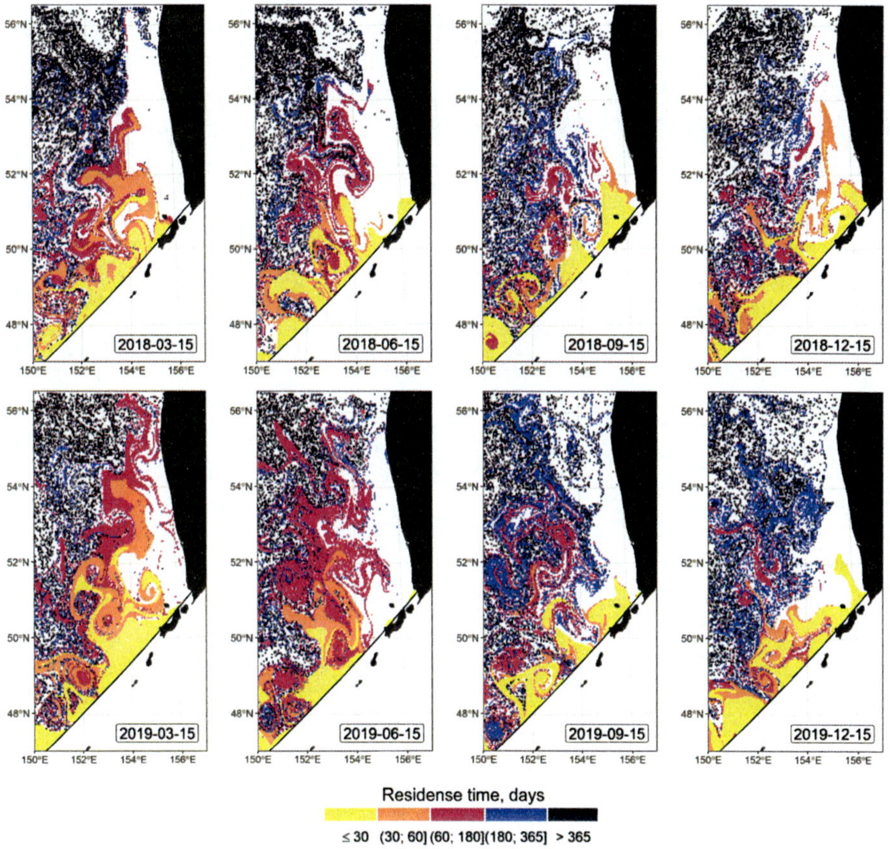

Fig. 2. Examples of Pacific-origin water distribution in the eastern Sea of Okhotsk on 15th of March, 15th of June, 15th of September, and 15th of December of 2018 (top row) and 2019 (bottom row). Color fill of each particle represents the time interval between crossing of conditional Kuril transect and the date of observation. The Kuril transect is marked with the black solid line. The figure is prepared in R [15] with ggplot2 package [19].

As discussed previously, Pacific-origin water should lose its physical, chemical, and perhaps biological properties with time while being transported in the sea by currents. This may happen because of diffusion and lateral mixing. Therefore in this paper we focus on decomposition of area index of Pacific-origin water with residence time from 61–180 days (hereinafter S_{61-180}, units 10^3 km^2) as this index should be of value in further works related to the biological effects of Pacific-origin water transport and distribution.

Figure 4 illustrates results of decomposition of S_{61-180} into oscillatory modes (IMFs) of different periods with CEEMDAN method. The seasonal component (IMF9) of the index is prevailing which is an anticipated result. This component is strongly correlated with initial time-series and explains about 80% of its variability ($R^2 = 0.796$, p-value less than machine epsilon). Addition of residual (trend) component increases the correlation ($R^2 = 0.874$, p-value less than machine epsilon). Comparison of the initial time-series

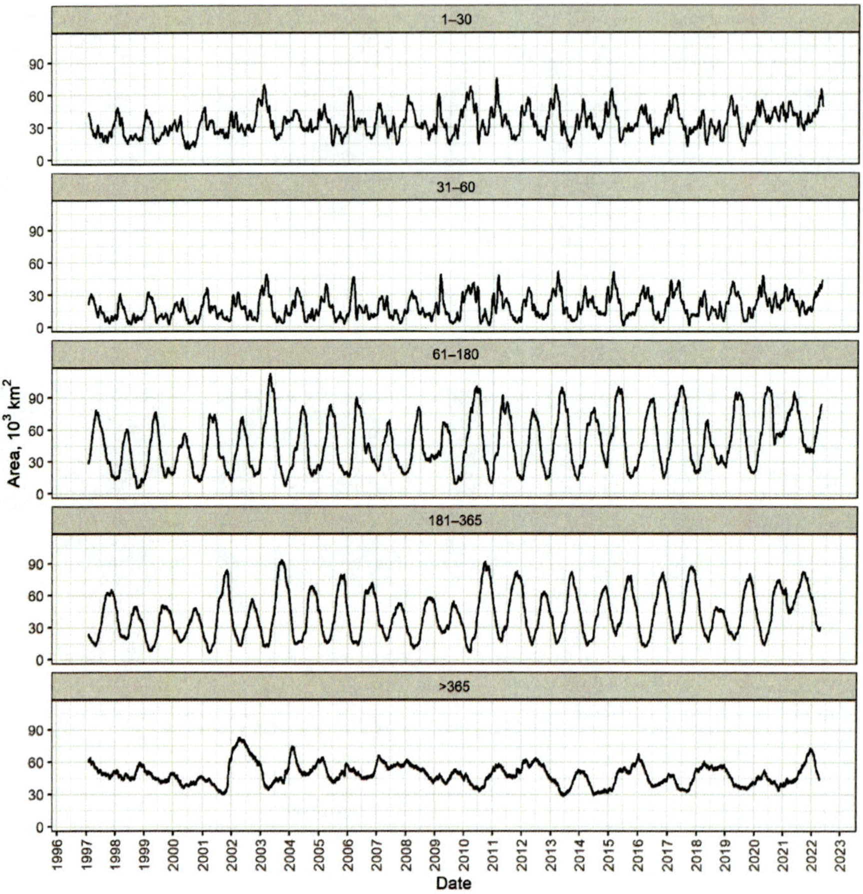

Fig. 3. Temporal variability of area (10^3 km^2) occupied by Pacific-origin water with different residence time. Only particles within region 50–56.5°N, 150–157°E are considered (dashed line in Fig. 1). Intervals of residence time (days) are same as in Fig. 2 and are marked at top side of each panel. The figure is prepared in R [15] with ggplot2 package [19]

with sum of seasonal component derived with CEEMDAN method and residual (trend) component is shown on Fig. 5.

At the same time, part of seasonal variability was assigned to the IMF8, especially between 2012 and 2016. Worth noting resulted infra-seasonal component (IMF8) was also correlated with initial time-series ($R^2 = 0.510$, p-value less than machine epsilon). Thus, CEEMDAN is a useful method for seasonal component decomposition in case of S_{61-180} index, however it has limitations and could not correctly return complete seasonal mode. Despite this, decomposed seasonal component of initial time-series is still valuable for investigation of variability of seasonality as its amplitude is considerably greater than amplitude of IMF8. Moreover, in cases when part of seasonal variability is attributed to IMF8, maximums and minimums of IMF8 and IMF9 occur on similar dates.

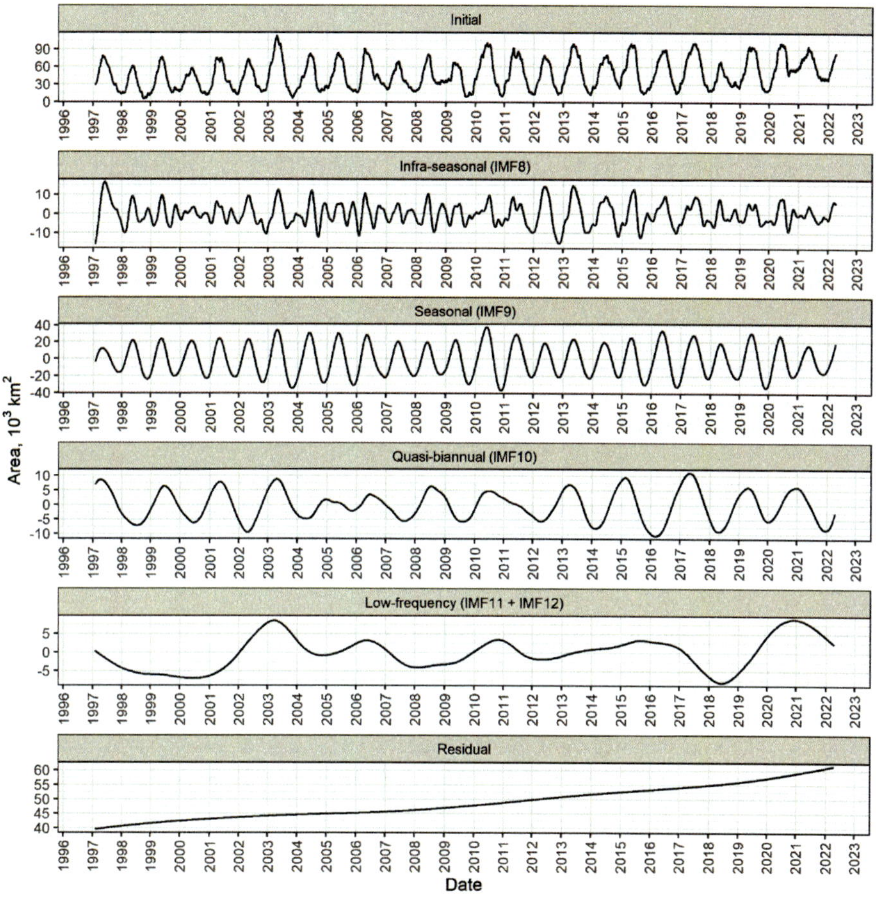

Fig. 4. Main variability modes of area of Pacific-origin water with residence time from 61–180 obtained by CEEMDAN method. Top panel shows initial time-series. Infra-seasonal (IMF8), seasonal (IMF9), quasi-biannual (IMF10), low-frequency variability (IMF11 + IMF12), and residual (trend) components are shown. High-frequency modes (IMFs from 1–7) are not shown. The figure is prepared in R [15] with ggplot2 package [19]

In order to investigate variability of seasonality of S_{61-180} index, values of local extrema of IMF9 and dates of their occurrence (Julian day of year; DOY_{max} and DOY_{min}) were calculated, and period was evaluated as time interval (days) between consecutive local maximums or minimums (P_1 and P_2, respectively). Amplitude (A) of each cycle was accessed as difference between local maximum and minimum values during each year. The statistics (mean and standard deviation) of those variables are presented in Table 1.

Maximum of S_{61-180} seasonal component occurred between last decade of April and first decade of June, on average in the middle of May, while minimum was observed between the middle of October and the middle of December, on average in the middle of November. The timing of annual minimum is more variable than the timing of maximum.

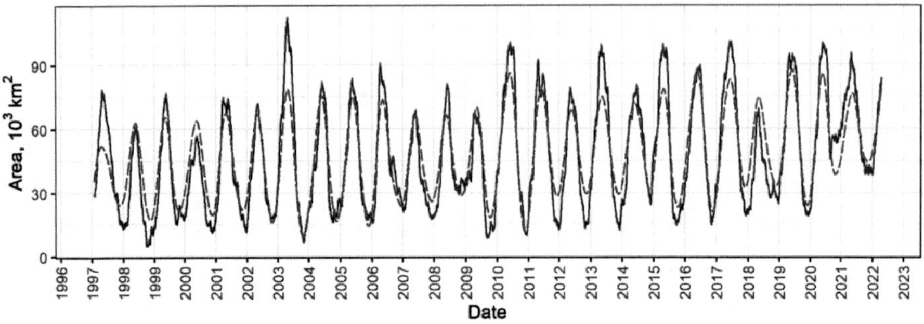

Fig. 5. Comparison of initial time-series of S_{61-180} index (grey solid line) and sum of IMF9 (the only IMF with apparent seasonal cycle) and residual component (blue dashed line). Initial time-series decomposition into IMFs and residual (trend) component was performed with CEEMDAN method. The figure is prepared in R [15] with ggplot2 package [19]

Standard deviation values of maximum, minimum, and annual amplitude reach about 22–24% of their mean values, so the inter-annual variability of those parameters is quite substantial. The period between consecutive maximums P_1 is 366 days on average, and the period between minimums P_2 is 365 days on average, so it is almost perfect annual cycle. However, variability of those two measures of period is also considerable (Table 1).

Table 1. Statistics describing variability of seasonality of S_{61-180} index. Seasonal component received with CEEMDAN method (IMF9) was used for analysis. Mean and standard deviation (SD) values are present for local extrema and their dates (day of year) as well as for the period and amplitude estimates

	DOY_{max}	DOY_{min}	max	min	P_1	P_2	A
Mean	138,1	320,6	24984	– 24728	366,1	364,8	49712
SD	13,1	17,3	6028	5822	18,8	22,8	11211

DOY_{max} – Julian day of year of local maximum;
DOY_{min} – Julian day of year of local minimum;
max – value of local maximum;
min – value of local minimum;
P_1 – period between consecutive maximums;
P_2 – period between consecutive minimums;
A – amplitude.

4 Conclusion

The work is aimed at analysis of seasonality of Pacific-origin water distribution in the Sea of Okhotsk. Additional objective of the study is characteristics of variability of this seasonality. The topic is important because of ecological and commercial value of the eastern Sea of Okhotsk. Lagrangian approach related to the life history of multiple water particles is used as the main method. For every date of the analysis, thousands of regularly distributed in the Sea of Okhotsk water particles were tracked for 1095 days back in time using satellite altimetry data. The condition of particle trajectory intersection with transect along the Kuril Islands was evaluated and residence time within the Sea of Okhotsk was estimated for each particle of Pacific origin. The analysis is done for every day in time interval between January 31, 1997, and April 17, 2022.

The results allowed to conclude that distribution of Pacific-origin water in the sea has substantial inter-annual variability. As the inflow of Pacific water into the sea vary with season, the distribution of Pacific-origin water in the sea also has strong seasonal component. Area of region occupied with water of Pacific origin with selected residence time was suggested as an index of Pacific-origin water volume variability within the region of interest. Several such indices were investigated. The index describing area of Pacific-origin water with residence time between 61 and 180 days was chosen for analysis of variability because of several reasons. First, the water residing in the sea for such limited time should still keep the initial physical, chemical, and biological properties while "older" water of Pacific-origin may already lose its specific characteristics due to mixing. Second, the water particles of this residence time form distinct polygons. This means that discussed water propagates in the sea as distinct water mass. Third, strong seasonality was observed in the area of this water.

The analysis of seasonality was carried out with CEEMDAN method. One of the Intrinsic Mode Functions returned by the method was apparently responsible for the seasonal component. This IMF was investigated in terms of annual local maxima values, their timing, period between consecutive maximums or minimums, and amplitude. The results revealed considerable inter-annual variability of those parameters.

Acknowledgements. Work of Budyansky M.V., Uleysky M.Y., and Prants S.V. is supported by Russian Science Foundation, project № 23-17-00068.

References

1. Kivva, K.K., Budyansky, M.V., Uleysky, M.Yu., Prants, S.V.: Analysis of spatio-temporal variability of Pacific water distribution in the Sea of Okhotsk based on Lagrangian approach. Trudy VNIRO 193 (In Russian, in Press)
2. Vanin, N.S., Yurasov, G.I.: Water exchange between the Okhotsk Sea and Pacific Ocean in Bussol and Krusenstern Straits. Russ. Meteorol. Hydrol. (7), 62–68 (1998)
3. Tanaka, Y., Hibiya, T., Niwa, Y.: Estimates of tidal energy dissipation and diapycnal diffusivity in the Kuril Straits using TOPEX/POSEIDON altimeter data. J. Geophys. Res. Oceans **112**(C10) (2007)
4. Tanaka, Y., Hibiya, T., Niwa, Y., Iwamae, N.: Numerical study of K1 internal tides in the Kuril straits. J. Geophys. Res. Oceans **115**(C9) (2010)
5. Matveev, V.I., Tikhomirova, E.A., Luchin, V.A.: Primary production of the Sea of Okhotsk in years with different temperature conditions. Russ. J. Mar. Biol. **41**(3), 176–185 (2015)
6. Katsumata, K., Yasuda, I.: Estimates of non-tidal exchange transport between the Sea of Okhotsk and the North Pacific. J. Oceanogr. **66**, 489–504 (2010)
7. Ohshima, K.I., Nakanowatari, T., Riser, S., Wakatsuchi, M.: Seasonal variation in the in-and outflow of the Okhotsk Sea with the north pacific. Deep Sea Res. Part II **57**(13–14), 1247–1256 (2010)

8. Fayman, P.A., Prants, S.V., Budyansky, M.V., Uleysky, M.Y.: Simulated pathways of the northwestern pacific water in the Okhotsk Sea. Izv. Atmos. Oceanic Phys. **57**(3), 329–340 (2021)
9. Zuenko, Yu.I., et al.: Recent changes in the Okhotsk Sea ecosystem (2008–2018). Izvestiya TINRO **197**(2), 35–61 (2019). (In Russian)
10. Zuenko, Yu.I., et al.: Marine Ecosystems of the North Pacific Ocean 2009–2016: Region 17 (Sea of Okhotsk), p. 30. North Pacific Marine Science Organization, Sidney, BC, Canada (2023)
11. Kolomeytsev, V.V.: Variability of the west Kamchatka current in winters of 1994–2019, by altimetry data. Izvestiya TINRO **200**(2), 412–426 (2020). (In Russian)
12. Budyansky, M.V., Kulik, V.V., Kivva, K.K., Uleisky, M.Y., Prants, S.V.: Lagrangian analysis of pacific waters in the Sea of Okhotsk based on satellite data in application to the walleye pollock fishery. Izvestiya Atmos. Oceanic Phys. **58**(12), 1427–1437 (2022)
13. Global Ocean Gridded L 4 Sea Surface Heights and Derived Variables Reprocessed 1993 Ongoing. https://doi.org/10.48670/moi-00148
14. Krovnin, A.S., et al.: Oceanographic conditions within the main area of walleye pollock. Trudy VNIRO **189**, 16–44 (2022). (In Russian)
15. R Core Team: R: A language and environment for statistical computing. R Foundation for Statistical Computing, Vienna, Austria (2022). https://www.R-project.org/
16. Torres, M.E., Colominas, M.A., Schlotthauer, G., Flandrin, P.: A complete ensemble empirical mode decomposition with adaptive noise. In: 2011 IEEE International Conference on Acoustics, Speech and Signal Processing (ICASSP), pp. 4144–4147 (2011)
17. Luukko, P.J., Helske, J., Räsänen, E.: Introducing libeemd: a program package for performing the ensemble empirical mode decomposition. Comput. Stat. **31**, 545–557 (2016)
18. Wickham, H., François, R., Henry, L., Müller, K., Vaughan, D.: dplyr: A Grammar of Data Manipulation. – R package version 0.4 (2023). https://dplyr.tidyverse.org. https://github.com/tidyverse/dplyr
19. Wickham, H.: ggplot2: Elegant Graphics for Data Analysis. Springer, New York (2016). https://doi.org/10.1007/978-0-387-98141-3. ISBN 978-3-319-24277-4. https://ggplot2.tidyverse.org

Turbulence Influence on the Thickness of the Mixed Layer in the Coastal Zone of the Black Sea

V. A. Korzhuev(✉)

Marine Hydrophysical Institute of RAS, Sevastopol 299011, Russia
genzak30@gmail.com

Abstract. Significant problem in present climate theory is the consideration of rapidly changing wind-dependent parameters when modeling the dynamics of the MBL depth. Perceptible errors in the simulation of the mixed layer are due to Langmuir and turbulent mixing. The paper presents a way to modernize the one-dimensional model of the mixed layer by Niiler-Kraus using a multiscale turbulence model, which takes into account three parameters of turbulence generation from wind: current velocity shear, surface wave nonlinearity and wave breaking. The verification of the model is presented by field data collected in the coastal zone of the Black Sea on a stationary oceanographic platform of the Marine Hydrophysical Institute of the Russian Academy of Sciences. The model working is evaluated with different variability of wind speed and direction.

Keywords: Turbulence · Marine Boundary Layer · Multiscale model · Surface layer · Generation mechanisms · Niiler-Kraus model

1 Introduction

Turbulent perturbations of the mixed ocean layer affect gas exchange with the atmosphere, heat and momentum transfer. The capability of the ocean to influence climate change by absorbing and accumulating heat and gas impurities largely depend on vertical exchange in the upper Marine boundary layer (MBL) [1, 2]. The mixed layer also affects deep-sea stratification and dynamics, from which it can be concluded that the MBL is crucial for determining the role of global ocean circulation in climate change [3].

Currently, one of the most effective methods of modeling climate change is global climate models or, as they are also called, general circulation models. Numerical modeling of the climate system with their help is a promising task of climate theory. However, turbulent mixing in the MBL of the ocean is not accurately parameterized in global climate models, which creates noticeable systematic errors in determining the depth and temperature of the mixed layer [4]. Considered that one of the reasons is that modern climate models do not accurately take into account the key processes of mixing by surface waves that cause turbulence and deepen the mixed layer faster than prolonged wind

T. Chaplina (Ed.): CIWO 2023, SPEES, pp. 69–79, 2023.
https://doi.org/10.1007/978-3-031-47851-2_9

exposure. According to seasonally averaged data from climate change simulation over 20 years in the HadGEM3 model, which are compared with Argo buoy data, the error in determining the depth of the MBL can be up to 100% [1].

This paper presents the results of the modernization of the Niiler-Kraus (N-K) mixed layer model [5] using a multiscale model [6] in order to improve the quality of modeling the thickness of the mixed layer for various modes of vertical exchange.

2 Material and Methods

One-dimensional mixed models are useful components of integrated ocean and climate models. Simplicity and the possibility of direct physical substantiation make it possible to provide realistic modeling of daily and seasonal temperature changes in selected areas of the ocean where there is a slight horizontal advection.

The one-dimensional Niiler-Kraus mixed layer model Insufficient correctly taking into account the parametrization of turbulent flows when calculating the vertical dynamics of the mixed layer. In this paper, we propose a method for compounding the N-K model with a multiscale turbulence model. The solution of the stated problem is carried out by the replacement method in the model of Eq. (1) parameter «D», responsible for the wind impact, into the sum of three components describing different mechanisms of turbulence generation: current velocity shear, surface wave nonlinearity and wave breaking, which are presented in Eq. (14).

The parametrization of heat fluxes is performed using the empirical formulas obtained by Shuleikin [7], their interaction is determined by the N-K model.

It is assumed that the thickness of the OSBL is an area bounded on two sides: from above - by the ocean-atmosphere boundary layer, from below - by the upper boundary of the seasonal thermocline, the depth of which can be calculated using the N-K model.

Measurements of empirical parameters and thermal profiles are made in the coastal zone of the Black Sea on a stationary oceanographic platform of the Marine Hydrophysical Institute of the Russian Academy of Sciences. Considered that the impact of the coast and the bottom affects the stirring only if the wind is directed from the landside. Assumed that the platform itself has no effect on mixing.

3 Niiler-Kraus Model and Heat Flows

The N-K model [5] is a somewhat cumbersome equation consisting of different unequal quantities, which makes it possible to simplify it for various processes occurring in the OSBL. In this work, the Nieler-Krauss model is used in the following form:

$$\underset{A}{\frac{dh}{dt}}\Big[\underset{B}{q^2} + \tfrac{1}{2}h_*^{\,2}N^2 - \underset{B'}{B_0 t} - \underset{C}{2su_*^4 f^{-2}h_*^{\,-2}(1-\cos ft)}\Big] = \underset{D}{mu_*^3} + \underset{E}{\tfrac{1}{2}B_0 h_*} \tag{1}$$

where $\frac{dh}{dt}$ is the MBL depth change; q is the energy value required to mix the entrained water; h_* is MBL depth at the zero moment of time; N is buoyancy frequency; B_0 is heat balance; t – time; s is the value corresponding to the Richardson number (Ri); f – Coriolis parameter.

A – rate of energy needed to agitate the entrained water;

B – work per unit time needed to lift the dense entrained water and to mix it through the layer;

C – rate at which energy of the mean velocity field is reduced by mixing across the layer base;

D – rate of working by the wind;

E – rate of potential energy change produced by penetrating solar radiation.

Heat balance according to N-K:

$$B_0 = \frac{g}{\rho}\left\{\frac{\alpha}{c}(R_0 - I_0 + H_0) - \beta P_0 S_0 + \left(\frac{\alpha A_c}{c} + \beta S_0\right) Q_0\right\} \quad (2)$$

where R_0 is the effective ocean surface radiation, I_0 is insolation, H_0 is latent fluxes (humidity), A_c is latent fluxes (evaporation), S_0 is salinity, Q_0 is evaporation, α is the thermal expansion coefficient, β is the salinity compression coefficient.

4 Multiscale Model

The near-surface layer of the sea, distinguished several turbulence generation mechanisms, belonging to different ranges of the turbulence spectrum, so it is logical to consider the influence of each of them on the corresponding site. In a sense, a similar idea of dividing the spectrum into zones in which energy is generated, and which are separated by inertial sections where the law «−5/3» is fulfilled, was developed by R.V. Ozmidov [8]. For the near-surface layer, three mechanisms of turbulent generation are considered to be the most important from an energy point of view. The first is due to the tangential stress of the wind on the surface, and the drift current created by it, the shear of which generates vortices of sufficiently large scales covering the entire thickness of the mixed layer. Turbulence caused by breaking and nonlinear effects of waves is more «fast» and has smaller spatial scales, but the intensity of the transformation of wave energy into turbulence can be very significant. The nonlinear effects of waves and their breaks have approximately the same scale, so they are considered to be close in terms of energy contribution. The influence of surface waves on the turbulence regime is most significant for the vertical transfer of kinetic energy by turbulent movements. The main role is performed by movements comparable in scale to the thickness of the wave layer [3].

Figure 1 shows schematically the sources of generation of turbulent energy by various mechanisms, its inertial transfer to the region of higher wave numbers and dissipation. It is believed that turbulent energy is produced in regions 1 and 2, which are characterized by the corresponding energy levels E_{p1} and E_{p2}, part of the energy from these regions is transformed with velocities ε_{p1} and ε_{p2} into the energy of pulsations of smaller scales, replenishing the energy in regions 2 and 3, respectively. In the range of wave numbers $k_2 \leq k \leq k_3$, energy is transferred at a rate of ε_{T3}, the energy level in this section is denoted by E_{T3}, this section is considered to be inertial, and turbulence is not generated in it. The fourth section of the spectrum is considered dissipative, it lies in the region of wave numbers $k \geq k_3$. In the first range of scales, turbulence is generated by a shear in the drift current velocity, in the second, by nonlinear effects of surface waves and their

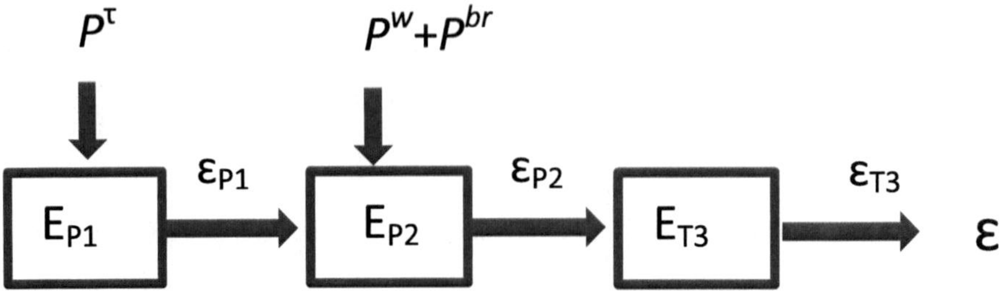

Fig. 1. Mechanisms of generation of turbulent energy [6]. P^τ is generation of turbulent energy by flow velocity shear; P^w is generation of turbulent energy by non-linearity of surface waves; P^{br} is generation of turbulent energy by breaking waves; ε – dissipation of turbulent energy

breaking. In the diagram, these energies are denoted by P^τ and $P^w + P^{br}$, respectively. It is assumed that the total energy of turbulence is distributed between Sects. 1–3, in the fourth section there is a small part of the energy that can be neglected. In the indices ε and E, in addition to the number, the letters «P» - generation and «T» - transfer.

The generation by flow velocity shear P^τ is determined by the expression:

$$P^\tau = -\frac{u_*^2}{\kappa}\left(\frac{dU}{dz}\right) \tag{3}$$

where u_* - friction velocity, κ - Karman's constant, U - mean flow velocity:

$$U(z) \approx U_0 - \frac{u_*}{\kappa}\ln\left(\frac{z+z_0}{z_0}\right) \tag{4}$$

where U_0 is the velocity on the surface, z_0 is the roughness parameter of the sea surface.

Dynamic speed:

$$u_* = \sqrt{\frac{\rho_a}{\rho_w}C_D V_{10}} \tag{5}$$

where ρ_a is the density of air, ρ_w is the density of seawater, V_{10} is the wind speed at a height of 10 m, C_D is the roughness coefficient of the water surface $\left(1.5 \cdot 10^{-3}\right)$.

Generation by non-linearity of surface waves:

$$P^w = \sigma_w u_* \frac{d}{dz}E_w \tag{6}$$

where the constant σ_w (0.2–0.5), E_w is the kinetic energy of the surface disturbance:

$$E_w = \frac{\overline{u_w^2}}{2} \tag{7}$$

where u_w is the velocity of the particles in the wave.

$$\frac{dE_w}{dz} = 2A^2k^2ge^{-2kz} \tag{8}$$

where k is the wave number, A is the amplitude. The parameters are determined from the energy spectrum of the perturbation.

$$k = \frac{\omega^2}{g} \; \omega = 2\pi f \tag{9}$$

Generation by breaking waves:

$$P^{br} = C_{br}\frac{u_0^3}{b_0}\left(1 + C_j\frac{z}{b_0}\right)^{-2.8} \tag{10}$$

where C_{br} is a constant $(1.5 \cdot 10^{-4})$, b_0 is the width of the crest of the crashing wave, C_j is a constant (6.75).

$$b_0 = \frac{8.5 \cdot 10^3 V_{10}^3}{2} \tag{11}$$

The generation of turbulence by breaking waves also depends on the value of the q – probability of breaking:

$$q = -2.47 + 32.87\left(u_*^a\right)^2 + 42.37\frac{u_*^a}{c_p} \tag{12}$$

where u_*^a is the dynamic velocity in the air, c_p is the phase velocity of the spectral peak of the perturbation:

$$c_p = \frac{g}{2\pi f_p} \tag{13}$$

where g is the acceleration of gravity, f_p is the frequency of the spectral peak of the disturbance, π is a mathematical constant.

Summing up all three generation mechanisms Eqs. (3), (6), (10) and taking into account the probability of breaking Eq. (12), we can proceed to a general expression describing the transition of mechanical wind energy into energy that sinks the OSBL, supplemented by the values g and α, scale coefficients necessary to bring mechanical energy and energy flows heat to the same units of measurement.

$$D = \frac{1}{g\alpha}\int_0^z\left[P^{\tau}(z) + P^{w}(z) + qP^{br}(z)\right]dz \tag{14}$$

where g is the acceleration of gravity, α is the coefficient of thermal expansion.

5 Modernized Model

From the Eqs. (1), (2), (13) we obtain:

$$\frac{dh}{dt}\left[q^2 + \frac{1}{2}h_*^2N^2 - B_0t - 2su_*^4f^{-2}h_*^{-2}(1 - \cos ft)\right] = D + \frac{1}{2}B_0h_* \tag{15}$$

If we assume that $\frac{dh}{dt} = \Delta h$ and solve Eq. (14) in finite differences, then we get a depth value corresponding to the thickness of the OSBL, which was established during the period of turbulence evolvement. This perion is an empirical characteristic based on the choice of time (in this paper – 20 min), during which the energy spectrum of the disturbance will be analyzed (Fig. 2). The obtained model curve is verified using the data of a distributed thermoprophilometer, which is a two-dimensional map of the temperature dependence on depth (Fig. 4, 5, 7).

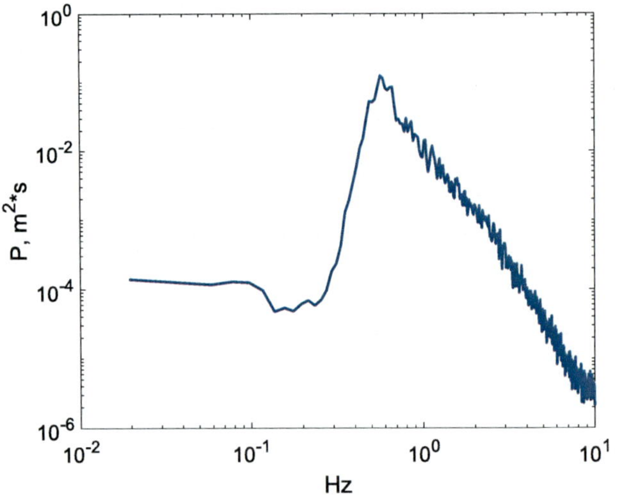

Fig. 2. An example of the experimental energy spectrum of the wave used in the calculations

6 Data Collection and Assumptions Used

In-situ data collection was carried out on the stationary oceanographic platform of the Black Sea Hydrophysical Subsatellite polygon of the Marine Hydrophysical Institute of the Russian Academy of Sciences (MHI RAS) in the Katsiveli in the period from June to October 2021. The observations were carried out with the help of specialized equipment, created mainly in the MHI.

The parameters of the wave perturbance were recorded using a six-string waveform, temperature profiles for model verification with full-scale data were recorded by the Gaisky P.V. thermoprophilometer [9], meteorological data (wind speed, air temperature, humidity, precipitation, insolation, evaporation) were collected using the «*Davis*» meteorological complex and provided by employees of the Department of Remote research Methods of the MHI.

The measuring area is located near the shore (Fig. 3), the distance from the sea surface to the bottom is 27 m, and there are local currents near the measuring area, which depend on the strength and direction of the wind. If the wind is directed at the measurement area from the shore, then a special hydrological situation should be created, which is not described by the model proposed in this paper.

Fig. 3. Stationary oceanographic platform (left) and the location of the platform (right)

Previously, it was experimentally shown that with wind and waves from the sea, the platform does not affect the waveform readings [10].

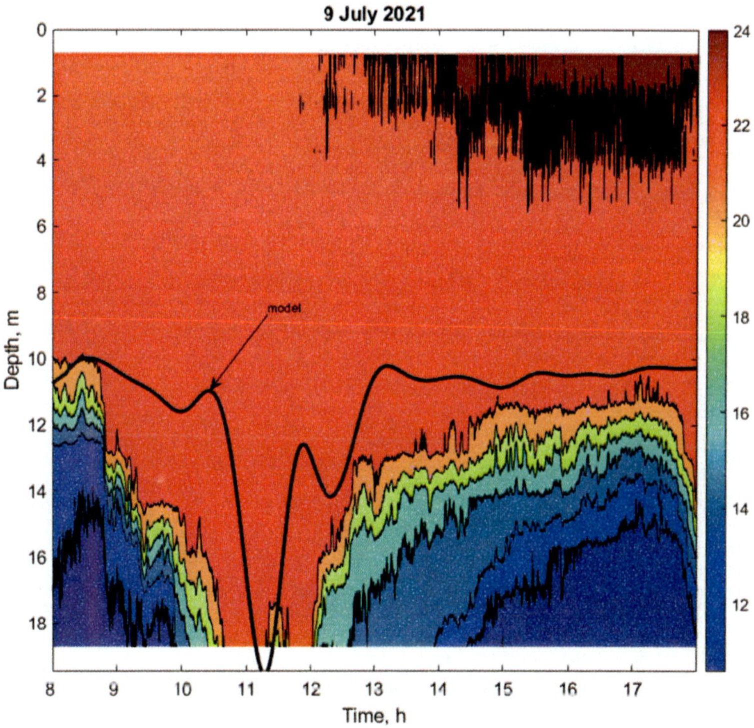

Fig. 4. Modeling of the thickness of the MBL using the Niiler-Kraus model. The wind speed is 10.3 m/s the wind direction is N-E. The correlation coefficient of model data to full-scale data is 0.75

7 Conclusion

The data analysis and verification modernized model are presented in Figs. 4, 5, 7. Revealed that the model performs well at high wind speeds. This is due to the fact that the main force stirring the MBL is rapidly developing turbulence, capable of deepening the seasonal thermocline by several meters in tens of minutes. The most representative simulation result is shown in Fig. 4. According to the thermoprophilometer, we can observe that the layer is homogeneous to a depth of 10 m; the wind in strength (10.3 m /s) and direction (N-E) almost does not change. Full-scale conditions provide an excellent opportunity to test the correctness of model work.

The situation presented in Fig. 5 is most interesting to present the operation of the model with high variability of direction (Fig. 6) and wind speeds. The absence of any serious fluctuations in the model data is caused by taking into account the frequency of buoyancy and the energy of reduction of the average velocity field due to mixing (parameters B and C in Eq. 1).

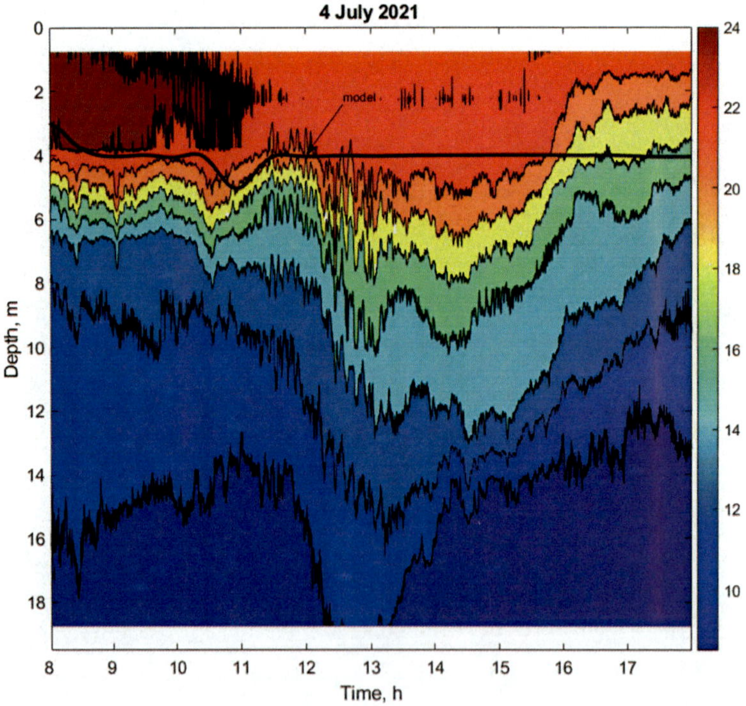

Fig. 5. Modeling of the thickness of the MBL using the Niiler-Kraus model. The wind speed is 5–16 m/s, the variability of wind speed and direction is high. The correlation coefficient of model data to full–scale 0.54

The simulation interval in Fig. 7 is shorter, because before 13 o'clock the wind was weak (less than 3 m/s). The model is not able to describe the dynamics of the MBL in conditions of weak wind action.

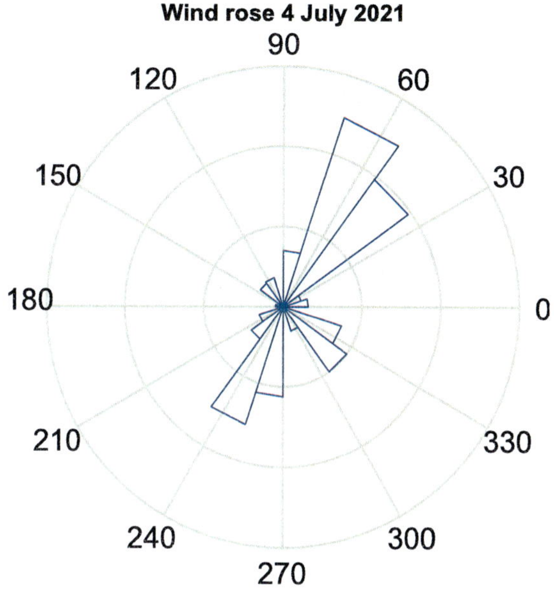

Fig. 6. The wind rose

Summing up the results of the work, we can say that the Niiler-Kraus model, coupled with a multiscale turbulence model, can be used to assess the dynamics of the thickness of the upper MBL under the influence of turbulent mixing processes and heat flows, and especially in strong winds. In addition, significant wind variability in the direction does not have a serious result on the calculations. Taking into account the fact that the Niiler-Kraus model is designed to simulate the processes of seasonal thermocline deepening in the ocean, where the depth of the mixed layer can reach 100 m, its application in the conditions of the shelf zone at shallow depths (up to 27 m) can create errors in the final results of modeling, which in this work could not be fully proved, due to the fact that some dynamic factors were not taken into account, for example, currents, as well as upwelling and downwelling. An important characteristic of the upgraded model is that all the mechanisms of energy exchange in the upper layers presented in this paper can be used individually and thus its refinement is an applicative procedure requiring consideration of methods for evaluating additional mechanisms that affect the dynamic processes occurring in the MBL. Summarizing, the following points can be distinguished:

- The Niiler-Kraus model, supplemented by a multiscale turbulence model, can be used to predict the dynamics of the thickness of the MBL in the coastal zone, which is confirmed by the obtained correlation coefficients of model and field data;
- The model behaves especially well at high wind speeds, while the variability of the wind slightly affects the results of calculations.;
- The discrepancy between observations and model calculations is apparently caused by the influence of unaccounted factors, in particular, the dynamics of local currents in the coastal zone. As well as upwelling and downwelling, which can be caused by

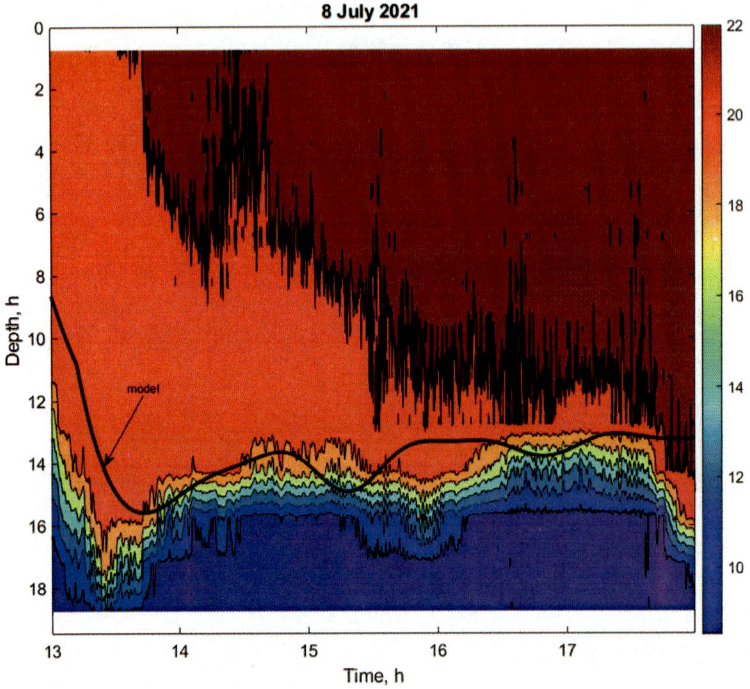

Fig. 7. Modeling of the thickness of the MBL using the Niiler-Kraus model. The wind speed is 9.6 m/s, the wind direction is N-E (from the sea), the variability is low. The correlation coefficient of model data to full-scale data is 0.61

a special dynamic structure of currents causing shear turbulence in the measurement area;

- It is also worth noting that the absence at the moment of a generally accepted method for calculating the contribution of excitement to turbulence reduces the objectivity of model calculations.

Acknowledgements. The research was supported by RSF (project No. 22-17-00150).

References:

1. Belcher, S.E., Grant, A.L., et al.: A global perspective on Langmuir turbulence in the ocean surface boundary layer. Geophys. Res. Let. **39**, 18605 (2012)
2. Rippeth, T.P., Palmer, M., Simpson, J.H., Fisher, N.R., Sharples, J.: Thermocline mixing in summer stratified continental shelf seas. Geophys. Res. Lett. **32**, L05602 (2005)
3. Kantha, L., Clayson, C.A.: Boundary Layers Ocean Mixed Layer. Academic Press, pp. 291–298 (2003)
4. Hewitt, H., Hewitt, C., et al.: Design and implementation of the infrastructure of HadGEM3: the next-generation met office climate modelling system. Geosci. Model Dev. Discuss. (2010)

5. Niiler, P.P., Kraus, E.B.: One-dimensional models of the upper ocean. In: Modelling and Prediction of the Upper Layers of the Ocean, vol. 1, pp. 143–172. Pergamon Press (1977)
6. Chukharev, A.M.: Multitime scale model of turbulence in the sea surface layer. Izv. Atmos. Ocean. Phys. **49**, 439–449 (2013)
7. Shuleikin, V.V.: Physics of the Sea, Nauka, Moscow (1968)
8. Ozmidov, R.V.: Certain features of the energy spectrum of oceanic turbulence. Dokl. Akad. Nauk SSSR **161**(4), 828–831 (1965)
9. Gaisky, P.V.: Algorithmic and software data registration of hydrological meters based on the distributed thermoprofilemeters. Ecolog. Saf. Coastal Shelf Zones Sea **3**, 128–141 (2022)
10. Dulov, V.A., el al.: Full-scale study of spatial uniformity of meteorological and holographic parameters in the coastal zone. To the problem of calibration of the radar BO «Sich-1M» as a tool for obtaining the speed of the driving wind. Mar. Hydrophys. J. (3), 31–43 (2005)

Heat Fluxes in a Multilayer Atmosphere-Snow-Ice-Water Environment Using the Gulf of Finland as an Example

K. V. Kravtsova(✉) and N. A. Podrezova

Russian State Hydrometeorological University, Saint Petersburg 192007, Russia
karkrav08.12@mail.ru

Abstract. This publication outlines and summarises the results of ice research in the eastern Baltic Sea conducted in February 2021 by oceanology students of the Russian State Hydrometeorological University and in February 2023. Namely, the vertical profile of the actual temperature, vertical distribution of hourly average temperatures in each layer was studied while setting up experiments with a thermometric streamer, and the purpose of the research was to calculate heat fluxes in a multilayer medium. Ice accretion from the ice surface through snow was calculated from the obtained fluxes. Currently there are not many scientific papers on the chosen topic and this is the relevance of this work, which can serve as a basis for more extensive directions in the future.

Keywords: Gulf of Finland · Ice cover · Snow cover · Multilayer environment · Heat fluxes

1 Introduction

The formation and disappearance of ice in the Gulf of Finland occurs annually, but the exact timing and extent depend on the severity of the winter. The ice usually forms first in bays, archipelagos and inland areas and spreads in an east-west direction. The ice period begins around mid-November, with maximum ice cover in late February to March. Ice thickness is usually relatively thin (0.1–0.3 m), but can reach 0.8 m and even 1 m in severe winters. The ice cover in the eastern part of the sea breaks up from west to east. In the Gulf of Finland, the continental ice begins to break up in late March and early April. The timing of ice break-up can vary by 60–70 days from year to year. In severe winters the ice melts completely in the second half of May, in mild winters in early May or even in the first or second decade of April. [0,2].

Under the influence of the heat flux from the sub-ice layer of water towards the ice surface, melting can occur from its lower boundary despite sub-zero air temperatures. This is why it is important to study the thermodynamics of the ice cover, especially in navigable areas of water areas prone to freezing.

T. Chaplina (Ed.): CIWO 2023, SPEES, pp. 80–94, 2023.
https://doi.org/10.1007/978-3-031-47851-2_10

2 Experiments and Data Representation

From 8 to 20 February 2021 in the eastern part of the Gulf of Finland in the area of Oranienbaum (Lomonosov city) there was an internship of second-year oceanology students of the Russian Hydrometeorological University. The aim of the training was to collect data on the vertical temperature distribution in the layered atmosphere-snow-ice-water environment. For this purpose, a thermometric streamer with four sensors was embedded in the frozen medium. The sensors were installed in such a way that 1 sensor was located at the upper boundary of ice and the lower boundary of snow cover, 2 and 3 sensors were located in a drilled hole and 4 sensor was located in the subglacial water layer. Temperature values were measured on February 18 and 19, that is to say two experiments were conducted in 2021. The values were recorded with a sampling rate of 30 s [1]. Scheme of sensor location on 19.02.21 is presented in Fig. 1.

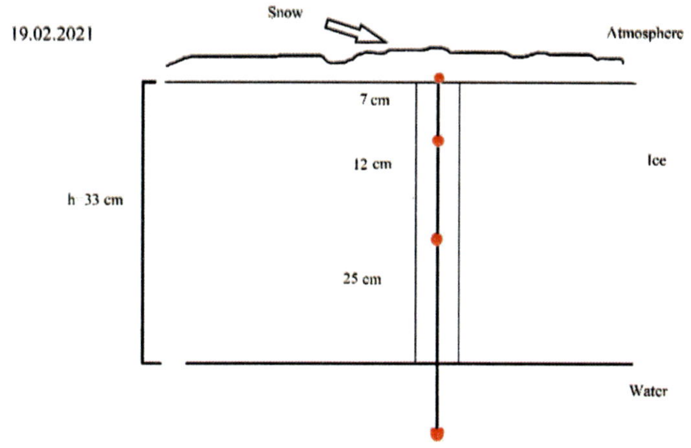

Fig. 1. Diagram of the thermometric tress sensor layout [1]

3 February 2021

Experiment 1, which took place on February 18th, had an average atmospheric temperature of −18 °C. There were no clouds in the sky and the wind was calm. The experiment was conducted in a closed terrain called the Sidorovsky channel. To protect the surface temperature sensor from the atmosphere, it was covered with snow, which had a thickness of 42 cm. The duration of the experiment was 3 h and 47 min. Throughout the experiment, the temperature changes observed at sensor 1, located in the snow, ranged from −9.0 °C to −4.0 °C on average. On the other hand, sensor 2's temperature stabilized within an hour and recorded values from 1.6 °C to 1.0 °C. The final sensor and sensor 3 recorded temperatures close to each other between 0.1 °C and 1.2 °C over the entire measurement period. The resulting temperature profile can be seen in Fig. 2a [1].

The graph shows the contrast between the small interval with temperatures below the presented values and the environment, as some anomalous area of the order of 0.5 °C, this area is followed by a warm zone with temperatures of 1.0–1.5 °C. This relatively cold interval can be explained by the fact that the sensors of the thermal tress are very sensitive to even minor tampering, perhaps someone walked next to it, or the device itself was checked to ensure it was registering correctly [1].

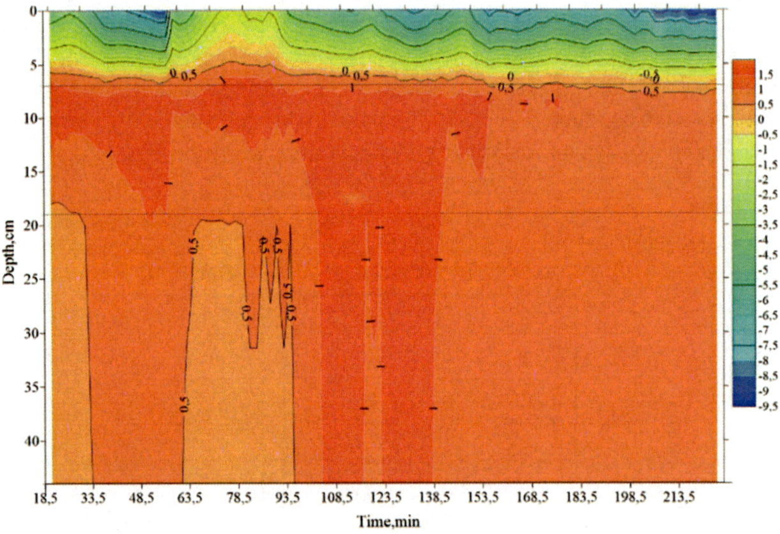

a)

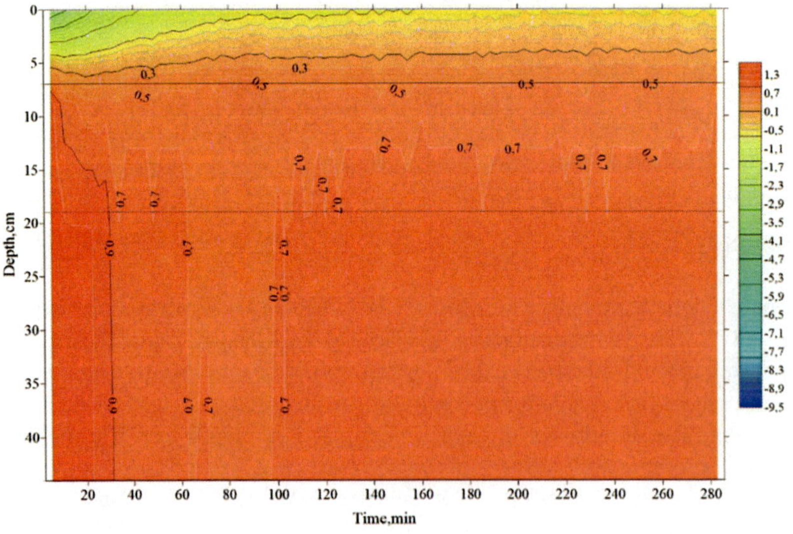

b)

Fig. 2. Vertical profiles of the actual temperature on 18.02.21 (a) and 19.02.21 (b) [1]

Experiment 2 was made on the 19th of February at an average atmospheric temperature of −10 °C. The cloudiness level was 8–9 points and the wind speed was 2 m/s. The experiment was conducted in the open area, on landfast ice. To protect the first temperature sensor from external environmental conditions, it was covered with snow. Ice thickness was 33 cm. The time of the experiment was 4 h and 45 min. The resulting temperature profile is shown in Fig. 2b [1].

Temperature stabilized at all sensors in less than an hour, about 30–45 min. The most contrasting temperature changes were noted at sensor T_1 at the beginning of the experiment, which is in the snow. Here we can note the temperature increase from −

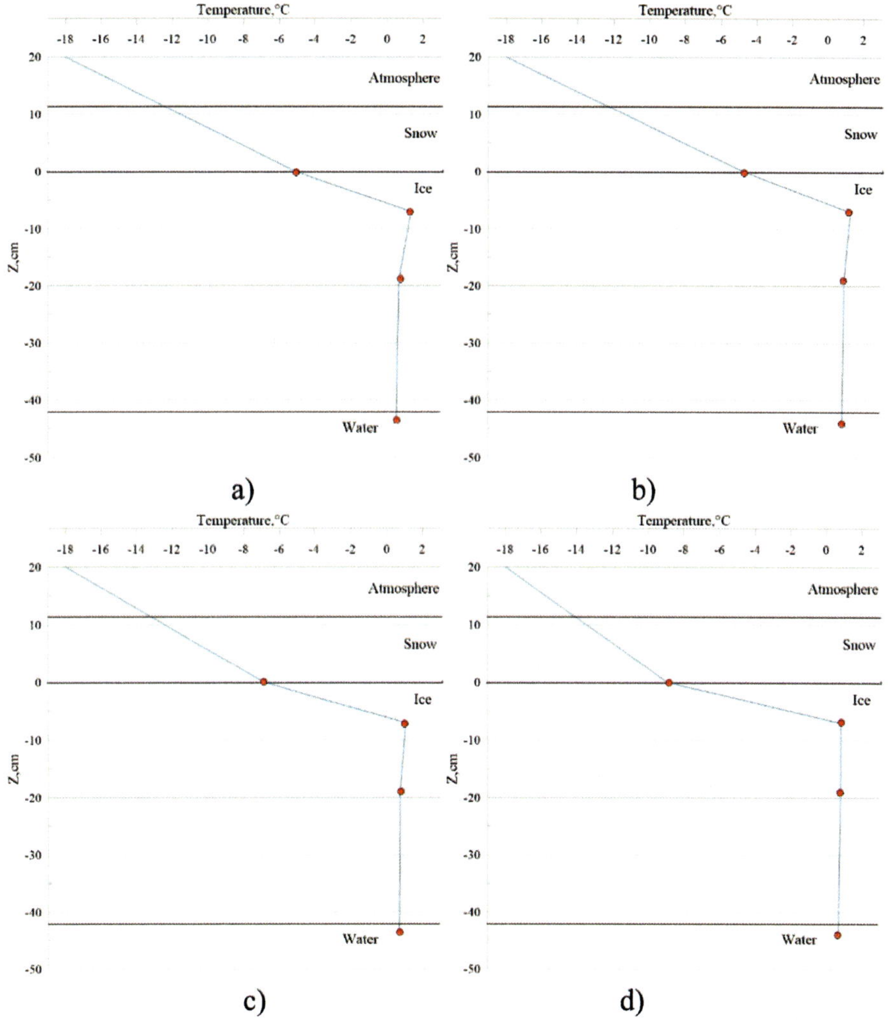

Fig. 3. The vertical distribution of the average water-ice-snow-atmosphere temperature for each hour (a-1st hour, b-2nd hour, c-3rd hour, d-last 47 min) 18.02.21

3.6 °C to −1.0 °C on average for an hour and a half and then steadily −1.0 °C [1]. Sensor T_2 recorded temperatures between 1.5 °C and 0.6 °C for 15 min, followed by small changes around 0.5–0.7 °C. T_3 and T_4 recorded temperature values ranging from 1.5 °C to 0.9 °C for approximately 30 min from the start of recording and then variations in the range 0.6 °C to 0.8 °C until the end of the experiment.

From the graph shown in Fig. 3a–d, it is clearly seen that the sub-ice water and the lower ice boundary had a temperature of about 0 °C at the beginning of the measurements and about 1 °C at the last hour [1]. Whereas the surface temperature was significantly different: it was about −5 °C and by the end of the experiment a lower value of about

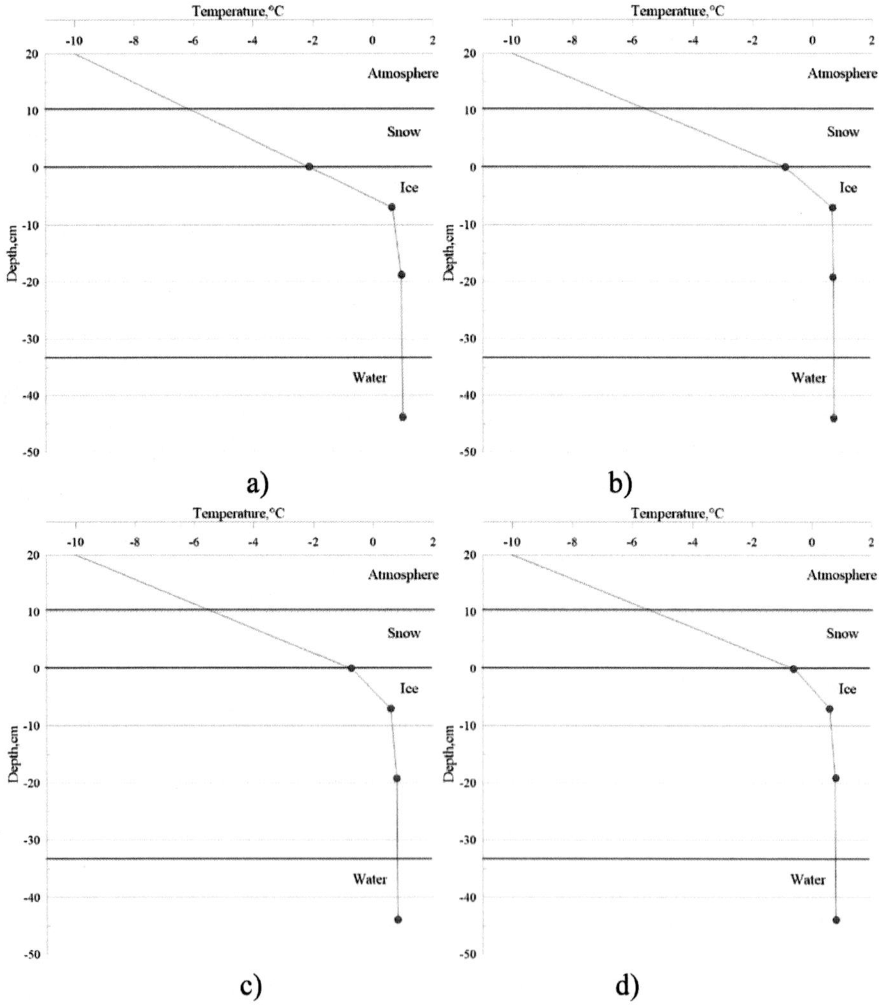

Fig. 4. The vertical distribution of the average water-ice-snow-atmosphere temperature for each hour (a-1st hour, b-2nd hour, c-3rd hour, d-4th hour) 19.02.21

−9 °C was noted on the surface sensor. It is important to note that the water is homothermal throughout the experiment.

From the graphs in Fig. 4a–d, it can be observed that the temperature of the sub-ice water and the lower ice boundary was 1 °C during the initial hour of measurements and slightly below 1 °C during the final hour. The temperature of the upper ice boundary and the lower boundary of the snow cover was −2 °C at the beginning of the experiment and approached −0.5 °C towards the end [1]. Thus, the water in the hole is also quite homogeneous.

The thermal conductivity coefficient of snow exhibits a wide range of values for the same snow density due to various factors, including the structural characteristics of the snow cover itself. The layering of the snow cover, influenced by meteorological conditions and metamorphism processes, affects its thermal resistance and the temperature distribution of the underlying surfaces. This indicates that snow is a good thermal insulating material, as it has poor heat conduction properties and acts as a protective layer for the ice surface [1, 2].

As the rules the heat flux through the snow cover is directly proportional to the temperature difference and inversely proportional to the thermal resistance of the snow cover. [1, 3]. It follows that the contributions of air temperature and the thermal resistance of the snow cover depend to a large extent on the thermal conductivity coefficient of the snow.

Equation was used to calculate the heat fluxes in each of the studied layers [1, 3, 4]:

$$F = \left(\frac{T_2 - T_1}{h}\right)\lambda, \left[W/m^2\right] \quad (1)$$

where T_1-temperature of overlying layer, T_2-temperature of underlying layer, h-distance between sensors [m], ρ-density of water [kg/m^3], which was determined from Oceanographic tables at salinity S = 2‰ (ρ_{wat} = 1001.61 kg/m), λ-thermal conductivity of water determined from reference tables [7] and varying slightly at different water temperatures between 0.569–0.5698 [W/m•°C]. The ice temperature was assumed to be 0 °C (Tables 1, 2 and 3).

Table 1. Required constant parameters for calculations

Parameter	18.02.2021	19.02.2021
h_{snow}, cm	11.5	10.25
ρ_{snow}, kg/m^3	259	221
T_a, °C	−18.0	−10.0
ρ_{wat}, kg/m^3	1001.61	1001.61
ρ_{ice}, kg/m^3	917.0	917.0

The heat transfer from water to ice is a natural occurrence that contributes to the reduction of frost degree-days, particularly when the ice is thicker. This gradual increase in heat flux from water to the atmosphere is evident in the temperature measurements

Table 2. Results of heat fluxes calculation on 18.02.2021

18.02.2021			
1	2	3	4
1st hour			
$F_{wat\text{-}ice}$, W/m^2	F_{wat} (s.3), W/m^2	F_{wat} (s.2), W/m^2	F_{snow}, W/m^2
−0.33	−3.21	10.77	89.06
2nd hour			
$F_{wat\text{-}ice}$, W/m^2	F_{wat} (s.3), W/m^2	F_{wat} (s.2), W/m^2	F_{snow}, W/m^2
−0.22	−1.68	10.50	89.06
3rd hour			
$F_{wat\text{-}ice}$, W/m^2	F_{wat} (s.3), W/m^2	F_{wat} (s.2), W/m^2	F_{snow}, W/m^2
−0.12	−1.47	8.51	89.06
4th hour (last 47 min)			
$F_{wat\text{-}ice}$, W/m^2	F_{wat} (s.3), W/m^2	F_{wat} (s.2), W/m^2	F_{snow}, W/m^2
−0.28	−0.05	6.86	89.06

Table 3. Results of heat fluxes calculation on 19.02.2021.

19.02.2021			
1	2	3	4
1st hour			
$F_{wat\text{-}ice}$, W/m^2	F_{wat} (s.3), W/m^2	F_{wat} (s.2), W/m^2	F_{snow}, W/m^2
0.12	1.38	5.30	55.51
2nd hour			
$F_{wat\text{-}ice}$, W/m^2	F_{wat} (s.3), W/m^2	F_{wat} (s.2), W/m^2	F_{snow}, W/m^2
0.08	0.25	5.24	55.51
3rd hour			
$F_{wat\text{-}ice}$, W/m^2	F_{wat} (s.3), W/m^2	F_{wat} (s.2), W/m^2	F_{snow}, W/m^2
0.07	0.99	4.77	55.51
4th hour			
$F_{wat\text{-}ice}$, W/m^2	F_{wat} (s.3), W/m^2	F_{wat} (s.2), W/m^2	F_{snow}, W/m^2
0.08	0.88	4.86	55.51

of the sub-ice water and lower ice boundary, which were initially at 1 °C and slightly below 1 °C towards the end. Similarly, the temperature of the upper ice boundary and lower boundary of the snow cover started at −2 °C and approached −0.5 °C as the experiment progressed. It is important to note that the thermal dynamics of the sea ice cover are influenced not only by meteorological and hydrological conditions but also by the properties of the snow covering its surface. In areas with negative air temperatures, the snow cover acts as a protective layer for the upper ice boundary [1, 6].

4 February 2023

In February 2023, as part of the winter ice practice, experiments were repeated to obtain data on the vertical distribution of temperature in the multilayer environment using a thermal stress. For the correctness of the data obtained, the conditions of the 2021 experiment were followed, i.e. the distances between the instrument sensors were maintained, and the surface sensor was covered with snow to minimise weathering. Sensors recorded the temperature with a discreteness of 30 s. Temperature values were measured on 10 and 11 February, i.e. 2 experiments were conducted. A diagram of the sensor locations on 10 February is shown in Fig. 5.

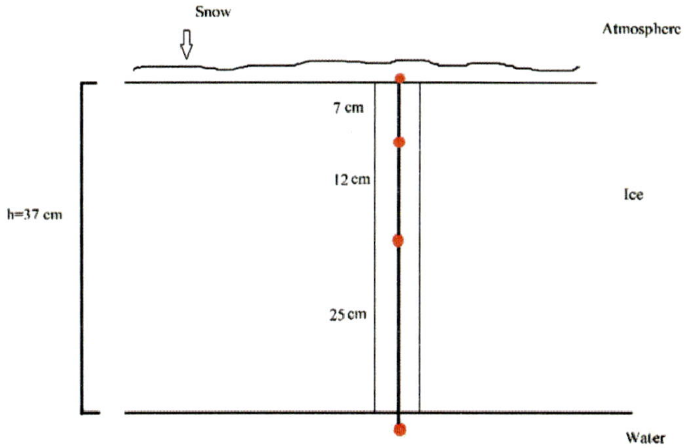

Fig. 5. Diagram of the thermometric tress sensor layout

Experiment 1 was conducted on 10 February under certain weather conditions. The average atmospheric temperature at the time of the experiment was −1 °C, cloudiness of 8–9 points, strong wind blowing from the west at a speed of 20–23 m/s. The experiment lasted 4 h 8 min in an open area on landfast ice. The ice cover thickness in this area was 37 cm. Figure 6a shows the temperature profile obtained during the experiment.

The temperature values equalised relatively quickly, namely in less than 15 min, due to the low negative air temperature. It can also be seen that the water in the drilled hole is fairly uniform, as the sensors recorded close values. The surface sensor recorded temperatures ranging from 1.0 °C to 0.6 °C for 25 min from the start and then changes in the range of 0.4–0.6 °C. Sensors T_2 and T_3 recorded temperatures between 1.0 °C to 0.5 °C for 25 min followed by 0.4–0.7 °C for the rest of the time. Sensor T_4 recorded values from 1.4 °C to 0.3 °C for 30 min, followed by small fluctuations of 0.4–0.8 °C for the rest of the measurements.

Experiment 2 took place on 11 February. The average atmospheric temperature was −3.17 °C. Cloudiness was 10, average wind speed was 2.20 m/s, direction was variable.

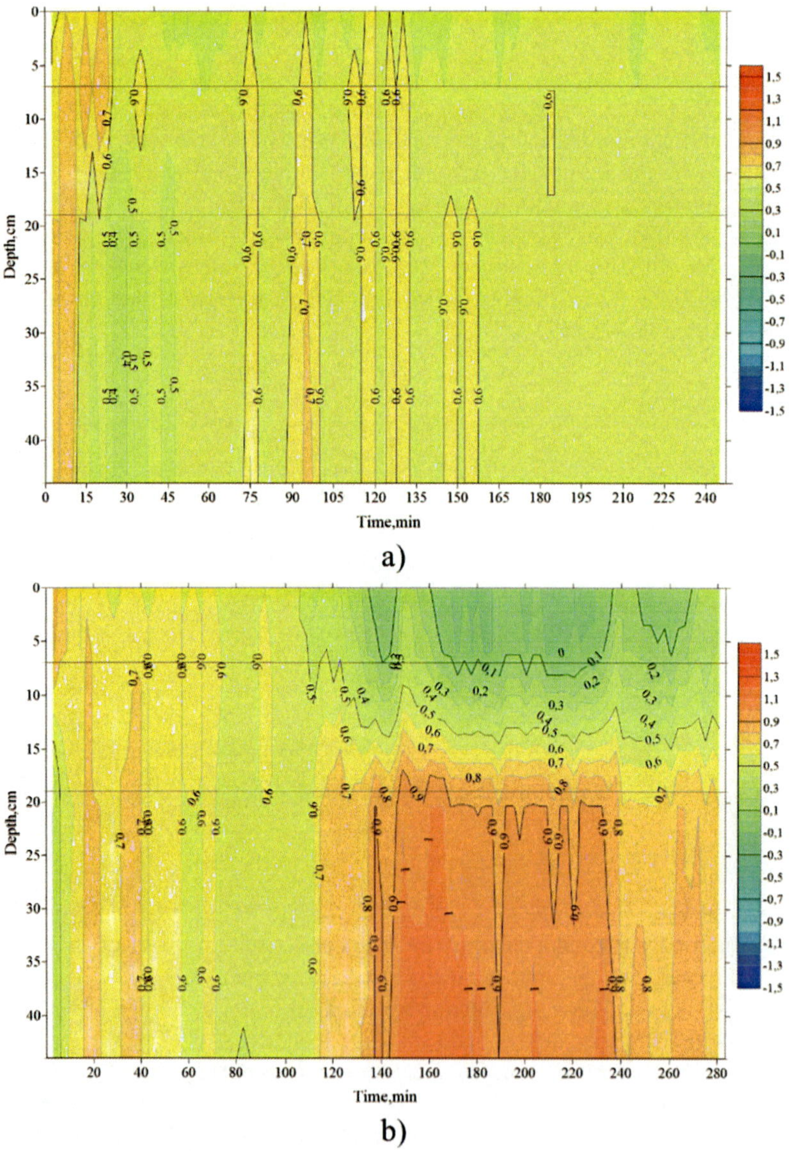

Fig. 6. Vertical profiles of actual temperature 10.02.2023 (a) and 11.02.23 (b)

The area was open and the time of the experiment was 4 h and 27 min. Sensors at a sampling rate of 30 s recorded temperature. The thickness of the ice cover was 38 cm. The resulting temperature profile is shown in Fig. 6b.

The values on the thermocouples stabilised quickly, within 10 min. From the vertical profile of the actual temperature, it can be noted that T_1 recorded temperature for two hours between 1.0 °C and 0.5 °C, and then to the end of the measurement in the range 0.5 to −0.1 °C. The T_2 thermocouple recorded temperature values between 0.9–0.4 °C for two hours and 0.6 °C to 0.0 °C until the end of the experiment and the T_3 sensor recorded

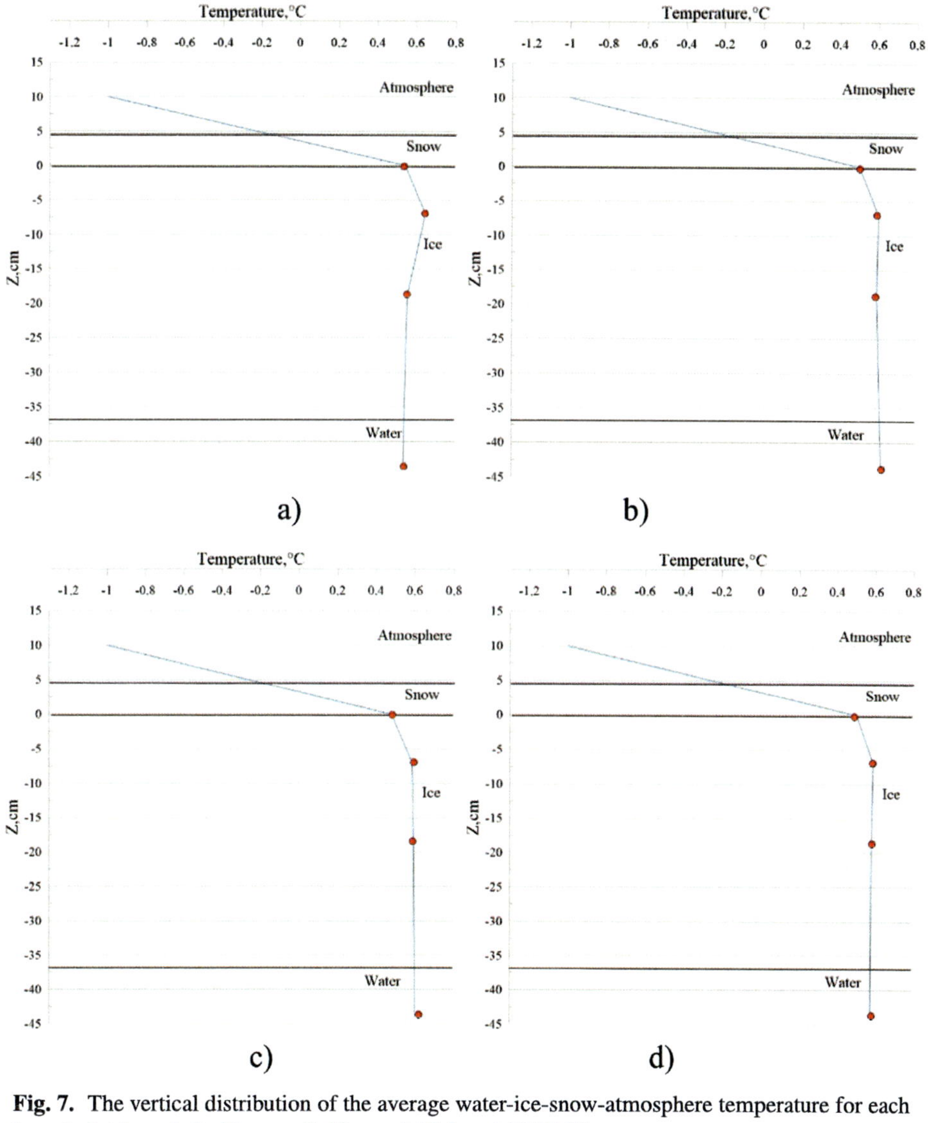

Fig. 7. The vertical distribution of the average water-ice-snow-atmosphere temperature for each hour (a-1st hour, b-2nd hour, c-3rd hour, d-4th hour) 10.02.23

temperature values between 0.4 °C to 0.7 °C for the first two hours and then 0.6–1.1 °C. The final sensor had the most stable temperatures without significant fluctuations for 2 h and was 0.5–0.7 °C and after that 0.7 °C to 1.2 °C.

Immediately noticeable is a prominent area dividing the profile into two parts, which appears as a decrease in temperature on T_1 and T_2, and an increase on T_3 and T_4, breaking the uniform temperature profile.

The graphs in Fig. 7a–d clearly show that the under-ice water and lower ice boundary had a temperature of about 0.5 °C in the first hour of measurements, by the end of the experiment it was about 0.6 °C. At the same time, the temperature of the upper ice

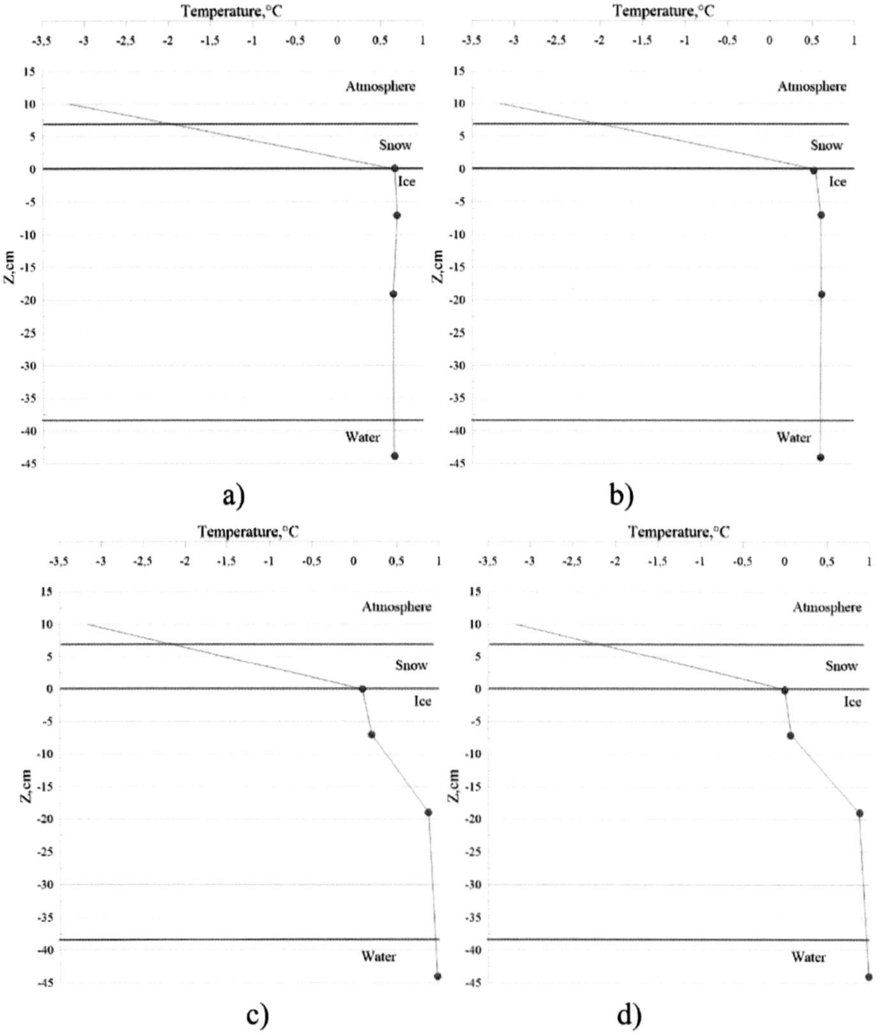

Fig. 8. The vertical distribution of the average water-ice-snow-atmosphere temperature for each hour (a-1st hour, b-2nd hour, c-3rd hour, d-4th hour) 11.02.23

boundary and the lower boundary of the snow cover at the beginning of the experiment was 0.5 °C and did not change over time.

From the presented graphs in Fig. 8a–d, it can be seen that the under-ice water and the lower ice boundary had a temperature of about 0.5–0.6 °C at the beginning of the experiment, and in the last hour 0.9–1.0 °C. Ice surface and lower snow cover boundary temperatures were also 0.5 °C at the start of measurements and after the first 2 h, there was a temperature drop to about 0 °C at the surface sensor (Table 4).

Similarly, using Eq. (1), the heat fluxes shown in Tables 5 and 6 were calculated from the measurements of the two experiments on 10.02.23 and 11.02.23.

Table 4. Required constants for the calculations

Parameter	10.02.2023	11.02.2023
h_{snow}, cm	4.75	7.0
ρ_{snow}, kg/m^3	285	229
Ta, °C	−1.0	−3.17
ρ_{wat}, kg/m^3	1001.61	1001.61
ρ_{ice}, kg/m^3	917.0	917.0

Table 5. Results of heat fluxes calculation on 10.02.23

10.02.2023			
1	2	3	4
1st hour			
$F_{wat\text{-}ice}$, W/m^2	F_{wat} (s.3), W/m^2	F_{wat} (s.2), W/m^2	F_{snow}, W/m^2
−0.05	−0.44	5.25	11.98
2nd hour			
$F_{wat\text{-}ice}$, W/m^2	F_{wat} (s.3), W/m^2	F_{wat} (s.2), W/m^2	F_{snow}, W/m^2
0.06	−0.04	4.87	11.98
3rd hour			
$F_{wat\text{-}ice}$, W/m^2	F_{wat} (s.3), W/m^2	F_{wat} (s.2), W/m^2	F_{snow}, W/m^2
0.01	0.04	4.77	11.98
4th hour			
$F_{wat\text{-}ice}$, W/m^2	F_{wat} (s.3), W/m^2	F_{wat} (s.2), W/m^2	F_{snow}, W/m^2
1	2	3	4
−0.01	−0.04	4.78	11.98

Table 6. Results of heat fluxes calculation on 11.02.23

11.02.2023			
1	2	3	4
1st hour			
$F_{wat\text{-}ice}$, W/m^2	F_{wat} (s.3), W/m^2	F_{wat} (s.2), W/m^2	F_{snow}, W/m^2
0.02	−0.22	5.62	25.77
2nd hour			
$F_{wat\text{-}ice}$, W/m^2	F_{wat} (s.3), W/m^2	F_{wat} (s.2), W/m^2	F_{snow}, W/m^2
−0.03	0.02	4.89	25.77
3rd hour			
$F_{wat\text{-}ice}$, W/m^2	F_{wat} (s.3), W/m^2	F_{wat} (s.2), W/m^2	F_{snow}, W/m^2
0.24	3.15	1.69	25.77
4th hour			
$F_{wat\text{-}ice}$, W/m^2	F_{wat} (s.3), W/m^2	F_{wat} (s.2), W/m^2	F_{snow}, W/m^2
0.26	3.81	0.63	25.77

With the available heat fluxes in mind, the idea arose to calculate ice growth from the ice surface through the snow using the equation [8, 9]:

$$\frac{dh}{dt} = \frac{F}{L\rho}, \quad [\text{cm/day}] \tag{2}$$

where ρ-density of water is 1001.6 [kg/m^3], L-specific heat of crystallization is $334 \bullet 10^3$ [J/kg], F-heat flux from the upper ice boundary through snow [W/m^2], received in calculations earlier. The ice temperature was assumed to be 0 °C.

The calculated values of ice formation on the available through snow using Eq. (2) for the average time of the experiments and the day are presented in Table 7.

The highest value of ice growth was received for 18.02.21, when the lowest atmospheric temperature of −18 °C and the maximum of the presented heat flux F through snow were observed, and the lowest value was for 10.02.23 with the lowest flux and the highest atmospheric temperature of −1 °C.

Table 7. Results of ice growth calculations

date	T_a, °C	V, m/s	h_{snow}, cm	F_{snow}, W/m2	dh/dt, cm/4 h	dh/dt, cm/day
18.02.2021	−18.0	0.0	11.5	89.06	0.383	2.30
19.02.2021	−10.0	2.0	10.25	55.51	0.239	1.43
10.02.2023	−1.0	20.0	4.75	11.98	0.052	0.31
11.02.2023	−3.17	2.2	7.0	25.77	0.111	0.67

5 Conclusions

As a result of this research work, vertical profiles of the actual temperature in the atmospheric-snow-ice-water multilayer medium were received, plots of the vertical distribution of hourly average temperatures in the multiphase medium were constructed, and calculations of heat fluxes for each hour of the experiments were performed and ice build-up was calculated for the average experiment time (4 h) and day.

The following conclusions can be made based on the results of the research work:

- After conducting thermal tress experiments, it became clear that the water in the well did not have time to freeze during the experiments (4–5 h), and as such, no "ice" medium was formed. During the experiments, a column of water with initial ice and snow types was observed in the hole.
- In all experiments conducted, the water in the excavated hole was characterised by temperature homogeneity with neutral density stratification. Thus, there was not convective mixing in the hole.
- Cooling was from the surface.
- The presence of snow cover on the ice surface smoothed its temperature course.
- The thickness of snow-covered ice is less than that of snow-free ice. Its temperature under the snow is higher and changes more slowly than that of bare ice. This reduction depends on the thickness of the snow layer and its thermal conductivity.

References

1. Zakharov, V.F.: Sea ice in climatic system. In: Problems of the Arctic and Antarctic, vol. 69, pp. 15–26. Gidrometeoizdat, Saint Petersburg (1995)
2. Podrezova, N.A., Kravtsova, K.V.: Ice research in the Gulf of Finland. Physical and mathematical modeling of earth and environment processes. In: Proceedings of the 7th International Conference Series Springer Proceedings in Earth and Environmental Sciences, pp. 271–279 (2022)
3. Kotlyakov, V.M., Sosnovsky, A.V., Osokin, N.I.: Estimation of snow thermal conductivity coefficient from its density and hardness in West Spitsbergen. Ice Snow **58**(3), 343–352 (2018)
4. Doronin, Y.P.: Ocean Physics. Textbook, p. 340. RSHU, Saint Petersburg (2000)
5. Doronin, Y.P., Lukyanov, S.V.: Laboratory Work on Ocean Physics, vol. 87. RSHI, Saint Petersburg (1993)

6. Aslamov, I.A.: Heat exchange at water-ice boundary and structure of subglacial water layer in Lake Baikal: specialty 25.00.27 "Land hydrology, water resources, hydrochemistry". In: Abstract for Candidate of Physical and Mathematical Sciences/Aslamov I.A. Institute of Hydrodynamics named after M.A. Lavrentyev RAS. M.A. Lavrentev RAS, Irkutsk, vol. 24 (2019)
7. Volkov, A.I., Zharsky, I.M.: Big Chemical Reference Book, vol. 608. Modern School, Minsk (2005)
8. Zavyalov, D.D., Solomakha, T.A.: Influence of freshly fallen snow on sea ice build-up and melting. Marine Hydrophysical Institute of the Russian Academy of Sciences, Sevastopol. Ice Snow **59**(1), 103–111 (2019)
9. Doronin, Y.P., Kheisin, D.E:. Sea Ice, vol. 320. Gidrometeoizdat, Leningrad (1975)
10. Chizhov, A.N.: Ice Cover Formation and Spatial Distribution of its Thickness, vol. 129. Gidrometeoizdat, Leningrad (1990)
11. Kravtsova, K.V.: Study of ice cover in eastern part of Gulf of Finland in February 2021. In: Proceedings of X International Scientific-Practical Conference Volume I (III) Marine Research and Education. MARESEDU-2021, p. 234. CMI MSU, Moscow (2021)
12. Kravtsova, K.V., Podrezova, N.A.: Heat fluxes in multilayer atmosphere-snow-ice-water medium by example of Gulf of Finland. In: Proceedings of XI International Scientific-Practical Conference Volume II (IV) Marine Research and Education. MARESEDU-2022, p. 241. CMI MSU, Moscow (2022)

Spatial Trend Analysis of Significant Wave Heights in the Caspian Sea

E. E. Kruglova[1,2](✉) and S. A. Myslenkov[1,2,3]

[1] Department of Oceanology, Faculty of Geography, Lomonosov Moscow State University, Leninskie Gory, GSP-1, Moscow 119991, Russia
lissavetaandin@gmail.com

[2] Shirshov Institute of Oceanology, Russian Academy of Sciences, Nakhimovsky Pr. 36, Moscow 117997, Russia

[3] Hydrometeorological Research Centre of Russian Federation, Bolshoy Predtechensky Lane, Moscow 123376, Russia

Abstract. The paper presents the analysis of a number of storms and the trends for average annual and maximum annual of significant wave height based on the results of wave model WAVEWATCH III and the Peak Over Threshold method. The mean, maximum and 95th percentile SWH were analyzed for the Caspian Sea. Significant positive trends were found for the number of storms with SWH > 3–4 m. A significant positive trend was found in total storm duration: every 10 years, the number of stormy days with SWH > 2 m increases by 4 days. Trends were calculated for the average annual and maximum annual SWH at each point of the Caspian Sea. The maximum significant positive trends—0.03 m/10 years, the minimum trend value −0.03 m/10 years. For the average and maximum annual SWH values, negative trends observed for the area of the border between the Middle and South Caspian. The maximum values of trends are observed in the western part of Middle Caspian Sea.

Keywords: Wind waves · Wave modeling · Number of storms · Spatial trend · WAVEWATCH III · Caspian Sea

1 Introduction

The Caspian Sea is a region of significant economic activity, including shipping, fishing, and oil and gas extraction [1]. Understanding climate change impacts on storm activity and long-term variations in significant wave height (SWH) is crucial for civil engineering, scientific research, marine infrastructure safety, and weather forecasting [2].

The Caspian Sea is a closed inland body of water with a maximum depth of 1025 m, while its northern part is relatively shallow, depths do not exceed 20 m [3]. The depth in this northern region limits the development of waves. In the deeper Middle and South Caspian Sea regions, the stretching of the basin plays a significant role in wave formation. Consequently, the development of waves under western and eastern wind directions is constrained by the available fetch [2]. Due to differences in seabed topography and

T. Chaplina (Ed.): CIWO 2023, SPEES, pp. 95–103, 2023.
https://doi.org/10.1007/978-3-031-47851-2_11

physical conditions, the Caspian Sea is commonly divided into the North, Middle, and South Caspian Seas [4]. General information about the hydrometeorological conditions of the Caspian Sea is presented in [5, 6], with several studies focusing on sea level variability and storm surge phenomena, particularly in the North Caspian region [3, 7–9].

The regime and extreme characteristics of wind waves were considered in a number of studies [10–13]. Earlier, our scientific team received some basic information the wind wave climate of the Caspian Sea [11, 14, 15]. Preliminary findings regarding storm occurrences across the entire Caspian Sea were presented in [11], while a joint analysis of storm waves and wind speed climate variability, along with detailed regional and seasonal analyses, was published in [15].

The primary objective of this study is to analyze the number of storms and trends in average annual and maximum annual significant wave heights from 1979 to 2020 in the Caspian Sea. To assess wave condition variability in the Caspian Sea, we employed a wave model.

2 Data and Methods

To analyze wind waves within the Caspian Sea, we employed the WAVEWATCH III wave model [16]. Input wind data were sourced from high-resolution NCEP/CFSR/CFSv2 reanalysis data spanning the years 1979 to 2020. The reanalysis data offers a spatial resolution of ~0.2–0.3 degrees and a time step of 1 h. The ice fields were extracted from the NCEP/CFSv2 reanalysis and the OSI-450 database. Calculations were carried out using an unstructured triangulation grid. Figure 1 illustrates that the grid spacing is roughly 10 km in the open expanse of the Caspian Sea and narrow down to 900 m in the coastal area, featuring a total of 17,529 nodes. The figure also highlights three points within different regions of the Caspian Sea: the North Caspian (NC), the Middle Caspian (MC), and the South Caspian (SC).

To generate waves in the model, we utilized the ST6 scheme [16]. A more comprehensive understanding of the model and its adaptation for the Caspian Sea, refer to [11, 14].

The model outputs include wind wave characteristics with a 3-h time interval spanning 1979 to 2020, totaling 42 years of data. SWH with a 3-h time step for storm activity and mean and max annual SWH for trends was used in this research.

To analyze storm activity within the Caspian Sea, we applied the Peak Over Threshold (POT) methodology, a technique previously detailed in [11, 15, 17]. The analysis process involved several steps: First, a specific criterion was established (in our case, we selected a wave height criteria from 2 to 5 m). Next if any node within the examined sea area exhibited a significant wave height (SWH) surpassing the chosen criterion, the event was categorized as a storm case. The event concluded when the SWH decreases below the specified criterion. The storm's duration was calculated by determining the time difference between the end and start times of the event. Storms were tallied across the entire expanse of the Caspian Sea water area. If the interval between two storms was less than 9 h, they were regarded as a single long storm. When the gap exceeded 9 h, the storms were counted as two separate events.

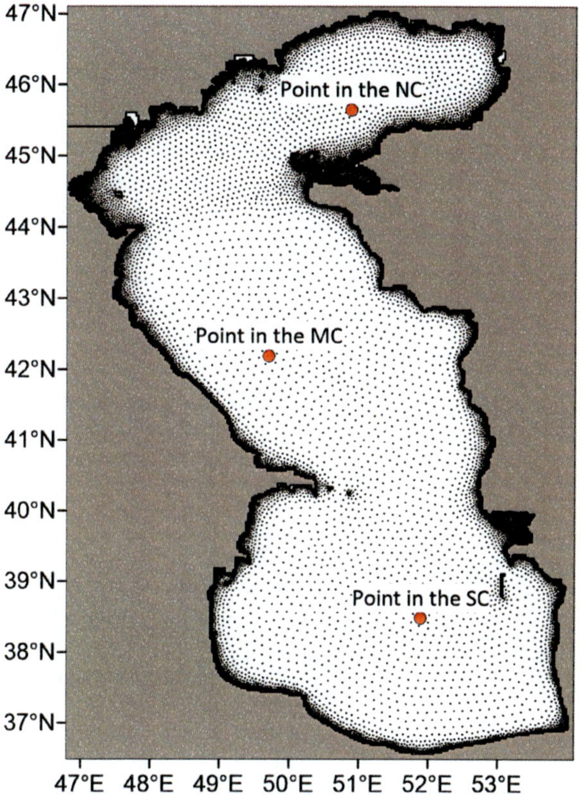

Fig. 1. The computational unstructured grid for the WAVEWATCH III model with points in the different parts Caspian Sea

Trend analysis based on average annual and maximum annual data which calculated for each model node.

We employed a linear regression model ($y = a1x + a0$) to calculate these trends. The trend value corresponds to the coefficient a1 of the linear trend and is expressed in the units of the characteristic y per unit of discreteness. To assess the significance of these trends, we analyzed the adequacy of the regression models through Fisher's test and examined the significance of the regression coefficients. For verification, we computed Student's criteria and determined the critical value of the Fisher's criterion at a specified significance level of 5%. Same methodology we used in [15].

3 Results and Discussion

Utilizing modeling data, were obtained maps displaying the mean and maximum multiyear significant wave height (SWH) values, as well as the distribution of the 95th percentile SWH, spanning the entire study period.

Figure 2 presents the distribution of maximum (a), mean (b), and 95th percentile SWH (c) over the entire study duration. The highest recorded maximum SWH was located in the Middle Caspian region, reaching a peak of 8.17 m. The North Caspian area did not witness maximum SWH values exceeding 5 m. Meanwhile, the South Caspian region exhibited a local maximum of approximately 7 m. The maximum mean SWH observed over the entire period was also in the Middle Caspian region, with values not surpassing 1.5 m. In both the North Caspian and South Caspian regions, mean SWH values for the entire study duration remained below 1 m. The 95th percentile SWH reached its maximum in the Middle Caspian region, it was more than 2.5 m. The 95th percentile SWH ranged from 1.5 to 2.2 m in the South Caspian and from 0.7 to 1 m in the North Caspian. [15] provides a detailed seasonal analysis of mean, maximum, and 95th percentile SWH values, further enhancing our understanding of these wave characteristics.

Also in [15] we conducted mean and maximum multiyear values of SWH, and the 95th percentile, SWH distribution maps for standard calendar seasons: winter (December–February), spring (March–May), summer (June–August), and autumn (September–November).

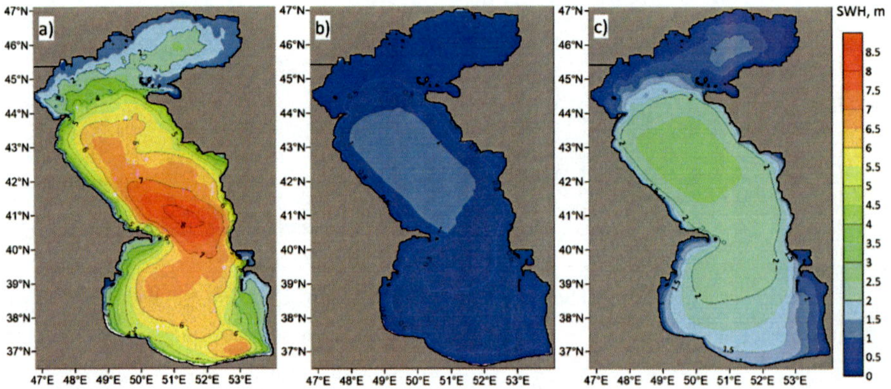

Fig. 2. The long-term maximum (a), mean (b), and the 95th percentile (c) SWH for the period 1979 to 2020.

To evaluate storm activity, we employed the POT method. Our analysis involved calculating the yearly count of storm events with SWH ranging from 2 to 5 m.

The multiyear mean number of storms with SWH > 2 m was 90, while those with SWH > 3 m totaled 45 (Fig. 3). Notably, the highest numbers of storms with SWH > 2 m occurred in 1985, 2009, and 2018 (101 events), while storms with SWH > 3 m were most prevalent in 2014 and 2016 (59 storms) and in 2015 (62 storms). In 2014, there was an absolute maximum of storms with SWH > 4 m, reaching 23 storms. Were revealed significant positive trends for storms with thresholds of 3 and 4 m, with the maximum trend value of 0.24 storms per year observed for storms with SWH > 3 m. This implies that every ten years, the number of storms with SWH > 3 m increases by 2.

Additionally, our analysis considered the total duration (in days) of storms with various SWH criteria for each year. Figure 3b demonstrates an increase in the duration of storms with SWH > 2 m. The mean number of stormy days featuring SWH > 2 m was ~114 days per year, with a maximum of ~145 days in 2016. The positive and significant trend was identified: every ten years, the number of stormy days with SWH ~ 2 m increased by 4 days. Furthermore, a significant positive trend was found for the total duration of storms with SWH > 3 m, with a trend value of 2 days per decade. However, trends for the number of stormy days associated with other SWH thresholds did not exhibit statistical significance.

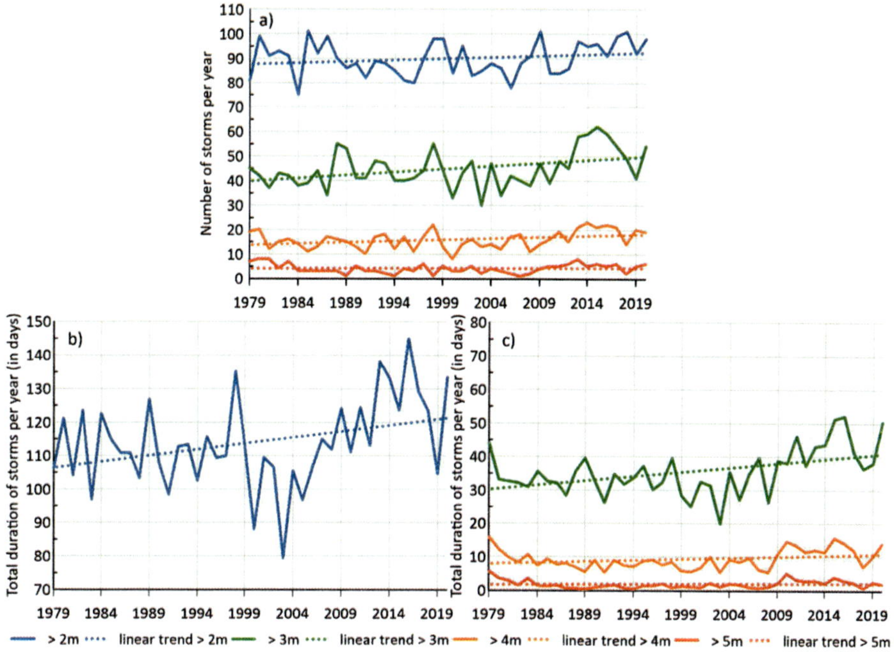

Fig. 3. The number of storms with different SWH thresholds per year (a), total duration of storms (b, c), and their linear trends for 1979 to 2020.

In prior research [15], we conducted separate calculations for the number of storms and their duration in distinct regions of the Caspian Sea. Moreover, we analyzed trends in the number of storms and duration across different areas and entire of the Caspian Sea and throughout various seasons of the year.

Trends were calculated in each node of the grid according to the obtained average annual and maximum SWH. The trend values are shows in Fig. 4. The black dots on Fig. 4 mark the grid nodes in which the trends found are significant. Spatial trend analysis shows the distribution of positive and negative trends for both average and maximum SWH is approximately the same. In the Middle Caspian Sea, there is an increase in the average annual wave height along the western coast, as well as negative trends on the border with the South Caspian. A similar distribution is typical for the maximum annual SWH. Using spatial trend analysis in the North and South Caspian, a decrease in the maximum annual SWH in most of the water area, while the average annual SWH increases. Changes in the average altitude in the North Caspian Sea can be explained by an increase in the number of milder winters, which is why this part of the sea has been ice-bound for less time in recent decades [18, 19].

Number of nodes with significant trends, for the average annual SWH 7167, and for the maximum—175. This is logical considering that the maximum wave heights at the points depend on one-time extreme events, while the average annual height responds to climatic changes and is less susceptible to unique events.

The maximum value of a significant trend for average annual values is 0.03 m/10 years (western part of the Middle Caspian), the minimum is –0.03 m/10 years (eastern part of the Middle Caspian). The maximum value of a significant trend for maximum annual values is 0.11 m/10 years (western part of the Middle Caspian Sea), the minimum is –0.13 m/10 years (near Cape Tyub—Karagan).

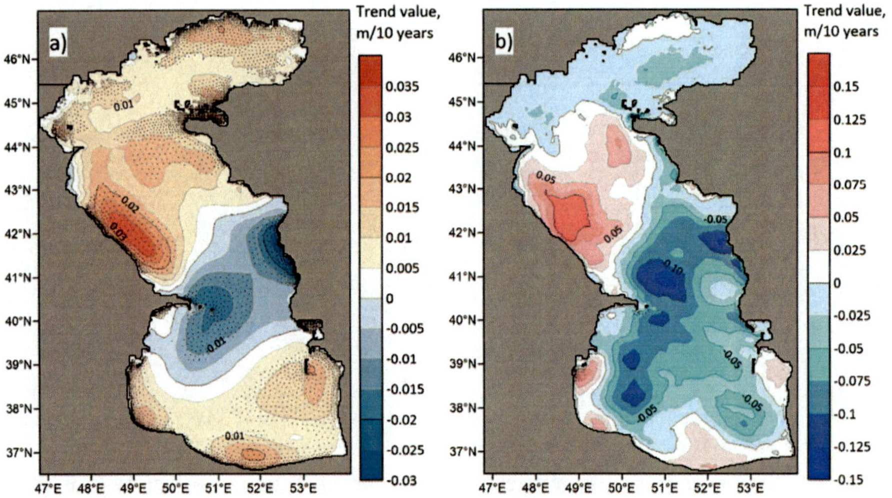

Fig. 4. Spatial distribution of the trends for average annual (a) and maximum annual SWH (m/10 years). The black dots mark the grid nodes in which the trends are significant.

Figure 5 shows the average annual (a) and maximum annual (b) SWH at three points located in different parts of the Caspian Sea. The location of the points (Fig. 1) was chosen based on the maps of the average and maximum SWH (Fig. 2). Trends were significant for the average annual SWH in Middle Caspian (trend value 0.02 m/10 years) and in the South Caspian (trend value 0.01 m/10 years). In the North Caspian, the trend for the average annual significant wave heights at the point is not significant. Trends for the maximum annual values insignificant for all selected points.

Pay attention to that at all three selected points, the trend is positive for the average annual SWH values, and negative for two of the three in the maximum annual SWH values. The positive trend in the Middle Caspian, as seen in Fig. 4, does not characterize the entire area, because the southeastern part of the Middle Caspian has negative trends. Therefore, drawing conclusions based on several points from the entire grid of the Caspian Sea can lead to errors.

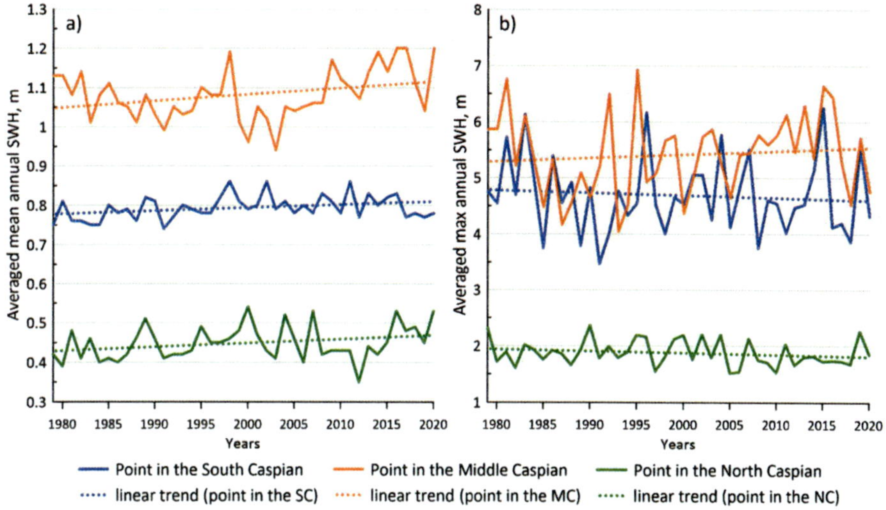

Fig. 5. Averaged mean (a) and max (b) annual SWH in the points Caspian Sea

4 Conclusion

The Middle Caspian recorded the highest maximum SWH, reaching 8.17 m, also in this part of Caspian Sea was observed the maximum mean SWH for the entire period but did not exceed 1.5 m. The maximum 95th percentile SWH during the entire study period exceeded 2.5m.

Employing the POT method, was analyzed a multiyear analysis the number of storm events with SWH ranging from 2 to 5 m across the entire Caspian Sea. This analysis revealed significant positive trends in the number of storms featuring SWH greater than 3–4 m. Additionally, a positive trend was found in the total storm duration, indicating that every ten years, the number of stormy days with SWH exceeding 2 m increases by 4 days.

Trends were found for the average annual and maximum annual SWH at each point of the Caspian Sea. The maximum significant positive trends were found for the western part of the Caspian Sea (0.03 m/10 years), the minimum trend value was found on the border between the Middle and South Caspian (–0.03 m/10 years).

Spatial trend analysis shows since most of the Caspian Sea has a positive trend in average annual SWH, greater number of significant trends are also positive. For the average annual SWH values, negative trends are characteristic only for the area of the border between the Middle and South Caspian. The maximum values of trends are observed in the Middle Caspian Sea.

For the maximum annual SWH, the largest modulo negative trends are also observed on the border between the Middle and Southern Caspian. The maximum trends are typical for the western region of the Middle Caspian Sea.

It is important to take into account that it is impossible to estimate the variability of SWH in the entire Middle or South Caspian Sea at one selected point, since both significant positive and negative trends are present in these areas. For a more correct and complete description, it is necessary to perform calculations for each node of the computational grid and make a spatial analysis of variability.

Acknowledgements. Data analysis was funded by the Ministry of Science and Higher Education of Russia, theme FMWE-2021–0002 E.E. Kruglova. This work was done with the support of MSU Program of Development, Project No 23-SCH07-33.

References

1. Akbulaev, N., Bayramli, G.: Maritime transport and economic growth: Interconnection and influence (an example of the countries in the Caspian Sea coast; Russia, Azerbaijan, Turkmenistan, Kazakhstan and Iran). Marine Policy **118**, 104005 (2020)
2. Leroy, S.A.G., Gracheva, R., Medvedev, A.: Natural hazards and disasters around the Caspian Sea. Natural Hazards **114**(3), 2435–2478 (2022)
3. Nesterov, E.S.: Water balance and level fluctuations in the Caspian Sea. In: Modelling and Forecasting: A Text-Book, p. 378. Triad Ltd., Piraeus (2016)
4. Kosarev, A.N.: Physico-geographical conditions of the Caspian Sea. In: The Caspian Sea Environment, pp. 5–31 (2005)
5. Baidin, S.S., Kosarev, A.N.: The Caspian Sea: hydrology and hydrochemistry. Nauka. Moscow **261**(272), 25 (1986)
6. Aladin, N., Plotnikov, I.: The Caspian Sea. In: Lake Basin Management Initiative Thematic Paper (2004)
7. Ivkina, N.I.: Climate and water balance changes in the Caspian region. In: A Method for Research and Forecasting Runup and Surge Phenomena in the Kazakh Part of the Caspian Sea, Taking into Account Changing Water Balance Characteristics, Astrakhan, pp. 22–23 (2010)
8. Malinin, V.N., Gordeeva, S.M.: Caspian Sea level as an indicator of large-scale moisture exchange in the "Ocean-atmosphere-land" system. Proc. Karelian Res. Cent. Russ. Acad. Sci. **4**, 5–20 (2020)
9. Serykh, I.V., Kostianoy, A.G.: The links of climate change in the Caspian Sea to the Atlantic and Pacific Oceans. Russ. Meteorol. Hydrol. **45**, 430–437 (2020)
10. Lopatoukhin, L.I., Yaitskaya, N.A.: Wave climate of the Caspian Sea, input wind data for hydrodynamic modeling, and some calculation results. Oceanology **59**, 7–16 (2019)
11. Pavlova, A., et al.: Storm surges and extreme wind waves in the Caspian Sea in the present and future climate. Civil Eng. J. **8**(11), 2353–2377 (2022)
12. Amirinia, G., Kamranzad, B., Mafi, S.: Wind and wave energy potential in southern Caspian Sea using uncertainty analysis. Energy **120**, 332–345 (2017)
13. Golshani, A., et al.: Wave hindcast study of the Caspian Sea. J. Mar. Eng. **1**, 19–25 (2005)
14. Myslenkov, S.A., et al.: Wave climate in the Caspian Sea based on wave hindcast. Russ. Meteorol. Hydrol. **4**(10), 670–678 (2018)
15. Kruglova, E., Myslenkov, S.: Influence of long-term wind variability on the storm activity in the Caspian Sea. Water **15**(11), 2125 (2023)

16. Tolman, H.: The WAVEWATCH III Development Group User Manual and System Documentation of WAVEWATCH III Version 6.07. Tech. Note 333, March 2019. NOAA/NWS/NCEP/MMAB 2019. https://www.researchgate.net/publication/336069899_User_manual_and_system_documentation_of_WAVEWATCH_III_R_version_607. Accessed 18 Dec 2020
17. Myslenkov, S.A., et al.: Increase in storm activity in the Kara Sea from 1979 to 2019: numerical simulation data. Dokl. Earth Sci. (Pleiades Publishing) **498**, 502–508 (2021)
18. Lavrova, O.Y., et al.: Interannual variability of ice cover in the Caspian Sea. J. Hydrol. X **17**, 100145 (2022)
19. Yaitskaya, N.A., Magaeva, A.A.: Ice regime of the Northern Caspian. In: Bulletin of the Moscow University. Series 5. Geography, no. 6, pp. 63–72 (2020)

An Influence of Meridional Oceanic and Atmospheric Heat Fluxes to the Atlantic Sector of the Arctic on Arctic Amplification

M. M. Latonin[1,2](✉), I. L. Bashmachnikov[1,2], and L. P. Bobylev[2]

[1] Saint Petersburg State University, Saint Petersburg 199034, Russia
m.m.latonin@spbu.ru

[2] Nansen International Environmental and Remote Sensing Centre, Saint Petersburg 199034, Russia

Abstract. One of the triggers of the Arctic amplification is the poleward heat transport; however, the relative contribution of its oceanic and atmospheric components remains poorly understood. In this study, we investigate the role of the Atlantic meridional oceanic and atmospheric heat fluxes to the Arctic for the emergence of the Arctic amplification. The meridional heat fluxes were computed using ORAS4 and ERA5 reanalyses for the Atlantic Water layer and over the lower troposphere. The horizontal and vertical boundaries of the sections in the ocean and atmosphere were selected to include only the areas, where the time mean heat transport is directed into the Arctic. It is found that the meridional heat fluxes and the regional Arctic amplification in the Eurasian Arctic are strongly coupled on the decadal time scales (10–15 years). We argue that the low-frequency variability of the Arctic amplification is regulated via a feedback, which elements include the oceanic heat transport, atmospheric heat transport and Arctic amplification. An increase (decrease) in the oceanic heat transport leads to a decrease (increase) in the atmospheric heat transport and to a decrease (increase) in the Arctic amplification. The atmospheric response to the ocean forcing occurs with a delay of three years and is attributed to the Bjerknes compensation mechanism. In turn, the atmospheric heat and moisture transport directly affects the magnitude of the Arctic amplification, with the latter lagging by one year. Thus, the variability of the oceanic heat transport at the southern boundary of the Nordic Seas might be a predictor of the magnitude of the Arctic amplification over the Eurasian Basin of the Arctic Ocean with a lead time of four years.

Keywords: Ocean-atmosphere interaction · Bjerknes compensation · Poleward heat transport · Nordic Seas · Variability · Eurasian Basin of the Arctic Ocean

1 Introduction

Present climate warming, which has begun in the 1980s, is global in nature, with higher warming rates in the Northern Hemisphere, a pronounced Arctic amplification and a very weak Antarctic amplification [1]. Arctic amplification of global warming or cooling is

T. Chaplina (Ed.): CIWO 2023, SPEES, pp. 104–110, 2023.
https://doi.org/10.1007/978-3-031-47851-2_12

an inherent and robust feature of the global climate system [2]. It was revealed based on proxy data from paleoclimatological studies in the past [3], is observed now, and is predicted in the future [4]. It is thought that the Arctic amplification phenomenon is a result of the climate feedbacks interplay in a nonlinear climate system and of variations in the meridional energy (heat and moisture) transport by the atmospheric and oceanic circulation [5, 6]. The current positive phase of the Arctic amplification is closely related to the reduction in the Arctic sea ice area, and positive surface albedo feedback was considered to be the main mechanism for a long time [7]. However, model experiments with a locked surface albedo feedback have shown that the Arctic amplification becomes only 15% weaker [8]. More recently, studies have emerged indicating that temperature feedbacks (lapse-rate feedback and Planck feedback) may be more important than the albedo mechanism [5, 9, 10]. However, the most recent studies again point to the leading role of positive surface albedo feedback. Initially, this mechanism works only in summer, when a large amount of shortwave solar radiation enters the Arctic. However, the true albedo mechanism appears in winter, when the heat stored in the Arctic Ocean is released into the atmosphere causing heating of the lower troposphere [11, 12]. Moreover, the positive lapse-rate feedback in the Arctic is most pronounced in the regions where sea ice is shrinking. Therefore, this mechanism might be a consequence of the Arctic amplification and not its cause [11–13]. However, these estimates of climate feedbacks were obtained from the top-of-atmosphere perspective. When assessing climate feedbacks from the surface perspective, the leading role of the albedo mechanism is still disputed. In particular, it was found that the positive surface albedo feedback is largely compensated by the negative cloud feedback. This, in turn, results in a lower increase in the ocean heat content in summer and lower heat release into the atmosphere in winter [14]. Moreover, water vapor might play a significant role in the formation of Arctic amplification, the increase of which over the Arctic leads to an increase in the downward longwave radiation [15]. This process can be decisive in the timing of sea ice melting in the Arctic [16]. At the same time, atmospheric circulation is the source of water vapor [17], which, in turn, can respond to oceanic circulation through the Bjerknes compensation mechanism [18].

Thus, at the moment, there is no clear understanding of the relative importance of the mechanisms causing the Arctic amplification [19]. Moreover, there is no consensus on how to correctly numerically estimate the magnitude of the Arctic amplification [20]. The aim of this study is to find new linkages between the variability of Arctic amplification on different time scales and the coupled oceanic and atmospheric meridional heat fluxes at the entrance to the Atlantic sector of the Arctic.

2 Data and Methods

The oceanic reanalysis ORAS4 (https://icdc.cen.uni-hamburg.de/daten/reanalysis-ocean/easy-init-ocean/ecmwf-ocean-reanalysis-system-4-oras4.html) and atmospheric reanalysis ERA5 (https://climate.copernicus.eu/climate-reanalysis) were used for the calculation of the meridional heat fluxes in the ocean and atmosphere. The following variables are necessary for the assessment of heat fluxes: seawater potential temperature data, salinity and meridional current velocity for the ocean and air temperature data, specific humidity, meridional wind velocity and geopotential heights for the atmosphere.

The Arctic amplification was estimated based on the surface air temperature (SAT) data from the ERA5 archive. The heat fluxes and Arctic amplification were computed for the period 1958–2017, and annual means were studied.

The meridional advective heat fluxes were estimated in the Atlantic domain across the following sections at 66.5°N: from −4.5°E to 13.5°E in the ocean and from −5°E to 80°E in the atmosphere (Fig. 1).

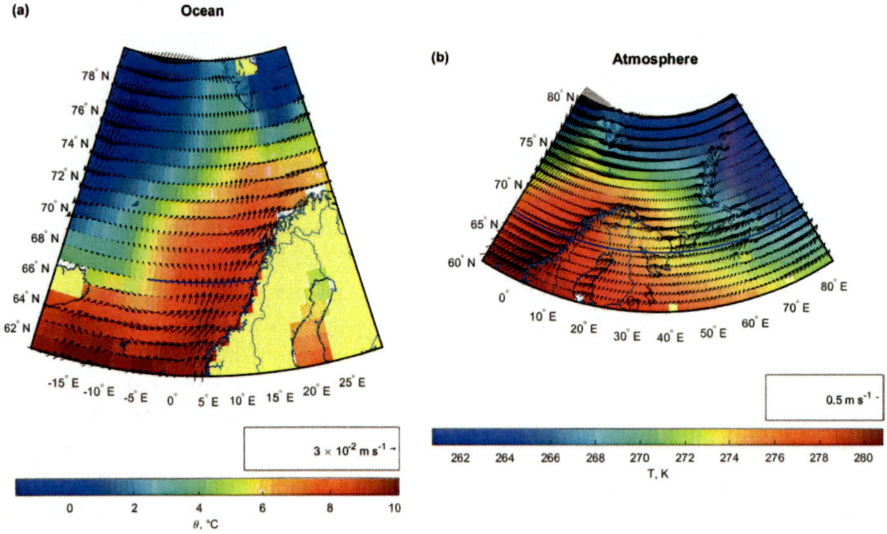

Fig. 1. Studied area with the sections for the heat transports' estimation displayed in blue. Climatological fields of two key parameters for the ocean and atmosphere are also added [18].

The oceanic heat flux was estimated in the Atlantic Water layer, whereas the atmospheric heat fluxes were estimated in the layer from 1000 hPa to 800 hPa. The horizontal and vertical boundaries of the sections were defined based on a condition to include only the heat fluxes directed predominantly northwards both in the ocean and atmosphere. The detailed explanations and equations can be found in [18].

The Arctic amplification was calculated for the area 60°–90°N using the original 0.25° × 0.25° ERA5 resolution. It was defined in the same way as in [18]. This approach has allowed to obtain the Arctic amplification time series per grid point in the Arctic region. Thus, the spatial inhomogenuity of the intensity of Arctic amplification has been highlighted. In contrast, previous studies considered single time series of Arctic amplification obtained on the basis of average SAT anomalies both in the Arctic and reference regions [21–23].

3 Results and Discussion

The time series of integrated oceanic and atmospheric heat transport across the studied sections (66.5°N, −4.5°E–13.5°E and 66.5°N, −5°E–80°E correspondingly) are shown in Fig. 2.

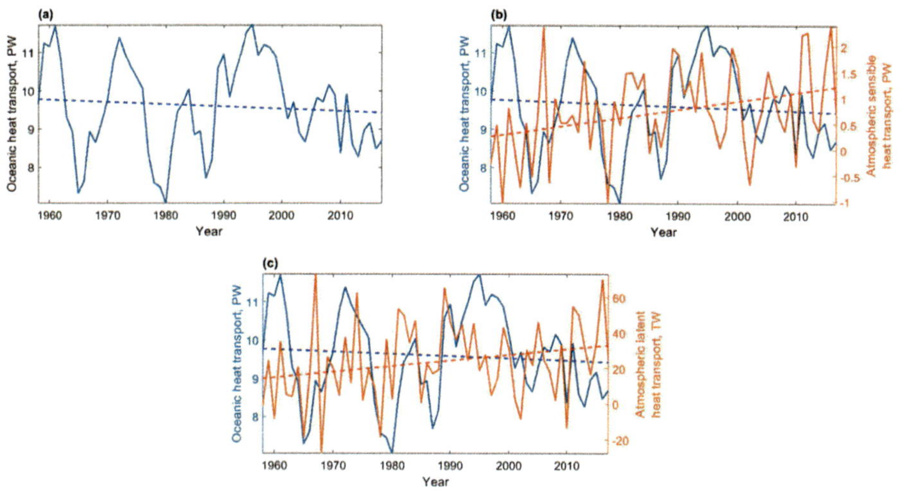

Fig. 2. Meridional integral heat fluxes at the entrance to the Atlantic sector of the Arctic [18]. Negative values indicate the southward direction

Figure 2a shows the dominance of the 10–15 years cycle in the ocean until the beginning of the 21st century. In contrast, the short-term variability prevails in the atmosphere (Fig. 2b,c). Due to a high coherence of the components of atmospheric heat transport, their sum was considered in the further analysis.

Figure 3 shows the maps of spatial variability of correlations of interannual variability of oceanic and atmospheric heat fluxes (across the sections shown in Fig. 1) with the mean annual Arctic amplification. The analysis was performed after the removal of the linear trends from the time series. The Arctic amplification lags from zero (a–b) to four years (i–j).

The strongest relationship between the Arctic amplification and the oceanic heat flux was found when the ocean leads by four years (Fig. 3i). This linkage is negative and concentrated in the Central Arctic. The same region was determined from the correlation maps with the atmospheric heat flux; however, with no delay in time (Fig. 3b). Based on this analysis, the mean values of Arctic amplification were found for the following region of the Central Arctic: 80°–90°N, 50°E–140°E (Fig. 4).

According to Fig. 4, there is a strong linkage between the Arctic amplification and atmospheric heat flux. This is especially pronounced on the decadal time scale (Fig. 4d). This relationship is positive, with the Arctic amplification lagging by one year. The oceanic heat flux also has a high coherence with the Arctic amplification in the low-frequency band (Fig. 4c). However, the correct phase relationship is determined from the cross-correlation analysis and after the application of the low-pass Butterworth filter with a threshold period of six years (Fig. 3i and Fig. 4b). The oceanic heat flux leads the Arctic amplification by four years, and this relationship is negative.

The similar analysis between the heat fluxes showed that the ocean leads the atmosphere by three years (Fig. 4b). As this linkage is negative, the following chain can be constructed: the ocean forces the atmosphere through the Bjerknes compensation, and

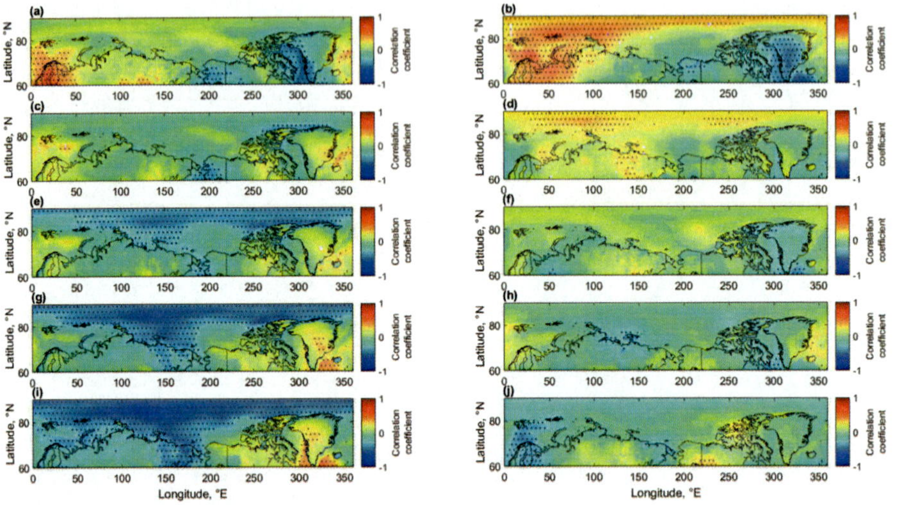

Fig. 3. Correlation maps showing the linkages between the studied heat transports and Arctic amplification [18]. The dotted areas highlight the highest correlations

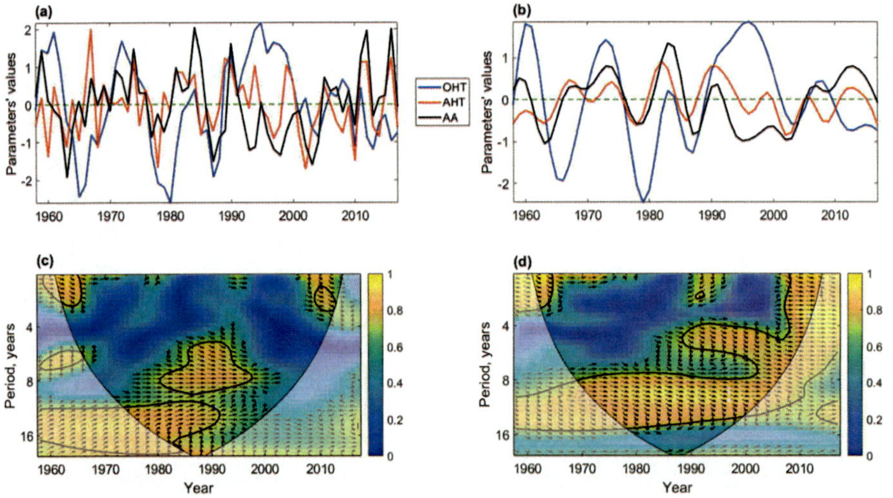

Fig. 4. Wavelet analysis of the heat fluxes (oceanic heat transport (OHT) and atmospheric heat transport (AHT)) and Arctic amplification (AA) [18]

the resulting response of atmospheric advection of heat and moisture affects the Arctic amplification. Therefore, the ocean, due to its influence on the transport of atmospheric heat and moisture, might induce the decadal variability of Arctic amplification.

4 Conclusion

This study has highlighted the strong relationship between the meridional heat transports in the subpolar North Atlantic and the Arctic amplification. According to [24], in a climate model, the highest anti-correlation between the meridional oceanic and atmospheric heat fluxes is detected when the ocean leads the atmosphere by 3–4 years. This is consistent with the findings in our paper based on the reanalyses.

Thus, the meridional oceanic heat flux at the southern boundary of the Nordic Seas might be a predictor of the decadal variability of the Arctic amplification over the Eurasian Basin of the Arctic Ocean with a lead time of four years. This is because the atmospheric heat flux leads the Arctic amplification by one year in the low-frequency band; at the same time, the atmosphere responds to the ocean forcing with a delay of three years.

Acknowledgements. The publication was funded by the Saint Petersburg State University, project № 94033410.

References

1. Allen, M.R., et al.: Framing and context. In: Masson-Delmotte, V., et al. (eds.), Global Warming of 1.5 °C. An IPCC Special Report on the Impacts of Global Warming of 1.5 °C above Pre-industrial Levels and Related Global Greenhouse Gas Emission Pathways, in the Context of Strengthening the Global Response to the Threat of Climate Change, Sustainable Development, and Efforts to Eradicate Poverty, pp. 49–91. IPCC, Geneva (2018)
2. Serreze, M.C., Barry, R.G.: Processes and impacts of Arctic amplification: a research synthesis. Glob. Planet. Change. **77**, 85–96 (2011). https://doi.org/10.1016/j.gloplacha.2011.03.004
3. Masson-Delmotte, V., et al.: Past and future polar amplification of climate change: climate model intercomparisons and ice-core constraints. Clim. Dyn. **26**, 513–529 (2006). https://doi.org/10.1007/s00382-005-0081-9
4. Collins, M., et al.: Long-term climate change: projections, commitments and irreversibility. In: Stocker, T.F., et al. (eds.) Climate Change 2013: The Physical Science Basis. Contribution of Working Group I to the Fifth Assessment Report of the Intergovernmental Panel on Climate Change, pp. 1029–1136. Cambridge University Press, Cambridge (2013). https://doi.org/10.1017/CBO9781107415324.024
5. Goosse, H., et al.: Quantifying climate feedbacks in polar regions. Nat. Commun. **9**, 1919 (2018). https://doi.org/10.1038/s41467-018-04173-0
6. Latonin, M.M., Bashmachnikov, I.L., Bobylev, L.P.: The Arctic amplification phenomenon and its driving mechanisms. Fundam. Prikl. Gidrofiz. **13**, 3–19 (2020). https://doi.org/10.7868/S2073667320030016
7. Screen, J.A., Simmonds, I.: The central role of diminishing sea ice in recent Arctic temperature amplification. Nature **464**, 1334–1337 (2010). https://doi.org/10.1038/nature09051
8. Graversen, R.G., Wang, M.: Polar amplification in a coupled climate model with locked albedo. Clim. Dyn. **33**, 629–643 (2009). https://doi.org/10.1007/s00382-009-0535-6
9. Pithan, F., Mauritsen, T.: Arctic amplification dominated by temperature feedbacks in contemporary climate models. Nat. Geosci. **7**, 181–184 (2014). https://doi.org/10.1038/ngeo2071

10. Henry, M., Merlis, T.M.: The role of the nonlinearity of the Stefan–Boltzmann law on the structure of radiatively forced temperature change. J. Clim. **32**, 335–348 (2019). https://doi.org/10.1175/JCLI-D-17-0603.1
11. Jenkins, M., Dai, A.: The impact of sea-ice loss on Arctic climate feedbacks and their role for Arctic amplification. Geophys. Res. Lett. **48**, e2021GL094599 (2021). https://doi.org/10.1029/2021GL094599
12. Jenkins, M.T., Dai, A.: Arctic climate feedbacks in ERA5 reanalysis: seasonal and spatial variations and the impact of sea-ice loss. Geophys. Res. Lett. **49**, e2022GL099263 (2022). https://doi.org/10.1029/2022GL099263
13. Boeke, R.C., Taylor, P.C., Sejas, S.A.: On the nature of the Arctic's positive lapse-rate feedback. Geophys. Res. Lett. **48**, e2020GL091109 (2021). https://doi.org/10.1029/2020GL091109
14. Lu, J., Cai, M.: Seasonality of polar surface warming amplification in climate simulations. Geophys. Res. Lett. **36**, L16704 (2009). https://doi.org/10.1029/2009GL040133
15. Cai, Z., et al.: Amplified wintertime Barents Sea warming linked to intensified Barents oscillation. Environ. Res. Lett. **17**, 044068 (2022). https://doi.org/10.1088/1748-9326/ac5bb3
16. Mortin, J., Svensson, G., Graversen, R.G., Kapsch, M.-L., Stroeve, J.C., Boisvert, L.N.: Melt onset over Arctic sea ice controlled by atmospheric moisture transport. Geophys. Res. Lett. **43**, 6636–6642 (2016). https://doi.org/10.1002/2016GL069330
17. Alexeev, V.A., Walsh, J.E., Ivanov, V.V., Semenov, V.A., Smirnov, A.V.: Warming in the Nordic Seas, North Atlantic storms and thinning Arctic sea ice. Environ. Res. Lett. **12**, 084011 (2017). https://doi.org/10.1088/1748-9326/aa7a1d
18. Latonin, M.M., Bashmachnikov, I.L., Bobylev, L.P.: Bjerknes compensation mechanism as a possible trigger of the low-frequency variability of Arctic amplification. Russ. J. Earth Sci. **22**, ES6001 (2022). https://doi.org/10.2205/2022ES000820
19. Semenov, V.A.: Modern arctic climate research: progress, change of concepts, and urgent problems. Izv. Atmos. Ocean. Phys. **57**, 18–28 (2021). https://doi.org/10.1134/S0001433821010114
20. Bekryaev, R.V.: Statistical aspects of quantitative estimation of polar amplification. Part 1: The ratio of trends. Russ. Meteorol. Hydrol. **47**, 419–427 (2022). https://doi.org/10.3103/S1068373922060012
21. Francis, J.A., Vavrus, S.J.: Evidence for a wavier jet stream in response to rapid Arctic warming. Environ. Res. Lett. **10**, 014005 (2015). https://doi.org/10.1088/1748-9326/10/1/014005
22. Davy, R., Chen, L., Hanna, E.: Arctic amplification metrics. Int. J. Climatol. **38**, 4384–4394 (2018). https://doi.org/10.1002/joc.5675
23. Latonin, M.M., Bashmachnikov, I.L., Bobylev, L.P., Davy, R. Multi-model ensemble mean of global climate models fails to reproduce early twentieth century Arctic warming. Polar Sci. **30**, 100677 (2021). https://doi.org/10.1016/j.polar.2021.100677
24. van der Swaluw, E., Drijfhout, S., Hazeleger, W.: Bjerknes compensation at high northern latitudes: the ocean forcing the atmosphere. J. Clim. **20**, 6023–6032 (2007). https://doi.org/10.1175/2007JCLI1562.1

Lagrangian Method for Studying Agulhas Leakage in the Water Mixing Area

A. A. Malysheva[1](✉), T. V. Belonenko[1], and M. V. Budyansky[2]

[1] St. Petersburg State University, St. Petersburg 199034, Russia
alinamalysheva97@mail.ru

[2] Pacific Oceanological Institute of the Russian Academy of Sciences, Vladivostok 690041, Russia

Abstract. This article is motivated by a vital component of the ocean circulation called the Agulhas leakage. The aim of the study is to examine the impact of the South Atlantic Gyre, Benguela Current, and Benguela upwelling on the thermohaline properties of mesoscale eddies generated in the Cape Basin. The research explored statistics by Eddy Detection Algorithm, Lagrangian analysis, and vertical thermohaline cross-sections. The findings of the research illustrate how the process of mixing the waters of the Agulhas Current with the waters of the South Atlantic Gyre, including the Benguela Current, occurs. The impact of the South Atlantic Gyre on the Cape Basin eddies is significant. The findings support the unique eddy structure illustrated by alternative studies.

Keywords: Agulhas · Agulhas leakage · Cape Basin · South Atlantic Gyre · Lagrangian analysis · AMEDA

1 Introduction

One of the most dynamically active areas of the World Ocean is the basin located to the south of Africa [1]. The simplified circulation scheme of the considered basin is shown in Fig. 1. This area includes the Agulhas Current, that transports warm and salty water from the Indian Ocean to the Atlantic Ocean by eddies. These eddies can pass through thousands of kilometers, hence, they are well-known as tracers of water exchange between two oceans [2, 3]. This phenomenon is the Agulhas leakage. Agulhas leakage eddies are generated in the Cape Basin. This is where the process of mixing the waters of the Agulhas Current with the waters of the South Atlantic Gyre and the Benguela Current occurs [4]. Although there is no unambiguous opinion about exactly what contribution Agulhas waters make to the Benguela Current or vice versa, there is a close connection between the power of the Benguela and the Agulhas eddies [5]. Moreover, Matano and Beier [6] found out that the Agulhas eddies provide a large contribution to the Benguela Current energy. Thus, the Benguela Current can be examined as a method of transporting Agulhas eddies into the Atlantic Ocean. It is also worth noting the influence of the South Atlantic Gyre waters on the structure of eddies formed in the Cape Basin. Guerra et al. [7] examined thermohaline characteristics of Agulhas eddies with the longest lifespan and

T. Chaplina (Ed.): CIWO 2023, SPEES, pp. 111–117, 2023.
https://doi.org/10.1007/978-3-031-47851-2_13

noticed two basic types of eddy water: I (salinity 35.6, temperature 16.2 °C), and type II (salinity 35.2, temperature 12.9 °C). In our opinion, the first type of water belongs to the Agulhas Current, and the second one - to the Benguela Current and the South Atlantic Gyre. The eddies created in the Cape Basin contain this two types of water (67%) [7]. Thus, the Agulhas leakage transports water to the Atlantic by Agulhas eddies and Cape Basin eddies. However, the contribution of the Benguela Current and the South Atlantic Gyre to the Agulhas leakage has not been studied. Thus, the purpose of this study is to examine the dimensional water particles distribution of different origins in the Cape Basin.

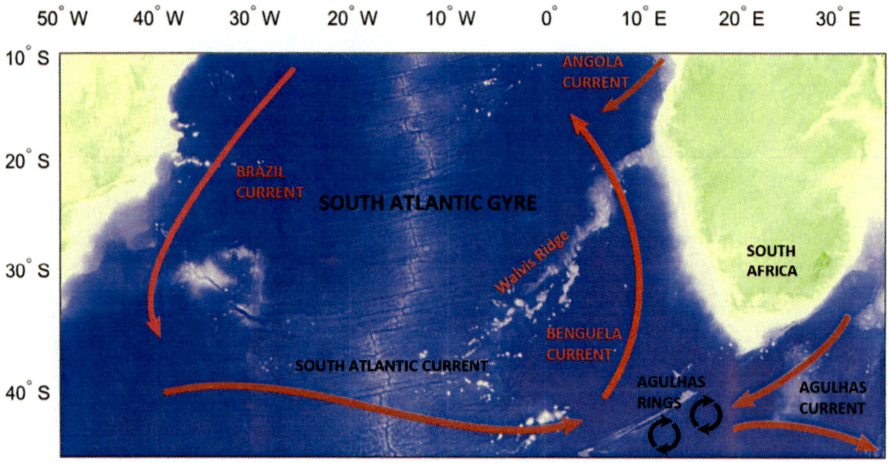

Fig. 1. The simplified circulation scheme of the considered area.

2 Material and Methods

In the current research we applied the GLORYS version 12v1 (Global Ocean Physics Reanalysis) data, that is available on the CMEMS (Copernicus Marine Environment Monitoring Service) portal. Temperature, salinity, velocities, sea surface height, and many other parameters are available on 1/12° at 50 vertical levels for 1993–2021.

Also we used "MESOSCALE EDDY TRAJECTORY ATLAS PRODUCT META 3.2 DT". This altimetric atlas was distributed by service AVISO+ (https://aviso.altimetry.fr). META3.2 DT identify anticyclones and cyclones in the World Ocean and their parameters based on altimetric information [8].

AMEDA (Angular Momentum Eddy Detection and Tracking Algorithm) is an optimized eddy identification based on the GLORYS12V1 reanalysis data. This algorithm uses velocity fields for detecting eddies.

To simulate water circulation in oceanic basins we use Lagrangian approach. This method is based on the computing trajectory of many passive artificial particles based on altimetry or a numerical-model velocity field. This allows to track the origin and "fate" of different waters, their transfer and mixing by solving equations of motion. With this method, we study the water masses transport in the Cape Basin. The Lagrangian maps are parameterized by the integration period found empirically (T = 365 days). Thus, many passive particles were distributed throughout the region on the given day and their trajectories were computed backward in time for our integration period.

3 Results and Discussion

We discovered 54496 eddies of the Agulhas Current (28018 cyclones and 26478 anticyclones) for 1993–2017. Only anticyclones have a lifespan exceeding a year. The study area was divided into 3 parts: part 1 (34–45° S, 0–13° E), part 2 (34–45° S, 13–24° E) and part 3 (20–34° S, 0–24° E). Statistical analysis was calculated for each individual part. A part of information is shown in Table 1. *S* is the eddy contours integral square (km^2) based on AMEDA.

Table 1. Number (*N*) of eddies with a lifespan > 60 days for the three parts. *AC* is for anticyclones and *C* is for cyclones.

Areas (*S*, km^2) of part 1–3	*N* (*AC*)	*N*/*S* (*AC*) × 10^{-4}	*C*	N/S (*C*) × 10^{-4}
1 (1356 × 10^3)	260	2	268	2
2 (1147.3 × 10^3)	160	1	155	1
3 (2468.1 × 10^3)	389	1.5	425	1.7

The next stage of the study is the analysis of Lagrangian maps (the area under consideration: 20–46°S, 0–20°E). The selected depth for the study is 260 m to exclude the influence of surface effects. Also this level of depth allows us to identify the mesoscale eddies cores that are located at a depth of 200–800 m [9, 10]. Thus, 6 types of daily Lagrangian maps were analyzed over a period of 10 years (2010–2020). Below we presented several Lagrangian maps, but only for certain dates.

Figure 2 presents the O-map (origin map) for chosen region at a depth of 260 m. This type of map shows the passive particles distributed over the region. The pink color refers to the Agulhas leakage water particles. These waters include the Agulhas water formed in the Indian Ocean. The violet color refers to the particles, that may belong to the southern branch of the South Atlantic Gyre and may also be part of the Antarctic Circumpolar Current waters. The blue color marks the particles, that travelled across the 0° meridian in the last 365 days. These particles most likely belong to the South Atlantic Gyre waters and influence the Benguela Current and the Agulhas leakage. The green color marks the particles that belong to the Angola Current. It is shown that only the northern part of the region under consideration is influenced by the Angola current. One can see white particles that spread over the area. They refer to particles that did not reach any of the region borders and the coastline during 1 year of counting in reverse time. White color means particles that have not reached the borders of the region or the coastline in the 365-day countdown.

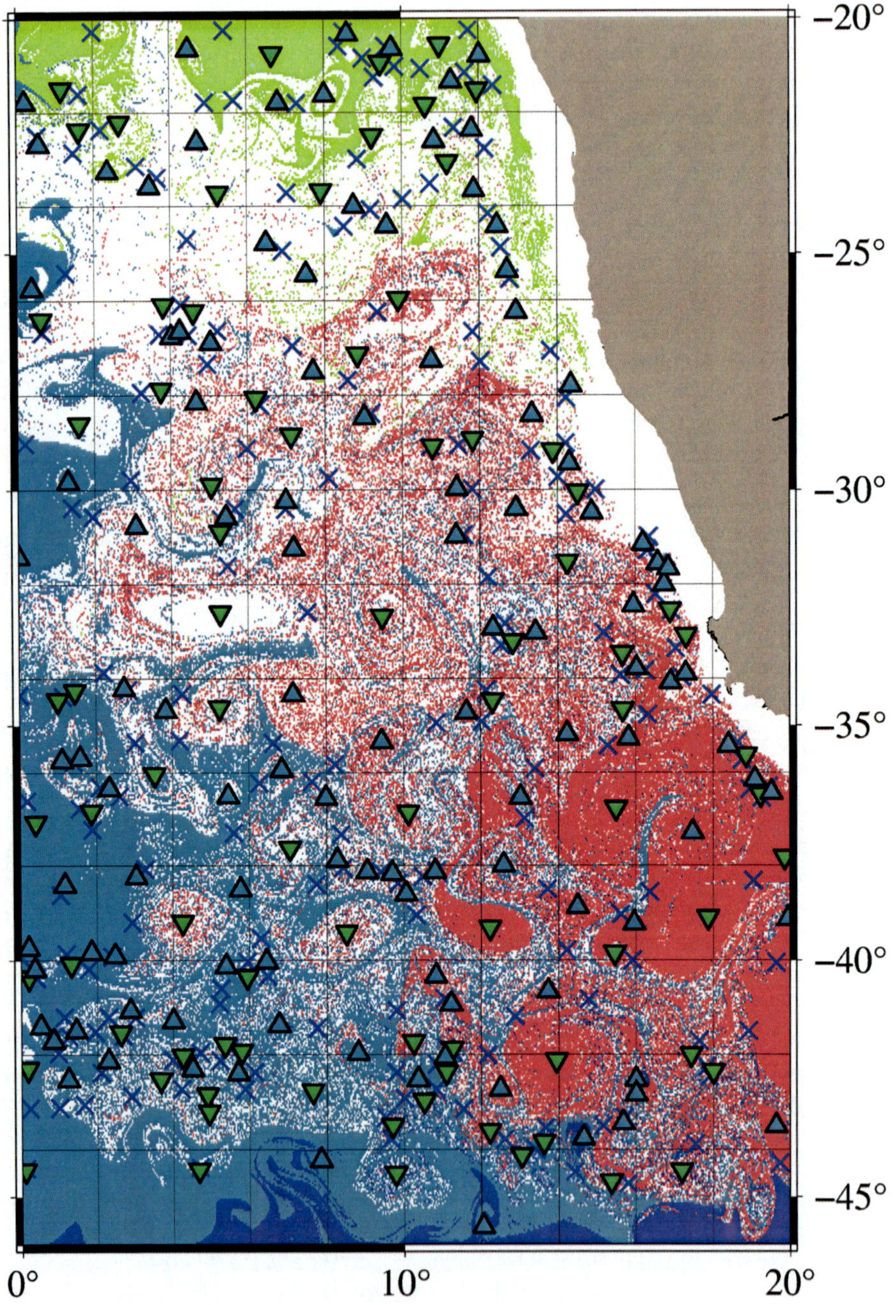

Fig. 2. O-map (Lagrangian map of the water masses origin) on 13 September 2013, at a depth of 260 m. Elliptical points are represented by the triangles: blue triangles refer to the centers of cyclones eddies shown △, and green triangles refer to the centers of anticyclones eddies shown ▽. Hyperbolic points (or instability points) are shown by the dark blue X.

Figure 3 demonstrates the T-map, that is an addition to the O-map. The T-map shows how old the water particles are. Thus, the T-map shows the border crossing time. According to the black-and-white scale, "young waters" that crossed the border 0–50 days ago are colored black, and "old waters" that crossed the border 200–365 days ago from the date are colored white It is shown, that the largest number of particles are old waters, they have not left the region even once in 365 days. "Young waters" refers to the Agulhas water, Angola Current, Benguela Current and Antarctic Circumpolar Current. Although many particles in the area painted white color do not have any obvious mobility.

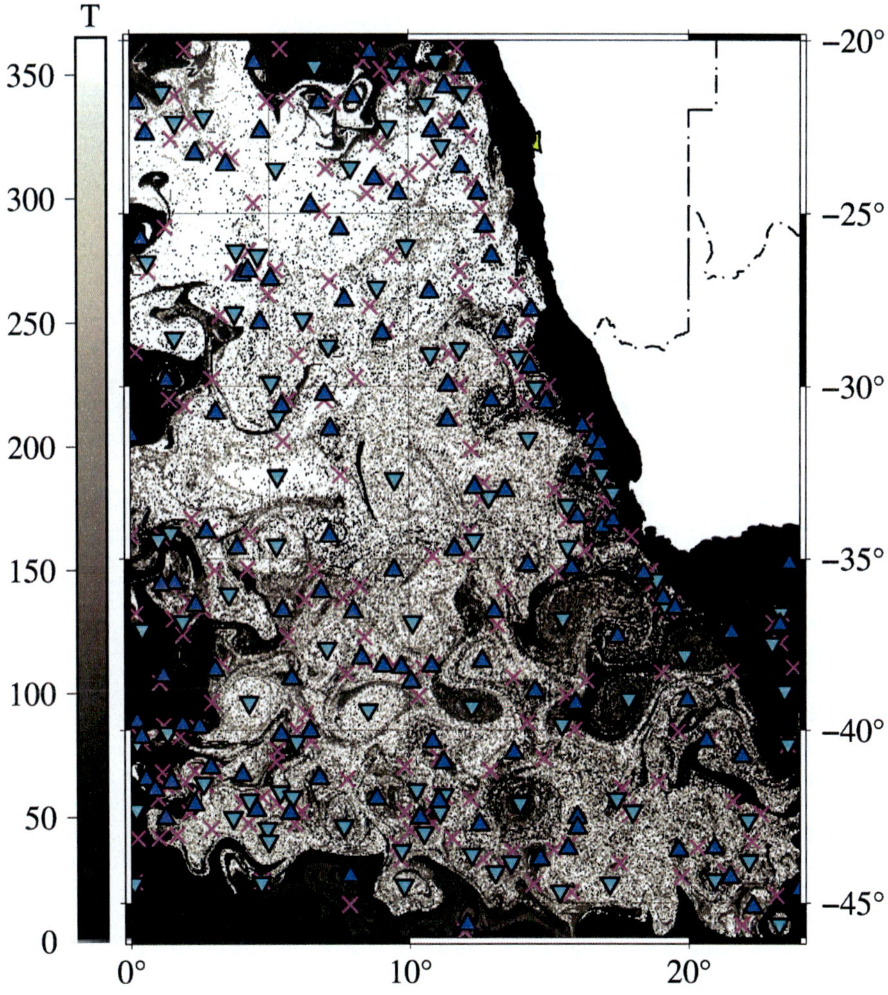

Fig. 3. T-map (Lagrangian map of the water masses age) on 13 September 2013, at a depth of 260 m. The color scale shows the border crossing time between putting particles into the water on the given date and the moment when particles reach one of the area boundaries. Other elements are the same as in the previous figure.

Thus, the Cape Basin is an area where three types of water from different sources mix and influence the Agulhas eddies, which is an important part of water exchange. Due to this mixing of the three types of water, the Agulhas eddies change their thermohaline characteristics from warm and salty Indian Ocean waters to fresher and colder. Lagrangian maps (Figs. 2 and 3) demonstrate the mixing of the waters in the region under consideration.

4 Conclusion

Using the GLORYS12V1 reanalysis and the altimetric Eddy Atlas, we obtained some statistics of eddies in the study region and note some differences between the parts of the divided study region. The area under consideration is one of the most dynamically active areas of the World Ocean. The main components of large-scale circulation as the Benguela Current, Benguela upwelling and South Atlantic Gyre make the main contribution to the high dynamic activity. In order to eliminate the effect of the Benguela upwelling on the eddies, we chose a depth of 260 m for our study.

In our study, we obtain a new result: quantitative estimates water mixing characteristics in the study region. We study the behavior of water particles in the region where the Agulhas waters mix with the waters of the Benguela Current and the South Atlantic Gyre. Mixing occurs generally in the Cape Basin. It was shown, that the South Atlantic Gyre, the Benguela current and the Agulhas leakage have a major impact on the waters of the region under consideration. Based on the Lagrangian analysis, it is shown that waters of different origins and ages mix in the study area. The South Atlantic Gyre waters interact significantly with the Agulhas waters. Thus, the two unique types of eddies explained in [7] may be a consequence of this effect. Not only the Agulhas Current is a source of particles in the Agulhas leakage but also and the Benguela Current the South Atlantic Gyre. Water particles travel hundreds of kilometers from the west and southwest and fill the region. It has also been shown that the South Atlantic Gyre has a greater influence on the Agulhas leakage than previously thought.

References

1. Olson, D.B., Evans, R.H.: Rings of the Agulhas current. Deep Sea Res. A: Oceanog. Res. Pap. **33**(1), 27–42 (1986)
2. Gordon, A.L., Weiss, R.F., Smethie, W.M., Jr., Warner, M.J.: Thermocline and intermediate water communication between the South Atlantic and Indian Oceans. J. Geophys. Res. Oceans **97**(C5), 7223–7240 (1992)
3. Donners, J., Drijfhout, S.S., Coward, A.C.: Impact of cooling on the water mass exchange of Agulhas rings in a high resolution ocean model. Geophys. Res. Lett. **31**(16), L16312 (2004)
4. Hutchings, L., et al.: The Benguela current: an ecosystem of four components. Prog. Oceanogr. **83**(1–4), 15–32 (2009)
5. Garzoli, S.L., Gordon, A.L.: Origins and variability of the Benguela current. J. Geophys. Res. **101**(C1), 897–906 (1996)
6. Matano, R.P., Beier, E.J.: A kinematic analysis of the Indian/Atlantic inter-ocean exchange. Deep-Sea Res. **II**(50), 229–250 (2003)

7. Guerra, L.A.A., Mill, G.N., Paiva, A.M.: Observing the spread of Agulhas Leakage into the Western South Atlantic by tracking mode waters within ocean rings. Front. Mar. Sci. **9**, 958733 (2022)
8. Pegliasco, C., Busché, C., Faugère, Y.: Mesoscale Eddy Trajectory Atlas META3.2 Delayed-Time all satellites: version META3.2 DT allsat (2022)
9. Sandalyuk, N.V., Belonenko, T.V.: Three-dimensional structure of the mesoscale eddies in the Agulhas current region from hydrological and altimetry data. Russ. J. Earth Sci. **21**, ES4005 (2021)
10. Malysheva, A.A., Belonenko, T.V., Iakovleva, D.A.: Characteristics of two eddies of different polarity in the Agulhas Current. Gidrometeorol. Ekol. [J. Hydrometeorol. Ecol.] **68**, 478–493 (2022). [In Russian]

Wind Profile Evaluation by Using Result of Wave Forecast Model

D. D. Romashchenko[1] (✉) and K. Yu. Bulgakov[2]

[1] Russian State Hydrometeorological University, St. Petersburg 195196, Russia
r0mashenkodan@yandex.ru
[2] Shirshov Institute of Oceanology RAS, Moscow 117997, Russia

Abstract. The paper presents method of using wind wave spectrum for evaluation of wind profile in the wave boundary layer. Result of experiments performed by the wave boundary layer model with the method are shown.

Keywords: Wave boundary layer · Wave spectrum · Wave impact flux

1 Introduction

The wave boundary layer is the lowest part of the planetary (or atmospheric) boundary layer where the intensity of the flux induced by wave is comparable with the turbulent one. Thickness of the wave boundary layer has order of magnitude of the significant wave height which equals average of the one third largest waves. The wave boundary layer in the atmospheric models is simulated by the methods developed for the boundary layer above solid surface. Wave impact on momentum, heat and water vapor exchange is parametrized by the changing of rough scale. Mostly Charnok formula is used.

$$z_0 = m\frac{v_*^2}{g} \tag{1}$$

where $v_* = \sqrt{\tau/\rho_a}$ - friction velocity, g - gravity acceleration, m - empirical coefficient, order of magnitude is 0.01 (τ - vertical momentum flux, ρ_a - air density). Length scale $Ch = v_*^2/g$ was proposed in [1].

This formula is suppose that waves are fully developed and can be described by the wind velocity or friction velocity v_*. This assumption is reasonable because the stronger wind the higher waves. Lack of this method due to large spread of data of coefficient m from 0.01 to 0.04. It can be expected that value of this coefficient has significant influence on accuracy of simulation of atmospheric circulation over the sea.

Exchange of momentum, heat and water wapour in the wave boundary layer is crucial for atmospheric and ocean circulation and for wind waves generation. Surface fluxes of this substance are boundary conditions for numerical models and parameters which define stratification in planetary boundary layer. Therefore, it's necessary to take into account impact of wave on fluxes in the wave boundary layer.

T. Chaplina (Ed.): CIWO 2023, SPEES, pp. 118–125, 2023.
https://doi.org/10.1007/978-3-031-47851-2_14

2 Wind Waves Interaction

The main properties of the processes of interaction between wind and waves in a wide range of wind speeds and wave height have not yet been fully investigated. Wind-generated waves are a complex process, the study of which presents significant difficulties associated with describing the generation, dissipation and interaction of waves both inside the wave fields themselves and with other processes in the ocean and atmosphere. Moreover, the main mechanisms of interaction between wind and waves are concentrated very close to the moving interface, which makes direct measurements difficult. Therefore, the development of a numerical model will allow us to study the wave boundary layer in the most optimal way and then consider a wide range of input parameters and obtain accurate results that are not inferior to field experiments.

Structure of the wave boundary layer are formed by the turbulence fluxes on the upper border and shape of surface that can be expressed in the term of wave spectrum which depends on the wave development. Impact of sea surface movement on structure of the wave boundary layer can be expressed through Wave Produced Momentum Flux (WPMF) which consists of momentum flux provided by wave fluctuations of turbulent stresses (term III), momentum flux provided by the pressure and momentum flux provided by the velocity field [3]. The vertical profiles of WPMF were investigated in detail in [3], where the numerical experiments with the coupled wind/wave model were described. It was shown that the part of the turbulent flow and the WOLF is constant in the wave boundary layer, where the WPMF is maximal near the surface and decreases to zero with height.

According to the linear theory [2] surface value of WPMF can be evaluated through beta-function which connects Fourier components of surface pressure with those of surface elevation. In [3] approximation of beta-function and WPMF dependence on height were derived and parametrization of WPMF was formulated. Description of one dimensional model of the wave boundary layer based on this parametrization was given in [3–5]. There was investigation of drag coefficient above sea by the model. It was shown that drag coefficient dependence not only on wind speed but also on reverse wave age which can be expressed as relation of 10 m wind velocity to wave peak phase velocity.

In this paper the model is used as a part of algorithm which allows to take into account WPMF in atmospheric circulation models. The possible impact of WPMF on dynamic characteristics is estimated.

3 Parametric Model of the Wave Boundary Level (WBL)

One-dimensional model [6, 7] (here and further 1D model) of the wave boundary layer consists of movement equation for u and v velocity components:

$$\frac{\partial u}{\partial t} = \frac{\partial}{\partial z}\left(k\frac{\partial u}{\partial z} + \tau_{wx}\right) \tag{2}$$

$$\frac{\partial v}{\partial t} = \frac{\partial}{\partial z}\left(k\frac{\partial v}{\partial z} + \tau_{wy}\right) \tag{3}$$

where τ_{wx}, τ_{wy} -x, y components of the WPMF, k - turbulence coefficient which was calculated through evolutionary equations of turbulence kinetic energy closed by turbulent length scale which was determined proportional to the height.

$$k = Kz\sqrt{e/C_1} \tag{4}$$

$$\frac{\partial e}{\partial t} = \frac{\partial}{\partial z}k\frac{\partial e}{\partial z} + P - \varepsilon \tag{5}$$

$$P = \frac{\partial u}{\partial z}\left(k\frac{\partial u}{\partial z} + \tau_{wx}\right) + \frac{\partial v}{\partial z}\left(k\frac{\partial v}{\partial z} + \tau_{wy}\right) \tag{6}$$

where ε - turbulence dissipation rate, P - rate of turbulence energy generation

$$\varepsilon = \frac{(e/C_1)^{3/2}}{Kz} \tag{7}$$

where K – Karman's constant 0.4; $C_1 = 3.7$ – constant.

Parameterization of the WPMF:

$$\tau_w(z) = \sum\nolimits_{j=0}^{M-1}\sum\nolimits_{i=0}^{N-1}\omega_i^2\beta(\Omega_{i,j})S(\omega_i,\theta_j)\exp\left(-G\left(\widetilde{\omega}_i\right)\frac{\omega_i^2 z}{g}\right)\Delta\omega_i\Delta\theta_j \tag{8}$$

where N – number of grid points, M – number of wave numbers, ω– frequency, S – two-dimensional wave spectrum, $\beta(\Omega_{i,j})$ – imaginary part of the beta function, Ω – apparent (calculated from the wind direction) dimensionless frequency, $G\left(\widetilde{\omega}\right)$ – weak function, where $\widetilde{\omega} = \Omega/\Omega_P$, Ω_P – apparent frequency per spectrum peak, $\Delta\theta_i$ – grid angle step, $\Delta\omega_j$ – grid frequency step.

Approximation of the imaginary part of the beta function:

$$\beta_{-k} = \begin{cases} \beta_0 + a_0(\Omega - \Omega_0) + a_1(\Omega - \Omega_0)^2\Omega_0 < \Omega \\ \beta_0 + a_0(\Omega - \Omega_0) - a_1(\Omega - \Omega_0)^2\Omega_0 > \Omega \end{cases} \tag{9}$$

where $a_1 = 0.09476$, $a_2 = -0.3718$, $a_3 = 14.8$, $\beta_0 = -0.02$, $\beta_1 = -148$, $a_0 = 0.02277$, $\Omega_0 = 0.58$, $\Omega_1 = 1.2$, $\Omega_2 = -18.8$, $\Omega_3 = 21.2$. $a_0 = 0.02277$.

4 Model Grids and Boundary Condition

The equations of the WBL model are approximated at the irregular stretched grid:

$$z_k = \frac{(\gamma_z - 1)\cdot H}{\gamma_z^{L-1}}\gamma_z^{k-1} k = 1, 2, \ldots, L \tag{10}$$

where γ_z – stretching coefficient. The height of the domain was assumed to be $H = 10$ meters with the number of vertical levels $L = 50$ and the stretching coefficient $\gamma_z = 1.1$, thickness of the lowest level was of the order of height of the viscous sub-layer $z_v = 60\eta/V_{*0}$ ($\eta = 1.5 \cdot 10^{-5}$ м^2/c – molecular viscosity coefficient; V_{*0} – modulus of local friction velocity).

Upper boundary conditions for Eqs. (2) and (3) are the wind velocity u_H, v_H. On the lower boundary conditions are the components of the tangential friction:

$$\left(k\frac{\partial u}{\partial z}\right)\bigg|_{z=0} = v_{*0x}^2; \left(k\frac{\partial v}{\partial z}\right)\bigg|_{z=0} = v_{*0y}^2 \tag{11}$$

where v_{*0x}^2; v_{*0y}^2 - near-wall values of the friction velocity.

Upper boundary condition for Eq. (5) is value of the turbulence energy e:

$$e_{z=H} = C_1\left(v_{*x}^2 + v_{*y}^2\right) \tag{12}$$

At the lower boundary, the condition of the absence of flux:

$$\left(k\frac{\partial e}{\partial z}\right)\bigg|_{z=0} = 0 \tag{13}$$

Equation (8) used 2D array of wave spectrum which is function of angle θ and frequency ω and are specified on the grid:

$$\Delta\theta = \frac{2\pi}{M}; \theta_j = (j - 1/2)\Delta\theta; j = 1, 2, 3, \ldots, M \tag{14}$$

$$\begin{cases} \Delta\omega_1 = \frac{1}{2}\omega_1 \cdot (\gamma_\omega - 1) \\ \Delta\omega_i = \frac{1}{2}\omega_i \cdot \frac{(\gamma_\omega - 1)}{\gamma_\omega} \quad i = 2, 3, \ldots, N - 1 \gamma_\omega = \left(\frac{\omega_N}{\omega_1}\right)^{\frac{1}{N-1}} \\ \Delta\omega_N = \frac{1}{2}\omega_N \cdot \frac{(\gamma_\omega - 1)}{\gamma_\omega} \end{cases} \tag{15}$$

that is, a uniform angle grid with M = 72 directions where the angle is within $0 < \theta < 2\pi$ and a stretched frequency grid ω with a stretching coefficient γ_ω is used.

5 Model Calculation Algorithm

1D models used results of run of wave forecast system which consists of WRF [8] and WAVEWATCH III [9] models. Input parameters for experiments were following data from the one grid: V_{σ_1} - value of the velocity at the first sigma level of the WRF model, and 2D wave spectrum.

The initial condition (profiles of (u, v) and $e(z)$) were defines by the Monin-Obukhov similarity theory [10] (logarithmic profile of velocity, constant value of kinetic turbulence energy). Firstly, friction velocity and rough scale were calculated

$$V_* - \frac{K \cdot V_{\sigma_1}}{\ln(\sigma_1/z_0)} \tag{16}$$

$$z_0 = \frac{0.0185 \cdot V_*^2}{g} + 1.59 \cdot 10^{-5} \tag{17}$$

The Eqs. (16) and (17) are transcendental, so they are calculated iteratively. Then profiles were evaluated:

$$\begin{bmatrix} u(z) \\ v(z) \end{bmatrix} = \begin{pmatrix} cos\psi \\ sin\psi \end{pmatrix} \cdot \frac{V_*}{K} \cdot \ln\frac{z}{z_0} \tag{18}$$

$$e(z) = C_1 V_*^2 \tag{19}$$

where ψ – wind direction angle.

At each time step, the prognostic equations of the model are calculated using a tridiagonal matrix algorithm, and then the values of the turbulent and WPMF are calculated. The height-averaged sum of these fluxes is considered to be the total momentum flux in the WBL. After each time step, the velocities are corrected at the upper boundary according by the following formulas:

$$\tau(H) = \overline{\tau_u(z) + \tau_w(z)} \tag{20}$$

$$V_* = \sqrt{\tau(H)} \tag{21}$$

$$\begin{bmatrix} u_H \\ v_H \end{bmatrix} = \begin{pmatrix} cos\psi \\ sin\psi \end{pmatrix} \cdot \left[\frac{V_*}{K} \cdot \ln\frac{10}{\sigma_1} + V_{\sigma_1} \right] \tag{22}$$

Integration over time is carried out until the criterion of constancy of the sum of fluxes with height is reached.

$$\frac{[\max(\tau(z)) - \min(\tau(z))]}{\max(\tau(z))} < 0.001 \tag{23}$$

6 Results of Experiments

Data of simulation of storm in Baltic sea which occurs (24.10.14) was used. One point was selected for the Gulf of Bothnia, the Gulf of Finland, the Gulf of Riga, as well as for the central and western parts of the Baltic Sea. 1D model were run with the wind values and spectrum fields from abovementioned points.

Wave spectrum for point (62.6N, 18.9E) is shown on the Fig. 1. Significant wave height in this case was 4 m.

Profiles of initial wind velocity and wind velocity in the end of 1D model run are shown on the left panel of Fig. 2. It can be said that one the first wind profile calculated without WPMF (purely logarithmic profile) and the second one with taking into account WPMF. Values on the second profile are less than on the first profile, that mean that WPMF slows down wind. Value of 10 m wind velocity and friction velocity for logarithmic profile are 18.66 and 0.51 m/s. The same value with WPMF are 17.6 and 0.8 m/s. Reverse wave age is 1.36.

Figure 3 shows a graph in polar coordinates where many cases are displayed for wave heights of 2–4.6 m. According to the graph, the following conclusions can be drawn: with increasing angle, the number of cases decreases and the excitement weakens; the degree of development of the excitement also affects the distribution of strong and weak cases.

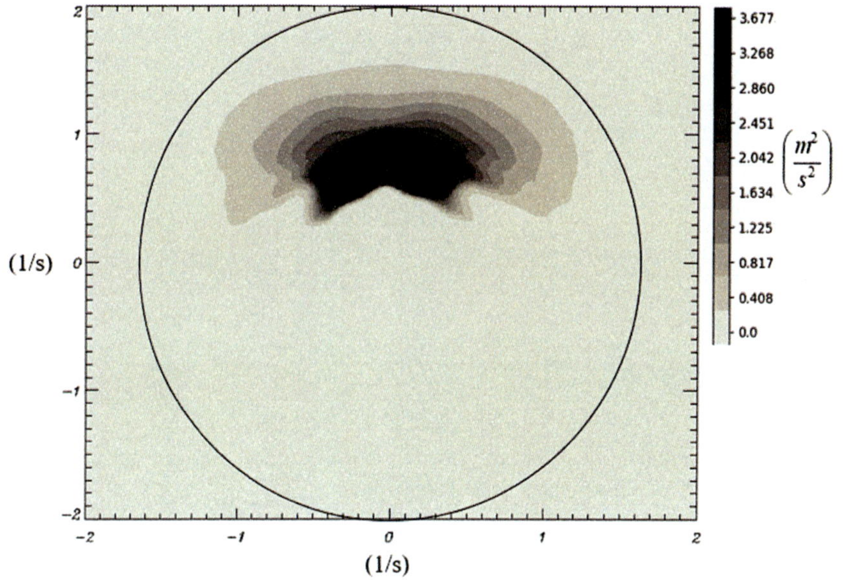

Fig. 1. Wave spectrum in polar coordinates

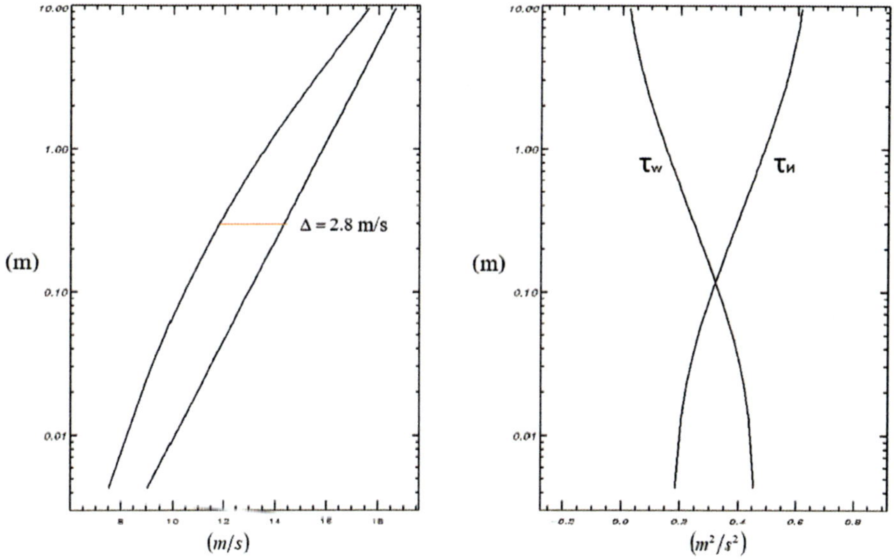

Fig. 2. Vertical profile of velocity (left), fluxes (right)

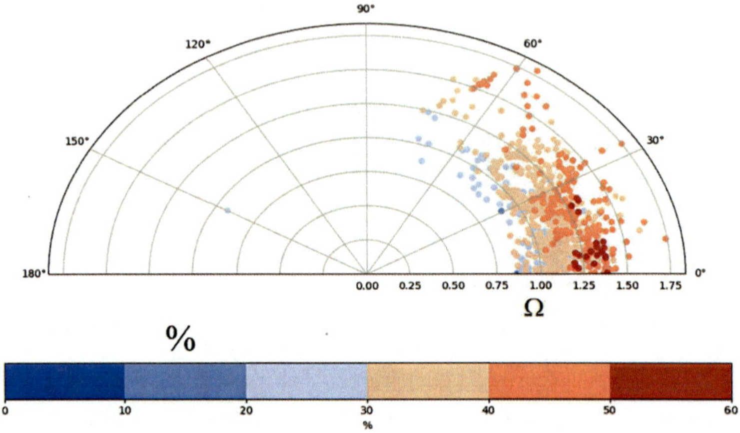

Fig. 3. Distribution of relative change of velocity friction as dependence of angle between wind direction and peak wave direction and reverse wave age

7 Conclusion

WPMF formed by a two-dimensional wave spectrum has a significant effect on one dynamic characteristic in the near-wave layer of the atmosphere - the friction velocity. Its significant change by 57% at high wind speeds and sufficiently energetically dense spectra changes proportionally the heat and moisture flows in the wave boundary layer.

The importance of taking into account the reverse wave age of the wave and taking into account the angle between the wind and the peak mode of the wave is confirmed.

Taking into account the WPMF in modeling would improve the modeling of wind-wave processes over water areas.

Acknowledgments. The work was carried out within the framework of the State assignment № FMWE-2021-0014.

References

1. Charnock, H.: Wind stress on a water surface. Quart. J. Roy. Meteorol. Soc. **81**, 639–640 (1955). https://doi.org/10.1002/qj.49708135027
2. Hristov, T., Miller, S., Friehe, C.: Dynamical coupling of wind and ocean waves through wave-induced air flow. Nature **422**(6927), 55–58 (2003). https://doi.org/10.1038/nature01382
3. Chalikov, D., Rainchik, S.: Coupled numerical modeling of wind and waves and the theory of the wave boundary layer. Boundary-Layer Meteorol. **138**, 1–41 (2010)
4. Chalikov, D., Babanin, A.: Parameterization of wave boundary layer. Atmosphere **10**(11), 686 (2019)
5. Chalikov, D.V., Bulgakov, K.: The structure of surface layer above sea. Fundam. Appl. Hydrophys. **12**(2), 50–65 (2019)
6. Bulgakov, K., Fokina, K.V.: Modeling the surface layer above sea with parameterization of stratification and the elements of the wind waves influence. Fundam. Appl. Hydrophys. **14**(2), 3–16 (2021)

7. Fokina, K.V., Bulgakov, K.: Coupled modelling of wind waves and wave boundary layer. Fundam. Appl. Hydrophys. **15**(1), 73–81 (2022)
8. Michalakes, J., et al.: The weather research and forecast model: software architecture and performance. In: Mozdzynski, G. (ed.) Proceedings of the 11th ECMWF Workshop on the Use of High Performance Computing in Meteorology, 25–29 October 2004, Reading (2004)
9. Tolman, H.L.: User manual and system documentation of WAVEWATCH III version 3.14. Technical Report. NOAA/NWS/NCEP/MMAB (2009)
10. Monin, A.S., Yaglom, A.M.: Statistical Fluid Mechanics: Mechanics of Turbulence, vol. 1, p. 769. The MIT Press, Cambridge (1971). https://doi.org/10.1119/1.10870

Features of the Ice Extent Dynamics in the Western Sector of the Arctic on the Example of the Archipelagos of Svalbard and Franz Josef Land

N. O. Rudenko(✉) and B. V. Ivanov

Arctic and Antarctic Research Institute, St. Petersburg 199397, Russia
norudenko@aari.ru

Abstract. The article is devoted to the analysis of interannual and seasonal variability of sea ice extent in the waters of the Svalbard and Franz Josef Land archipelagos for the period of 1979–2020. Quantitative results are based on the analysis of trends in surface air temperature, sea surface temperature and sea ice extent in the water areas of both archipelagos are obtained, indicating the current climate warming. In the course of the study, global warming trends were confirmed and evaluated, which are expressed in the reduction of ice cover, increase in surface air temperature and surface sea temperature in the investigated region.

Keywords: Franz Josef Land · Spitsbergen · Sea ice cover · Sea surface temperature · Surface air temperature · Climate change and variability

1 Introduction

Sea ice in the Arctic is the most sensitive and responsive component of the global climate system. The Arctic region is known for its "relatively dramatically" changes compared to other regions due to strong positive feedback mechanisms. First of all, modern climate change most clearly expressed in the reduction of the region's ice cover [1–3].

Currently, many domestic and foreign experts are paying attention that the polar climate system has moved from a relatively "stable state" of the middle of the 20th century to the current state, which is characterized by intense warming, which is most pronounced in the Arctic region. These changes affect the entire "ocean-sea ice-atmosphere" system [4]. The main reason of these changes is modern global warming. The rate of warming in the Arctic exceeds the global average by two to three times. This phenomenon is called "Polar amplification", of which the "Arctic amplification" is a special case [5]. The process of warming in the Arctic is intensifying every year, which is manifested in an increase in the average annual surface air temperature (SAT), sea surface temperature (SST) and a record reduction in the sea ice extent (SIE) of the Arctic seas, especially in recent decades [6–10]. As known, the term ice cover or SIE means the ratio of an area of ice of any concentration to the total area of sea surface within some large geographic local [11].

T. Chaplina (Ed.): CIWO 2023, SPEES, pp. 126–134, 2023.
https://doi.org/10.1007/978-3-031-47851-2_15

One of the hotspots of Arctic warming is located in the northern Barents and Kara seas. This region is experiencing the strongest declines in winter sea ice concentration and most rapid surface warming in the entire Arctic [12]. It was found that the heating of the Barents atmosphere plays an important role in "Arctic amplification" [13]. In the scientific article [14], devoted to the study of heat transfer by the ocean based on the global climate model, the researchers concluded that an increase in the Atlantic waters (AW) input into the Barents Sea (BS) has a strong effect on the SIE of this water area by reducing the time when the water area is covered with ice and increasing the ice-free period during summer.

Another aspect pointing to the relevance of the study is the fact that for more than half a century, oceanologists have been studying the mechanisms of the influence of the North Atlantic (NA) on the Earth's climate system, in particular, on the climate of the polar regions. Scientists note an important contribution of an increase in the inflow of AW into the Arctic Basin (AB) and also AW temperature to a decrease in the sea ice area, the so-called "Atlantification" of the Arctic [15–21].

In addition the water area of the Western sector of the Arctic (WSA) is a dynamically active region in which the sea ice extent is most sensitive to climate change [22]. The largest seasonal and long-term changes in the area of the sea ice cover are observed here [23]. These features make the WSA an important and demonstrative region for studying the interaction of components in the "ocean-sea ice-atmosphere" system. In order to assess the features of regional variability of hydrometeorological conditions in the water surrounding the Svalbard and Franz Josef Land (FJL) archipelagos. The region is located at the junction of the Arctic and Atlantic air and water masses [22]. According to the study by [10], the average annual SAT in the northern part of the BS over the past 20–40 years has been increasing at a rate of 2.7 °C/decade (in terms of linear trend), with a maximum in the autumn season (up to 4 °C/decade), which in generally consistent with the warming rates (2.2–2.9 °C/decade) that were previously established for the FJL. For the period 1990–2017 the increase in annual SAT at FJL was 6.3 °C [9].

The main aim of this study is to find out the general patterns and features of SIE characteristics of Svalbard and FJL and also assessed the influence of the NA on the features of climate change in the WSA on the example of said archipelagos.

2 Material and Methods

To establish relationships between the components of the "ocean-sea ice-atmosphere" system, equally discrete series of the same length were used. The assessment of long-term changes in SIE in the water areas of Svalbard and FJL was carried out on the basis of an analysis of satellite observation from multichannel microwave radiometers SSMR-SSM/I-SSMIS (1979–2020), collected by the Arctic and Antarctic Research Institute [24]. Spatial resolution of satellite data is 25 × 25 km. The analysis of the seasonal and interannual changes in SIE was conducted for quasi-homogeneous regions around the archipelagos shown in Fig. 1 [24]. The rationale for the principle of selecting these regions was described in paper [25]. In general, on the basis of natural conditions observed in the Arctic, two main seasons have been singled out: summer (May-October) and winter (November-April).

In order to find out the long-term variability of SAT, composite data series were used. Information about meteorological observations (daily and monthly averages) at the Krenkel Observatory (Hayes Island) was collected from various sources. A detailed procedure for obtaining a long homogeneous series is described by [9]. As for the Svalbard's SAT, most of the data was obtained from the RIHMI-WDC archives [26], some data were calculated by the method of linear regression for the monthly average SAT at the meteorological site of the HMO "Barentsburg" from the average monthly SAT at Finneset cape (Green Harbour), part of the data was obtained from Longyearbyen and Isfjord Radio. Also, one of the periods was calculated according to the 20CRv3 reanalysis. The method for obtaining a long composite series is also detailed in the article [27].

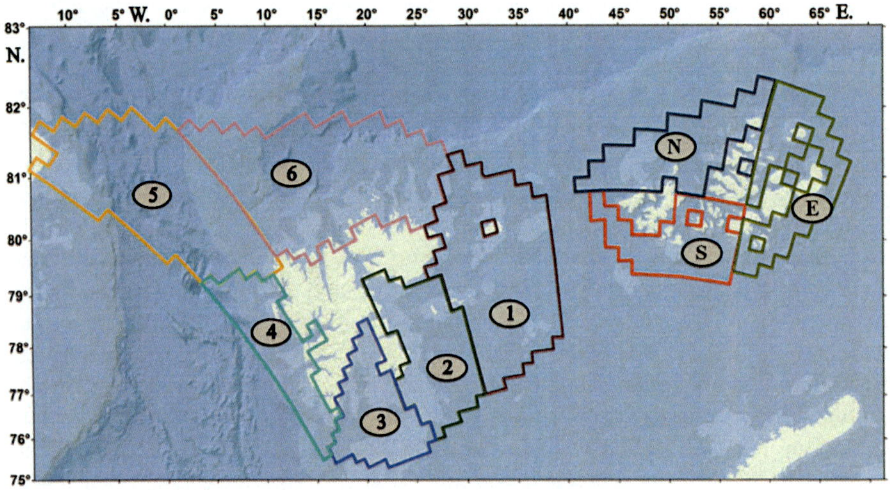

Fig. 1. Areas around the Svalbard and Franz Josef Land: Svalbard: 1 – Eastern, 2 – Barents Sea region, 3 – Southern, 4 – Western, 5 – Fram Strait, 6 – Northern; Franz Josef Land: 7 – Southern, 8 – Northern, 9 – Eastern

SST information was received from the unified information system on the Great Oceans situation (UISGO) [28].

The impact of the NA on the regional features of climate change in the WSA was estimated by finding statistical connections with the Atlantic multidecadal oscillation (AMO) and water temperature along the Kola Section. The Kola Section is a unique standard oceanographic section not only because of the long duration of observations, but also because of their high frequency. These measurements are extremely useful for analyzing the depth distribution of oceanic parameters and establishing the contribution of AW to BS warming. In this work, we used data on the average water temperature in the 50–200 m layer for stations 3 to 7. The choice of the 50–200 m layer for the analysis of water temperature changes on the Kola Section was due to the fact that the upper mixed layer is greatly influenced by the atmosphere and the seasonality of the SST, as a result, the average depth temperature can be significantly distorted, while the near-surface (intermediate) layer is much less susceptible to the above described signals. Data series with information on water temperature are freely available on the website of the PINRO [29].

The method for determining the effect of AW on the interannual variability of ice cover in the water area around the FJL archipelago is based on a statistical approach, taking into account the main factors and mechanisms. In this case, the mechanism of influence of fluctuations in the flow of AW into the BS with their subsequent transfer by warm currents on the SIE variability in the water area around the FJL was considered. Since the AMO index characterizes the thermal regime of the water surface in the North Atlantic, it can be assumed that the influx of warmer and saline surface waters into the Arctic should cause a decrease in ice cover. Monthly AMO indexes are obtained from the NOAA PSL website [30] as unsmoothed long series in standard format. The spatial resolution of the grid is $5^\circ \times 5^\circ$ and consists of monthly SST anomaly values from 1856 to the present.

For statistical processing of data series, regression analysis and calculation of the cross-correlation function were used. Adhering to the provisions on the possible limiting shift of two series relative to each other [31], a reliable cross-correlation result can be obtained if the shift is not exceeded more than 1/10 of the length of the entire data series. In our case, the limit shift was 4 years. Also, for all the obtained relationships, the calculation of the level of statistical significance (P-Value) was carried out in order to make sure that the obtained dependencies were not random.

3 Results and Discussion

In the articles of authors who dealt with a similar topic earlier [3, 25, 32], preliminary studies were carried out on the seasonal and interannual variability of the SIE in the area of both archipelagos. It was established that from November to April in the water area surrounding Svalbard [32], the maximum variability of ice conditions was observed, while in the FJL area this variability was the least. At the same time, there are some features that are characteristic of both regions. For example, the temporal structure of interannual changes in SIE has a very similar character, which may indicate a general external cause of the observed processes. Figure 2 shows seasonal and long-term changes in SIE in the eastern (E), northern (N) and southern (S) areas around of the FJL archipelago.

Obviously, in recent years there has been a noticeable reduction in SIE in all three areas. Since 2004, the southern area has almost always been ice-free in September, with the exception of 2014. In the northern and eastern areas, situations with complete ice-free water began to be observed only since 2011–2012. In addition, such events were not observed there every year. Nevertheless, it is clearly noticeable that for each of the areas, approximately since 2004, not only the absolute SIE values decrease, but also the frequency of occurrence of minimum values in different months increases. The identified features are in good agreement with the fact that after 2002–2004, the central Arctic basin began to experience the predominance of first-year ice compared to multi-year ones [33].

The determination of the moment of "ice regime change" for each selected area around the archipelagos was carried out separately. For Svalbard the change of a relatively

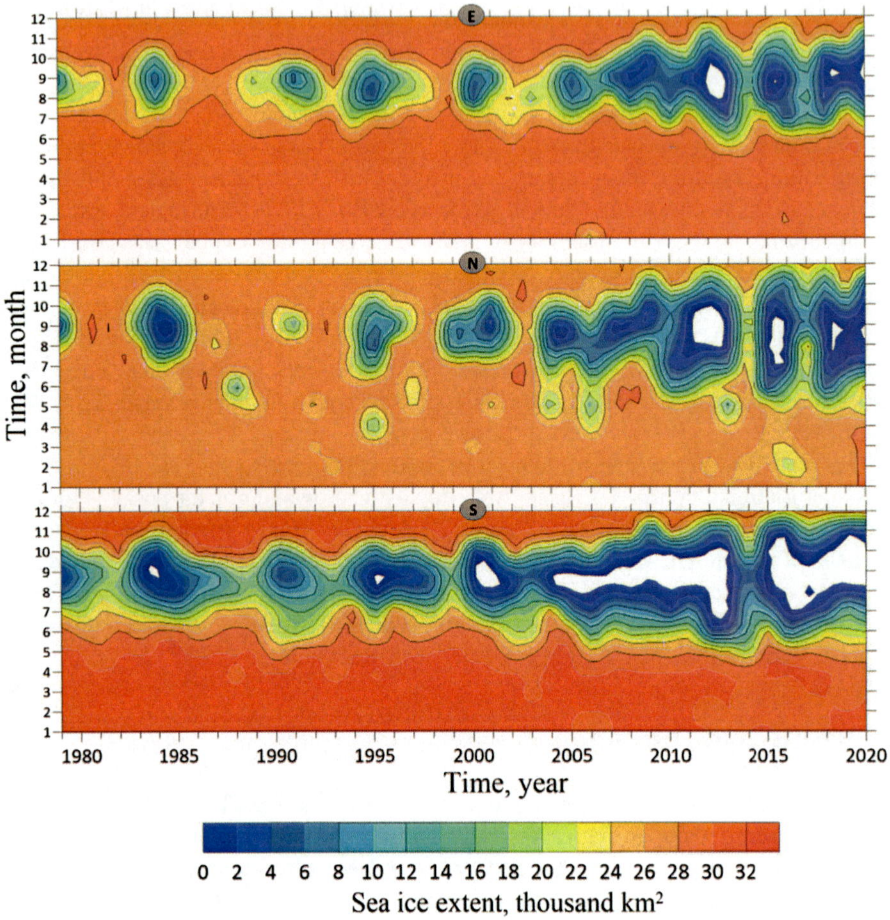

Fig. 2. Temporal distribution of average monthly sea ice extent in the eastern (E), northern (N) and southern (S) areas of the Franz Josef Land archipelago (white color indicates the absence of ice)

stable ice regime to a regime with a high level of interannual variability, which occurred in 2004–2006, was established. The reduction in the SIE (average annual values) occurs at almost the same rate in all areas, with the exception of "Fram Strait" area [25]. Since 2006, the amplitude of interannual fluctuations in the SIE has increased significantly in each of the identified areas of the FJL, as well as in areas 1 and 2 of the Svalbard. Also, in the period from August to September in the area of the Svalbard archipelago, the difference between the maximum and minimum values of SIE is two times less than in the FJL area.

For areas E, N, S, the "ice regime change" occurred in 2000, 1995 and 1996 respectively. For Svalbard, this process has been characteristic since 2006. In general, the western and northeastern regions of the BS are characterized by a steady transition to a new state of the SIE after 2005 (the rate of SIE reduction has almost doubled compared to the previous period).

One of the objectives of this paper was to refine estimates of climate change on the example of the FJL archipelago, which, according to [34], corresponds to the beginning of warming in Russia. We have found that the rate of increase in the average annual SAT for the period 1977–2020 was about 1.5 °C/decade, and warming over the entire period above was 6.5 °C. For the period 1977–2000 the rate of change in the SAT was 0.7 °C/decade, while for the period 2001–2020, it was almost 2.2 °C/decade, which is three times higher. For the summer season over the entire observation period, the SAT increase rate was at the level of 0.6 °C/decade, for the winter season it was 2.35 °C/decade. Changes in seasonal SAT values for summer and winter for the period 1977–2020 were 2.8 °C and 10.3 °C, respectively. Over the same period, the SAT in Svalbard increased by almost 3.5 °C in terms of a linear trend. The "autumn" isotherm -6 °C, previously observed in the first half of September, since 2005–2006 "passed over" to October, and in 2016 it was noted already in November, which should affect the process of ice formation in the FJL water area. A stable trend towards mitigate the climate in the area of Svalbard corresponds to the well-known concept of "rapid warming" of the Arctic [7]. All these linear trends are statistically significant at $P < 0.05$.

SST is an important indicator that affects the potential for ice formation in the Arctic Ocean. Since the end of the 1990s, the SST in the FJL has been increasing not only in absolute values, but also in the frequency of recorded maxima. Thus, starting from 2004, the process of warming began most intensively. It is curious that in the winter months the trend of SST values is negative, while in the summer it is positive. The average monthly SST values in August were always negative, however, in 2010, 2013 and 2016 cases of previously unrecorded record-breaking warm SST indicators were consistently established. Based on the available data, it can be said that, starting from 2015, a fairly stable transition of the August SST values through 0 °C occurred in the FJL water area. It is also worth noting that the maximum SST corresponds to cases of elevated FJL SIE values. The SST distribution is shown in Fig. 3.

It was found that between 1995 and 2000 there have been significant changes in the average annual ice cover in the water area of FJL. This led to an "ice change regime" and the transition of the climate system to a quasi-stable state of reduced SIE. The study revealed negative trends in changes in SIE in certain large areas of the BS and in the water areas around the both archipelagos. An original approach has been applied, which consists in describing changes in the "ocean-sea ice-atmosphere" system on the SIE-SAT and SIE-SST planes for two ice regimes. Graphically they appear as closed curves, an example for the eastern region of the FJL is shown in Fig. 4. These curves make it possible to assess the potential for climate warming before and after the "change of ice regime" in a particular area by the area enclosed inside the figure and its shape.

With regard to the influence of the AW, statistically significant correlations between SIE and the AMO and water temperature along the Kola Section appear with a delay of 3 and 1 year in the Southern region of FJL with correlation coefficients 0.72 and 0.76 respectively. The maximum correlation coefficients with the AMO for Svalbard are observed in the winter season, and in the FJL region – in summer, which may be due to the final velocity of the AMO signal.

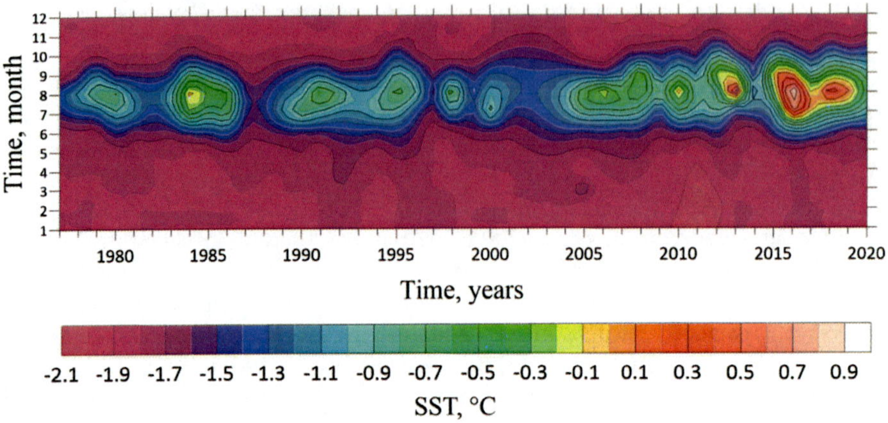

Fig. 3. Temporal distribution of average monthly sea surface temperature according to the data of the Krenkel Observatory (Hayes Island)

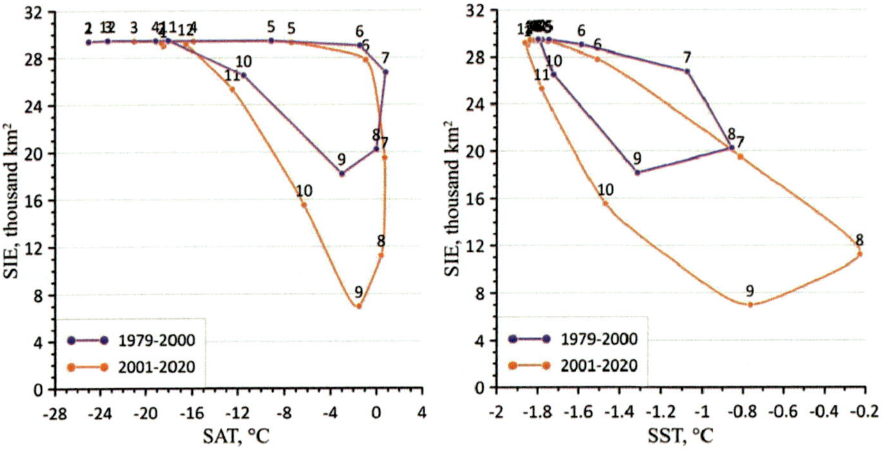

Fig. 4. Relationship between SST/SAT and SIE in the eastern area of the Franz Josef Land (1,2,3,…,12 – months of the year)

4 Conclusion

In the course of the study, global warming trends in the WSA were analyzed and evaluated using the example of the archipelagos of Svalbard and FJL, which are expressed in a reduction in the area of SIE, an increase in SAT and SST.

Quantitative results are obtained that indicate the modern climate warming. We have established that since 2000 there have been significant changes in the average annual SIE in the WSA (between 1995 and 2000 for FJL and around 2006 for Svalbard). This led to the transition of the climate system to a quasi-stable state of reduced ice cover. Statistically significant relationships between the FJL SIE and the AMO index and water temperature in the Kola Section appear with a delay of 3 and 1 year, respectively.

In addition, an original approach was applied, which consists in describing changes in the "ocean-sea ice-atmosphere" system on the SIE-SAT and SIE-SST planes for before and after the "change of ice regime". On the basis of the obtained curves, in the future it is planned to formulate a new climate index – the "climate anomaly index", which characterizes the regional features of warming.

The quantitative assessments of changes and variability of individual parameters of the "ocean-sea ice-atmosphere" system revealed in the course of the study indicate a significant restructuring of the processes of interaction between these environments, which, apparently, are global in nature and affects large areas of the Arctic Ocean. The restructuring of the ice regime, observed in other regions of the WSA, suggests that this process is of a large-scale nature and may indicate global changes in the circulation of the atmosphere and/or ocean.

Acknowledgments. This work was carried out in accordance with the Roshydromet project "Monitoring of the state and pollution of the environment, including the cryosphere, in the Arctic basin and in the areas of the research base «Ice Base Cape Baranova», Tiksi Hydrometeorological Observatory and the Russian Science Center at Svalbard Archipelago".

References

1. Meier, W.N., et al.: Arctic sea ice in transformation: a review of recent observed changes and impacts on biology and human activity. Rev. Geophys. **52**(3), 185–217 (2014)
2. Alekseev, G.V., et al.: Arctic sea ice cover in connection with climate change. Izv. Atmos. Ocean. Phys. **51**(9), 889–902 (2015)
3. Tislenko, D.I.: et al.: Seasonal and long-term changes in ice coverage in the area of the Spitsbergen archipelago. In: The Current State of the Natural Environment of the Spitsbergen Archipelago: A Collective Monograph, Under the General Editorship of Dr. Savatyugin L.M., pp. 95–103. AARI, St. Petersburg (2020). [In Russian]
4. AMAP Arctic Climate Change Update: Key Trends and Impacts Arctic Monitoring and Assessment Programme (AMAP), p. 2021. Tromsø, Norway (2021)
5. Francis, J.A., Vavrus, S.J., Cohen, J.: Amplified Arctic warming and mid-latitude weather: new perspectives on emerging connections. WIREs Clim. Change **8**(5), e474 (2017)
6. Ivanov, V.V., et al.: Does arctic ocean ice cover become seasonal? Explorat. Earth Space **4**, 50–65 (2013). [In Russian]
7. Alekseev, G.V.: Manifestation and intensification of global warming in the Arctic. Fundam. Appl. Climatol. **1**, 11–26 (2015). [In Russian]
8. Przybylak, R., Wyszyński, P.: Air temperature changes in the Arctic in the period 1951–2015 in the light of observational and reanalysis data. Theoret. Appl. Climatol. **139**(3), 75–94 (2020)
9. Ivanov, B., et al.: Assessment of long-term changes in the surface air temperature from the High Arctic archipelago Franz Joseph Land from 1929 to the present (2017). Czech Polar Rep. **11**(1), 114–133 (2021). [In Russian]
10. Isaksen, K., et al.: Exceptional warming over the Barents area. Sci. Rep. **12**, 9371 (2022). https://doi.org/10.1038/s41598-022-13568-5
11. World Meteorological Organization. WMO sea-ice nomenclature, terminology, codes and illustrated glossary. In: WMO/OMM/BMO (1970–2014), vol. 259, pp. 147 (2014)
12. Lind, S., Ingvaldsen, R.B., Furevik, T.: Arctic warming hotspot in the northern Barents Sea linked to declining sea-ice import. Nat. Clim. Chang. **8**(7), 634–639 (2018)

13. Carmack, E., et al.: Toward quantifying the increasing role of oceanic heat in sea ice loss in the new Arctic. Bull. Am. Meteor. Soc. **96**(12), 2079–2105 (2015)
14. Smedsrud, L.H., et al.: The role of the Barents Sea in the Arctic climate system. Rev. Geophys. **51**(3), 415–449 (2013)
15. Sandø, A.B., Gao, Y., Langehaug, H.R.: Poleward ocean heat transports, sea ice processes, and Arctic sea ice variability in NorESM1-M simulations. J. Geophys. Res. Oceans **119**(3), 2095–2108 (2014)
16. Ivanov, V.V., Alekseev, V.A., Repina, I.A.: Increasing impact of Atlantic waters on the ice cover of the Arctic Ocean. In: Turbulence, Atmospheric and Climate Dynamics: Proceedings of the International Conference in Memory of Academician A.M. Obukhov, pp. 267–273 (2014). [In Russian]
17. Alekseev, G.V., et al.: The influence of the North Atlantic on climate variations in the Barents Sea and their predictability. Russ. Meteorol. Hydrol. **41**(8), 544–558 (2016). [In Russian]
18. Makhotin, M.S., Ivanov, V.V.: Distribution of Atlantic water masses in the Barents Sea according to observations and numerical modeling. Proc. Hydrometeorol. Res. Center Russ. Feder. **361**, 169–191 (2016). [In Russian]
19. Polyakov, I.V., et al.: Greater role for Atlantic inflows on sea-ice loss in the Eurasian Basin of the Arctic Ocean. Science **356**(6335), 285–291 (2017)
20. Aksenov, P.V., Ivanov, V.V.: "Atlantification" as a probable cause of the reduction of the sea ice area in the Nansen basin in the winter season. Probl. Arctic Antarctic **64**(1), 42–54 (2018). [In Russian]
21. Bashmachnikov, I.L., et al.: Seasonal and interannual variations of heat fluxes in the Barents Sea region. Bullet. Russ. Acad. Sci. Phys. Atmosph. Ocean **54**(2), 239–250 (2018). [In Russian]
22. Dumanskaya, I.O.: Ice conditions of the seas of the European part of Russia. In: Hydrometeorological Center of Russia, 608p (2014). [In Russian]
23. Zubakin, G.K.: Large-scale variability of the state of the ice cover of the seas of the North European basin. In: Gidrometeoizdat, 160p (1987). [In Russian]
24. Arctic and Antarctic Research Institute World Sea Ice Data Center (WDC SI AARI) Global Sea Ice Databank [Web source]. http://wdc.aari.ru
25. Shapkin, B.S., et al.: Long-term changes in ice coverage in the area of the Svalbard (Spitsbergen) and Franz Josef Land archipelagos. Ice Snow **61**(1), 128–136 (2021). [In Russian]
26. RIHMI-WDC [Web source]. http://www.meteo.ru
27. Demin, V.I., Ivanov, B.V., Revina, A.D.: Reconstruction of air temperature series at Russian station in Barentsburg (Svalbard). Russ. Arctic. **9**, 30–40 (2020). [In Russian]
28. UISGO [Web source]. http://esimo.ru
29. PINRO [Web source]. http://www.pinro.vniro.ru
30. AMO Index time series database [Web source]. https://psl.noaa.gov/data/timeseries/AMO/
31. Grigorkina, R.G., Guber, P.K., Fuks, V.R.: Applied methods of correlation and spectral analysis of large-scale oceanological processes. In: LSU, 172p (1973). [In Russian]
32. Tislenko, D.I., et al.: Seasonal and long-term changes of sea ice extent in the Svalbard archipelago area during 1979–2015. Probl. Arctic Antarctic **109**(3), 50–59 (2016). [In Russian]
33. Yulin, A.V., Vyazigina, N.A., Egorova, E.S.: Interannual and seasonal variability of ice area in the Arctic Ocean according to satellite observations. Russ. Arctic **7**, 28–40 (2019). [In Russian]
34. Gruza, G.V., Rankova, E.Y.: Observed and Expected Climate Changes in Russia: Air Temperature. RIHMI-WDC, Obninsk, p. 194 (2012). [In Russian]

Estimation of Ice Age Categories Composition in the East Siberian Sea at the End of Winter Period

M. V. Sharatunova(✉) and V. V. Ivanov

Arctic and Antarctic Research Institute (AARI), Saint Petersburg 199397, Russia
mvsh@aari.ru

Abstract. The article is dedicate to the study of ice age categories composition in the East Siberian Sea. The calculation results showed that thick first-year ice (over 120 cm) prevails in drifting ice and fast ice in the sea by the end of the winter period (May). The transition of ice from one age gradation to an older one in fast ice occurs earlier than in drifting ice, by ten-day period 2–4. In the western part, thick first-year ice forms on average a month earlier than in the eastern part. The amount of drifting thick first-year ice in the East Siberian Sea in May was about 50%.

Keywords: East Siberian Sea · ice conditions · of ice age categories composition · fast ice · drifting ice

1 Introduction

In winter, the most important characteristic of the ice regime of the Arctic seas is the ice thickness. In addition, there is a spatial uneven distribution of ice by development stages (ice thickness), associated with climatic and hydrometeorological conditions [1]. The East Siberian Sea is the most ice extent of the seas of the Russian Arctic. Ice conditions in the western and eastern parts of the sea differ from each other and have their own characteristics [2].

In the western part of the sea and in the Novosibirsk Straits, there are ice mainly of local origin. About 2/3 of fast ice is located in the western part of the sea. The ice cover in the eastern part consists of first-year ice, residual ice and multi-year, brought from the Arctic Basin [3]. The increase in ice thickness is very intense and continues on average until the end of May.

It is planned to switch to year-round navigation along the eastern direction of the Northern Sea Route (NSR) in the near future. Considering that, it is necessary to have accurate information on the age characteristics of the ice cover in order to carry out marine operations.

T. Chaplina (Ed.): CIWO 2023, SPEES, pp. 135–140, 2023.
https://doi.org/10.1007/978-3-031-47851-2_16

2 Materials and Methods

As initial data, digital regional ice maps of the Center "Sever" of the State Scientific Center "AARI" for the period 1997–2022 from end October to May, with ten-day period discreteness were used. The methodology is based on the calculation of the area of a homogeneous ice zone of a certain age (km^2) using GIS-technologies. This area (km^2) is then converted to a percent (%) of the total ice area East Siberian Sea [4]. The boundaries of the studied seas and their areas are presented in the manual "Boundaries of oceans and seas" [5]. Other authors [6–8] published similar articles on development of ice age categories composition of ice in the Arctic seas previously.

3 Results and Discussion

Within the framework of this work, characteristic distributions of drifting ice and fast ice of various ages were obtained for the East Siberian Sea *(western (1) and eastern (2) parts*), as well as for the local area of the sea - region 23 are shown in Fig. 1. This region is located in central zone eastern part of sea, in accordance with the Rules for navigation in the water area of the Northern Sea Route [9]. There is some of the main obstacles are the Ayon Ice Massif. The selection of routes in the central zone is defined by the presence of drifting ice and shallows, which may be dangerous to sailing.

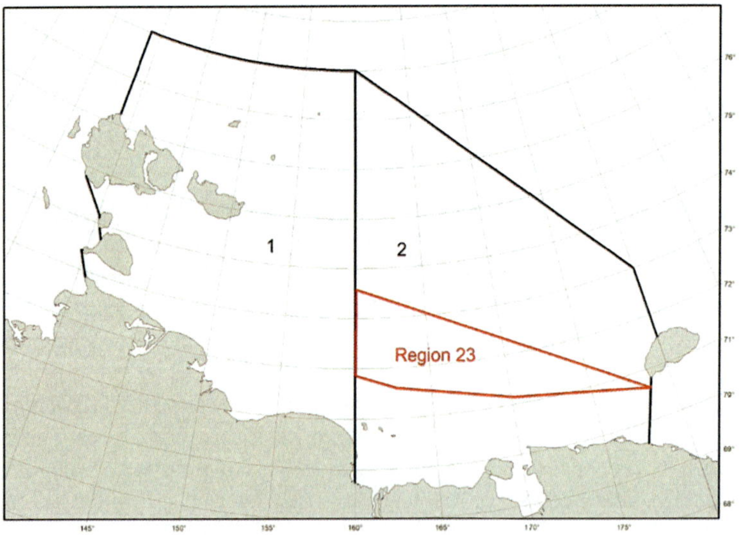

Fig. 1. Boundaries of the East Siberian Sea

4 Western Part of the East Siberian Sea

Drifting ice. Change in the of ice age categories composition of drifting ice in the western part of the East Siberian Sea are complex. The constant drift of ice, ice formation and ice growth in polynya zones, and ice advection make changes in the of ice age categories composition of drifting ice not as pronounced as in fast ice. Figure 2 shown the change in the amount of drifting ice from October to May. The maximum amount of thin first-year ice (30–70 cm) falls on the third ten-day period of November - the first ten-day period of December and is 45%. Their predominance continues until the end of December. By the beginning of December, medium ice (70–120 cm) begins to form, by the beginning of January they make up about 30% and prevail until mid-March. In the process of increasing thickness, part of the ice passes into the age gradation of thick first-year ice with a thickness of more than 120 cm. By the mid-January, thick first-year ice begins to appear (3%), and by the end of March is make up 18%. By the end of May, they make up about 23% of the western part of the East Siberian Sea. When analyzing the seasonal variability of the age composition of ice in the western part of the East Siberian Sea, it is necessary to take into account all types of ice. Because all types of ice located in the area, since are areas of active navigation.

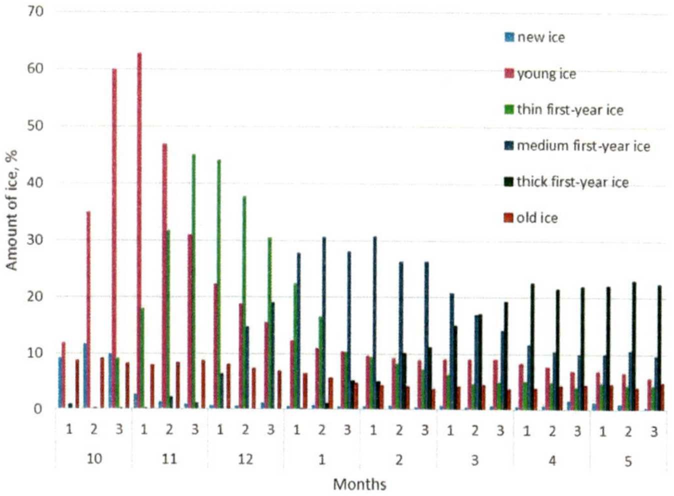

Fig. 2. The change the amount of drifting ice from October to May

Fast Ice. The change the amount of fast ice from October to May is shown in Fig. 3.

When fast ice formation is beginning, in the second ten-day period of October, gray and gray-white ice prevails. By the end of October in fast ice, young ice passes into the stage of thin first-year ice. Rapidly growing, by the end of November, thin first-year ice occupies about 13%. In early December, the appearance of medium first-year ice is noted, in the second ten days period they make up more than 10%, and in the second ten-day period of January - 27%. The transition of medium first-year ice to thick first-year ice starts from the second ten-day period of January. However, the predominance of medium first-year ice is continuing for 5 ten-day period. By the first ten-day period of February, in fast ice is dominated thick first-year ice (20%), then until mid-May, its amount increases to 47%. Also according to estimates, fast ice contains about 1.5% of old ice. The thickness of the fast ice play the main role in determining the timing of fast ice destruction by the end of the winter period. However, the variability of fast ice destruction periods can reach 2 months.

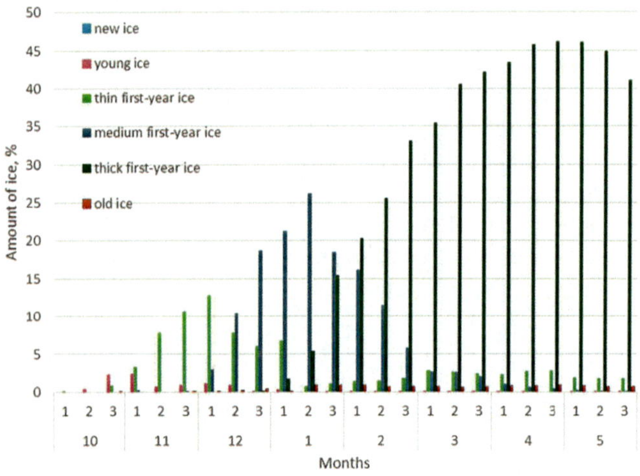

Fig. 3. The change the amount of fast ice from October to May

In drifting ice in the western part of Sea, the transition to an older age gradation occurs on average 1 month later than in fast ice, depending on the age of the ice.

Young ice passes into the gradation of thin first-year ice at 0.5 months later than in fast ice, into the gradation of medium first-year ice at 1 month later, into the gradation of thick first-year ice 1 month later.

This feature is a good prognostic sign for determining the timing of the transition of drifting ice to older age gradations according to an accomplished phenomenon in fast ice [10].

5 Eastern Part of the East Siberian Sea

When analyzing the seasonal variability of ice age categories composition of ice in the eastern part of the East Siberian Sea, it is necessary to take into account only drifting ice located in the water area of the region. This is due to the fact, that fast ice in the eastern part of the sea is located at shallow depths along the coast, occupies on average no more than 15% of the water area and is not a zone of active navigation.

Figures 4, 5 show the successive transition of ice age categories composition from new ice to thick first-year ice in eastern part and local area. In the eastern part, the transition of thin first-year ice to the medium first-year ice gradation occurs in early January, then they transitions to the thick first-year ice gradation in early March.

In mid-November, in the local area, the amount of young ice increases rapidly and reaches its maximum of 57%; for the entire eastern part, this value is 47%. By the second half of November, the transition of ice to the thin first-year ice stage occurs evenly for the entire eastern part of the sea. The maximum value the thin first-year ice is about 50%

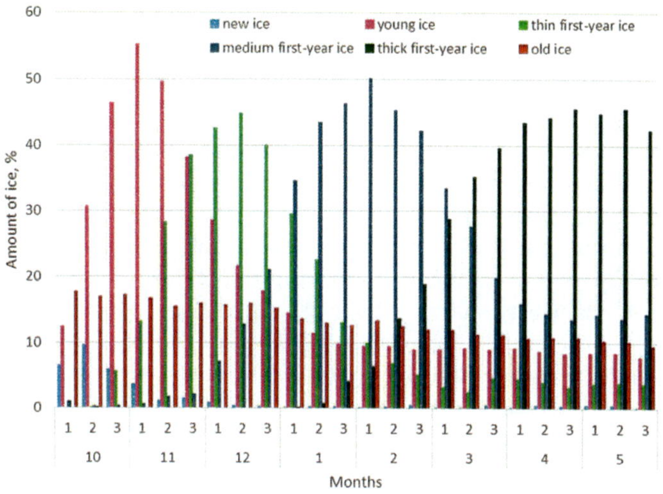

Fig. 4. The change the amount of drifting ice from October to May, eastern part

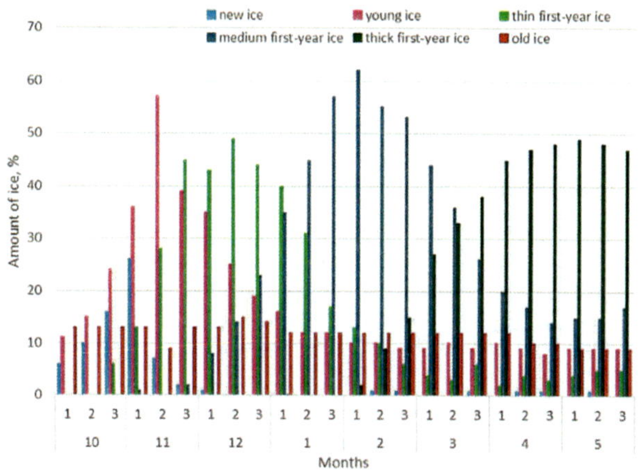

Fig. 5. The change the amount of drifting ice from October to May, local area

in mid-December. In the local area 23, medium first-year ice prevails from the second ten-day period of January (maximum 62% in early February).It is on one ten-day period later than in the east of the sea (at the end of January about 50%). Thus, the transition from medium to thick ice in a local area occurs one ten-day period later (the values are close to 30–35%). By mid-May, in some areas and in the entire eastern part of the sea, thick first-year ice reaches its maximum of about 50%.

Just like first-year ice, about 5–7% of new ice and young ice observed in the ice cover in the eastern part of the sea during the entire winter period, which is formed in the areas of formation of fast ice polynyas, as well as in fractures and cracks among the ice cover.

6 Conclusion

An estimation of the ice age categories composition of the ice cover by the end of the winter period (in May) showed the predominance of thick first-year ice in drifting and fast ice. For the transition of ice to the next age gradations, it is necessary to accumulate a certain amount of ice. All types of young and first-year ice are present in the ice cover until the end of the ice cover formation. However, the predominant types of ice are medium first-year ice and thick first-year ice. It is important to note the peculiarity of the difference in the increase in ice thickness in fast ice and drifting ice. In fast ice, the thickness of ice increases faster than in drifting ice. The transition of fast ice from one age gradation to an older one occurs 2–4 ten-day period earlier than in drifting ice. Consequently, when analyzing the local area under consideration, the spatial non-uniformity of the distribution of the ice cover by age is confirmed here. The obtained data on the age composition of ice in the East Siberian Sea in winter complement the regime characteristics and allow us to assess the ice conditions to ensure the safety of navigation of ships along the NSR route.

Acknowledgements. The authors wish to express their thanks to Alexander V. Yulin, head of laboratory "Long-term ice forecasts" OLRiP and specialists of Center "Sever" AARI for expert advice in working with data sets of ice age categories composition.

References

1. Seas of the Russian Arctic in modern climatic conditions, p. 360. AARI, St. Petersburg (2021)
2. Observations of the Ice Conditions: Textbook, p. 360. AARI, St. Petersburg (2009)
3. Karelin, I.D., Karklin, V.P.: Fast ice and fast ice polynyas of the Arctic seas of the Siberian shelf at the end of the 20th - beginning of the 21st centuries, p.180. AARI, St. Petersburg (2012)
4. JCOMM Expert Team on Sea Ice. Sea-Ice Nomenclature: snapshot of the WMO Sea Ice Nomenclature No. 259, p.121. WMO-JCOMM, Geneva, Switzerland (2014)
5. The boundaries of the oceans and seas, p. 206. GUNIO MO RF, St. Petersburg (2000)
6. Hotchenkov, S.V.: Stages of sea ice development in the Laptev Sea, vol. 4, no. 114, pp. 5–15. Arctic and Antarctic Research (2017)
7. Karklin, V.P., Hotchenkov, S.V., Yulin, A.V., Smolyanitsky, V.M.: Seasonal changes in the stages of sea ice development in northeast part of the Kara Sea during the autumn and winter period, vol. 4, no. 110, pp. 41–50. Arctic and Antarctic Research (2016)
8. Karklin, V.P., Hotchenkov, S.V., Yulin, A.V., Smolyanitsky, V.M.: Formation of the stages of sea ice development composition in the south-western part of the Kara Sea during autumn-winter season, vol. 3, no. 113, pp. 16–26. Arctic and Antarctic Research (2017)
9. Rules for navigation in the waters of the Northern Sea Route, p. 34. RF GR, Moscow (2020)
10. Sharatunova, M.V., Ivanov, V.V.: The variability of ice age composition in the winter period of the East Siberian Sea. In: Proceedings of the II All-Russian Conference «Hydrometeorology and Ecology: Achievements and Development Prospects», pp. 692–696. KhIMIZDAT, St. Petersburg (2018)

Surface Manifestations of Short-Period Internal Waves in the Sea of Okhotsk and the Kuril-Kamchatka Region of the Pacific Ocean

E. I. Svergun[1,2,3](✉), A. V. Zimin[1,2,3], K. V. Kulik[1,2,3], and N. S. Frolova[1,2,3]

[1] Shirshov Institute of Oceanology RAS, Moscow 117997, Russia
egor-svergun@yandex.ru
[2] Saint Petersburg State University, Saint Petersburg 199034, Russia
[3] Russian State Hydrometerological University, Saint Petersburg 192007, Russia

Abstract. The paper considers the characteristics of surface manifestations of short-period internal waves in the Sea of Okhotsk based on the analysis of Sentinel-1 radar images for the summer period of 2019. The features of the field of wave manifestations in the Sea of Okhotsk and the Kuril-Kamchatka region of the Pacific Ocean are also compared. It was found that in the Sea of Okhotsk, thrice fewer wave manifestations are recorded than in the Kuril-Kamchatka region, while the differences in the characteristics of the manifestations are insignificant. The manifestations of the Sea of Okhotsk are characterized by the formation of isolated clusters, while in the Kuril-Kamchatka region, manifestations are recorded everywhere. The position of the wave manifestations partially coincides with the potential generation sites of diurnal internal tidal waves, which may indicate the generation of short-period waves under the influence of the disintegration of the diurnal internal tide.

Keywords: short-period internal waves · satellite radar images · surface manifestations · internal tide · tidal body force · shelf · sea of Okhotsk · Kuril-Kamchatka region of the Pacific Ocean

1 Introduction

Short-period internal waves (SIW's) are an important component of the dynamics of waters on the shelf, but due to the small space-time scale, they remain insufficiently studied in many areas of the World Ocean. The Sea of Okhotsk is no exception – a marginal sea of mixed mainland-marginal type, which is separated from the Pacific Ocean by the Kuril Ridge. Under the influence of the tidal wave of the Pacific Ocean, a pronounced tide arises, which is diurnal in most of the water area of the Sea of Okhotsk [1]. In the north of the Sea of Okhotsk there is a continental shoal, and in the rest of the sea there is a continental slope with many irregularities of bottom relief. The complex bottom relief of the in combination with pronounced tidal dynamics is a preconditions for the generation of SIW's.

T. Chaplina (Ed.): CIWO 2023, SPEES, pp. 141–149, 2023.
https://doi.org/10.1007/978-3-031-47851-2_17

For the Sea of Okhotsk, only separate evidence of satellite registration of surface manifestations of SIW's on the eastern shore of Sakhalin Island, in the Aniva and Sakhalinsky bays is available [2, 3], as well as the results of episodic measurements of wave characteristics by in situ methods [4, 5]. Measurement data show that intense SIW's with a height of up to 11 m are recorded in the shelf zone of the Sea of Okhotsk [5]. However, there are still no studies identifying areas of frequent occurrence of SIW's manifestations for the entire water area of the Sea of Okhotsk based on the analysis of modern satellite data for the summer period. Identification of these areas is very relevant, given the commercial importance of the Sea of Okhotsk, since SIW's are capable of local strengthening of vertical mixing [6], redistribution of biogens [7]. According to [8], for the contiguous Kuril-Kamchatka region, based on the analysis of the archive of Sentinel-1 satellite images for the summer of 2019, a wide presence of SIW's manifestations was established in almost the entire water area of the region. In [9] it was found that the main reason for the generation of SIW's in the Kuril-Kamchatka region is the disintegration of diurnal internal tidal waves (ITW's) beyond the critical latitude. Given that the diurnal tide dominates in the Sea of Okhotsk, it can be assumed that this generation mechanism will also work effectively. However, in the Kuril-Kamchatka region, the average bottom slopes are significantly higher than in the Sea of Okhotsk, which may affect the intensity of SIW's generation.

Thus, the purpose of this work is to analyze the characteristics of SIW's manifestations and the features of SIW's generation in the Sea of Okhotsk according to radar observations for the summer of 2019 in comparison with the Kuril-Kamchatka region of the Pacific Ocean.

2 Materials and Methods

536 Sentinel-1A and Sentinel-1B satellite images with HH and VV polarization and a spatial resolution of up to 40 m were used to analyze the characteristics of SIW's manifestations in the Sea of Okhotsk in the period from June 1, 2019 to August 31, 2019. In the Kuril-Kamchatka region, 205 images were available during the same period. Satellite images were selected and uploaded from the Alaska Satellite Facility (ASF) Data Search website [10]. The coverage of the water area with satellite data is shown in Fig. 1.

Figure 1 shows that the central part of the Sea of Okhotsk is not covered with satellite data, and the coastal part of the study area is covered with images fairly evenly. It is should be noted that the manifestations of SIW's were analyzed only from the Pacific side of the Kuril Ridge and the Kamchatka Peninsula in order to emphasize the differences in the features of the manifestations of SIW's in the oceanic and marine parts of the region.

The manifestations of SIW's were recorded in the ESA SNAP software as alternating curved bands of amplification and attenuation of the radar signal. An example of the manifestations of SIW's in this region is demonstrated in [9]. The coordinates of the manifestations, the number of waves in the packet (n in Fig. 1b), the wavelength (λ in Fig. 1b), the length of the leading crest (Cr in Fig. 1b) and the direction of propagation (dir in Fig. 1b) were determined, similar to the method presented in [9]. The spatial and temporal variability of wave characteristics was estimated from maps and diagrams. In

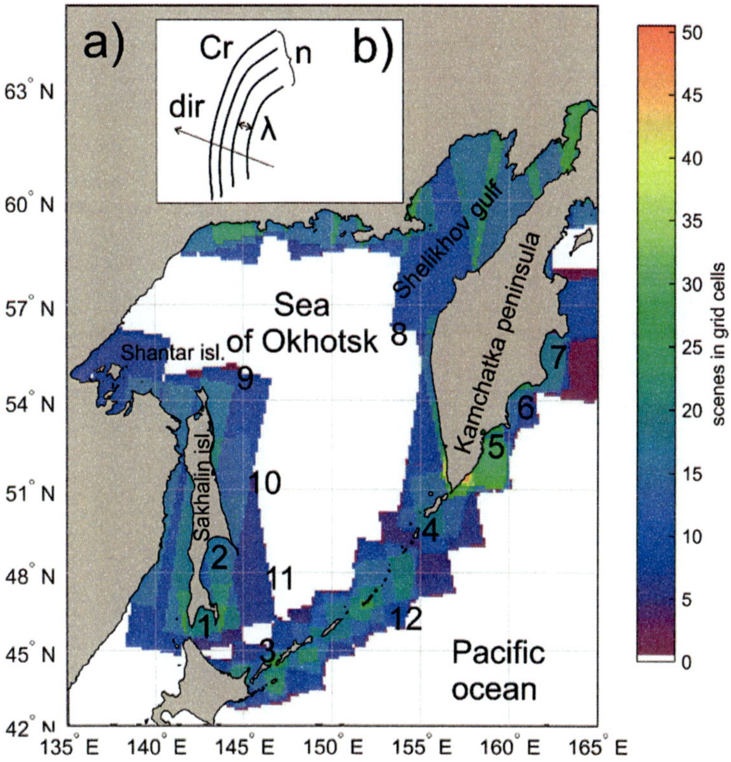

Fig. 1. a) geographical location of the research region and its coverage with satellite radar images: 1 – Aniva Bay, 2 – Terpeniye Bay, 3 – Kunashir Island, 4 – Onekotan Island, 5 – Avacha Bay, 6 – Kronotsky Bay, 7 – Kamchatka Bay., 8 – TINRO basin, 9 – Deryugin basin, 10 – Okhotsk basin, 11 – Kuril basin, 12 – Kuril-Kamchatka trench; b) schematic representation of the manifestation of SIW's (see the text for designations).

order to identify areas of regular registration of SIW's manifestations, the frequency of their occurrence was calculated as the ratio of the number of wave manifestations to the number of images in the cell.

To analyze the causes of SIW's, the average monthly values of the depth-integrated tidal body force [11] were calculated as a criterion for the intensity of ITW's generation for the K1 harmonic. Tidal body force is a function of barotropic tidal flow, stratification and topography. The criterion was calculated using the TPXO9 tidal atlas [12], MERCATOR system reanalysis [13] and ETOPO1 bathymetry [14]. In this work, it was assumed that the areas where the values of the tidal body force criterion exceed 5 $m^2\ s^{-2}$ are potential generation sites of internal tide. The number of manifestations of SIW's in these areas was calculated.

3 Results and Discussion

In the Sea of Okhotsk in the summer of 2019, 346 manifestations of internal waves were detected on 536 images. The maximum was observed in June – 165 manifestations (48% of the total number of waves), the minimum in August – 73 manifestations (21% of the total number of waves). In the Kuril-Kamchatka region, during this period, 927 manifestations of SIW's were registered on 205 images, which is thrice more than their number in the Sea of Okhotsk. The minimum number of manifestations was registered in June (289 manifestations or 31%), and the maximum – in July (341 or 36%).

In the Sea of Okhotsk, the length of the leading crest varied in the range of 2–60 km, the average value was 12 km. The wavelength ranged from 60 to 2000 m, with an average value of 450 m. During the study period, internal waves contained from 2 to 26 waves in a train, an average of 7. In the Kuril-Kamchatka region, the front of the leading crest measured from 2 to 70 km, averaging 14 km. The images revealed trains containing from 3 to 18 waves. The wavelengths inside the train ranged from 80 to 1900 m, with an average value of 400 m. Figure 2 shows frequency histograms of the characteristics of SIW's manifestations in the Sea of Okhotsk and the Kuril-Kamchatka region.

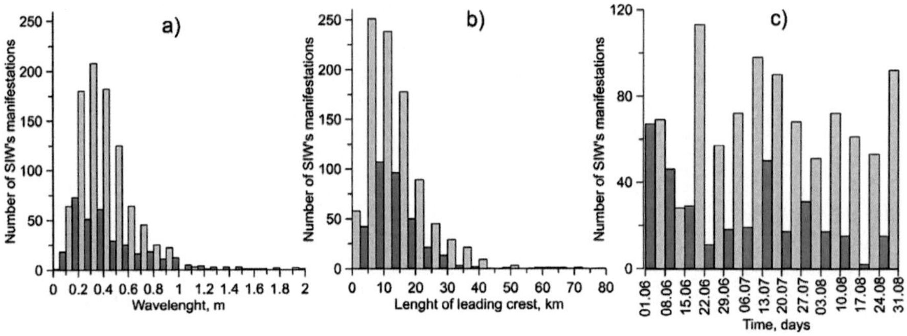

Fig. 2. Frequency histograms of the characteristics of wave manifestations in the Sea of Okhotsk (green) and in the Kuril-Kamchatka region (gray): a) the wavelength, b) the length of the leading crest, c) the number of manifestations by week

Both in the Sea of Okhotsk and in the Kuril-Kamchatka region, the manifestations with the length of the leading crest from 5 to 10 km had the greatest qrequency. The wavelength of the greatest frequency in the Sea of Okhotsk was 200–300 m, and in the Kuril-Kamchatka region – 300–400 m. Figure 2c shows in more detail the intra-seasonal variability of the number of registered manifestations: in the Sea of Okhotsk, the maximum number of manifestations occurred in the first week of June and the third week of July, and in the Kuril–Kamchatka region - in the third week of June, the second week of July and the 4th week of August.

Thus, despite the significant difference in the number of registered manifestations of SIW's and in the dates of registration of maxima, in the Sea of Okhotsk and the Kuril-Kamchatka region, their geometric characteristics are very similar.

Figure 3 shows maps of the spatial distribution of the leading crests of SIW's manifestations and their frequency of occurrence.

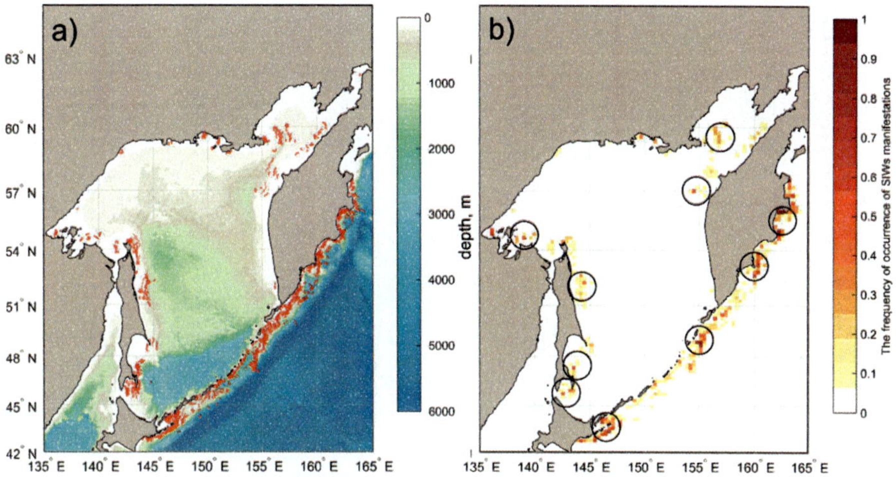

Fig. 3. Maps of spatial distribution: a) the leading crests of SIW's manifestations for the summer of 2019, combined with ETOPO1 bathymetry, b) the frequency of occurrence of SIW's manifestations for the summer of 2019 (black circles are the areas of maxima of frequency of occurrence)

From Fig. 3a it can be seen that, in general, the manifestations of SIW's are confined to areas of irregularities of the bottom relief, but in the Kuril-Kamchatka region, manifestations are recorded almost everywhere, and in the Sea of Okhotsk in the form of local isolated clusters.

In the Sea of Okhotsk, the most frequent internal waves were recorded on the Sakhalin Island shelf in the Aniva and Terpeniye bays, on the border of the Deryugin and Okhotsk basins, east of the Shantar Islands, in the central part of Shelikhov gulf and on the border of the TINRO basin (Fig. 3b). In the Kuril-Kamchatka region, the areas with the highest frequency of occurrence are located south of the islands of Kunashir and Onekotan, in the Avacha, Kronotsky and Kamchatka Bays. It is worth noting that the areas of frequent occurrence, characteristic of the whole summer as a whole, are stable in some months. Figure 4 shows maps of the spatial distribution of the averages in the cell of the wavelength and the length of the leading crest.

It was found that, in general, during the study period, for the areas of frequent occurrence of SIW's manifestations in the Sea of Okhotsk, the average wavelength is higher than for the areas of frequent occurrence of the Kuril-Kamchatka region (1400 m vs. 1000 m). At the same time, the average length of the leading crests in areas of frequent occurrence has similar values both in the Sea of Okhotsk and in the Kuril-Kamchatka region, and is about 40 km.

As mentioned earlier, in the Sea of Okhotsk the diurnal tide is dominate. The wave of the main diurnal harmonic K1 is sub-inertial (the tidal frequency is lower than the inertial one), internal tidal waves cannot propagate as free [15], therefore, the disintegration of ITW's in the conditions of critical latitudes into SIW's packets will be a possible mechanism for generating SIW's in the Sea of Okhotsk. At the same time, the areas of frequent occurrence of SIW' will be close or even coincide with the of generation sites of ITW's. Thus, it is possible to use the tidal body force criterion to identify potential

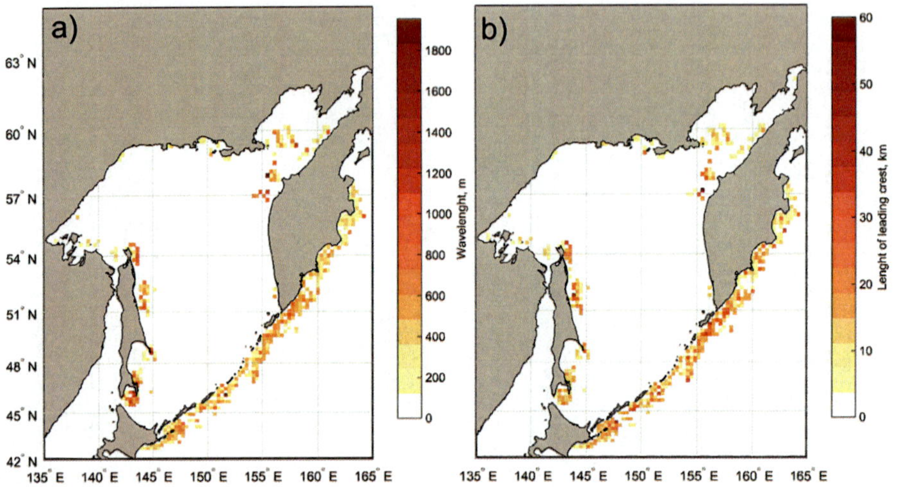

Fig. 4. Spatial distribution for the summer of 2019: a) wavelength, b) length of the leading crest

areas of SIW's generation. Figure 5 shows a composite map of the tidal body force criterion of the K1 harmonic, calculated for stratification in June (since the number of manifestations of SIW's in the Sea of Okhotsk is maximum), with the leading crests of SIW's manifestations.

As a result of the analysis of the composite map, it was revealed that potential generation sites of ITW's in the Sea of Okhotsk are located northeast of Sakhalin Island, in Shelikhov Bay, on the borders of the TINRO and Kuril basins, and in the Kuril-Kamchatka region - around the Kuril Islands, in the Avacha, Kronotsky and Kamchatka bays. The main areas of frequent occurrence of SIW's manifestations (in the Sea of Okhotsk in the Terpeniye bay, near the Deryugin basin, and in Shelikhov Gulf, and in the Kuril-Kamchatka region near the islands of Kunashir and Onekotan) partially coincide with potential generation sites of ITW's. The percentage of SIW's manifestations in ITW's generation areas in the Sea of Okhotsk is 52% (83 out of 159 manifestations), and in the Kuril-Kamchatka region – 32% (95 out of 289 manifestations). Thus, about half of the manifestations of SIW's in the Sea of Okhotsk may be associated with the disintegration of diurnal ITW's. However, in the Sea of Okhotsk, there are large generation sites of ITW's on the border of the Kuril and TINRO basins, as well as northeast of Sakhalin Island, where no manifestations of SIW's have been recorded. Figure 1 shows that these regions are not covered with Sentinel-1 satellite data, therefore it is necessary to apply additional multisensory satellite observations in these areas to identify the connection of generation sites of ITW's with SIW's manifestations.

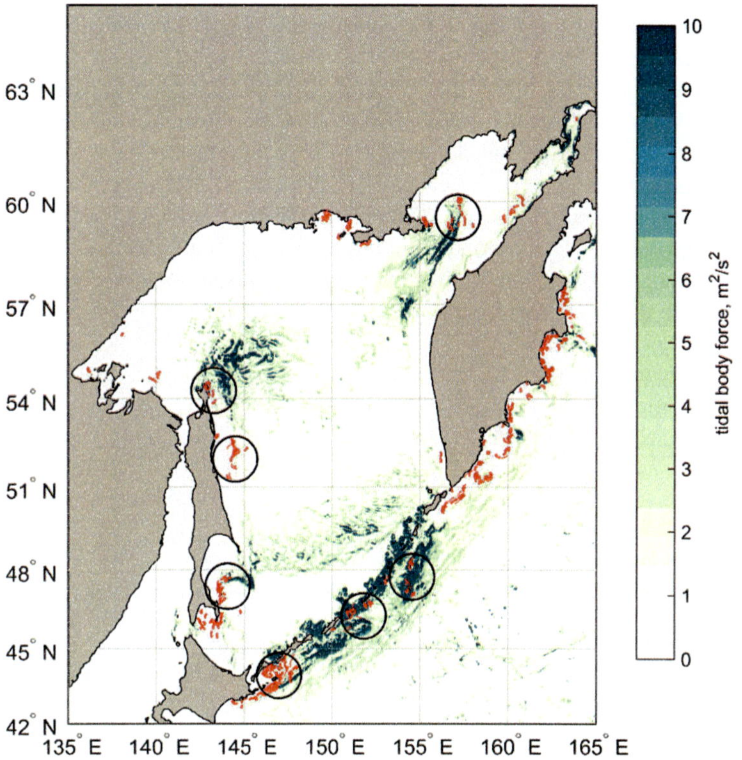

Fig. 5. Composite map of the depth-integrated tidal body force criterion for the K1 harmonic and the position of the leading crests of the SIW's manifestations. Black circles indicate the maximum number of SIW's manifestations in the generation sites of ITW's

4 Conclusion

The paper analyzed the characteristics of SIW's manifestations in the Sea of Okhotsk based on the results of processing the archive of Sentinel-1 satellite images for the summer of 2019 and compared the features of SIW's manifestations in the Sea of Okhotsk and the Kuril-Kamchatka region of the Pacific Ocean.

It was found that the geometric characteristics of wave manifestations in the Sea of Okhotsk and the Kuril-Kamchatka region have no significant differences in both average values and values of the higher frequencies. However, in the Sea of Okhotsk, thrice fewer manifestations of SIW's were registered than in the Kuril-Kamchatka region. The feature of SIW's manifestations in the Sea of Okhotsk is their registration in the form of isolated clusters, while in the Kuril-Kamchatka region, manifestations are recorded almost everywhere. This is probably due to the fact that there is a steep continental slope along the entire Kuril-Kamchatka region, and isolated irregularities of the bottom relief in the Sea of Okhotsk.

For the first time, areas of frequent occurrence of SIW's manifestations in the Sea of Okhotsk were demonstrated, located in the Aniva and Terpeniye bays, on the border of the Deryugin basin, east of the Shantar Islands, in Shelikhov gulf and on the border of the TINRO basin.

A comparison of the positions of SIW's manifestations with potential generation sites of internal tide revealed that about half of the manifestations in the Sea of Okhotsk are associated with the disintegration of diurnal ITW's, however, satellite data coverage does not allow to establish a connection between SIW's manifestations and ITW's generation sites in the areas northeast of Sakhalin Island, the TINRO and Kuril basins.

As a continuation of this work, a more detailed study of the seasonal variability of the characteristics of manifestations of internal waves and the mechanisms of their generation with the use of consequence satellite data is proposed.

Acknowledgements. The work was supported by the Russian Science Foundation grant No. 23–17-00174.

References

1. Glukhovsky, B.H., Gidrometeoizdat, L. (eds.): Hydrometeorology and hydrochemistry of the seas of the USSR. T.9, Hydrometeorological conditions. Sea of Okhotsk, Issue 1, p. 318 (1998)
2. Mitnik, L.M., Dubina, V.A.: Spatial-temporal distribution and characteristics of internal waves in the Okhotsk and Japan Seas studied by ERS-1/2 SAR and Envisat ASAR, pp. 23–27. In: Proc. Envisat Symposium (2007)
3. Atlas for working with the database of observations of internal waves - electronic data. https://lmnad.nntu.ru/ru/igwatlas_map/
4. Nagovitsyn, A.P., Pelinovsky, E.N.: Observations of internal wave solitons in the coastal waters of the Sea of Okhotsk. Meteorologiya i hydrologiya **4**, 124–126 (1988)
5. Zimin, A.V., Svergun, E.I.: Short-period internal waves in the shelf areas of the White, Barents and Okhotsk Seas: estimation of the extreme heights occurrence and dynamic effects in the bottom layer. Fundamentalnaya i Prikladnaya Gidrofizika **11**(2), 66–72 (2018). https://doi.org/10.7868/S2073667318040081
6. Purwandana, A., Cuypers, Y., Bouruet-Aubertot, P.: Observation of internal tides, nonlinear internal waves and mixing in the Lombok Strait Indonesia. Cont. Shelf Res. **216**, 104358 (2021)
7. Garwood, J.C., Musgrave, R.C., Lucas, A.J.: Life in internal waves. Oceanography **33**(3), 38–49 (2020)
8. Svergun, E.I., Zimin, A.V., Lazutkina, E.S.: Characteristics of manifestations of short-period internal waves of the Kuril-Kamchatka region based on satellite observations in summer. Fundamentalnaya i Prikladnaya Gidrofizika 14(1), 106–115 (2012). https://doi.org/10.7868/S2073667321010111
9. Svergun, E.I., Sofina, E.V., Zimin, A.V., Kruglova, K.A.: Seasonal variability of characteristics of nonlinear internal waves in the Kuril-Kamchatka region by Sentinel 1 data. Cont. Shelf Res. **259**, 104986 (2023). https://doi.org/10.1016/j.csr.2023.104986
10. Alaska Satellite Facility (ASF) Data Search. https://search.asf.alaska.edu/#/
11. Baines, P.G.: On internal tide generation models. Deep Sea Res. Part **A29**(3), 307–338 (1982). https://doi.org/10.1016/0198-0149(82)90098-X
12. Egbert, G.D., Erofeeva, S.Y.: Efficient inverse modeling of barotropic ocean tides. J. Oceanic Atmos. Technol. **19**, 183–204 (2002). https://doi.org/10.1175/1520-0426(2002)019%3c0183:EIMOBO%3e2.0.CO;2

13. ETOPO1 bathymetry. https://www.ncei.noaa.gov/access/metadata/landing-page/bin/iso?id=gov.noaa.ngdc.mgg.dem:316
14. Global Ocean Physics Reanalysis. https://resources.marine.copernicus.eu/product-detail/GLOBAL_REANALYSIS_PHY_001_030/INFORMATION
15. Vlasenko, V., Stashchuk, N., Hutter, K.: Baroclinic Tides: Theoretical Modelling and Observational Evidence, p. 351. Cambridge University Press, New York (2005). https://doi.org/10.1017/CBO9780511535932

Topographic Waves of the Aleutian Trench

S. P. Khudyakova[1(✉)], V. S. Travkin[1,2], and T. V. Belonenko[1]

[1] St. Petersburg State University, St. Petersburg 199034, Russia
khydyakova.s@gmail.com
[2] N.N.Zubov's State Oceanographic Institute, Roshydromet, Moscow 119034, Russia

Abstract. Based on the mesoscale eddy trajectory data, the propagation of topographic Rossby waves south of the Aleutian Ridge in various areas of topography is described. It is noted that anticyclonic eddies propagate along the shelf zone at depths of 1–2 km in the form of shelf waves, and cyclonic eddies drift westward along the trench with depths of 5–7 km in the form of trench waves. Using the example of the trajectories of individual long-lived anticyclonic eddies, it is shown that the separation of eddies from isobaths occurs under the action of a current. The amount of the mesoscale eddies and the number of mesoscale eddies generation are analyzed. It was found that more anticyclonic eddies are formed in the area of the Aleutian Ridge than cyclonic ones, and the maximum number of anticyclonic eddies per cell 0.25° in latitude by 0.5° in longitude is observed south of the Near and Rat islands. The mechanisms influencing the propagation of mesoscale eddies of the region are compared: the β-effect, the meridional gradient of the zonal velocity shear, the topographic factor and the combined action of the flow and topography. The places of eddy "separation" from the flow and topography are determined. Anticyclonic eddies stop following isobaths approximately between 171–176° E.

Keywords: topographic Rossby waves · mesoscale eddies · satellite altimetry · aleutian Trench

1 Introduction

The eddies of deep-sea trenches are of great interest to researchers. In the northern part of the Pacific Ocean, the Japanese, Kuril-Kamchatka and Aleutian Trench are distinguished, which have a depth exceeding 7500 m. This special relief is reflected in the generation, evolution and movement of the corresponding eddies. The eddies associated with significant decreases in the relief of the seabed include the Aleutian eddies. By definition [13] Aleutian eddies are mesoscale anticyclonic eddies formed in the area of the passage of the Alaskan Stream south of the Aleutian Ridge. The Alaskan Stream is one of the most eddy-active zones not only in the North Pacific Ocean, but also in the entire World Ocean [6]. For the first time, the formation of anticyclonic eddies south of the Aleutian Ridge according to satellite observations was noted in the article [10]. They form directly above or near the Aleutian Trench and stagnate there for some time [4]. After separating from the Alaskan Stream, they move either southwest or northwest

T. Chaplina (Ed.): CIWO 2023, SPEES, pp. 150–159, 2023.
https://doi.org/10.1007/978-3-031-47851-2_18

to Kamchatka, crossing the northern part of the Imperial Seamount chain [13, 16]. The crossing of seamounts is accompanied by eddy deformation, according [12] most of the Aleutian eddies interact with the Detroit Rise with a horizontal size of 165 × 70 km. The baroclinic radius of Rossby deformation in the area of the Aleutian Trench does not exceed 15 km [3, 5].While the radius of the Aleutian eddies is estimated at 50 km or more [4], which allows them to be attributed to mesoscale.

Figure 1 shows the bathymetry of the study area. To the south of the Aleutian Ridge there is a significant decrease in relief – the Aleutian Trench. It is confined to the Aleutian Island Arc and stretches for 3400 km from the Alaska Peninsula to the Kamchatka Peninsula with a width of 40–100 km along an isobath of 5500 m. The maximum depth according to various estimates reaches from 7670 to 7855 m and is located in the central part of the Trench. Also in the middle part at a depth of about 1800 m there is a wide (up to 50 km) underwater terrace. The slope of the bottom to the island part is estimated from 4° to 12°. At the western border in the area of the Commander Islands, it passes into the Kuril-Kamchatka Trench, which has a south-westerly direction.

It is known that mesoscale eddies in the ocean appear as Rossby waves in the form of coherent structures [7]. The propagation of two systems of topographic waves in the Kuril region is described in the article [15]. It is shown that on the shelf anticyclones move to the southwest in the form of shelf waves, and on the sea side of the Trench - in the form of Trench waves propagating to the northeast. On the inner side of the Trench

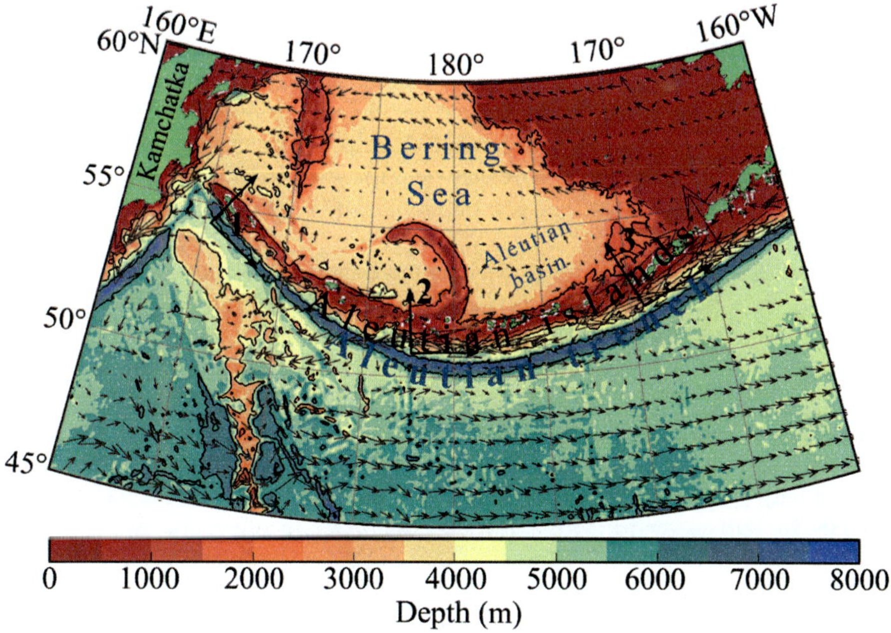

Fig. 1. Bathymetric map of the area according to ETOPO1. The coordinate systems 1, 2 and 3, oriented across the Trench, are marked with black arrows. The gray arrows show the surface velocity averaged over the period 1993–2021 according to GLORYS12V1

and on the slope of the topographic elevation behind the Trench, topographic waves appear mainly in the form of cyclones moving to the southwest.

The paper investigates topographic Rossby waves in the area of the Aleutian Ridge, appearing as chains of mesoscale anticyclonic and cyclonic Aleutian eddies. The purpose of this work is to identify and compare the contribution of various factors determining the propagation of mesoscale eddies in the Aleutian Trench. Mesoscale eddies are an important part of an integral dynamic system in the ocean, because they are able to transfer shape, mass, heat and biochemical characteristics over long distances. The Aleutian eddies form over the Aleutian Trench, move southwest after separating from the Alaskan Stream and pass through the Western Subarctic Gyre, transferring the transformed waters of the Gulf of Alaska to the western part of the Pacific Subarctic [1]. In addition to the water exchange between the Aleutian Islands region and the western and central parts of the Pacific Subarctic, the eddies of the Alaskan Stream can affect the water exchange between the North Pacific Ocean and the Bering Sea [17].

2 Material and Methods

META3.2 DT

Atlas of mesoscale eddies META3.2 DT is a product of the French agency AVISO (https://www.aviso.altimetry.fr/en/data/products/value-added-products/global-mesoscale-eddy-trajectory-product/meta3-2-dt.htm). It is based on the use of satellite altimetry data to identify and determine the trajectories of mesoscale eddies of the World Ocean. Eddies are identified as closed structures with a diameter of 100–300 km. After filtering, extremes are detected in the field of sea level anomalies to determine the location and characteristics of the eddy. The trajectories of long-lived (more than 10 days) anticyclonic and cyclonic eddies and their main parameters (amplitude, radius, area, orbital velocity, displacement velocity and lifetime) were used. The time period considered in the work is almost 30 years: from January 1, 1993 to December 31, 2021.

GLORYS12V1

The GLORYS12V1 database is a global oceanic reanalysis covering satellite altimetry data since 1993. It is a product of the European Center for Marine Forecasts (English Copernicus Marine Environment Monitoring Service—CMEMS, http://marine.copernicus.eu). It is based on the CMEMS real-time global forecasting system. A component of the model is the NEMO platform controlled on the ECMWF ERA-Interim surface, then the ERA5 analysis is used for recent years. Observations are assimilated using a reduced-order Kalman filter. The model implementation is based on the assimilation of data not only of sea level anomalies by satellite altimetry, but also jointly assimilated satellite (sea surface temperature, sea ice cohesion) and in situ data (vertical profiles of temperature and salinity). The product includes daily and monthly average values of temperature, salinity, components of the flow velocity, sea level, depth of the mixed layer and parameters of the ice cover. The output data is displayed on a standard regular grid with a spatial resolution of 1/12° (approximately 8 km) and at 50 standard levels. The daily and monthly average values of the zonal (u) and meridional (v) components of the surface flow velocity were used for the entire available period from 1993 to 2021.

ETOPO1

The ETOPO1 Global Relief Model combines topography, bathymetry, and shoreline data from regional and global datasets for a comprehensive display of information in high resolution. ETOPO1 is available on the website of the American division of the National Oceanic and Atmospheric Administration (NOAA) (https://www.ncei.noaa.gov/products/etopo-global-relief-model). The data is linked vertically to sea level, and horizontally to the World Geodetic System of 1984 (WGS 84). The cell size is one angular minute. In this work, a set of bathymetric information displayed on a spatial coordinate grid with a step of 500 m was used.

Effective β-effect

Rossby waves are classified as gradient-vortex waves, and their propagation is carried out according to the law of conservation of a potential vortex:

$$\frac{d_h}{dt}\left(\frac{\zeta + f}{H}\right) = 0$$

where $\frac{d_h}{dt} = \frac{\partial}{\partial x} + U\frac{\partial}{\partial x} + V\frac{\partial}{\partial y}$ x and y are the zonal and meridional variables in a rectangular coordinate system, t is time, U and V are zonal and meridional components of the velocity, f is the Coriolis parameter, H is the depth. On the shelf, potential vorticity occurs due to an increase in depth in the direction perpendicular to the large-scale topographic irregularity. We accept the designations adopted in the monograph [9], and assume that the depth increases from the ocean surface to the bottom.

Let us apply the theory described in the work [8] on the calculation of the topographic effect of Rossby waves on the shear flow. We introduce the expression of the effective β-effect (β*):

$$\beta^* = \beta\cos(\theta) - U_{yy} - \frac{fH_y}{H} + \frac{U_y H_y}{H} \tag{1}$$

Here f is the Coriolis parameter, H is the depth, β = df/dy, k and l are zonal and meridional wave numbers, U is the zonal component of the flow velocity. Expression (1) is valid for the area where the Alaskan Stream and the Aleutian Trench have a rotation relative to the original coordinate system – coordinate systems 1 and 3 by an angle θ (Fig. 1). For the area where the Stream is zonal (coordinate system 2 – Fig. 1) we use a simplified expression. The individual terms of Eq. (1) will describe various mechanisms affecting the propagation of mesoscale eddies of the studied region: the β-effect, the meridional gradient of the zonal velocity shear, the topographic factor and the combined action of the flow and topography.

3 Results and Discussion

Analysis of mesoscale eddies trajectories of the Aleutian Trench

Let us consider the trajectories of long-lived mesoscale anticyclonic and cyclonic eddies formed in the area of the Aleutian Ridge during the period 1993–2021. Eddies formed to the south of the Aleutian Ridge drift mainly to the west along the islands of the archipelago, moreover, individual anticyclonic eddies migrate to the southwest, crossing the Trench, which is observed in almost the entire considered area from 170° E to 170° W (Fig. 1). For cyclonic eddies, it is characteristic the distribution is zonal to the west along the isobaths. There is a separate cyclonic eddy that reached the Commander Islands (in the area of 162° E) and migrated south along the Kuril-Kamchatka Trench. Only one cyclonic eddy breaks away from the isobaths and crosses the northern part of the Imperial Seamount chain in the area of 170° E, while almost all anticyclonic eddies that broke away from the isobaths and flow west of 173° E crossed the Imperial Seamount chain. The direction of the eddies formed to the north of the Aleutian Ridge is determined mainly by the eastern direction. There are also anticyclonic and cyclonic eddies directed to the north, especially near the Amchitka and Middle Straits.

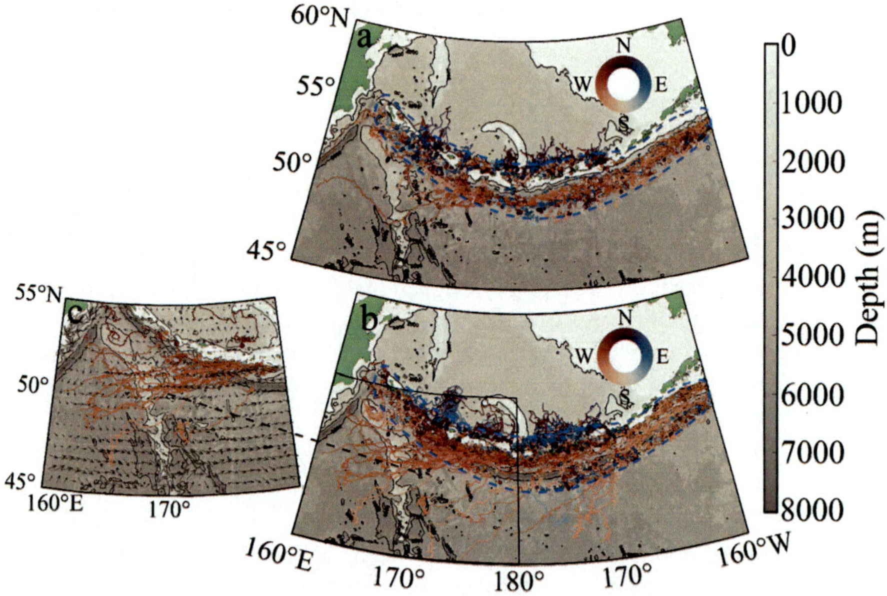

Fig. 2. Trajectories of mesoscale a) cyclonic, b) anticyclonic eddies and c) the place of separation of anticyclonic eddies from isobaths according to META3 data.2 DT for the period 1993–2021. The points mark the places where the eddies are generated, the color indicates the direction of the eddies. The isobaths are shown with black lines, the flow velocities on the surface are plotted with black arrows according to GLORYS12V1 data. The blue dotted line shows the boundary of the selected eddies

Anticyclonic eddy trajectories of that have traveled more than 300 km from the moment of origin shows that most of them cross the chain of the Underwater Imperial Mountains and only a part of them reaches the Kuril-Kamchatka Trench (Fig. 3). Two separate eddies were selected for consideration, characterizing the most typical trajectories for long-lived anticyclonic eddies. Thus, an anticyclonic eddy with the serial number 154217 was formed between the Near and Rat islands in the area of 52° N 175.5° E. After its origin, it crosses the Trench line in the area of 175° E, following the southwest in accordance with the direction of the current on the surface, overcomes the chain of the Imperial seamounts at 50° N 167.7° E and dissects to the west of them at 49.5° N 162.5° E.

The eddy with the serial number 19917, was formed to the east – in the area of 52° N 178° W. to the south of Amchitka Island. The eddy describes a similar trajectory: after its origin, it drifts westward in the direction of the current at a depth of 541 m. It crosses the Aleutian Trench and dissipates in the area of 51.5° N 171° E. The contour of the eddy on the day under consideration (the position of the center of the eddy is marked with a red dot in Fig. 3b is visible in the velocity field, at the periphery its velocities reach 0.3–0.45 m/s, while the flow velocity at a depth of 541 m for the rest of the water area does not exceed 0.1–0.2 m/s.

The amount of the mesoscale eddies and the number of mesoscale eddies generation.

The largest number of anticyclones (from 80 to 100 eddies per cell) is observed south of the Near and Rat Islands approximately between 173° – 180° E, as well as south of Unimak Island in the area of 163° W (Fig. 4). Cyclonic eddies in the amount of 70 to 80 per cell are located south of Rat Islands in spatial coordinates 170° – 162° E. Moreover, it is noted that anticyclonic eddies are localized in the shelf zone closer to the islands of the Aleutian Archipelago, and cyclonic ones are located in a narrow strip along the line of the Aleutian Trench.

The maximum number of anticyclonic eddies generation is 40–50 per cell and is fixed between the Near and Rat Islands (176° E) and south of Tanaga Island (178° W). Also, a large number of anticyclones (from 30 to 40 per cell) is formed at Rat Islands between 177° – 180° E and Atka Island in the area of 175° W. The generation of cyclonic eddies is less active, the maximum number of generations reaches 20 eddies per cell and is located to the east of Agattu Island, approximately at 52° N 175° E.

Evaluation of the terms in the effective β-effect.

Figure 5 shows three components of the generalized β-effect: the meridional gradient of the zonal velocity shear, the topographic factor and the combined action of the flow and topography, the module of which was divided by $\beta cos\theta$. It can be seen that the greatest influence of the flow can be traced in the area of the passage of the Alaskan Stream – south of the Aleutian Ridge from the Gulf of Alaska to the central part of the ridge. However, the effect of the flow probably does not have such a strong effect on the separation of eddies in comparison with the other parameters considered, since its effect is local: only small zones of predominance of the meridional gradient of the zonal velocity shear over the β-effect are distinguished. This may be due to the use of the average values of the zonal velocity component for the considered period from 1991 to 2021, and not for a specific day, as shown in Fig. 3. When comparing the topographic factor with the β-effect, small areas are distinguished where the topographic factor and the β-effect are comparable, which indicates the possibility of eddies in this part to cross isobaths. For the greater part of the water area (except for the deep-water zone of the Bering Sea), the topographic factor is an order of magnitude higher than $\beta cos\theta$ due to the significant heterogeneity of the seabed relief in the region under consideration. The combined contribution of topography and flow probably has little effect on the propagation of mesoscale eddies in the region, the value of the fourth term in the effective β-effect exceeds $\beta cos\theta$ only in small areas. However, there is an area south of the Aleutian Ridge between 171° – 176° E, where there is a weakening of the combined action of the current and topography. In this part, anticyclonic eddies "break away" from the current and topography, as can be seen from Fig. 2a, and follow the southwest.

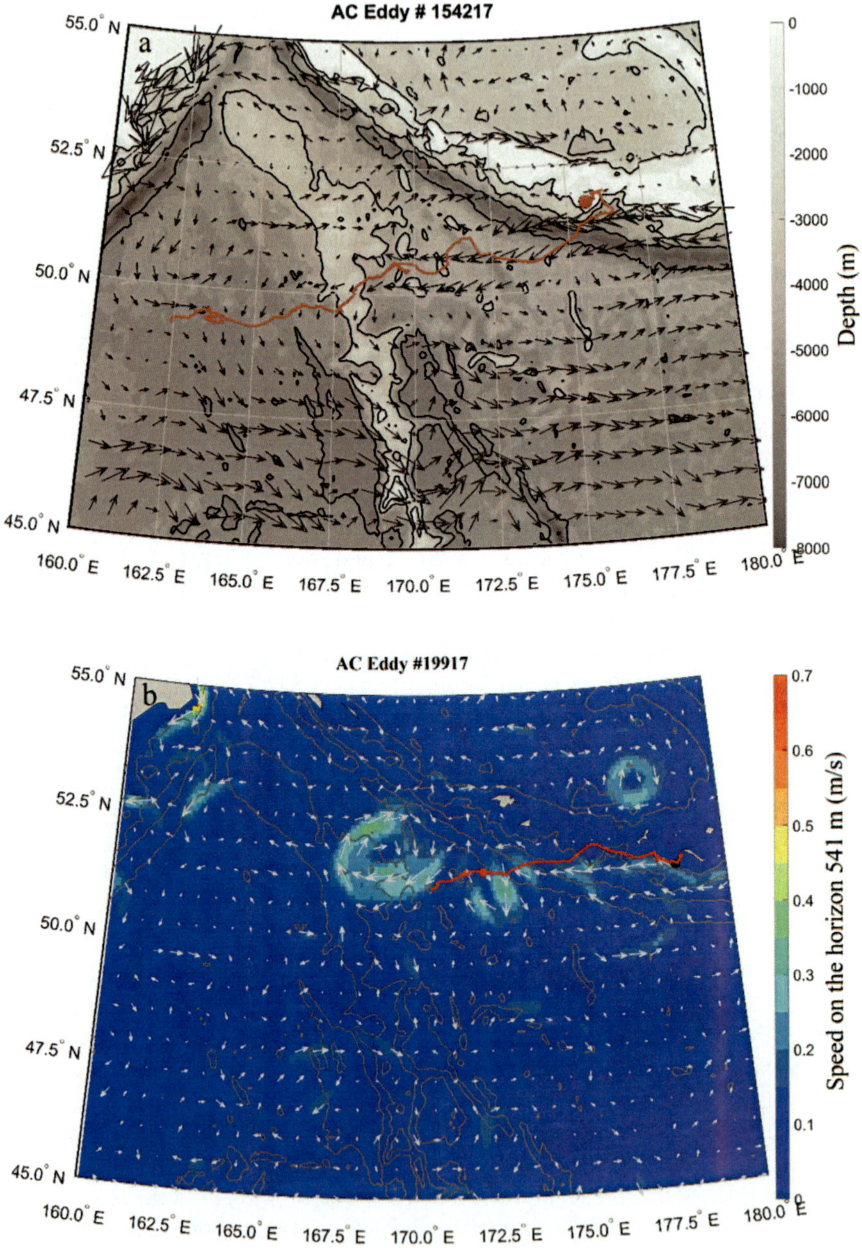

Fig. 3. Trajectories of individual long-lived anticyclonic eddies that have traveled a distance of more than 300 km from the day of formation. The red lines represent the trajectories of eddies according to META 3.2 DT data. a) Eddy 154217. A red dot marks the place of generation. The isobaths are shown with black lines, the flow velocities on the surface are plotted with black arrows according to GLORYS12V1 data. b) Eddy 19917. A black dot marks the place of generation, a red dot marks the position of the eddy on the day in question. Gray lines represent isobaths, gray arrows represent the flow velocity at the horizon of 541 m according to GLORYS12V1

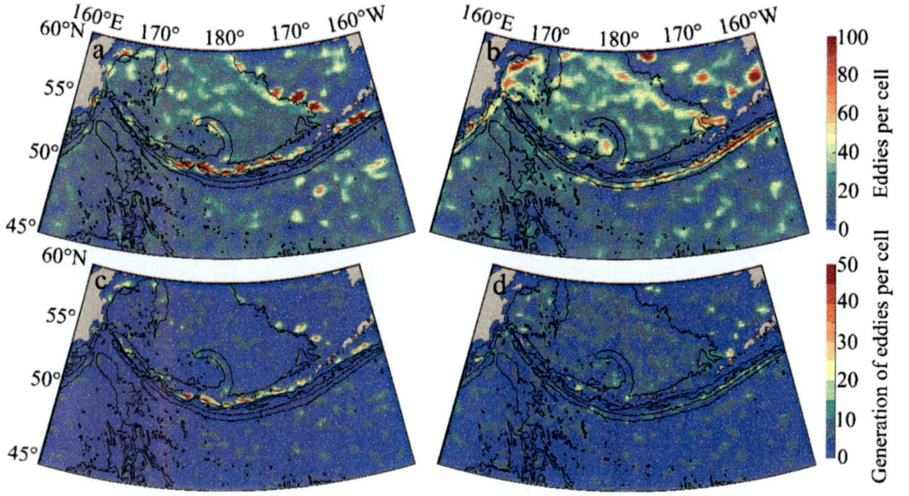

Fig. 4. The amount of mesoscale a) anticyclonic, b) cyclonic eddies and the number of generations of mesoscale c) anticyclonic, d) cyclonic eddies per cell (0.25° latitude by 0.5° longitude) according to META3.2 DT for the period 1993–2021. Isobates are displayed in black

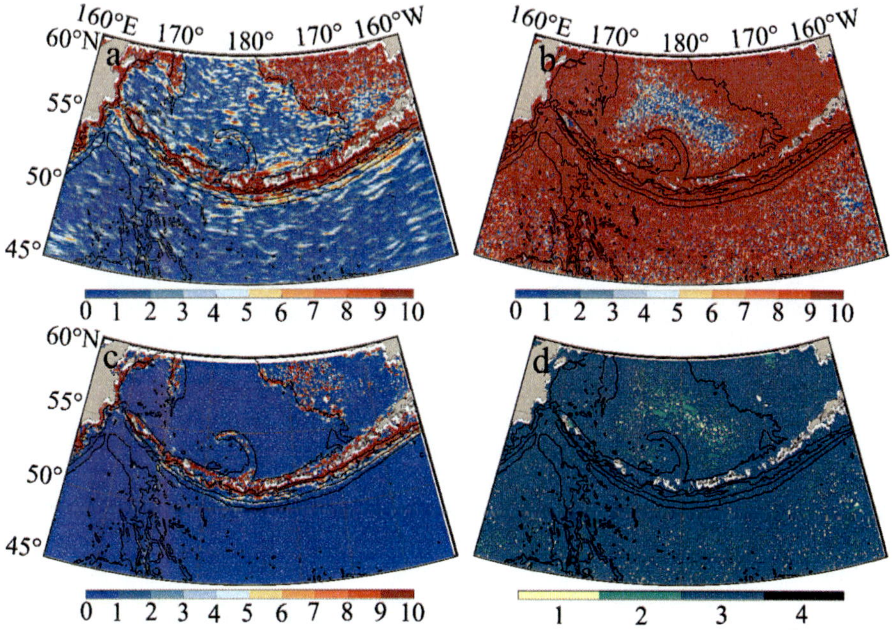

Fig. 5. Terms of the effective β-effect modulo: a) U_{yy}, b) fH_y/H, c) U_yH_y/H divided by $\beta cos\theta$. c) The dominant term. The values are normalized to $\beta cos\theta$

4 Conclusion

The propagation of mesoscale eddies after separation from isobaths is determined mainly by the western, and in some cases by the south-western direction. Using the example of the individual long-lived anticyclonic eddies trajectories, it is shown that the separation of eddies from isobaths occurs under the action of a current.

The characteristic distribution of anticyclonic and cyclonic eddies in various sections of the topography of the studied area is described. It is noted that anticyclonic eddies propagate northward along the Trench at depths of no more than 1–2 km along the shelf. Cyclonic eddies move westward along the central part of the Trench at depths from 5 to 7 km, spreading in the form of Trench waves.

It is found that more anticyclonic eddies are formed in the area of the Aleutian Ridge than cyclonic ones, and the maximum number of anticyclonic eddies per cell 0.25° in latitude by 0.5° in longitude is observed south of the islands of Near and Rat. The generation of cyclonic eddies is less active, the maximum number of generations is observed to the east of Agattu Island.

The predominance of the topographic factor over the β-effect in certain areas of the water area is noted, which indicates the tendency of eddies to leave the isobaths and follow the southern, south-western and other directions. The meridional gradient of the zonal velocity shear and the combined action of the flow and topography have a lesser effect on the propagation of eddies in the region in comparison with the factor determining the action of topography. It is possible that such estimates were obtained due to the calculation of the meridional gradient of the zonal velocity shear not by daily values, but by the average for the entire period under consideration from 1991 to 2021. An area of weakening of the combined action of the flow and topography is highlighted, in which anticyclonic eddies cease to follow isobaths.

Acknowledgements. This research was funded by Russian Science Foundation, grant number 22-27-00004, St. Petersburg State University, grant number 94033410.

References

1. Andreev, A.G.: Aleutian eddies and their impact on temperature and dissolved oxygen in the Western Subarctic Pacific. J. Oceanological Res. **48**(3), 109–122 (2020)
2. Andreev, A.G.: Water Circulation in the North-Western Bering Sea Studied by Satellite Data. Issledovanie Zemli iz Kosmosa (4), 40–47 (2019)
3. Belonenko, T.V., Kubryakov, A.A.: Temporal variability of the phase velocity of Rossby waves in the North Pacific. Sovrem. Probl. Distantsionnogo Zondirovaniya Zemli Kosmosa **11**(3), 9–18 (2014)
4. Budyansky, M.V., Prants, S.V., Uleysky, M.Y.: Odyssey of Aleutian eddies. Ocean Dyn. **72**(6), 455–476 (2022)
5. Chelton, D.B., et al.: Geographical variability of the first baroclinic Rossby radius of deformation. J. Phys. Oceanogr. **28**(3), 433–460 (1998)
6. Chelton, D.B., Schlax, M.G., Samelson, R.M.: Global observations of nonlinear mesoscale eddies. Prog. Oceanogr. **91**(2), 167–216 (2011)
7. Efimov, V.V. (ed.): Waves in the border areas of the ocean. Hydrometeoizdat (1985)

8. Gnevyshev, V.V., Frolova, A.V., Belonenko, T.V.: Topographic effect for Rossby waves on non-zonal shear flow. Water Resour. **49**(2), 240–248 (2022)
9. Mysak, L.A., LeBlond, P.H.: Waves in the Ocean. Elsevier, Amsterdam (1978)
10. Okkonen, S.R.: The shedding of an anticyclonic eddy from the Alaskan stream as observed by the Geosat altimeter. Geophys. Res. Lett. **19**(24), 2397–2400 (1992)
11. Prants, S.V., et al.: Lagrangian study of mesoscale circulation in the Alaskan Stream area and the eastern Bering Sea. Deep Sea Res. Part II: Topical Stud. Oceanogr. **169**, 104560 (2019)
12. Prants, S.V.: Trench eddies in the northwest pacific: an overview. Izvestiya, Atmos. Oceanic Phys. **57**, 341–353 (2021)
13. Saito, R., et al.: Subsurface hydrographic structures and the temporal variations of Aleutian eddies. Ocean Dyn. **66**(5), 605–621 (2016)
14. Strub, P.T., James, C.: Altimeter-derived surface circulation in the large–scale NE Pacific Gyres.: part 1. seasonal variability. Prog. Oceanogr. **53**(2–4), 163–183 (2002)
15. Travkin, V.S., Belonenko, T.V., Kochnev, A.V.: Topographic waves in the Kuril region. Sovrem. Probl. Distantsionnogo Zondirovaniya Zemli Kosmosa **19**(5), 222–234 (2022)
16. Ueno, H., Crawford, W.R., Onishi, H.: Impact of Alaskan stream eddies on chlorophyll distribution in the North Pacific. J. Oceanogr. **66**(3), 319–328 (2010)
17. Ueno, H., Freeland, H.J., Crawford, W.R., Onishi, H., Oka, E.: Anticyclonic eddies in the Alaskan Stream. J. Phys. Oceanogr. **39**(4), 934–951 (2009)

Probability Distributions for Finite Ensembles of Irregular Waves

A. S. Tregubov[1,2](✉), E. G. Didenkulova[1,2], and A. V. Kokorina[2]

[1] National Research University Higher School of Economics, Nizhny Novgorod 603014, Russia
andrey.tregubov2001@mail.ru

[2] Institute of Applied Physics RAS, Nizhny Novgorod 603024, Russia

Abstract. The data of wind waves in the area of Cape Svobodny in the Sea of Okhotsk have been obtained and processed. For comparison, numerical simulation of waves with similar parameters was carried out, and probability distribution functions for the appearance of waves of a certain magnitude were constructed. Particular attention was paid to the anomalously large waves that arise in experiments, the so-called "freak waves".

Keywords: Freak waves · Numerical simulation of wind waves · Long-term wave records · Wave spectrum

1 Introduction

Freak waves - the name of unexpected abnormally high waves in the sea [1–3]. Such waves, as a rule, arise suddenly for no apparent reason, and have a short lifetime from several minutes to tens of seconds. In oceanography, freak waves are more precisely defined as waves whose height is more than twice the significant wave height, which in turn is defined as the average of a third of the largest waves in the wave record. Therefore, freak waves are not necessarily the biggest waves on the water; rather, they are unusually large waves for a given state of the sea. Such waves pose a serious danger to ships, oil and gas platforms and coastal infrastructure therefore they are currently an object of study.

As a result of extensive observations, the statistics of freak waves are gradually accumulating, which makes it possible to study the mechanisms of their formation, their characteristic shape, lifetime, and the probability of their occurrence under different conditions. Numerous measurements of surface waves in various water areas demonstrate the high prevalence of freak waves in the World Ocean [4–8]. This work is devoted to numerical modeling of the wind wave field using two different wave spectra, obtaining their probability distributions, and analyzing long-term records of surface waves near Sakhalin Island.

The paper collected and analyzed the statistics of long-term wave records: model and natural ones, investigated the freak waves arising in them, and compared the probability of the appearance of such waves with theoretical values.

T. Chaplina (Ed.): CIWO 2023, SPEES, pp. 160–174, 2023.
https://doi.org/10.1007/978-3-031-47851-2_19

2 Material and Methods

In numerical experiments, a random wave field, as a rule, is specified using a superposition of spectral harmonics with random phases [2]:

$$\eta(x, t) = \sum_{i=1}^{N} a_i \cos(\omega_i t - k_i x + \varphi_i), \tag{1}$$

where spectral amplitudes $a_i = \sqrt{2S(\omega)\Delta\omega}$, $\Delta\omega$ is the sampling interval of the spectrum, $\Delta\omega = \frac{2\pi}{L}$, L is the record length, $L = N\Delta t$, $\Delta t = \frac{2\pi}{\Omega}$, N is the total number of harmonics, $\omega_i = i\Delta\omega$, $\Omega = \max(\omega_i)$, $k_i = \frac{\omega_i^2}{g}$ is dispersion ratio for deep water waves. The phases φ_i are uniformly distributed and are set using a random number generator ranging from 0 to 2π. $S(\omega)$ is the wave energy spectrum. The field contains approximately $N_w \approx \frac{\omega_0}{\Delta\omega}$ of individual waves, ω_0 is the abscissa of the peak value of the spectrum.

Each field is considered at the point $x = 0$. In this work, two spectra are used to model wave fields: the Pearson–Moskowitz spectrum derived from the real wind waves, and the model spectrum of Gauss.

To begin with, we will use the Pearson-Moskovitz spectrum, it is given by an equation with some empirical parameters:

$$S_{PM}(\omega) = \alpha g^2 \omega^{-5} e^{-\beta g^4 \omega^{-4}}, \tag{2}$$

where the parameter α is responsible for the height of the spectrum; the wind speed dependent parameter β is responsible for the peak value of the spectrum and its width; g is the free fall acceleration.

Significant wave height H_s in the record is defined as the average of one third of the largest waves in the record and is estimated as $H_s = 4\sigma$, where σ^2 is the dispersion of sea level fluctuations, which is the integral of the spectrum:

$$\sigma^2 = \int_0^\infty S(\omega) d\omega \tag{3}$$

Then we will carry out a similar numerical experiment using a Gaussian spectrum and compare the results. The shapes of the two presented spectra are shown in Fig. 1.

$$S_G(\omega) = Q e^{-\frac{(\omega-\omega_0)^2}{2K^2}}, \ Q = \frac{\sigma^2}{K\sqrt{2\pi}} \tag{4}$$

The parameter K is responsible for the width of the spectrum and was chosen to be $K = 0.04$. The choice of the peak frequency ω_0, equal to the peak frequency of the previous spectrum.

Further, according to Eq. (1), realizations of wave fields are considered. The values of Ω for the spectra are determined so that the sampling interval Δt allows each individual wave to be described by a sufficient number of points per period. Figure 2 shows realization built using the Pearson-Moskowitz spectrum for different Ω, from which it is obvious that small values of Ω do not allow a good description of each individual wave: small sampling intervals give "cut off" waves, which makes it impossible to estimate their real height.

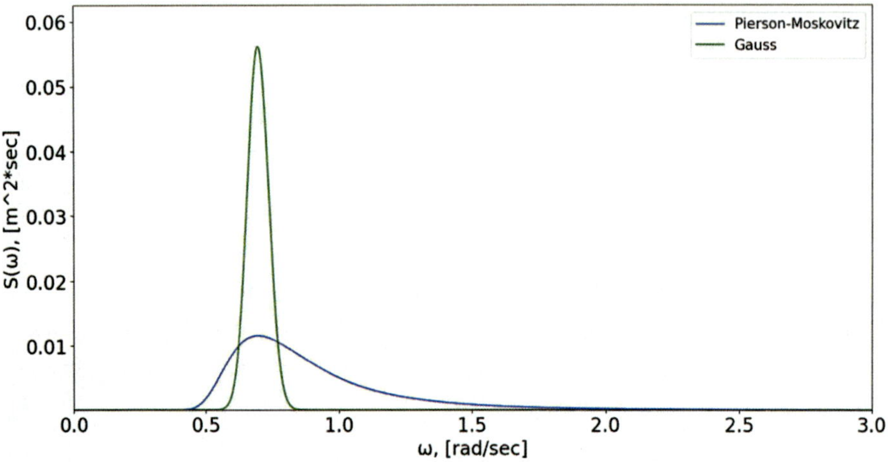

Fig. 1. Pearson-Moskowitz spectrum and Gaussian spectrum

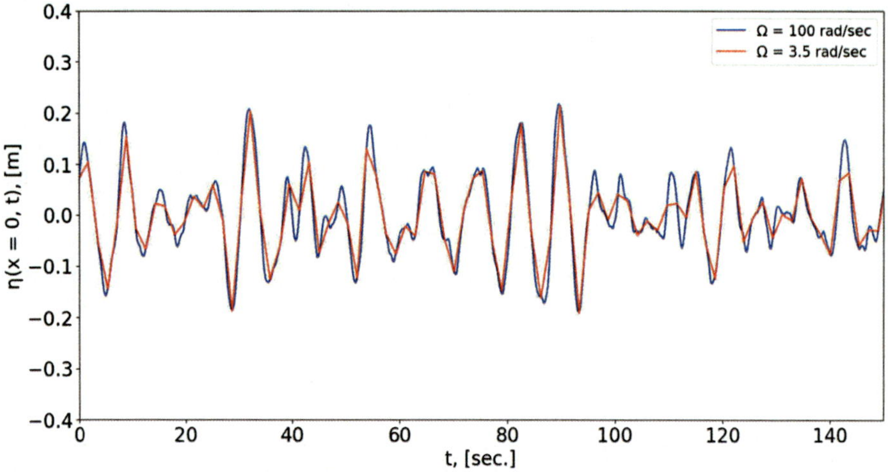

Fig. 2. Displacements of water surfaces realized by different number of points

One of the realizations of the wave field at the point $x = 0$ with each of spectra is shown in Figs. 3 and 4.

Individual waves in such records are determined by the zero-crossing method. For representative statistics of waves of different sizes and, in particular, freak waves, it is necessary to process a large number of individual waves. A wave is considered a freak wave if its height is $H > 2H_s$ according to the altitude criterion, or if its amplitude is $a > 2a_s$ according to the amplitude criterion (a_s is a significant wave amplitude and is determined similarly to a significant height, as the average of one third of the largest amplitudes).

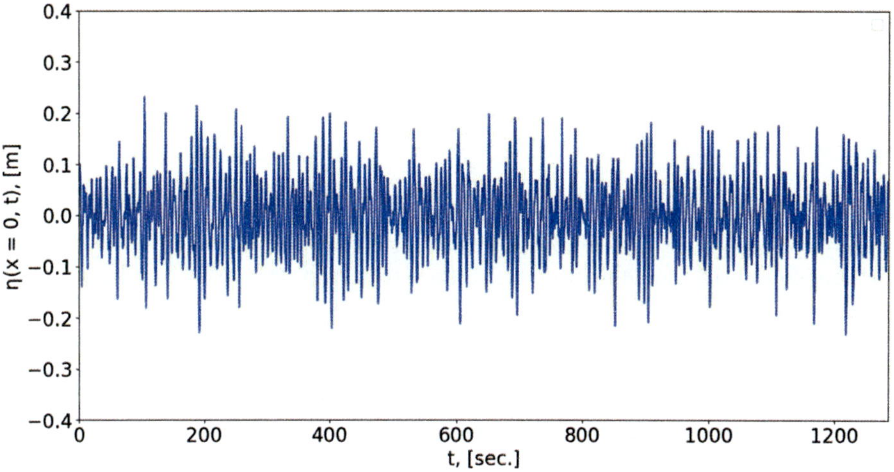

Fig. 3. Field of wind waves with Pearson-Moskowitz spectrum

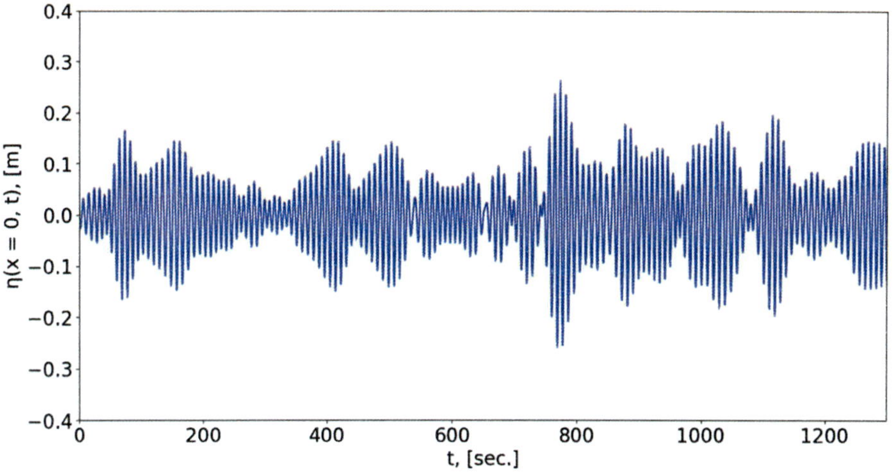

Fig. 4. Field of wind waves with a Gaussian spectrum

From the point of view of the calculation time of realizations, it is more convenient to consider a large number of realizations N_r with a small number of individual waves N_w in them. For example, one realization containing N harmonics requires N^2 operations, and two realization containing half as many harmonics require $2\cdot\left(\frac{N}{2}\right)^2 = \frac{N^2}{2}$ operations, and since $N_w \approx \frac{L}{L_0}$, $L_0 = \frac{\omega_0}{\Delta\omega}$, then the number of individual waves in one realization of size L and in two realizations of size $\frac{L}{2}$ is approximately the same.

We define the standard deviation and the average field for wave records as:

$$\sigma = \sqrt{\frac{1}{L}\int_0^L (\eta(t))^2 dt - \langle\eta\rangle^2}, \langle\eta\rangle = \frac{1}{L}\int_0^L \eta(t)dt, \tag{5}$$

As is clearly seen from Eq. (1), the displacement of the water surface at any time is a superposition of a large number of unrelated harmonic waves. Thus, each component of the sum is a random variable with a finite mean $\langle\eta\rangle$ and a finite nonzero standard deviation σ, all components of the sum are mutually independent, therefore, for a sufficiently large value of N, the conditions of the central limit theorem are satisfied and the displacement of the water surface is a random a process whose distribution density $f(\eta)$ coincides well with the normal distribution density. In order to choose the optimal value of N (and, consequently, N_r), graphs were plotted on a logarithmic scale (Fig. 5) of the normal distribution density in comparison with the distribution density of displacements η, normalized to the standard deviation σ. The density of the normal distribution in this case is given by the Eq. (6):

$$f(\eta) = \frac{1}{\sigma\sqrt{2\pi}} e^{-\frac{(\eta - \langle\eta\rangle)^2}{2\sigma^2}}, \qquad (6)$$

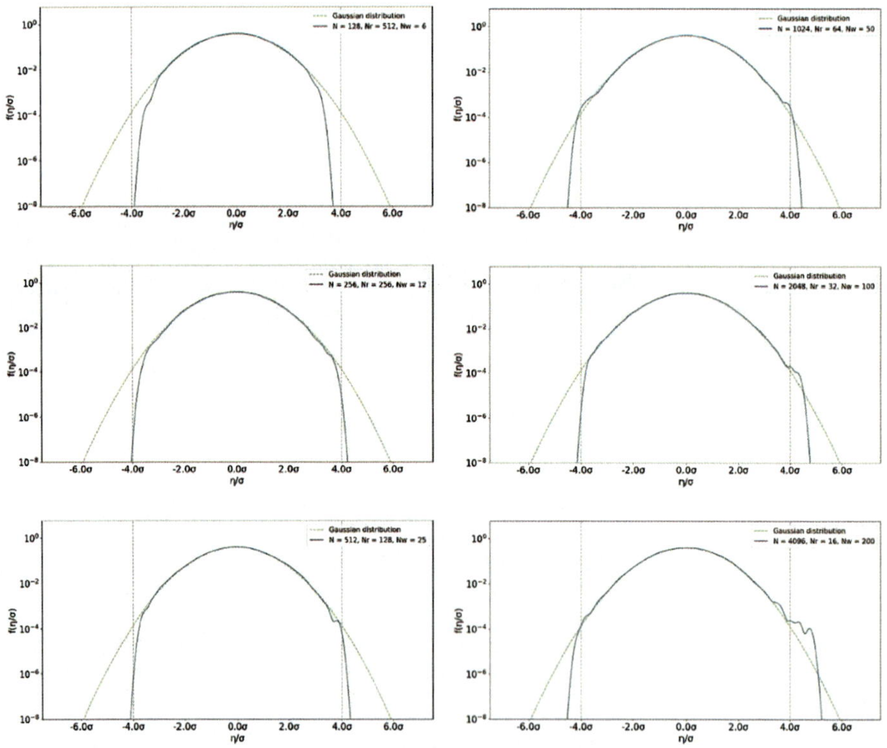

Fig. 5. Water surface distribution density depending on the number of realizations and harmonics

A good agreement between the distribution density and the Gaussian one in the entire area is already achieved when considering 16 realizations (containing a total of 3200 individual waves) obtained by a superposition of 4096 harmonics each. Further division into a larger number of realizations, apparently, does not allow one to describe displacements above 4σ, since the domain sizes become too small to obtain a freak wave in them (even with phasing of the maxima of all harmonics, it will not be possible to observe anomalously high waves in short recordings).

Next, we analyze the obtained wave fields with previously determined parameters from the point of view of the probability of occurrence of waves of certain amplitudes and heights in them. Figures 6, 7 and 8 show on a logarithmic scale the "tails" of the amplitude distribution functions of the wave fields realized by the Gaussian spectrum, averaged over the realizations and normalized to a_s. Since it is obviously impossible to obtain representative statistics with only one realization, we will increase the number of them. For all distributions, a comparison is made with the Rayleigh distribution:

$$F_r(a) = \exp\left[-2\frac{a^2}{(a_s)^2}\right], F_r(H) = \exp\left[-2\frac{H^2}{(H_s)^2}\right], \tag{7}$$

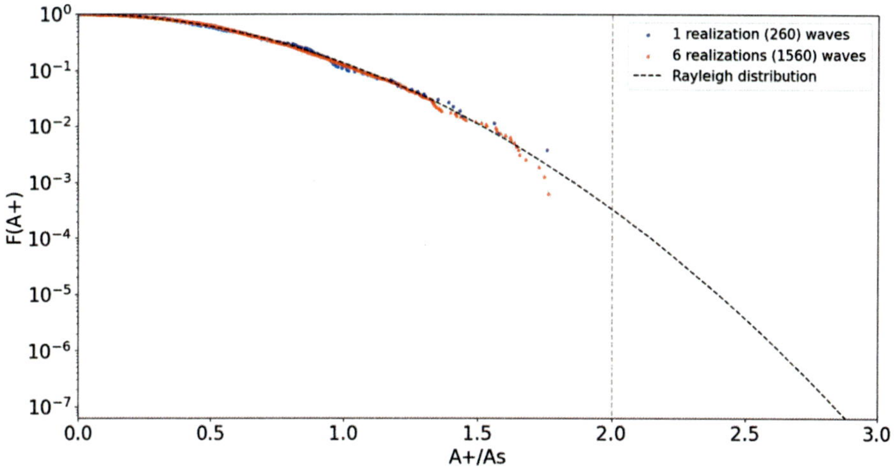

Fig. 6. Distribution functions of amplitudes, averaging over 1 and 6 realizations

When considering 520 thousand individual waves, we have representative statistics up to values of 2.2 A_s. There is also a slight overestimation by the Rayleigh distribution of the probability of the appearance of the largest waves.

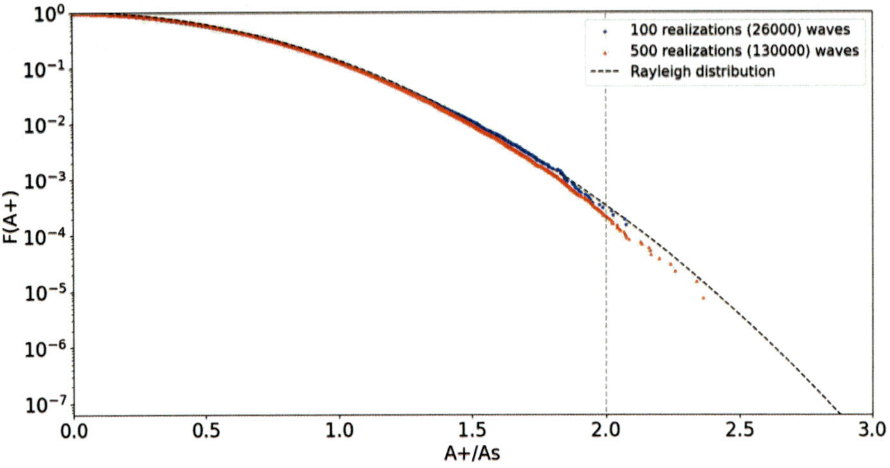

Fig. 7. Distribution functions of amplitudes, averaging over 100 and 500 realizations

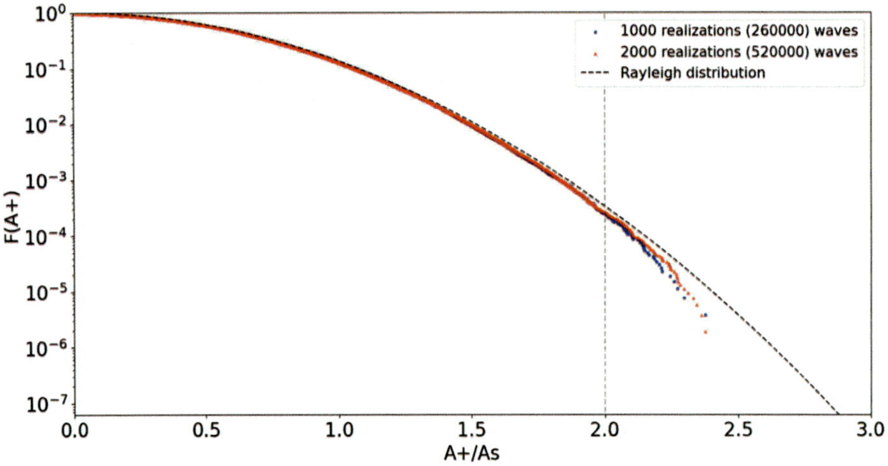

Fig. 8. Distribution functions of amplitudes, averaging over 1000 and 2000 realizations

Further on Figs. 9, 10, 11 and 12 on a linear and logarithmic scale, the distributions for wave fields in the case of the Pearson-Moskowitz spectrum are given, using the previously obtained parameters N_r and N_w, and they are compared with the case of the model spectrum and with the Rayleigh distribution.

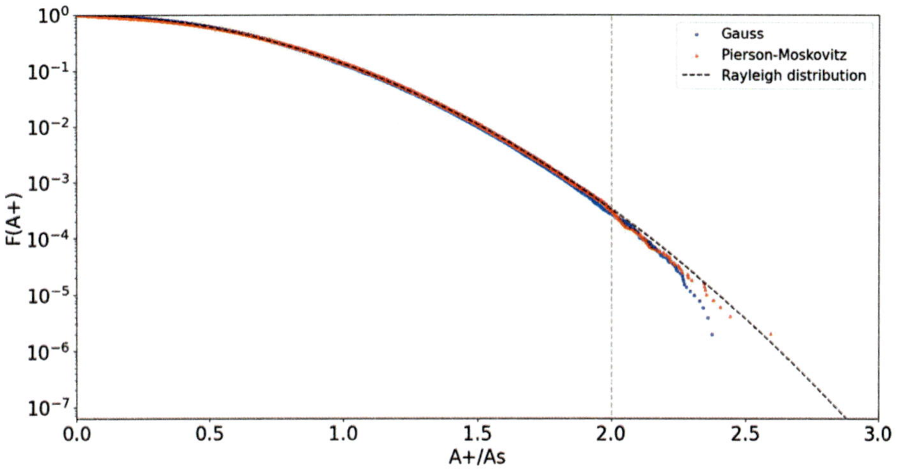

Fig. 9. Comparison of distribution functions of positive amplitudes of two spectra, logarithmic scale

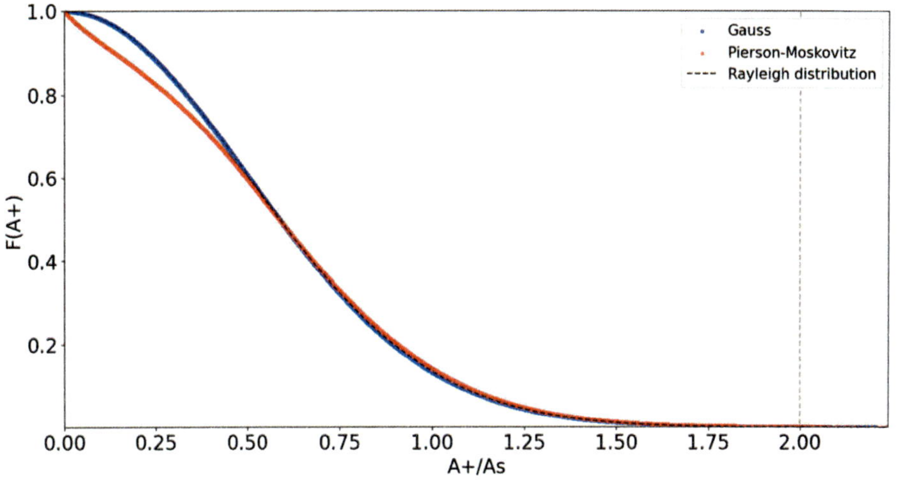

Fig. 10. Comparison of distribution functions of positive amplitudes of two spectra, linear scale

A similar behavior is observed for negative amplitudes.

Since the Pearson-Moskowitz spectrum is wide, the deviation from Rayleigh is significant: it is higher in the area of small amplitudes and in the area of high and medium heights.

The next stage of work is the processing of natural data. In this work, data obtained from September 2012 to May 2015 by autonomous bottom hydrostatic pressure recorders [9, 10] installed in the Sea of Okhotsk near Cape Svobodny. The devices measure bottom pressure fluctuations induced by surface waves.

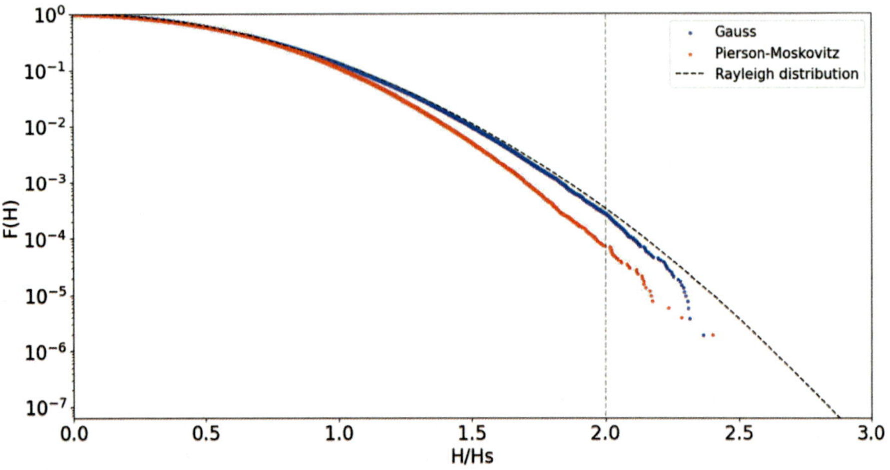

Fig. 11. Comparison of distribution functions of wave heights of two spectra, logarithmic scale

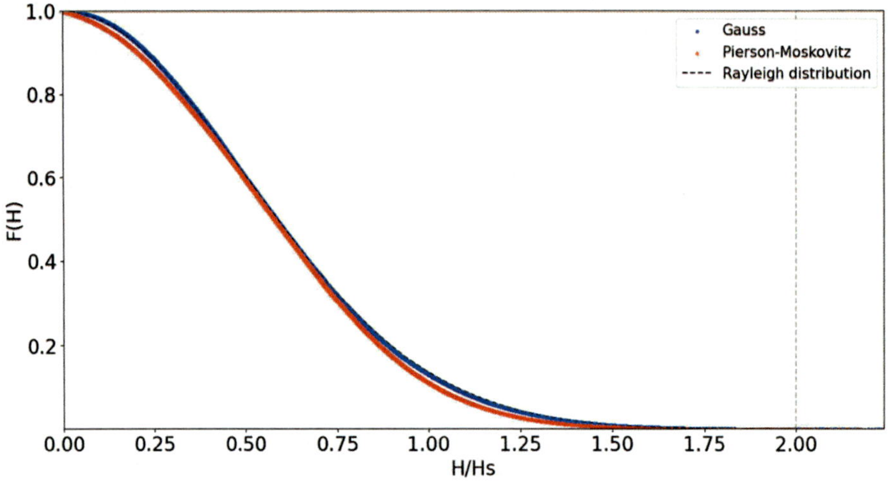

Fig. 12. Comparison of distribution functions of wave heights of two spectra, linear scale

The sensor immersion depth is $h = 9\ m$, measurement resolution is one second [10]. The wind waves are shallow, and the hydrostatic conversion formula can be used to approximate them:

$$p = p_{atm} + \rho g(h + \eta), \tag{8}$$

where p is the pressure on the sensor, p_{atm} is the atmospheric pressure, ρ is the density of water, and η is the displacement of the water surface. Under these conditions, the values of p_{atm} and h are not constant and can vary significantly.

Consequently, fluctuations in the level of atmospheric pressure and the level of immersion of the sensor h in the expansion of the displacement of the water surface in a Fourier series correspond to long harmonics with a large oscillation period T and a small frequency ω.

From the visualization of semi-annual record from September 2012 to March 2013 (Fig. 13) it can be noted that there is a significant heterogeneity: there are episodes of storm winds and periods in which water surface is covered with ice, fluctuations in the average level $\langle \eta \rangle$ of the sea surface are visible. Due to the relatively low recording frequency of the equipment, in some cases, when describing individual waves, "cut off" waves appear, the absolute value of the amplitude in this case can be underestimated.

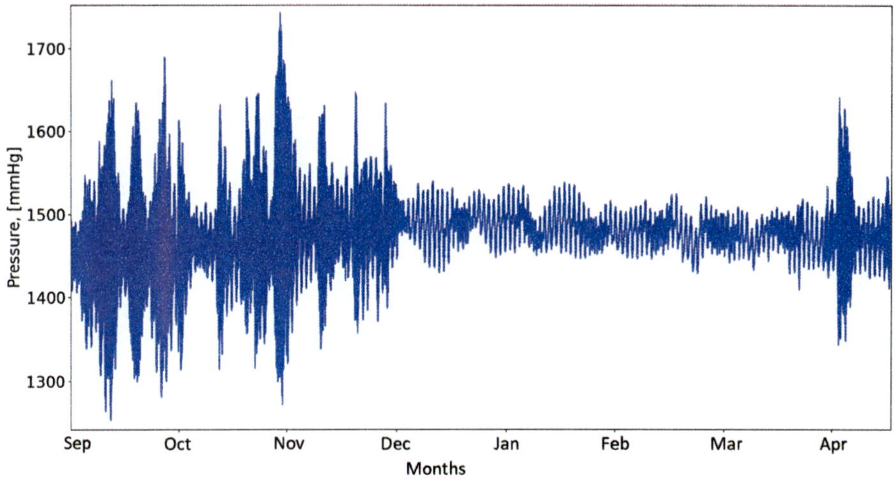

Fig. 13. Semi-annual wave record

After applying the hydrostatic scaling formula, we use the direct and inverse Fourier transform method to improve the quality of the data. We can find the frequency Fourier spectrum of each semi-annual record using FFT (Fast Fourier Transform), their shape is shown in Fig. 14.

It is necessary to remove from the spectra the components of the left part of the spectrum (waves with a period of more than 10 min); they correspond to tidal and seasonal fluctuations in the water level. Then we perform the inverse Fourier transform. We obtain records containing wind waves with an average level close to zero. Figure 15 shows a comparison of two recordings, in blue - the original, orange - after the removal of low-frequency components. This makes it possible to correctly determine each individual wave by the zero-crossing method. Next, apply the spline interpolation method (Cubic Spline) to improve the quality of the records, this method allows to describe waves with a large number of points per period by complementing the right side of the Fourier spectrum, thereby solving the problem of "cut off" waves. Figure 16 shows an example of a wave recording with and without spline interpolation.

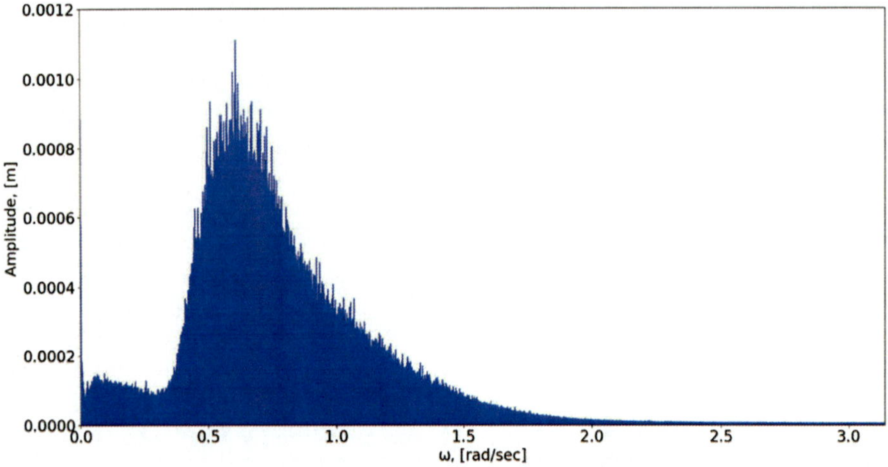

Fig. 14. Wave frequency spectrum of semi-annual records

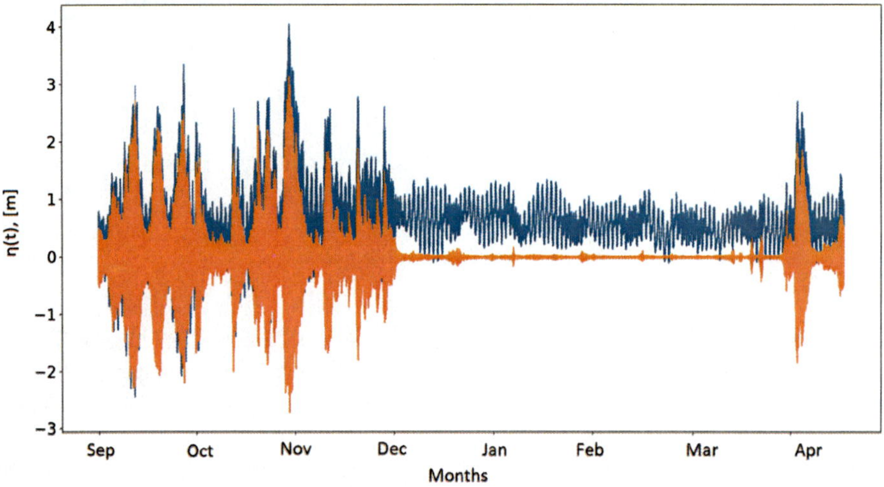

Fig. 15. Comparison of two recordings: before and after the removal of low-frequency components

Then, to solve the problem of heterogeneity, we divide the records into 20-min statistically homogeneous fragments, it is assumed that within such a period of time no qualitative changes in the nature of the wave can occur. In total, we have about 30 thousand records, each of which can now be classified. Then we calculate for each such record the practical values a_s^*, H_s^*, T_z. In this case, it is convenient to calculate significant amplitudes and heights through the standard deviation values calculated directly for each 20-min record: $a_s^* = 2\sigma^*$, $H_s^* = 4\sigma^*$. The zero-crossing period T_z is calculated by the formula:

$$T_z = \frac{T_r}{N_z}, \tag{9}$$

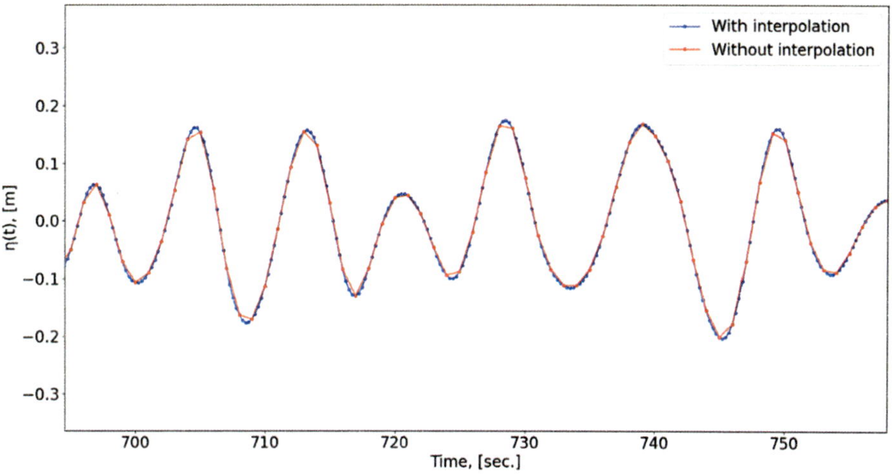

Fig. 16. Applying spline interpolation, approximating 20-min recording, the number of data points is increased by 4 times

where T_r is the length of the record, and N_z is the number of zero crossings in it. Now we can carry out the selection of records by wave modes, based on the values of H_s^* and T_z in each of them (Fig. 17). The numbers in the cells are the number of records matching the parameters specified on the axes.

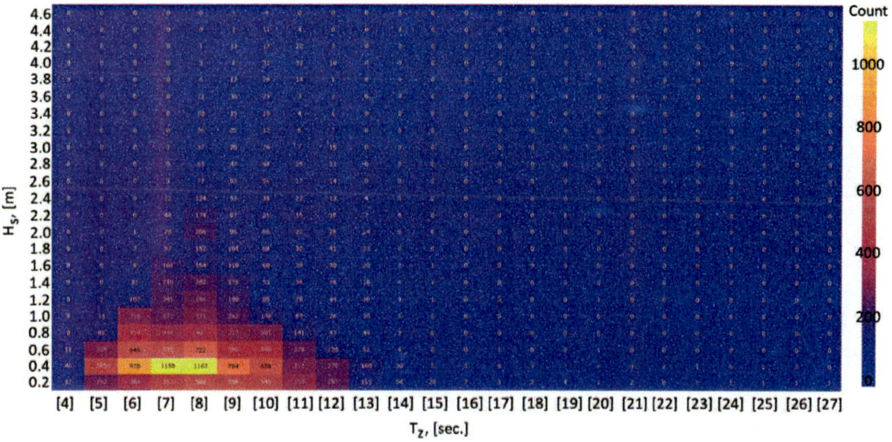

Fig. 17. Selection of records by wave modes

After the numerical simulation has been performed and nature data have been processed, one can begin to compare the results obtained. It is most interesting and visual to look at the statistics of the probability of occurrence of waves of various sizes (both for heights and for amplitudes), that is, to consider a comparison of distribution functions. On Figs. 18, 19 and 20 these functions are built on a logarithmic scale.

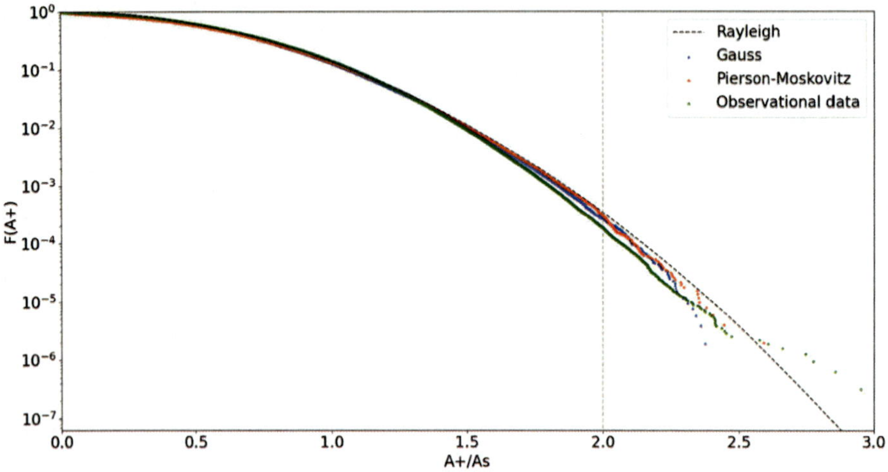

Fig. 18. Comparison of distribution functions of positive wave amplitudes of numerical experiments and field data, logarithmic scale

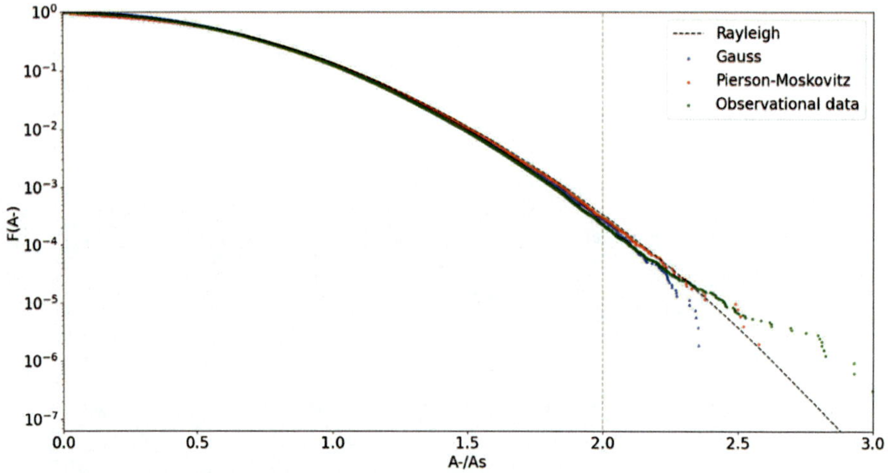

Fig. 19. Comparison of distribution functions of negative wave amplitudes of numerical experiments and field data, logarithmic scale

The "tails" of the distributions up to $2.2A_s$ lie below the Rayleigh and below the curves obtained as a result of computer simulation, but in the area above $2.2A_s$ there is a qualitative change in the upward behavior of the curve, as a result of which the set of points turns out to be above the Rayleigh.

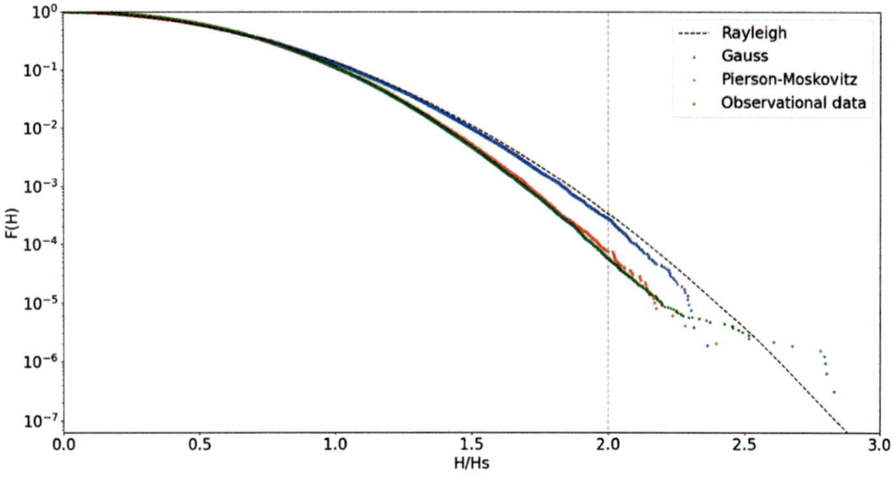

Fig. 20. Comparison of wave height distribution functions of numerical experiments and field data, logarithmic scale

3 Results and Discussion

In the course of the work, it was found that in order to obtain representative statistics of large waves, the optimal length of one realization is 200 individual waves, and about 2000 realizations are needed in total.

As part of a numerical experiment, when comparing the distributions obtained by both presented spectra, it was established how and where exactly the deviation from the Rayleigh curve occurs in the cases of a narrow and wide spectrum: a slight downward deviation in the area of large Amplitudes and heights in the case of a fairly narrow Gaussian spectrum and a significant deviation in the area of small amplitudes, high and medium heights in the case of the Pearson-Moskowitz spectrum.

As part of field measurements, about 3 million individual waves were obtained, 300 of which are killer waves, as a result of selection by wave modes, a characteristic wave height of about half a meter was obtained, and the characteristic period of zero crossing was 7–8 s.

Despite the difference in wave parameters between the simulation and the natural experiment (for example, the model considers the case of deep water and constant wind, and the measurements were carried out in shallow water with a strongly varying wind), when considering dimensionless values, one can note a good agreement between the natural experiment and the numerical, realized within the Pearson-Moskowitz spectrum, however, the proposed model is not able to describe the area of freak waves, which is especially interesting to us, and its estimates greatly underestimate the probability of the appearance of waves larger than $2.2H_s$ or A_s. This may be due to the linearity of the model, which does not take into account many real non-linear effects.

4 Conclusion

In this article, numerical modeling of wind wave fields with two spectra is carried out, model parameters are established, distribution functions for wave heights and amplitudes are constructed, and a comparison with the theoretical distribution is made. Also, a method for processing field data obtained from bottom pressure sensors is presented and a comparison of model data with field data is made.

Acknowledgements. This work was supported by the Russian Science Foundation (grant no. 22-17-00153).

References

1. Massel, S.R.: Ocean surface waves: their physics and prediction. Adv. Ser. Ocean Eng. **11**, 508 (1996)
2. Kurkin, A.A., Pelinovsky, E.N.: Rogue Waves: Facts, Theory and Modeling: Monograph, 2nd edn., p. 178. Direct-Media, Berlin (2015)
3. Kharif, C.H., Pelinovsky, E., Slunyaev, A.: Rogue Waves in the Ocean, p. 216. Springer, Cham (2009)
4. Mori, N., Liu, P., Yasuda, T.: Analysis of freak wave measurements in the Sea of Japan. Ocean Eng. **29**, 1399–1414 (2002)
5. Stansell, P.: Distributions of freak wave heights measured in the North Sea. Appl. Ocean Res. **26**, 35–48 (2004)
6. Christou, M., Ewans, K.: Field measurements of rogue water waves. J. Phys. Oceanogr. **44**, 2317–2335 (2014)
7. Häfner, D., Gemmrich, J., Jochum, M.: Real-world rogue wave probabilities. Sci. Rep. **11**, 10084 (2021)
8. Didenkulova, I., Anderson, C.: Freak waves of different types in the coastal zone of the Baltic Sea. Nat. Hazards Earth Syst. Sci. **10**, 2021–2029 (2010)
9. Zaitsev, A.I., Malashenko, A.E., Pelinovsky, E.N.: Abnormally large waves near the southern coast of Sakhalin. Fundam. Appl. Hydrophys. **4**(4), 35–43 (2011)
10. Kuznetsov, K.I., Zaitsev, A.I., Kostenko, I.S., Kurkin, A.A., Pelinovsky, E.N.: Observations of rogue waves in the coastal zone of Sakhalin. Ecol. Syst. Devices **2**, 33–40 (2014)

Model of the Schooner "Saint Anna" Drift with the Usage of NOAA-CIRES-DOE 20CR and ERA-20C Reanalises

S. V. Tsedrik(✉)

Saint Petersburg State University, Saint Petersburg, Russia
sofikotse@gmail.com

Abstract. "Saint Anna" disappeared in the Arctic Ocean in the middle of spring 1914 after two years drift with drifting ice. During this period the air pressure, daily mean air temperature, wind direction and velocity were measured, and now correlation analyze between schooner's data and data from 20th century reanalysis were made. Based on its results model of "Saint Anna's" movements after April, 1914 was suggested.

Keywords: "Saint Anna" · NOAA-CIRES-DOE 20CR · ERA-20C · sea ice drift · modelling

1 Introduction

"Saint Anna" with the captain G.L. Brusilov and 23 crew members left Saint-Petersburg on the 10th of August, 1912 with an aim to go through North-East passage. On 15th of September "Saint Anna" travelled through Yugorsky Schar and on the 17th went into Kara Sea, but in the middle of October of the same year the schooner was frozen in Kara Sea and pinned down to the shore of the Yamal peninsula on 71°45´N. On the 28th of October fast ice was damaged by south-east wind, and the schooner started its movement to the North with the drifting ice. After two winters spent in ocean, ship's steersman, V. I. Albanov, and thirteen crew members left schooner on the 23rd of April, because the vessel obviously wasn't able to get free from ice that year. They tried to reach the Franz-Yosef land, but only two of them, Albanov and one of the sailors, A.E. Konrad, stayed alive and were saved by the "Saint Foka"'s expedition and brought ship's documents to Saint-Petersburg. Despite the tragic end of the expedition, it has an enormous scientific value. During the voyage and the drift vessel's coordinates were determined, in the log book information about weather conditions and the state of the ice field surrounding the vessel was fixed, tables of meteorological parameters were filled in by daily triplicate determinations of air pressure and wind speed and direction and average daily air temperature counts and the results of depth measurements – daily depth measurements by lot – which made it possible to collect a significant array of data reflecting hydrometeorological conditions in the Kara sea and the Arctic ocean. For example, using the meteorology observations in Arctic Ocean and measurements of

T. Chaplina (Ed.): CIWO 2023, SPEES, pp. 175–180, 2023.
https://doi.org/10.1007/978-3-031-47851-2_20

the sea depth the Saint Anna's chute and Vize's Island were discovered. By the time of Albanov's leaving the drift had been lasted for 542 days with mean speed about 2,84 n.m. per day and drift the length of the drift had reached 1500 nautical miles [1, 2]. From Brusilov's log book we now have data about daily mean air temperature, air pressure and wind speed and direction, and also text descriptions of weather and sea ice conditions and can compare them to reanalysis data.

2 Materials and Methods

"Saint Anna"'s log book consists table of meteorological parameters – daily mean air temperature, air pressure and wind speed and direction for three terms - 7:00, 13:00 and 21:00 for each of drift days. For the entire duration of the drift array of 1731 pressure measurements and 576 daily mean temperatures had been collected [3].

For the correlation analyze two 20th century reanalysis, NOAA-CIRES-DOE 20CR and ERA-20C [8, 9], were used. NOAA-CIRES-DOE 20CR gives term, daily mean and monthly mean data for the period from 01.01.1836 to 31.12.2015 for 46 vertical levels. ERA-20C provides data for the same time intervals between 01.01.1900 and 31.12.2010 for 150 vertical levels. For both reanalysis nominal resolution is 1°x1°.

Scatter plots were used to identify if there is a correlation between reanalysis and Brusilov's data and Pirson's correlation coefficient as a linear correlation coefficient representing the mathematical expression of the statistical relationship between two quantities (1).

$$r = \frac{cov_{xy}}{\sigma_x \sigma_y} = \frac{\sum(x-\overline{x})(y-\overline{y})}{\sqrt{\sum(x-\overline{x})^2 \sum(y-\overline{y})^2}} \quad (1)$$

where $\overline{x}$, $\overline{y}$– mean values of the samples, σ_x, σ_y – mean square deviation.

Also, for both reanalysis samples mean absolute error and root mean square error regarding Brusilov's sample were calculated.

To detect correlation between vectors method from the paper [4] was used. For each pair of data samples set of four parameters was calculated. Two of them are correlation coefficients -r|| for characterizing collinear parts correlation and r_|_ for orthogonal parts. r|| varies from −1 to 1, with minimum when oppositely directed vectors are correlated and maximum in case of coincidence of vector directions. r_|_ also varies from −1 to 1 and is positive if one of the vectors is rotated predominantly to the right relative to the other, and negative if the rotation is predominantly left. Parameters should be analysed only together for correct interpretation.

The above indicators are supplemented by vector correlation indicator μ, which is determined by formula (2)

$$\mu = \sqrt{r||^2 + r\_|\_^2}, \quad (2)$$

that is quite convenient for because it allows select spatial areas (layers) in vector fields (or vertical profiles) with strong and weak correlations, as well as intermediate situations that subsequently analyzed using the first two indicators, and the angle α_μ, on which is necessary to rotate the coordinate system to achieve the greatest μ values. For more convenient calculations year was divided on two seasons – cold (October – April) and warm (May–September) (Table 1).

Table 1. Correlation coefficients between reanalysis' and Brusilov's data.

		r‖	r_⊥_	μ, °	α
Cold season	Brusilov - ERA	0.38	0.09	0.39	12.85
	Brusilov - 20CR	0.64	0.15	0.65	13.02
	ERA - 20CR	0.61	0.01	0.61	0.79
Warm season	Brusilov - ERA	0.32	0.07	0.33	13.11
	Brusilov - 20CR	0.55	0.11	0.56	11.63
	ERA - 20CR	0.62	−0.02	0.62	−1.62

Also, when analyzing vectors, additional indicators were calculated, including the magnitude and direction (clockwise from north) of the error vector (M_{bias} and D_{bias}), found as a vector difference between measured and predicted vectors, as well as the major and minor semiaxes ($\sqrt{\lambda_1}$ and $\sqrt{\lambda_2}$) and the direction of the major axis of the ellipse random error (α_1), limiting the area in which the end of the vector is located with 95% chance (Table 2).

Table 2. Additional indicators.

		M_{bias}, m/s	D_{bias}, °	$\sqrt{\lambda_1}$, m/s	$\sqrt{\lambda_2}$, m/s	α_1, °
Cold season	Brusilov- ERA	1.09	125.07	4.31	4.28	−56.41
	Brusilov- 20CR	0.43	63.95	2.78	2.58	−56.48
	ERA- 20CR	0.96	148.30	3.48	3.40	39.28
Warm season	Brusilov- ERA	0.22	47.36	4.42	3.91	85.48
	Brusilov- 20CR	0.58	−44.70	2.89	2.74	73.48
	ERA- 20CR	0.63	114.56	3.24	2.86	−81.83

For all coefficients significance levels were counted. During the analyse, it was found that all correlation coefficients are significant, because of sufficient length of data series for their determination.

3 Results

The obtained values of vector correlation indicators (0.64 and 0.55; 0.38 and 0.32, respectively, for the cold and warm months of the 20CR and ERA-20C reanalysis) indicate the possibility of using the 20CR reanalysis data to create a model of forced drift of an ice bound ship after departure of the Albanov group.

The drift model of the schooner "Saint Anna" is based on the ice wind drift formulas: the ship's drift speed was equal to the wind speed multiplied by the wind coefficient, the drift direction was determined by the wind direction, taking into account the wind drift deviation angle. The ship's drift trajectory was calculated using the Lagrangian description of hydrodynamics according to the explicit Euler scheme with daily discreteness. At each step, information from the 20CR atmospheric reanalysis was interpolated to the position of the ship, and the speed and direction of the ship's drift and the coordinates of the new position of the ship for the next time step were calculated from it. The initial conditions were the last known coordinates of the ship (82°55′30″ N 60°45′ E, 04/23/1914).

The model has a number of limitations: it does not take into account the internal interaction in the ice cover, the influence of coasts on ice drift, sea level slope and under-ice currents. But even such a rough model allows us to get an idea of the position of the ship after April 1914.

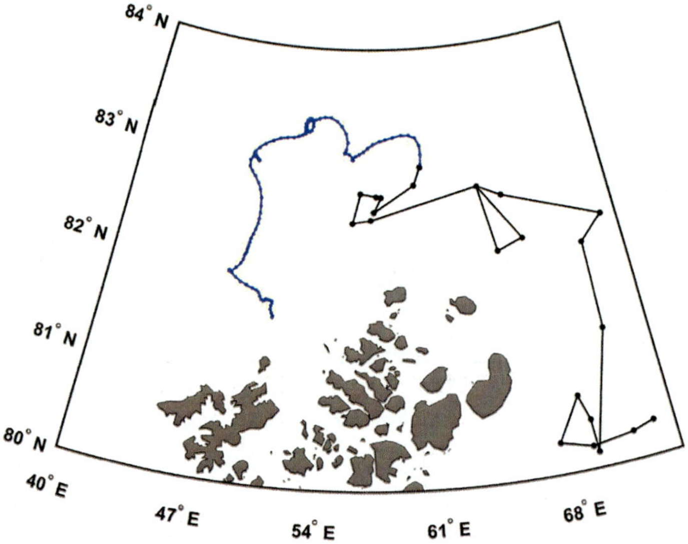

Fig. 1. Supposed "Saint Anna" drift after the Albanov's group leaving. Drift parameters by V.U. Vize [5, 7].

Figure 1 and 2 show the two implementations of the model with different drift parameters. On Fig. 1 wind coefficient and deviation angle calculated by V.Y. Vize in his works about "Saint Anna" drift and are same for the entire modelling period. Trajectory shown on Fig. 2, otherwise, was calculated based on coefficients and angles from Z.M. Gudkovich studies. Parameters varies from month to month, but, as it can be seen comparing the figures, two trajectories are practically not differed one from another. That's because month variations are tiny regarding to coefficient and angle quantities, so parameters can be constant.

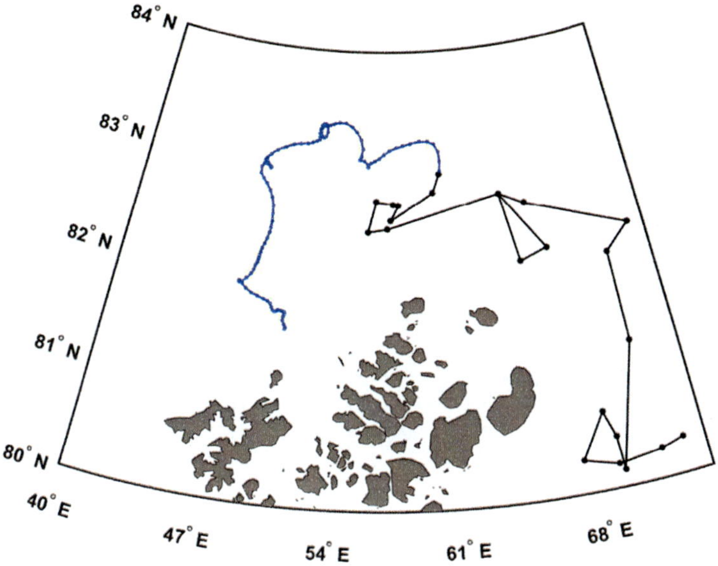

Fig. 2. Supposed "Saint Anna" drift after the Albanov's group leaving. Drift parameters by Z.M. Gudkovich [6].

According to the results obtained in the implementation of the model using the 20CR reanalysis, during the year after the departure of Albanov's group, the "Saint Anna" continued to move in a north-westerly direction along a broken trajectory until it reached 83°12'N. 55°39′ E (beginning of June 1914). Further, the drift trajectory was dominated by southern directions, the ship approached Franz Josef Land (mid-late June 1914). Thus, our study shows that, most likely, traces of the schooner should be sought to the north of Franz-Yosef Land.

4 Conclusion

Correlation analyzes showed that data collected during the "Saint Anna" drift is well in twentieth century reanalysis, especially in 20CR. This justifies the use of these data in drift modelling.

The drift model should be supplemented in the course of further work, but even such an easy model can be used for predicting the fate of the schooner and its crew after April, 1914.

The data obtained during the modeling can not only be used to study the ice drift of the Arctic Ocean, but also be of interest to historical science as material for reconstructing the future fate of the expedition.

Acknowledgements. This study was supported by the Russian Science Foundation (grant for Project No. 23-19-00039).

References

1. Savatugin, L.M., Dorozhkina, M.V.: And with the help of God, everything will be fine. Arctic Antarctic probl. **3**(92), 110–118 (2012)
2. Vize, V.U.: Seas of the Soviet Arctic: essays on the history of research. Leningrad, p. 496 (1936)
3. Breytfus, L.: The polar expedition of Lieutenant G. L. Brusilov on the schooner "St. Anna". Typography of the Maritime Department (1914)
4. Ivanov, N.E.: On the characteristics of the correlation of wind speeds, sea currents and ice drift. Meteorol. Hydrol. **8**, 61–73 (2004)
5. Vize, V.U.: On surface currents in the Kara Sea. Proc. Central Hydrometeorol. Bureau **3**, 86–101 (1924)
6. Vorobyov, V.N., Gudkovich, Z.M.: On intra-annual variability in ice drift and currents of the Arctic Basin. In: Proceedings of the AARI, vol. 303, pp. 76–88 (1971)
7. Vize, V.U.: On the issue of islands in the northern part of the Kara Sea. Res. Seas USSR **14**, 155–169 (1931)

Data

8. https://psl.noaa.gov/data/20thC_Rean/
9. https://www.ecmwf.int/

The Spatial and Temporal Distribution Patterns of Icebergs in the Barents Sea Based on Satellite Radar Data

T. S. Ustinova[1,2](✉)

[1] Nansen International Environmental and Remote Sensing Centre, Saint Petersburg 199034, Russia
tanya2001ustinova@gmail.com
[2] Marine Hydrophysical Institute of RAS, Sevastopol 299011, Russia

Abstract. In northern latitudes, icebergs frequently cross shipping routes and impair marine operations. To estimate the distribution of icebergs in the Barents Sea, this work explores the capabilities of an algorithm that detects and identifies icebergs based on sentinel-1 mission radar data. Synthetic Aperture Radar (SAR) is one of the main tools for studying the Arctic ice cover: it not dependent on daylight and largely independent of weather conditions. The applied algorithm uses the «Blob detection» method, based on searching for local bright areas of the image with subsequent classification using a Variable Selection Networks (VSN). The algorithm has shown its effectiveness in detecting icebergs on open water. Based on the obtained data, maps of iceberg density distribution in the Barents Sea for the ice-free period of 2021 were produced.

Keywords: Remote sensing · SAR · sea ice · icebergs · machine learning

1 Introduction

The observation of ice conditions is necessary for conducting safe and cost-effective maritime operations. The primary element of the iceberg management plan is the detection and subsequent mapping of the location of ice and icebergs, as this provides the fundamental basis for making all subsequent decisions, such as changing navigation routes or suspending operations.

Sources of information about icebergs have been filled by collecting data from aviation and shipborne ice reconnaissance. Multi-year data allows understanding the spatial distribution, assessing the risk of iceberg occurrence in a specific area, and tracking the variability of characteristics such as shape, size, and the southern boundary of ice formations [1]. However, information from such sources is heavily tied to travel routes and therefore uneven in time and space. Satellites are free from these drawbacks: operational Earth remote sensing data reflect the current state of the ice conditions, allowing timely adjustment of ship routes and preventing collisions with engineering structures. This work explores capabilities and limitations of an automated iceberg identification

T. Chaplina (Ed.): CIWO 2023, SPEES, pp. 181–190, 2023.
https://doi.org/10.1007/978-3-031-47851-2_21

algorithm based on SAR products. This work allows solving the most pressing problem in applied polar oceanography – the task of operational search for dangerous ice formations.

2 Data and Methods

Radar Data

Synthetic Aperture Radar (SAR) is one of the most important tools for monitoring ice in high latitudes due to its ability to provide images at any time of day, in cloudy or foggy conditions, and at different wind speeds. The Sentinel-1 mission is a group of two radar satellites in polar orbit equipped with a Synthetic Aperture Radar (SAR) for imaging in the C-band. The Extra-Wide Swath (EW) mode provides a spatial resolution of 40 m and a range of 400 km (250 miles). The mode is designed for observing sea ice, polar regions, monitoring oil spills, and security needs.

Optical Data

The joint use of SAR and optical images is necessary for creating a neural network training set and validating algorithm decryption results. The 10-m resolution allows for reliable identification of an iceberg and distinguishes it from other formations. Overall, a visible range image provides more information about the object but only in the absence of clouds and during daylight hours. Therefore, visible range images are only used for validation – confirming an object's membership in a particular class. The Sentinel-2A visible range has a spatial resolution of 10 m [2].

Blob Detection Method (Scikit-Image Library)

The blob detection method aims to identify areas on an image that differ in their properties, such as brightness or color, compared to adjacent areas. The family of methods [3] includes several ways that can be used to detect such areas:

1) Laplacian of Gaussian (LoG), which takes the Laplacian of a smoothed Gaussian image. The method has increased sensitivity and gives many false positives.
2) Determinant of Hessian (DoH), which takes the maximum value in the Hessian determinant matrix. It is characterized by low sensitivity and gives some artifacts along the edge of the image.
3) Difference of Gaussian (DoG), which takes the difference between two Gaussian smoothed images. The method provides a relatively more homogeneous picture and is used in this work.

Variable Selection Networks

Variable Selection Networks (VSN) is a machine learning technique used for classification tasks [4]. It aims to automatically select the most relevant features or variables from a given dataset to improve the accuracy and efficiency of the classification model. The utilization of VSNs enables the model to effectively eliminate any superfluous noisy inputs that may have a detrimental effect on performance. These techniques, when combined, aid in enhancing the ability of deep neural network models to learn and process information.

The backscatter from icebergs depends on both the spatial and temporal changes in the properties of the ice surface and the technical features of the radar. Due to the strong variability of ice properties, it is challenging to obtain a training dataset that can fully cover the diversity of its properties. Neural networks used for pattern recognition have the ability to generalize, which is an important advantage if the training dataset does not cover all possible states of objects. Another advantage of the neural network approach, unlike traditional classifiers, is its ability to solve problems based on noisy, distorted, internally inconsistent information [5].

3 Interpretation of SAR Images

Icebergs are often visible as bright points on a dark background in SAR images, but this is not always the case. There are three situations for iceberg detection: in open water, in fast ice and in drifting ice. In calm open water or undisturbed sea ice conditions, the contrast between iceberg and background signatures (reflection values) can be high enough [6]. Success of detection also depends on the spatial resolution of the radar and the size of the iceberg.

The resulting backscatter from the iceberg is influenced by both surface reflection and volume scattering [7]. The depth of penetration of the radar signal in the C-band (5.3 GHz) ranges from 3 to 14 m depending on the dielectric properties and volume structure of the ice (e.g., presence of air bubbles or other inclusions) [8, 9]. Thus, the

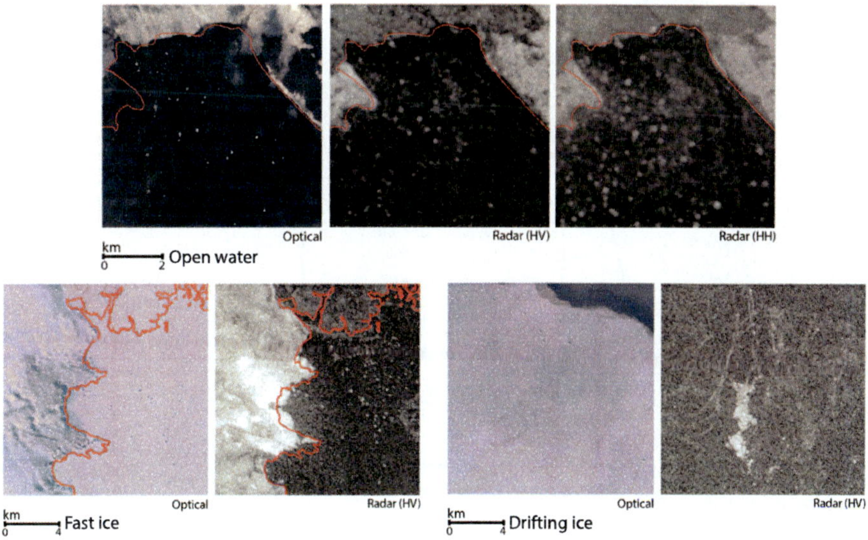

Fig. 1. Situations for iceberg detection

resulting brightness of the object and its probability of detection depend on the following factors [10–12]:

1. Orientation of the local surface of the iceberg relative to the angle of incidence of the radar beam;
2. Shape and texture of the surface, ratio of iceberg size to radar spatial resolution;
3. Presence of fresh water on the surface;
4. State of the sea or ice cover;
5. Air bubbles and other impurities inside the iceberg.

In high wind speeds and, as a result, rough surface conditions, the contrast between open water and icebergs in SAR images decreases [9, 13]. Figure 1 illustrate all three situations for iceberg detection.

4 Iceberg Detection Algorithm

The task of identifying objects in an image can be divided into two main subtasks: object detection and classification. The algorithm (Fig. 2) is based on the fact that the desired objects have increased brightness compared to the surrounding space in radar data. In this case, the task is reduced to the classical computer vision problem of determining objects based on intensity thresholds. The task is implemented using «blob detection»

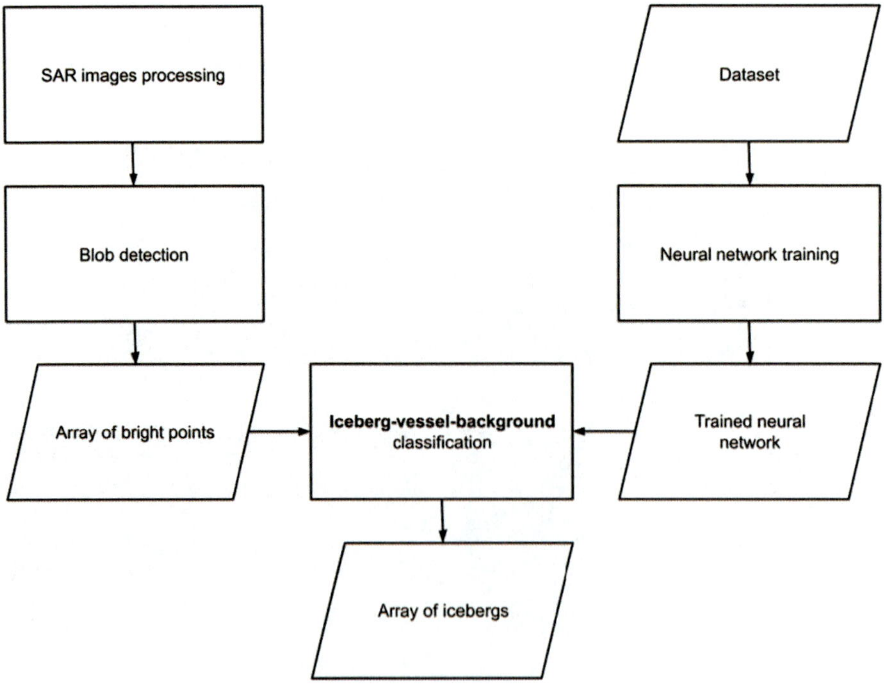

Fig. 2. Block diagram of the iceberg detection algorithm

operations – a stage of finding an array of areas that are highly likely to contain iceberg or vessel features. This approach allows for much faster image processing.

The first branch of the algorithm includes processing the target images of the studied region: pre-processing, searching for «bright spots» – local areas in images, and forming an array of the regions of interest. The second branch is responsible for selecting the neural network model, data augmentation of the training dataset if necessary, and training the classifier model. These operations prepare for the classification stage of «vessel-iceberg-background», which outputs an array of points with probabilities of belonging to one of the three classes.

5 Training Sample Data

In machine learning tasks, the formation of a training sample data (dataset) plays an important role. In general, the dataset includes a set of object samples of each class. One of the main requirements for the program is universality, so object selection was carried out in different places. Each region, each glacier is characterized by typical sizes and shapes of icebergs, and accordingly, numerical features reflected in the dataset. The inability to exclude the «angular» effect entails distorting the dataset in other cases: the resulting signal from an iceberg depends not only on its morphological and dielectric features but also on the angle of incidence of the radar beam on its surface. The way out of the situation is to collect data at all angular ranges of the image. The sample of icebergs was collected manually and included 1500 icebergs, more than 800 fragments of ships, as well as 1500 fragments of the «open water» background, selected randomly. Having data on classes, coordinates and images, it is possible to programmatic collection of samples. An important part of the sample collection algorithm is the use of the same image processing pipeline as during classification, ensuring the absence of additional distortions due to different display parameter settings. Each class sample in the dataset includes:

- the name of the object class;
- fragments of satellite images, including the object itself and some neighborhood around it (Fig. 3, 4 and 5).

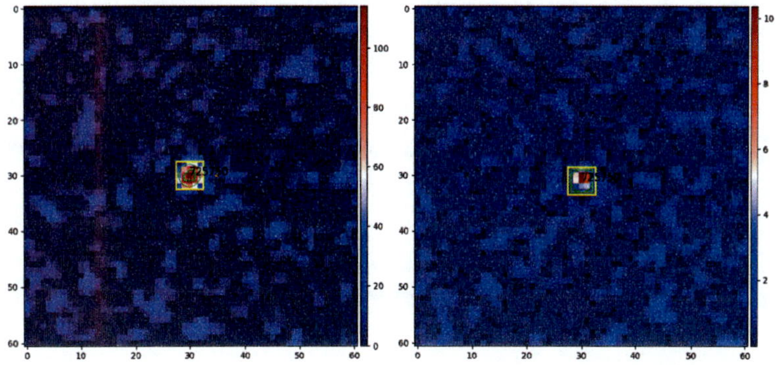

Fig. 3. Iceberg in open water. Fragment of the training sample (HH, HV)

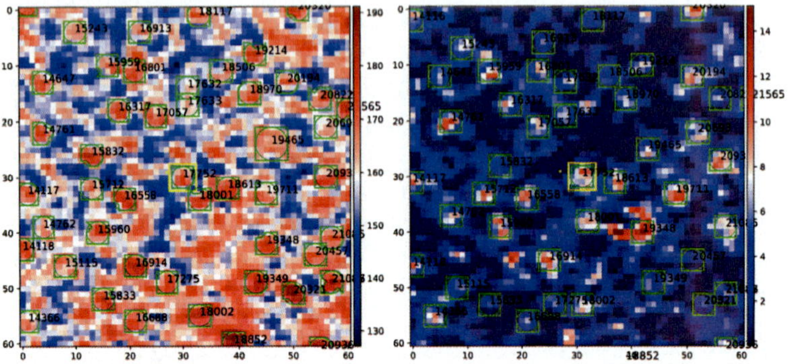

Fig. 4. Open water. Fragment of the training sample (HH, HV)

– calculated statistical characteristics of the image fragment.

Icebergs are bright spots or «points». The following characteristics can be distinguished for such objects (separately in HH and HV channels):

1) Object size (radius);
2) Maximum and average brightness of the «point» within the selected radius;
3) Maximum and average brightness of neighborhoods – areas around the point within 30 pixels;

Based on the above features, derivative characteristics can be calculated:

4) Contrast – the ratio of the average brightness of the point to the average brightness of neighborhoods;
5) Polarization ratio (HH/HV);
6) Noise – the number of adjacent bright «points» in neighborhoods;
7) Uniqueness – the number of neighboring objects that differ in other characteristics by no more than 20%.

These characteristics are received at the input of the neural network model.

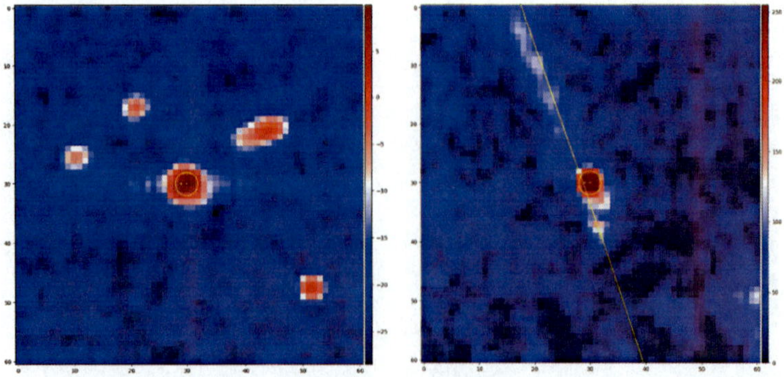

Fig. 5. «Prirazlomnaya» oil drilling platform (left). Vessel with a track line (right)

6 Iceberg Spatial Density Maps

Heatmaps are a valuable tool in geospatial analysis for displaying the density or intensity of events in data. They are particularly effective in identifying patterns and correlations in large datasets that may not be obvious from traditional visualization methods. The built-in «Heatmaps» module of the QGIS program was used to create the maps.

The estimates of the spatial distribution density of icebergs were carried out in the ice-free period: the number of objects refers to the area of the circle. The observed water area around the vessel's route, which depends on meteorological visibility conditions, was taken as circle's area. The value of maximum visibility distance, 10 nautical miles, was used for the mapping (circle area ~ 1000 km^2) (Fig. 6).

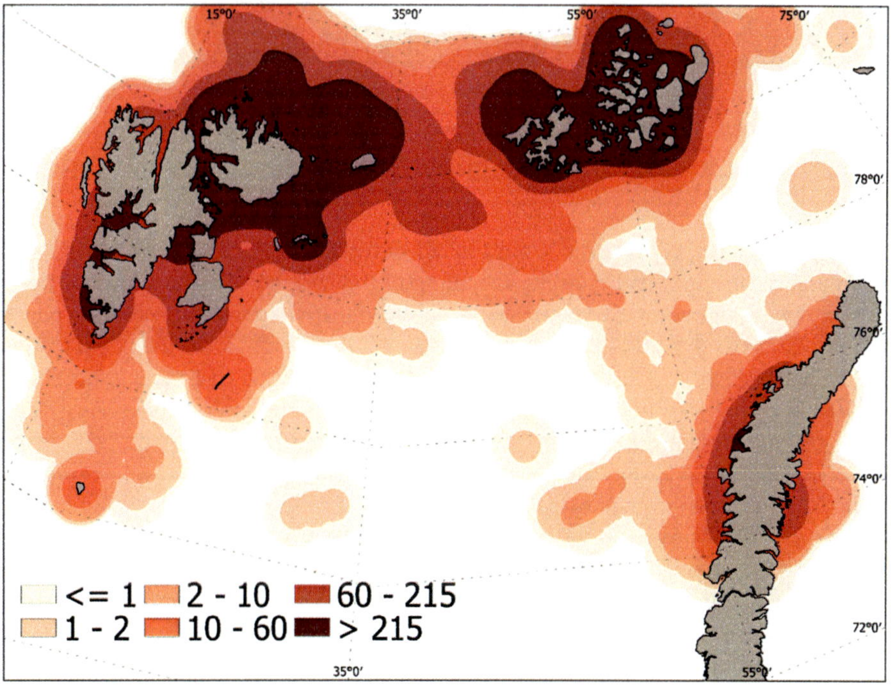

Fig. 6. Spatial density of icebergs (Number of icebergs/1000 km^2; Jul–Nov)

The highest concentration of icebergs is observed in the northern part of the Barents Sea, near the archipelagos of Svalbard, Franz Josef Land, and along the coasts of Novaya Zemlya. The density of clusters in glacier areas can indicate the level of iceberg productivity of the glaciers. Figure 7 shows the spatial density of distribution of objects classified as icebergs near glaciers. Thus, a relatively high concentration is observed in the following areas:

1) Franz Josef Land: Western part of Wilczek Island, Tindal Glacier; central and northern straits; southern part of Gallya Island and Mac-Clintock Island (Simoni Glacier in the RGI database); south of George Land.

2) Svalbard: Glaciers of Northeast Land: Austfonna and Braasvellbreen; Negribreen glacier. Objects classified as icebergs near King Charles Land were false positives – small islands were not covered by the coastline mask.
3) For the coast of Novaya Zemlya: near the Nordenskoid and Vize glaciers, Borzov and Chernyshev.

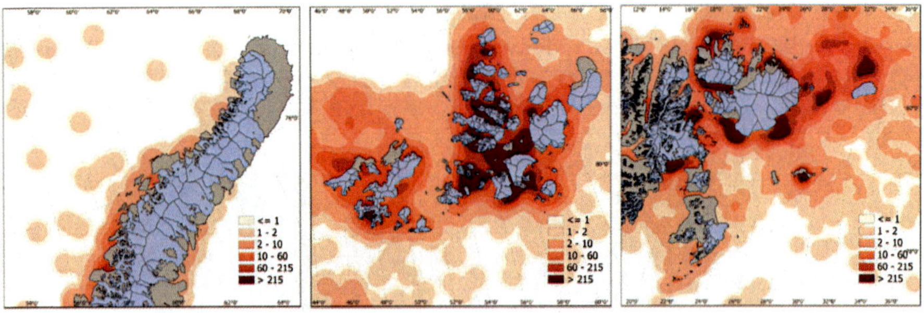

Fig. 7. Spatial density of icebergs near glaciers (Number of icebergs/1000 km^2; Jul-Nov)

The distribution of icebergs is not limited to these areas: icebergs can drift under the influence of winds and currents according to the scheme of the main water flows and spread throughout the Barents Sea area (Fig. 8). It is known that currents play a major role in the dynamics of iceberg movement. The character of water circulation in the Barents Sea is determined by many factors, including large-scale atmospheric circulation, inflow of water from neighboring seas, relief of the bottom, tides. As a result of the joint action of these forces, a complex and variable pattern of currents is observed in this region. Arctic basin waters, depending on the place of entry into the sea, spread to the south-southwest or practically to the west and follow to and along the eastern slopes of the Svalbard Bank, where they join the waters of the East Svalbard Current (Fig. 8). The contour of iceberg distribution corresponds to this scheme: it is elongated in a southwest direction. This leads to the fact that icebergs can be found in large numbers in the central part of the sea up to 77°N.

The western part of the sea is characterized by an increased concentration of icebergs in the area of the Svalbard Bank up to 74°. This phenomenon is caused by the Medvezhinsky Current, which carries a significant amount of ice and icebergs to the south: it is formed in the area of Hopen Island and recirculates when it reaches Bear Island (Bjørnøya). Further, it continues its movement to the north in the form of a narrow strip between the western slope of Svalbard and the Atlantic waters of the West Svalbard Current. The southern boundary of distribution covers the coastal areas of Novaya Zemlya, reaching 74°N. Further penetration of icebergs is hindered by opposing currents.

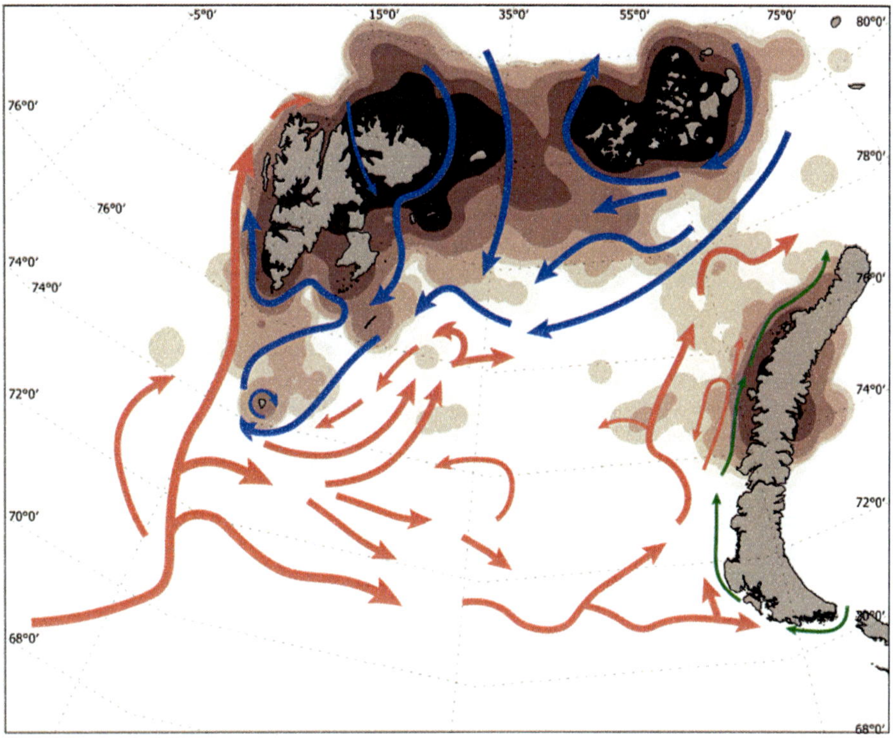

Fig. 8. The scheme of large-scale circulation in the Barents Sea according to [14]. Warm currents are represented by red arrows, cold currents by blue.

7 Conclusion

The spatial distribution of icebergs in the Barents Sea is the result of the interaction of various factors, including iceberg production from outlet glaciers, synoptic conditions, and water circulation. The overall distribution pattern is heterogeneous and variable over time. Coastal areas have a higher concentration, but icebergs can drift throughout the northern part of the Barents Sea, including areas near the shelves of Svalbard and Novaya Zemlya, thus hindering hydrocarbon development in the area. The non-zero probability of icebergs appearing on shipping routes also complicates maritime transport operations.

No two icebergs in the world are the same: as a natural object, it is constantly affected by various hydrometeorological factors, which determine its morphometric and dielectric characteristics. Analysis of the results confirms that the algorithm allows for identifying dangerous ice formations, assessing their spatial variability in open water conditions. Collecting a training dataset is an important stage in developing the algorithm, which has a significant impact on the quality and generalization ability of the model. The dataset takes into account the diversity of shapes and sizes, the position of icebergs relative to the radio incidence angle, and the peculiarities of their manifestation in two polarization channels.

This study explores the potential opportunities and limitations of modern approaches applicable to satellite data, and the development of an automated algorithm based on SAR products solves the issue of operational search for dangerous ice formations.

Acknowledgments. The author declare no conflicts of interest. The Sentinel-1 data used in this analysis is distributed by the European Space Agency.

References

1. Abramov, V.: Atlas of arctic icebergs: the Greenland, Barents, Kara, Laptev, East-Siberian and Chukchi seas and the arctic basin (1996)
2. Drusch, M., et al.: Sentinel-2: ESA's optical high-resolution mission for GMES operational services. Remote Sens. Environ. **120**, 25–36 (2012)
3. Vander Walt, S., et al.: scikit-image: image processing in Python. PeerJ (2014)
4. Lim, B., et al.: Temporal fusion transformers for interpretable multi-horizon time series forecasting. Int. J. Forecast. **37**(4), 1748–1764 (2021)
5. Osovsky, S.: Neural networks for information processing. Finan. Statist. (2004)
6. Gill, R.S.: Operational detection of sea ice edges and icebergs using SAR. Can. J. Remote Sens. **27**(5), 411–432 (2001)
7. Haykin, S., et al.: Remote Sensing of Sea Ice and Icebergs, vol. 13. John Wiley & Sons, Hoboken (1994)
8. Lewis, E.O., Currie, B.W., Haykin, S.S.: Detection and classification of ice (1987)
9. Power, D., et al.: Iceberg detection capabilities of RADARSAT synthetic aperture radar. Can. J. Remote Sens. **27**(5), 476–486 (2001)
10. Willis, C.J., et al.: Iceberg detection using ERS-1 synthetic aperture radar. Int. J. Remote Sens. **17**(9), 1777–1795 (1996)
11. Hannevik, T.N.: Literature review on ship and ice discrimination (2017)
12. Wesche, C., Dierking, W.: Iceberg signatures and detection in SAR images in two test regions of the Weddell Sea Antarctica. J. Glaciol. **58**(208), 325–339 (2012)
13. Silva, T.A.M., Bigg, G.R.: Computer-based identification and tracking of Antarctic icebergs in SAR images. Remote Sens. Environ. **94**(3), 287–297 (2005)
14. Loeng, H.: Features of the physical oceanographic conditions of the Barents Sea. Polar Res. **10**(1), 5–18 (1991)

Small Eddies Observed in Ladoga and Onega Lakes Using SAR Data

A. V. Zimin(✉), E. A. Blagodatskikh, and O. A. Atadzhanova

Shirshov Institute of Oceanology RAS, Moscow 117997, Russia
zimin2@mail.ru

Abstract. This article presents the results of analysis of observations of spatial and temporal variability of small eddy structures from May to October 2021–2022 in Ladoga and Onega lakes. As input data 626 high resolution Sentinel - 1A/B images were processed. There were 279 surface appearances of eddy structures of which 235 in Lake Ladoga and 44 in Lake Onega. In both lakes the cyclonic type of eddies were more common. The average diameter of the eddy structures in Lake Ladoga was 1.5 km, in Lake Onega 4.8 km. In Onega Lake the eddies are concentrated in the central and western parts of the basin. In Lake Ladoga - in the northern and central parts. More than 80% of the registered structures belong to the submesoscale range of variability.

Keywords: SAR imaging · Sentinel-1 · Eddies · Submesoscale · Lake Ladoga · Lake Onega

1 Introduction

The major large-scale characteristics of seasonal and synoptic circulation variability in Ladoga and Onega lakes are well represented through many years of research [1, 2]. However, studies of smaller-scale eddy dynamics with characteristic sizes from hundreds of metres to several kilometers and lifetimes from hours to days have been fragmentary to the present. This is due to the difficulty of observing such phenomena because of the small lifetime and the complexity of the mathematical description, as they do not fully advise the quasi-geostrophic theory and at the same time are not fully isotropically three-dimensional and also non-hydrostatic [3]. Such eddy structures, if they correspond to the baroclinic Rossby radius, belong to the sub-mesoscale interval of hydrophysical field variability.

Although the study of small eddy structures in the seas and oceans is increasing rapidly [4–8], their study in lakes from satellite data is still patchy.

One of the most comprehensive studies of eddy structures for lakes is the work [9] in which 45 small eddies were recorded in Lake Superior based on 361 ERS-1.2 SAR images in the warm season for the period 1992–1998 and their characteristics (position, diameter, eddy type: cyclonic/anticyclonic) were obtained. The authors found that eddies in the lake are often encountered in areas with a steep temperature gradient of 3–5 °C/3 km. In addition, there are known studies of eddies of the same scale (kilometer)

T. Chaplina (Ed.): CIWO 2023, SPEES, pp. 191–197, 2023.
https://doi.org/10.1007/978-3-031-47851-2_22

from satellite data in Lake Baikal, but there the researchers' attention was focused on eddies in ice fields [10].

It should be noted that information about them in Ladoga and Onega lakes is limited only to stating the fact of their existence [1]. Even though, small eddies can significantly influence the intensity of exchange through the thermobar and cause warm water to be thrown into the "cold" zone of the thermobar [11]. Estimates of the baroclinic Rossby radius for the lakes in question do not exceed 5 km [12].

The purpose of this work is to quantify the characteristics of the spatial and temporal variability of the manifestations of small eddy structures of Ladoga and Onega Lakes according to spaceborne SAR images from May to October 2021–2022.

2 Material and Methods

To investigate eddy signatures in Ladoga and Onega Lakes we analyzed SAR C-band images acquired by Sentinel-1 A/B during May–October 2021–2022. SAR-images taken in interferometric wide swath and extra-wide swath modes with spatial resolution of 20 m and 90 m, respectively. Sentinel-1 A/B data were obtained from Copernicus Open Access Hub [13].

In total, we analyzed for Ladoga Lake 454 SAR-images (2021 - 303, 2022 - 151), the maximum coverage of area was 214 radar images as illustrated in Fig. 1a. The maximum coverage for Lake Ladoga in 2021 was 140 pcs, mainly concentrated in the central part of the lake, between 31° and 32° E. The largest coverage for 2022 (more than 75 images) is concentrated in the central part of the lake.

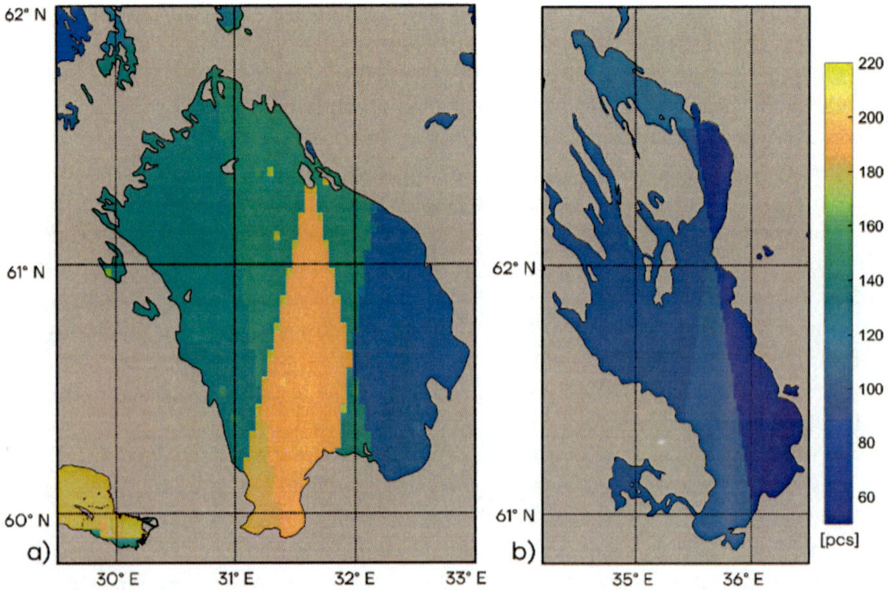

Fig. 1. Spatial coverage of Ladoga (a) and Onega (b) Lakes by SAR data during May–October 2021–2022

The total number of radar images for Lake Onega was 254 pcs (2021 - 194, 2022 - 60). Figure 1b shows the coverage of Lake Onega for two years was almost uniform over the water area and amounted to about 100 SAR-image, the maximum coverage was 127. The maximum number of image coverage of Lake Onega in 2021 was 97, mainly concentrated in the northern part of the reservoir, in the Povenets Bay. In 2022, there were already much fewer images, a maximum of 45, with an average of 30 images covering most of the western part of the Lake.

Based on the analysis of the surface manifestations of eddies in SAR images the coordinates of eddy center, their diameter, sign of rotation were found. Characteristics of eddy structures were determined based on the technique described in [6]. Data analysis and processing of results were carried out in ESA SNAP and Mathworks ©MatLab. On based characteristic we calculated different statistical parameters.

3 Results and Discussion

Analysis of SAR images made it possible to register 279 surface occurrences in the water areas of Ladoga Lake (235 eddies) and Onega Lake (44 eddies) from May to October 2021–2022. Figure 2a–b shows the locations of the eddies centers. The eddies in Lake Ladoga are mostly located in the central and northern parts of the lake. In areas with depths more than 50 m. Eddies were rarely observed in the southern, relatively shallow part of the lake. Eddies in Onega Lake were registered, as a rule, in the central part and

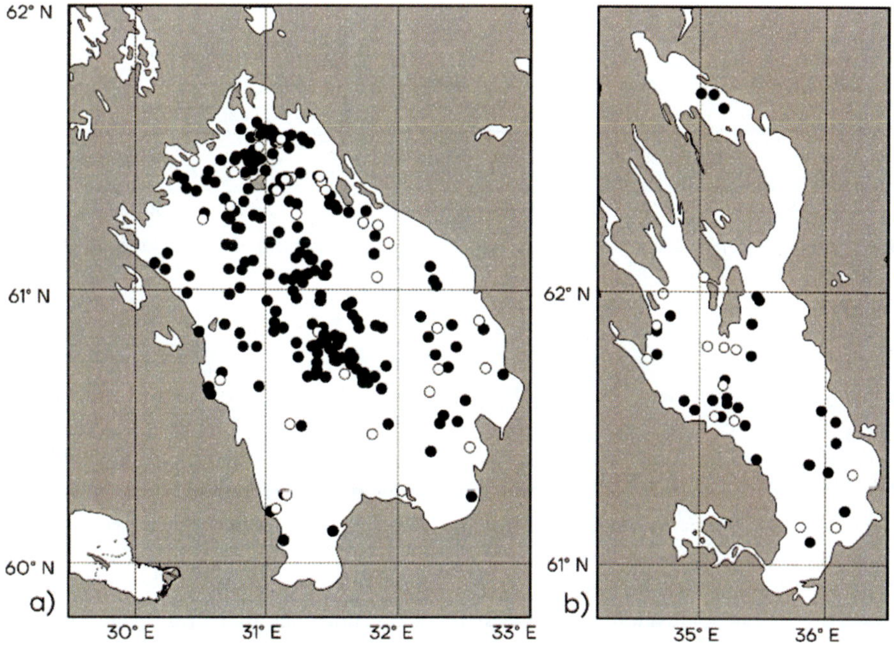

Fig. 2. Locations of the eddies identified at Ladoga (a) and Onega (b) Lakes by SAR data during May–October 2021–2022. Black (white) circles are mark cyclonic (anticyclonic) eddies.

in the vicinity of the western shore mainly in areas with depths of 30–50 m. In both lakes only isolated eddy structures were recorded close to the shore at depths less than 10–15 m.

The spatial distribution of eddies diameters per grid unit is shown in Fig. 1 and Fig. 3 The grid cell size for Ladoga is 123 km^2 and for Onega is 75 km^2. In Lake Ladoga, eddies with a diameter of 1–3 km were mostly recorded. Large eddies (with diameters > 6 km) were recorded as isolated events in the central part, north of Konevec Island and west of Cape Volchiy Nos. In Lake Onega, eddies with small diameters were recorded predominantly in the northwest of the lake, with large ones in the south. Eddies with a size of less than 3 km were most often recorded in the basin of both lakes.

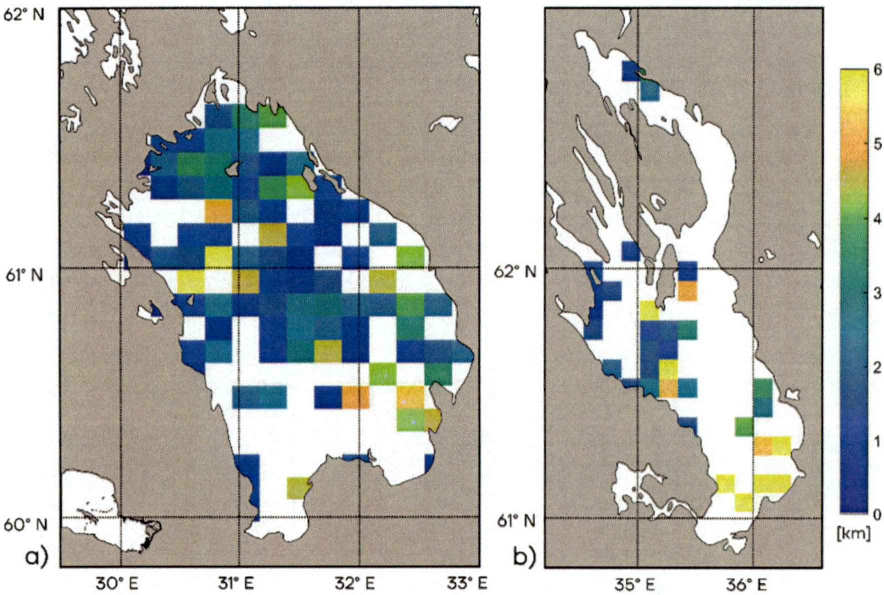

Fig. 3. Spatial distribution of the mean diameters of the eddies during May–October 2021–2022 in kilometers for Ladoga (a) and Onega (b) Lakes.

Table 1 presents annual variability of eddy characteristics in the lakes. In 2022 there were almost 2 times less SAR images, which has affected the amount of interannual variability data obtained. The number of cyclonic occurrences is always several times greater than that of anticyclonic ones. In Lake Ladoga in both years the average diameters of eddies vary from 1 to 2 km. There are no significant differences between the average diameters of cyclonic and anticyclonic eddies. The maximum diameters have structures of the cyclonic type. In Lake Onega, the number of eddies in 2022 was rather small for separate reliable statistical estimates. On average for two years, cyclonic eddies occur more often than anticyclonic eddies. At the same time, the diameter of anticyclonic eddies is larger. The average diameter of eddy structures in Lake Onega is 1.5–2 times larger than in Lake Ladoga.

Table 1. Summary of eddies detection during May–October 2021–2022 for Ladoga and Onega Lakes.

Lake	Number of Eddies				Mean Diameter (min-max), km		
	Year	C	Ac	SUM	C	Ac	SUM
Ladoga	2021	133	30	163	1.8 (0.3–7.5)	1.3 (0.2–4.6)	1.7
	2022	62	10	72	1.3 (0.2–8.4)	1.1 (0.4–1.9)	1.3
	Total	195	40	235	1.5	1.2	1.5
Onega	2021	28	10	38	3 (0.4–8.1)	3.3 (1–15.9)	3.1
	2022	2	4	6	2.9 (1.5–4.3)	8.3 (1.2–17.2)	6.5
	Total	30	14	44	3	5.8	4.8

An idea of the frequency of occurrence of eddies of different diameters can be obtained from Fig. 4a–b In Lake Ladoga, 74% of the recorded structures had a diameter up to 2 km. Both types of eddy structures (cyclonic and anticyclonic) had the maximum

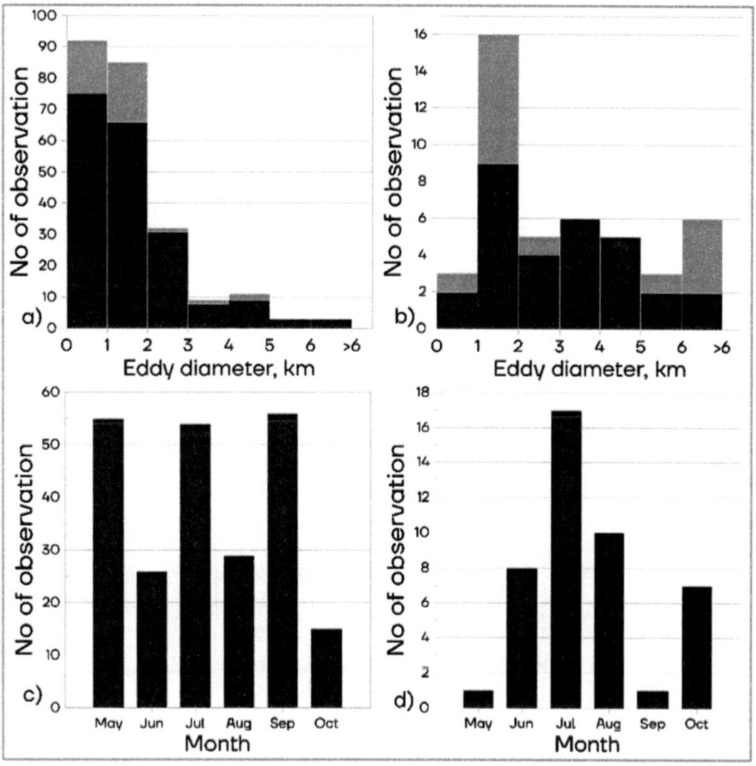

Fig. 4. (a, b) Histogram distributions of the number of eddies as a function of eddy diameter. Black and grey colors correspond to cyclones and anticyclones, respectively. (c, d) Monthly variability of the normalized number of eddies during May–October 2021–2022 for Ladoga (a, c) and Onega (b, d) Lakes.

occurrence in this interval. Larger eddies were much fewer and predominantly had the cyclonic type of rotation. In Lake Onega, 48% of one third of vortices had a diameter of up to 2 km. The maximum occurrence was noted in the interval from 1 to 2 km for both eddy types. If its maximum estimate of 5 km is taken as the boundary of the submesoscale interval, then for both lakes more than 80% of the registered structures can be classified within this range.

Figure 4c–d give an overview of the variability of the number of eddy occurrences on the lakes during the warm season. In Lake Ladoga, the maximum of occurrences is recorded in May, July and September. In Lake Onega, the maximum number was in July.

4 Conclusion

In this work for the first time for the Ladoga and Onega lakes, radar images for several years in the warm period are analyzed. The analysis has shown that eddies of the size of about one kilometer are a common phenomenon in the studied lakes. They are registered predominantly in areas with depths of more than 30–50 m. On Lake Ladoga, eddies are registered 5–6 times more often than in Lake Onega. The eddies in Lake Onega are on average larger.

Further studies will be aimed at extending the time interval of the study and detecting the dominant mechanisms of generation of eddy structures in each of the lakes.

Acknowledgements. This research was funded by the Ministry of Science and Higher Education of the Russian Federation, state assignments FMWE-2021-0014.

References

1. Lake Ladoga and the sights of its coast. In: Rumyantsev, V.A. (ed.) Atlas, p. 200. Nestor-History, St. Petersburg (2015). (in Russian)
2. Onega lake. In: Filatov, N.N. (ed.) Atlas, Revised edn., p. 151. Karelian Scientific Center of the Russian Academy of Sciences, Petrozavodsk (2010). (in Russian)
3. Thomas, L.N., Tandon, A., Mahadevan, A.: Submesoscale processes and dynamics. In: Hecht, M.W., Hasumi, H. (eds.) Ocean Modeling in an Eddying Regime/Geophysical Monograph Series **177**, 17–38 (2008). https://doi.org/10.1029/177GM04
4. Karimova, S., Gade, M.: Improved statistics of submesoscale eddies in the Baltic Sea retrieved from SAR imagery. Int. J. Remote Sens. **37**(10), 2394–2414 (2016)
5. Zimin, A.V., Atajanova, O.A., Romanenkov, D.A., Kozlov, I.E.: Submesoscale Eddies in the White Sea according to satellite radar measurements. Issledovanie Zemli iz kosmos **1**(2), 129–135 (2016). (in Russian)
6. Atadzhanova, O.A., Zimin, A.V.: Analysis of the characteristics of the submesoscale eddy manifestations in the Barents, the Kara and the White Seas using satellite data. Fundamentalnaya i Prikladnaya Gidrofizika **12**(3), 36–45 (2019)
7. Kozlov, I.E., Artamonova, A.V., Manucharyan, G.E., Kubryakov, A.A.: Eddies in the Western Arctic Ocean from spaceborne SAR observations over open ocean and marginal ice zones. J. Geophys. Res. Oceans **124**(9), 6601–6616 (2019)
8. Ji, Y., Xu, G., Dong, C., Yang, J., Xia, C.: Submesoscale eddies in the East China Sea detected from SAR images. Acta Oceanol. Sin. **40**, 18–26 (2021). https://doi.org/10.1007/s13131-021-1714-5

9. Mckinney, P., Holt, B., Matsumoto, K.: Small Eddies observed in Lake Superior using SAR and sea surface temperature data. J. Great Lakes Res. **38**, 786–797 (2012). https://doi.org/10.1016/j.jglr.2012.09.023
10. Kouraev, A., et al.: Giant ice rings on lakes and field observations of lens2 like eddies in the Middle Baikal (2016–2017). Limnol. Oceanogr. **64**, 2738–2754 (2019)
11. Kondratiev, S.A., Shmakova, M.V.: Mathematical modeling of mass transfer in the system catchment area - watercourse – reservoir, p. 246. Nestor-Istoriya, St. Petersburg (2019). (in Russian)
12. Pokazeev, K.V., Filatov, N.N.: Hydrophysics and ecology of lakes, vol. 1. Hydrophysics M. MGU. Faculty of Physics, p. 276 (2002). (in Russian)
13. Copernicus Open Access Hub. https://scihub.copernicus.eu. Accessed 1 Dec 2022

Antarctic Bottom Water in the Vema Channel

O. A. Zuev(✉) and A. M. Seliverstova

Shirshov Institute of Oceanology, RAS, Moscow 117997, Russia
qillois@gmail.com

Abstract. Based on complex data obtained in two IO RAS cruises, the distribution of the Antarctic Bottom Water flow at the exit of Vema Channel was studied. The distribution paths of two flow jets were confirmed and a third one was found. For the first time, synchronized measurements of hydrochemical and hydrophysical parameters were made, which allowed us to see small-scale changes inside the channel. Long-term measurements of bottom temperature at the main sill of the channel were continued.

Keywords: Antarctic bottom water · Vema Channel · abyssal flow · temperature and velocity profiles · dissolved oxygen · silicate

1 Introduction

The Earth's climate is largely determined by the ocean circulation, an important component of which is the Global Ocean Conveyor (GOC) that connects the thermocline and the abyssal. The lower depth branch of the GOC is of Antarctic origin - the densest water forms in a few areas of the Southern Ocean and then spreads in the bottom layer over much of the world ocean as Antarctic Bottom Water (AABW). There are several definitions of the AABW boundaries, including those depending on the study area; in the present work we follow the classical definition according to which [1] AABW is a water mass with a potential temperature of less than 2 °C. The Vema Channel is the main pathway of AABW from the Argentine Basin to the Brazilian Basin. The bathymetry of the area is shown in Fig. 1. It was washed out by the AABW flow itself about 35 millions years ago. It is more than 700 km long, 15–20 km wide with a depth relative to the terraces forming it of about 500 m [2]. The main sill of the channel is located at 31°12' S, 39°20' W and its depth is 4614 m [3]. Frequent researches have been conducted at this location for 50 years, including bottom water temperature measurements. Since the 1990s, there has been a steady trend of warming water at an average rate of about 0.002 °C per year [4, 5]. Most researches are concentrated in the area of the main saddle, but there is still no accurate picture of the distribution of AABW fluxes at the outlet of the Vema Channel at the place of its transition into the Brazilian Basin. In situ data are limited, and numerical models [6] have low resolution. In addition, the main direction of work in this area is hydrophysical and sometimes geological researches. However, measurements of hydrochemical parameters simultaneously with hydrophysical parameters are few and rather old, e.g. [7, 8]. The present work is intended to eliminate this gap as well.

T. Chaplina (Ed.): CIWO 2023, SPEES, pp. 198–204, 2023.
https://doi.org/10.1007/978-3-031-47851-2_23

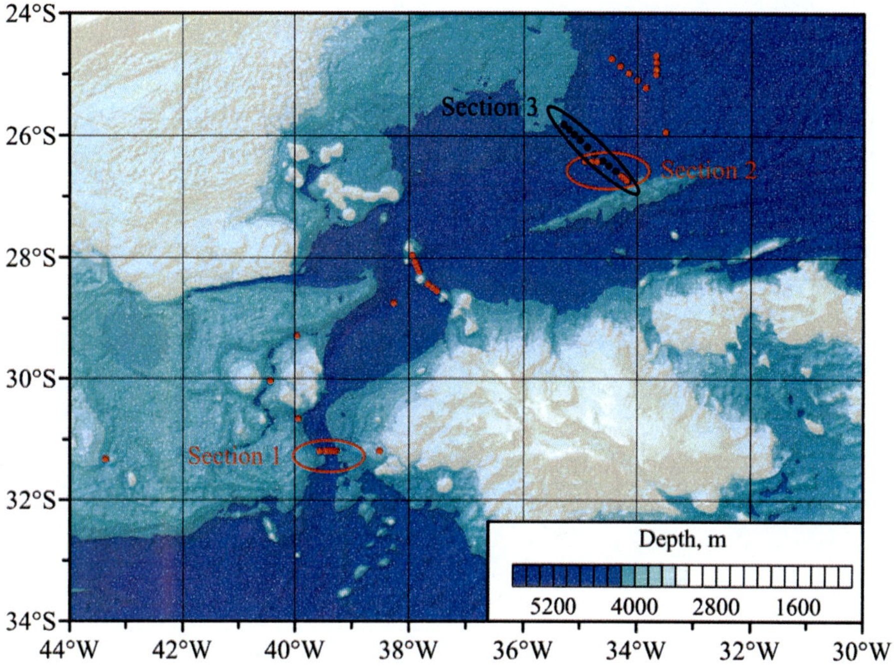

Fig. 1. Bottom topography in the region of the Vema Channel. Stations in 2020 are indicated with red dots and those in 2022 with black dots. Here and below the bottom relief is shown according the GEBCO2022 database.

2 Data and Methods

In 2020 and 2022, two cruises to Antarctica were carried out onboard the R/V "Akademik Mstislav Keldysh" (AMK79, AMK87) with associated work in the Vema Channel. There were 49 and 14 hydrophysical stations with measurements of temperature, salinity and current velocity (CTD/LADCP). Also hydrochemical parameters such as silicate, dissolved oxygen, nutrients were measured at most of the stations. The locations of the stations are shown in Fig. 1. The data were processed using standard methods: for CTD by SbeDataProcessing software package; for LADCP by LDEO Software version IX.10 [9] with additional consideration of tidal forces using software described in [10]; hydrochemical parameters in accordance with methods accepted in modern hydrochemistry [11, 12].

3 Results and Discussion

Two AABW flows are known to exist in the Vema Channel: the lower one runs along the deepest and narrowest part of the channel and is formed by the longitudinal pressure gradient of the accumulating abyssal water in the Argentine Basin, while the upper one is held by the geostrophic balance above the western terrace of the channel, located 300–400 m above the lower one [13]. The cores of both streams contain water with a potential temperature less than 0.2 °C. Figure 2 shows a transect on the main sill of the Vema Channel made in 2020. It also shows both AABW cores with minimum potential temperatures −0.11 °C and −0.06 °C, respectively, and flow velocities more than 25 cm/s. The lower core is located in the channel itself and is slightly displaced towards its eastern wall, while the upper core passes over the western wall, partly outside the channel itself. Such a distribution is typical of the Vema Channel. Total volume transport of AABW through the main sill is 1.6 Sv for $\theta < 0.2$ °C, which agrees well with literature estimates [5].

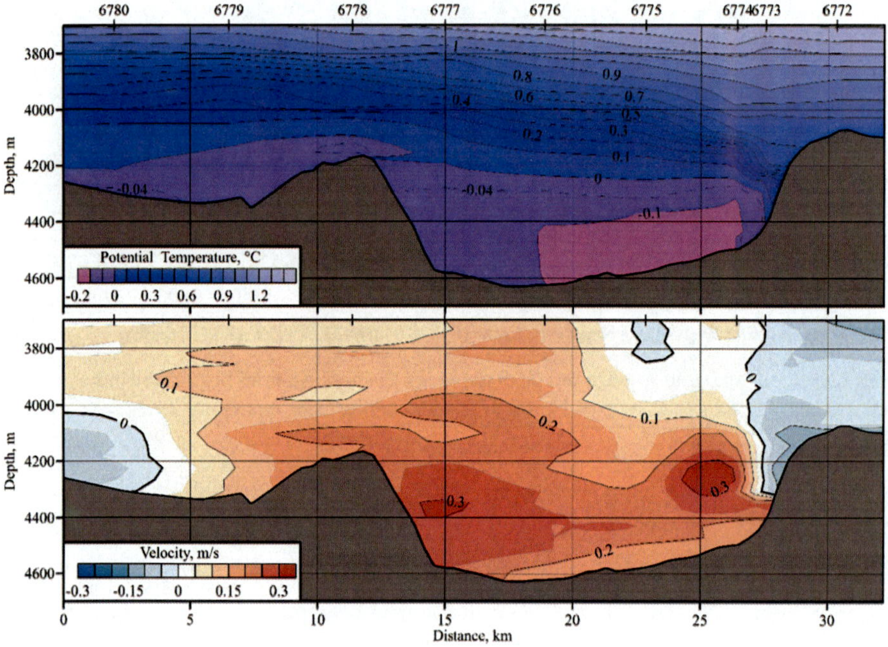

Fig. 2. Section of potential temperature (upper panel) and current velocities (bottom panel) in main sill of Vema Channel. Positive velocity direction north. Here and below black lines and numbers above indicate location of stations.

One of the tasks of the 2020 cruise was to study the behavior of both AABW jets at the exit of the Vema Channel - at 26°30' S, 34°30' W. Figure 3 shows two transects, each of which revealed AABW cores with $\theta < 0.2$ °C and significant northeastward flow velocities of up to 20 cm/s and 12 cm/s in the lower and upper streams, respectively. The lower core was located near the bottom at depths of 4700–4800 m, and the upper core was 150 m above the bottom at depths of 4100–4200 m; thus, the vertical distance between the cores was about 600 m. The lower branch of the current is more eastward, whereas the upper branch is more northward. Based on previous measurements as well as historical data, it can be assumed that the lower flow cannot overcome the significant elevation at 33°30' W and also turns northward along two narrow channels in front of it. Both channels only reach 26°S where the last potential temperatures below zero are observed, then the bottom potential temperature increases sharply. Thus, the coldest waters of the AABW can enter the Brazilian Basin only western - along the upper stream.

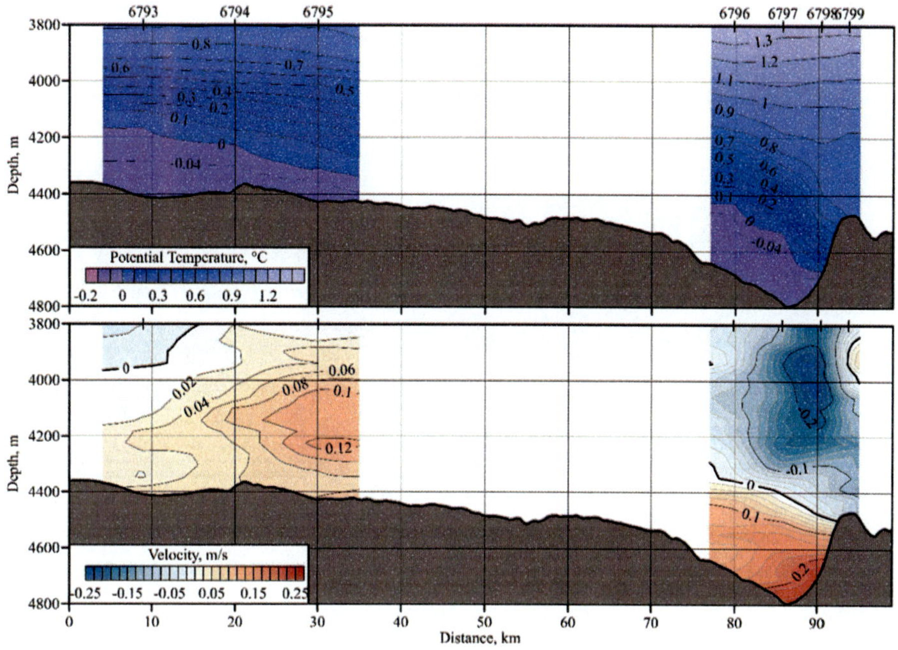

Fig. 3. Section of potential temperature (upper panel) and current velocities (bottom panel) in exit of Vema Channel in 2020. The section consists from two parts. Positive velocity direction northeast.

In 2009 and 2018, a repeat CTD transect was performed by British scientists at 24°S. Both times, the transect showed two jets with low bottom potential temperatures: from −0.04 °C to −0.05 °C at 33°50' W (at approximately 4650 m depth) in the first; and −0.03 °C to −0.04 °C at 31°25' W (in the 5000–5200 m depth range) in the second. It is likely that the areas of relatively cold bottom temperature may be related to the continued upward flow of bottom water from the Vema Channel. There are no other sources of such cold water, and the downstream AABW from the Vema Channel is located deeper and could not have risen to the indicated isobaths.

In the 2022 cruise, a longer and more detailed transect was made in the area of the Vema Channel Extension to search for possible sources of cold water on the British transects. The distribution of potential temperature and current velocities is shown in Fig. 4. The main objective was to verify the hypothesis that there was another AABW jet located higher up the slope and to confirm previous results. Thanks to a more frequent station pattern and an updated bottom topography base, it was possible to detect water with $\theta < 0.2$ °C at depths up to 4200 m. Three cores were identified - in the channel itself (−0.068 °C, 4700 m), at the beginning of the slope at the point where the upper jet was detected in 2020 (0.002 °C, 4400 m) and further west (35°12' W, 25°53' S,

0.037 °C, 4200 m). In all jets, current were mostly northward, northeastward, and reached 25 cm/s, 12 cm/s, and 15 cm/s, respectively. In turn, a weak southwest-directed current was observed between these jets. Thus it can be assumed that both cores of cold water found at 24°S were brought by two upper streams located above the western terrace of the Vema Channel, spreading along isobaths of 4200–4400 m one and deepening to 5000 m the other. The paths of spreading of all the streams on the basis of numerous historical data are described in more detail in [14].

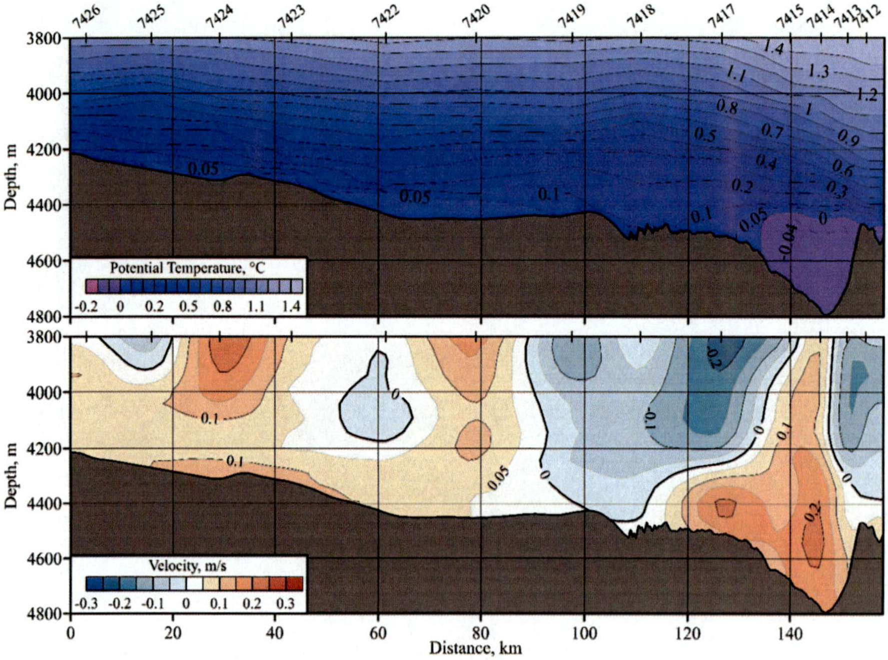

Fig. 4. Section of potential temperature (upper panel) and current velocities (bottom panel) in exit of Vema Channel in 2022. Positive velocity direction northest.

Additionally, Fig. 5 shows the distributions of dissolved oxygen and silicate. Such measurements were not carried out in this area and it is not possible to compare the obtained values with in-situ data. However, based on general considerations of the distribution of hydrochemical parameters in the bottom layer of the Atlantic Ocean, as well as comparison with hydrophysical parameters, we can conclude that these results are correct. The most important point is that the hydrochemical structure also shows the division of the AABW flow into three jets. High silicate values (125 μM, 115 μM and 117 μM from right to left) and local maxima of dissolved oxygen (5.14 ml/l, 5.13 ml/l and 5.13 ml/l, respectively) are observed at the respective stations. Thus we find further confirmation that the AABW flow at the exit of the Vema Channel separates, and follows on as single independent jets.

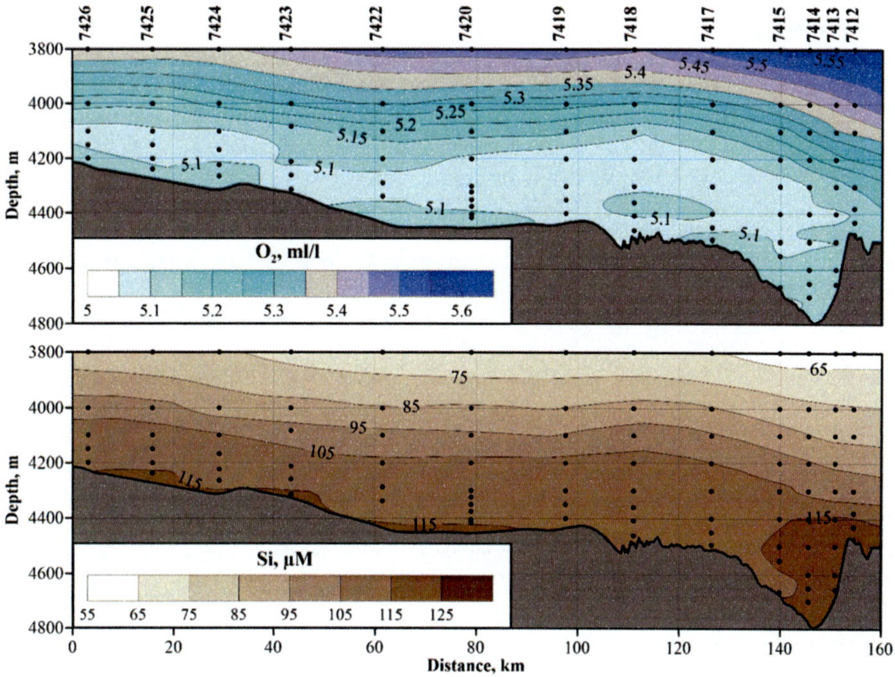

Fig. 5. Section of dissolved oxygen (upper panel) and silicate (bottom panel) in exit of Vema Channel in 2022. Black dots and numbers above indicate sampling levels.

4 Conclusion

The result of the work during two cruises was the solution of several tasks at once. First, observations of bottom water temperature changes were continued at the main sill of the Vema Channel. The long-term trend of AABW warming at this location continues. We also found and confirmed the existence of a third AABW jet passing further west, probably along the 4000–4200 m isobath, and additionally clarified the location and thickness of the main and upper AABW jets. Their location was observed from the results of complex synchronized measurements of temperature, salinity and current velocities, as well as silicate and dissolved oxygen. Finally, numerous hydrochemical parameters were measured for the first time for the Vema Channel, the results of which still need to be analyzed, but partially they already confirm such an important result as the separation of the AABW flow into separate jets at the exit of the Vema Channel.

Acknowledgments. The CTD and LADCP data processing were supported with Russian Science Foundation grant 21-77-20004.

References:

1. Wüst, G.: Schichtung und Zirkulation des Atlantischen Ozeans, Das Bodenwasser und die Stratosphäre. In: Wüst, G. (ed.) Wissenschaftliche Ergebnisse, Deutsche Atlantische Expedition auf dem Forschungs - und Vermessungsschiff "Meteor" 1925–1927, vol. 6(1), p. 411. Walter de Gruyter & Co., Berlin (1936)
2. Speer, K.G., Zenk, W.: The flow of Antarctic Bottom water into the Brazil Basin. J. Phys. Oceanogr. **23**, 2667–2682 (1993)
3. Zenk, W., Speer, K.G., Hogg, N.G.: Bathymetry at the Vema Sill. Deep-Sea Res. **40**, 1925–1933 (1993)
4. Zenk, W., Morozov, E.G.: Decadal warming of the coldest Antarctic bottom water flow through the Vema Channel. Geophys. Res. Lett. **34**, L14607 (2007)
5. Morozov, E.G., Tarakanov, R.Y., Frey, D.I.: Bottom Gravity Currents and Overflows in Deep Channels of the Atlantic Ocean: Observations, Analysis, and Modeling, p. 483. Springer, Dordrecht (2021). https://doi.org/10.1007/978-3-030-83074-8
6. Frey, D.I., Morozov, E.G., Fomin, V.V., Diansky, N.A., Tarakanov, R.Y.: Regional modeling of Antarctic Bottom Water flows in the key passages of the Atlantic. J. Geophys. Res. Oceans **124**(11), 8414 (2019)
7. Thunell, R.C.: Carbonate dissolution and abyssal hydrography in the Atlantic Ocean. Mar. Geol. **47**(3–4), 165–180 (1982)
8. Mantyla, A.W., Reid, J.L.: Abyssal characteristics of the World Ocean waters. Deep Sea Res. Part A Oceanogr. Res. Pap. **30**(8), 805–833 (1983)
9. Visbeck, M.: Deep velocity profiling using lowered acoustic Doppler current profilers: bottom track and inverse solutions. J. Atmos. Oceanic Technol. **19**(5), 794–807 (2002)
10. Egbert, G.D., Erofeeva, S.Y.: Efficient inverse modeling of barotropic ocean tides. J. Atmosp. Oceanic Technol. **19**(2), 183–204 (2002)
11. Bordovskii, O.K., et al. (ed.): Modern Methods of Hydrochemical Studies of the Ocean. (Inst. Okeanol. Akad. Nauk SSSR, Moscow) (1992). (in Russian)
12. Grasshoff, K., Kremling, K., Ehrhardt, M. (ed.): Methods of Seawater Analysis. Wiley, New York (2009)
13. Morozov, E.G., Frey, D.I., Tarakanov, R.Y.: Flow of Antarctic bottom water from the Vema Channel. Geosci. Lett. **7**(1), 1–9 (2020)
14. Morozov, E.G., Zuev, O.A., Frey, D.I., Krechik, V.A.: Antarctic bottom water jets flowing from the Vema Channel. Water **14**(21), 3438 (2022)

Ocean Biology

Trophic Ecology of Chaetognaths in the Barents Sea Revealed by Nitrogen Stable Isotope Analysis (δ^{15}N)

G. M. Artemev(✉), D. V. Zakharov, G. A. Batalin, B. I. Gareev, G. Z. Mingazov, R. M. Sabirov, and A. V. Golikov

Department of Zoology, Institute of Fundamental Medicine and Biology, Kazan Federal University, Kazan 420008, Russia
glebaartemjew@gmail.com

Abstract. Stable isotope analysis of nitrogen (δ^{15}N) was applied to investigate the trophic ecology of the three chaetognath species inhabiting the Barents Sea (*Parasagitta elegans*, *Eukrohnia hamata* and *Pseudosagitta maxima*). For the first time, we provide detailed information on the inter- and intra-specific variability of their isotopic composition and trophic levels. Different size-related and depth-related trends were found among the studied species, which indicated differences in their feeding strategies.

Keywords: Zooplankton · Chaetognatha · Trophic ecology · Stable isotope analysis · Arctic · Barents Sea

1 Introduction

Chaetognaths (arrow worms) constitute up to 7–18% of the total zooplankton biomass in the Arctic Ocean [1]. Understanding their trophic ecology is important for modeling food webs and energy flows in the pelagic communities. They have been considered obligatory predators for a long time [2], but nowadays there is strong evidence that some Arctic species may potentially feed on detritus and algae [3]. This means that the ecological role of arrow worms in pelagic food webs should be redefined.

Stable isotope analysis (SIA) is a widely applied technique that provides temporal and spatial resolution of the trophic relationships among the organisms [4]. Therefore, this approach may be especially useful for those organisms whose diet is difficult to characterize directly, such as chaetognaths. Previous isotope studies already indicated differences in the feeding strategies of the two most common chaetognaths in the Arctic, i.e. *Parasagitta elegans* and *Eukrohnia hamata*, with the former considered more predatory and the latter more omnivorous [3, 5, 6]. At the same time, the feeding strategy of the largest polar species, *Pseudosagitta maxima*, remains mainly unexplored and is of particular interest.

In this study, we used SIA of nitrogen (δ^{15}N) to investigate the trophic ecology of chaetognaths in the Barents Sea, where all three species coexist. Our main objectives

T. Chaplina (Ed.): CIWO 2023, SPEES, pp. 207–216, 2023.
https://doi.org/10.1007/978-3-031-47851-2_24

were: 1) to asses inter- and intra-specific variability in their $\delta^{15}N$ values; 2) to estimate their trophic levels (TLs); 3) to assess the effect of size and habitat depth on their $\delta^{15}N$ values and TLs.

2 Material and Methods

The samples were collected in four cruises of the Polar branch of the FSBSI "VNIRO" ("PINRO" named after N. M. Knipovich) on R/Vs M-0102 "Vilnus" and M-0662 "Nansen" during November and December 2013–2014 (Fig. 1). Zooplankton was caught by a trawl net (inlet diameter 50 cm, gas № 140) near the bottom and processed according to the unified method adopted by "PINRO" [7]. Samples were stored in 4% buffered formalin diluted with fresh water.

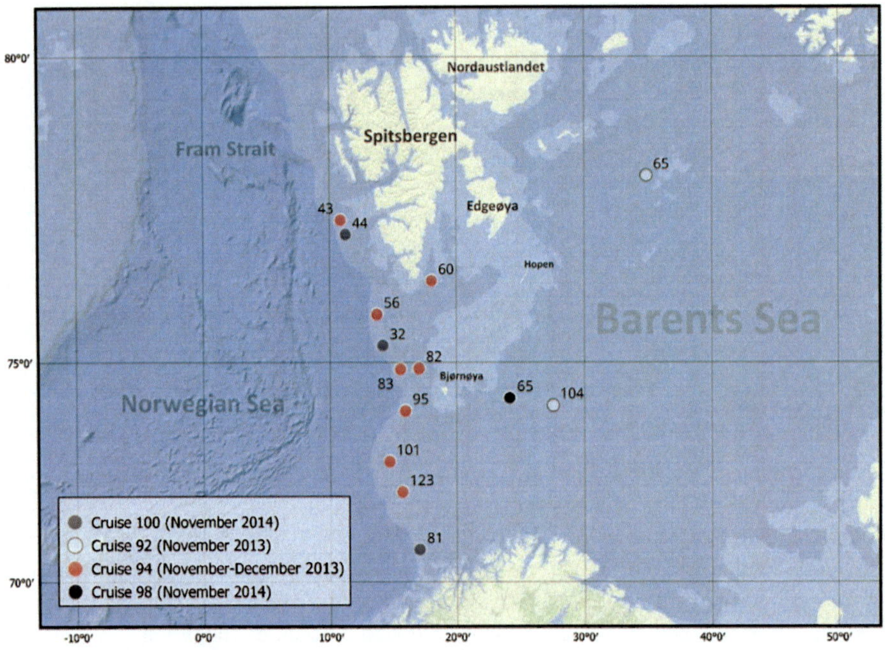

Fig. 1. Sampling stations in the Barents Sea (map created in QGIS 3.28.2)

Sample processing was conducted in 2021 in the Laboratory of Hydrobiology and Molecular Systematics (Kazan Federal University). In total, three species of chaetognaths were collected from 14 stations: *Parasagitta elegans* (175 ind.), *Eukrohnia hamata* (175 ind.) and *Pseudosagitta maxima* (33 ind.). Each selected individual was measured to the nearest millimeter from the beginning of the head to the tip of the tail, excluding the caudal fin. Then the specimens were washed in distilled water and dried at a temperature of 60 °C.

The stable isotope analysis was carried out in the Laboratory of Isotopic and Elemental Analysis (Kazan Federal University) with Flash HT series elemental analyzer coupled online via a ConFlo IV interface to a Delta V Plus mass spectrometer (TermoFisher Scientific, Germany). The ratios of nitrogen stable isotopes ($^{15}N/^{14}N$) in the samples were determined as a relative value delta (δ), which is measured in ppm (‰) as the difference between the ratios of isotopes in the test object and the standard:

$$\delta^{15}N\ (‰) = [(^{15}N/^{14}N_{sample} - {}^{15}N/^{14}N_{standart}) / {}^{15}N/^{14}N_{standart}] \times 1000$$

The isotope ratios were expressed relative to atmospheric nitrogen (AIR). The accuracy of determining the values of $\delta^{15}N$ on this mass spectrometer was ±0.2‰.

Trophic levels of chaetognaths were derived using the classical equation [4]:

$$TL_{consumer} = TL_{baseline} + (\delta^{15}N_{consumer} - \delta^{15}N_{baseline})/TEF$$

Filter-feeding copepods *Calanus glacialis* and *C. finmarchicus* were sampled from the same 14 stations and used as a baseline, so the values of $\delta^{15}N_{baseline}$ were unique for each station. We assumed that their diet is predominantly herbivorous and they correspond to the primary consumers ($TL_{baseline} = 2.0$). Among several known trophic enrichment factors (TEFs), we adopted the classical (= 3.4‰). It is known that formalin does not significantly affect $\delta^{15}N$-ratios [8], so no additional corrections were performed due to preservation.

Comparisons of $\delta^{15}N$ and TLs among the three species of chaetognaths were made using non-parametric Kruskal-Wallis and Wilcoxon rank sum tests. A regression analysis was used to determine the effect of size and habitat depth on their $\delta^{15}N$ values and TLs. All tests were performed using a significance value of $\alpha = 0.05$. Statistical data processing was performed in Rstudio based on R 4.0.3 software. A package "ggplot 2" was used to visualize the data.

3 Results

The three studied species of arrow worms had significant differences in both $\delta^{15}N$ values and TLs (Kruskal-Wallis rank sum test: $\chi^2 = 236.07$, df = 2, $p < 0.001$ for $\delta^{15}N$; $\chi^2 = 193.88$, df = 2, $p < 0.001$ for TLs). Further pairwise comparisons with the Bonferroni correction also showed significant differences among all species (Wilcoxon rank sum test, $p < 0.001$). The highest average values of $\delta^{15}N$ and TLs were found in *Parasagitta elegans*, followed by *Pseudosagitta maxima* and *Eukrohnia hamata* (Table 1, Figs. 2a,b). The largest ranges of $\delta^{15}N$ and TLs were observed respectively in *P. elegans* and *E. hamata*, while the smallest range was found in *P. maxima*.

No differences in $\delta^{15}N$ values and TLs were found between 2013 and 2014 in *E. hamata* and *P. maxima* (Wilcoxon rank sum test, $p > 0.05$). However, *P. elegans* showed significant differences between two years in $\delta^{15}N$ values, but not in TLs (Wilcoxon rank sum test: W = 4226, $p < 0.001$ for $\delta^{15}N$; W = 2837, $p > 0.05$ for TLs). In addition, there was a large variability in $\delta^{15}N$ values of all three species among 14 stations (Fig. 3). In most cases, they showed the general trend with *P. elegans* having the highest values, followed by *P. maxima* and *E. hamata*.

Table 1. Summary of $\delta^{15}N$ values and trophic levels of chaetognaths and copepods from the Barents Sea

Species	*n*, ind	Size, mm			$\delta^{15}N$, ‰			Trophic level		
		min	Max	mean (± SE)	min	max	mean (± SE)	min	max	mean (± SE)
Parasagitta elegans	175	12	36	23 ± 0	7.6	13.4	10.8 ± 0.1	2.0	3.6	2.7 ± 0.0
Eukrohnia hamata	175	11	36	22 ± 1	6.3	11.5	8.4 ± 0.1	1.2	3.2	2.1 ± 0.0
Pseudosagitta maxima	33	21	77	51 ± 2	7.8	11.5	9.4 ± 0.2	1.8	3.2	2.4 ± 0.1
Calanus glacialis/finmarchicus	14*	–	–	–	6.2	9.4	8.3 ± 0.2	2.0 (baseline)		

* 10–50 individuals per sample

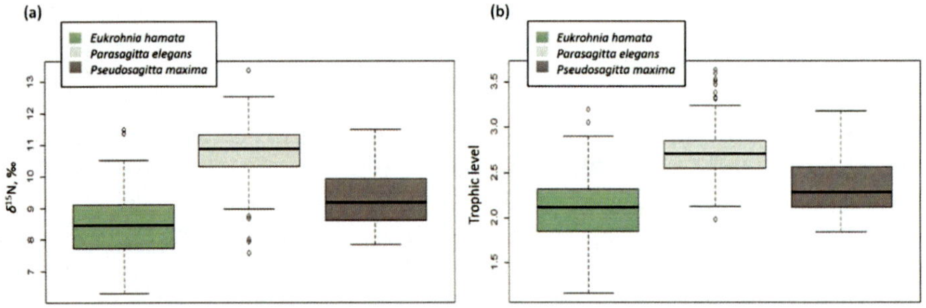

Fig. 2. Comparisons of $\delta^{15}N$ values (a) and trophic levels (b) among the three species of chaetognaths

Regression analysis revealed different size-related and depth-related trends in the three species (Table 2, Figs. 4 and 5). Only *P. maxima* had a significant positive correlation of both $\delta^{15}N$ values and TLs with size (Figs. 4e,f). *Parasagitta elegans* had a significant negative correlation of TLs and size, but not of $\delta^{15}N$ values and size (Figs. 4a,b). In *E. hamata*, no correlation of either $\delta^{15}N$ values or TLs with size was found (Figs. 4c,d).

Regression models of $\delta^{15}N$ and TLs relationship with station depth also showed different trends. In *P. elegans* and *P. maxima*, no correlations of $\delta^{15}N$ values and TLs with depth were observed (Figs. 5a,b,e,f). In *E. hamata*, a significant negative correlation of $\delta^{15}N$ values with depth was found, but there was no correlation of TLs with depth (Figs. 5c,d).

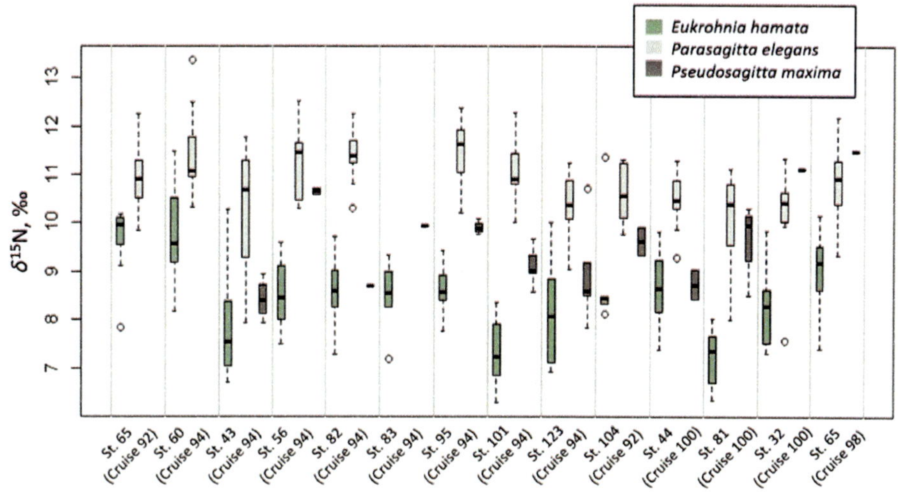

Fig. 3. Variability of δ^{15}N values in the three species of chaetognaths among 14 stations. At some stations (Cruise 92: St 65; Cruise 94: St 60, St 83), only two species were sampled

Table 2. Summary statistics for regression analysis of the effect of size and habitat depth on δ^{15}N values and trophic levels of chaetognaths. Statistically significant results are in bold

Species	Regression model	Intercept (± SE)	Slope (± SE)	Multiple R^2	*F*-statistic	df	*p*-value
Parasagitta elegans	δ^{15}N ~ size	11.079 (± 0.252)	−0.117 (± 0.107)	0.01	1.18	1, 173	0.28
	TL ~ size	2.967 (± 0.080)	−0.108 (± 0.034)	0.06	10.08	1, 173	**0.0018**
	δ^{15}N ~ depth	11.193 (± 0.365)	−0.001 (± 0.001)	0.11	1.38	1, 11	0.27
	TL ~ depth	2.611 (± 0.129)	0.0002 (± 0.0002)	0.08	0.94	1, 11	0.35
Eukrohnia hamata	δ^{15}N ~ size	8.189 (± 0.261)	0.115 (± 0.111)	0.01	1.06	1, 173	0.30
	TL ~ size	2.059 (± 0.094)	0.017 (± 0.040)	0.001	0.18	1, 173	0.67
	δ^{15}N ~ depth	9.895 (± 0.379)	−0.002 (± 0.001)	0.55	14.76	1, 12	**0.0023**

(*continued*)

Table 2. (*continued*)

Species	Regression model	Intercept (± SE)	Slope (± SE)	Multiple R^2	*F*-statistic	df	*p*-value
	TL ~ depth	2.238 (± 0.206)	−0.0003 (± 0.0003)	0.06	0.75	1, 12	0.40
Pseudosagitta maxima	$\delta^{15}N$ ~ size	7.908 (± 0.600)	0.294 (± 0.115)	0.18	6.60	1, 31	**0.0152**
	TL ~ size	1.773 (± 0.208)	0.115 (± 0.040)	0.21	8.33	1, 31	**0.0070**
	$\delta^{15}N$ ~ depth	9.193 (± 1.042)	0.001 (± 0.002)	0.03	0.26	1, 10	0.62
	TL ~ depth	2.073 (± 0.358)	0.001 (± 0.001)	0.10	1.16	1, 10	0.31

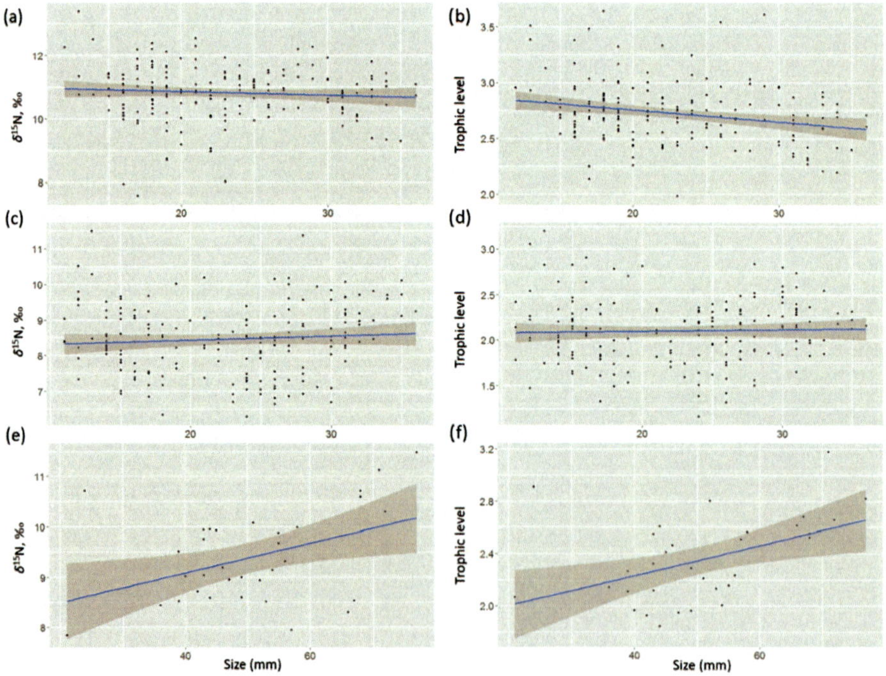

Fig. 4. Relationship of $\delta^{15}N$ values and trophic levels with size in the three species of chaetognaths: a, b – *Parasagitta elegans*; c, d – *Eukrohnia hamata*; e, f – *Pseudosagitta maxima*. The dark area surrounding the regression line corresponds to the confidence intervals

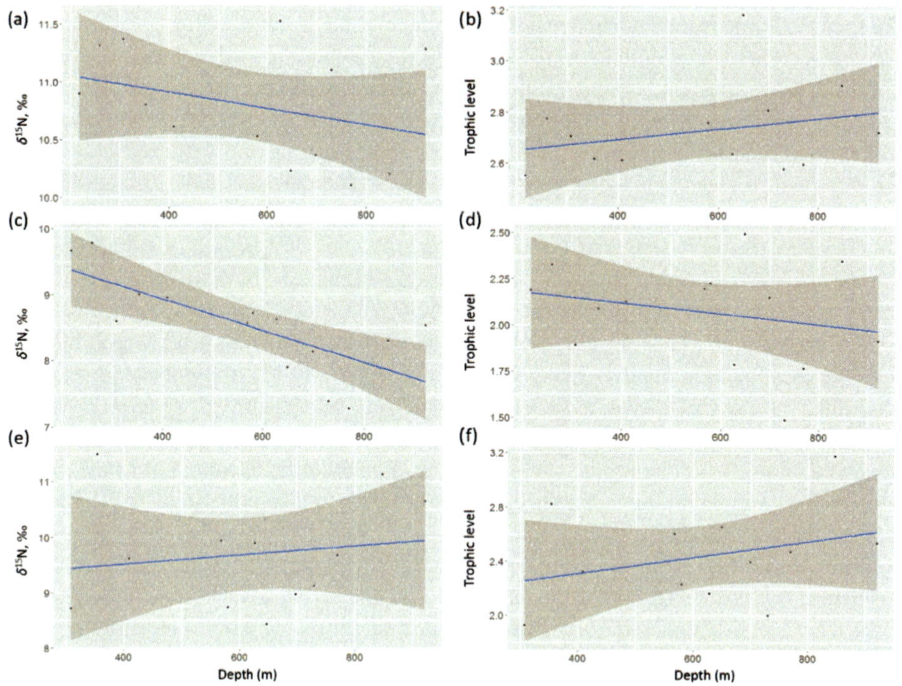

Fig. 5. Relationship of δ^{15}N values and trophic levels with station depth in the three species of chaetognaths: a, b – *Parasagitta elegans*; c, d – *Eukrohnia hamata*; e, f – *Pseudosagitta maxima*. The dark area surrounding the regression line corresponds to the confidence intervals

4 Discussion

Only several studies have focused on the inter-specific differences in the isotopic composition of arrow worms prior to this study [3, 9, 10]. Among them, only one is devoted to the ecology of two Arctic species [3]. In all these studies intra-specific isotope variability was largely overlooked due to the use of pooled samples instead of individual samples [3, 9, 10]. Therefore, our research represents the first attempt of a detailed analysis of both inter- and intra-specific variability in the isotopic composition of chaetognaths.

It should be emphasized that the material in the present study was collected during the polar night, when many zooplankters exhibit a shift in their dietary preferences [11]. Arrow worms belong to non-visual predators, so the absence of light should not considerably affect their feeding behavior. No significant changes in their diet during the polar night have been observed elsewhere, although a reduced feeding activity was proposed [12]. The low TLs of chaetognaths obtained in this research can be explained by the fact that herbivorous copepods (*Calanus* spp.), which we chose as a baseline, may switch to a more omnivorous diet in winter [13]. Thus, their actual trophic position is higher than we assumed in the equation.

The results of our study demonstrate large differences in the trophic ecology of the three species of chaetognaths from the Barents Sea. We observed great variability in their δ^{15}N values and TLs, as well as different size-related and depth-related trends among the species. Some researchers have already compared the trophic ecology of *Parasagitta elegans* and *Eukrohnia hamata* in the Arctic [3] and Subarctic [14] waters, highlighting the differences in their isotopic composition, lipid profile and gut content. Hence, in order to explain the differences among all three species, a comprehensive consideration of their ecology is required.

Parasagitta elegans is a predominantly epipelagic species up to 40–50 mm in size. The diet of *P. elegans* in the Barents Sea in early summer consisted mainly of nauplii, small copepods (early stages of *Calanus*, *Pseudocalanus* and *Oithona*) and appendicularians [15]. In the present study, *P. elegans* has the highest $\delta^{15}N$ values and TLs among the three species, which is consistent with previous isotope research [3, 5, 6].

Eukrohnia hamata is an eurybathic species up to 40 mm in size, which usually inhabits mesopelagic depths. According to a recent study [3], the guts of *E. hamata* from the Amundsen Gulf were dominated by detritus and diatom algae throughout the year. Only small numbers of copepods (*Pseudocalanus* spp. and *Oithona* spp.) were observed, along with one report of cannibalism. In our study, *E. hamata* has the lowest $\delta^{15}N$ values and TLs among the three species, which is also consistent with previous isotope research [3, 5, 6].

Pseudosagitta maxima is an eurybathic and one of the largest species in the group (up to 90 mm), which usually inhabits meso- and bathypelagic depths. The trophic ecology of *P. maxima* is poorly understood, with only limited data from the Arctic [16] and Antarctic [17] being available. The diet of this species in polar waters includes large copepods (*Metridia* spp., *Euchaeta* spp.), appendicularians, polychaetes and chaetognaths [16, 17]. In the present study, *P. maxima* has intermediate $\delta^{15}N$ values and TLs among the three species. To our knowledge, there are only a few other reports on the isotopic composition of this species [18, 19].

Our results show that the inter-specific differences in $\delta^{15}N$ values and TLs of chaetognaths cannot be explained only in terms of their size, as has been previously suggested [9, 10]. The three species also appear to differ in metabolic activity [20], which is closely related to their lifestyles and feeding strategies. Among them, *P. elegans* is known to have high food requirements [14] and respiration rates [20]. It is assumed that this carnivorous species is able to actively search for prey during its diel vertical migrations [21]. In contrast, predominantly omnivorous *E. hamata* shows lower food requirements [14] and respiration rates [20], but has the ability to store reserve lipids. Finally, *P. maxima* has one of the lowest levels of metabolic activity in the group [20]. This flaccid species with reduced musculature is assumed to be an ambush predator using the "sit-and-wait" strategy [21].

There are several studies devoted to the effect of body size on the prey composition of chaetognaths [22, 23], and they demonstrated the expected positive relationship between the prey and predator sizes. If simplified, this should have led to an ontogenetic increase in $\delta^{15}N$ values and TLs with body length, as it is usually observed in carnivores. In our study, we can see such size-related trend only in *P. maxima*, perhaps due to its larger size range, but not in the other two species. The reasons for these differences remain unclear and require further research.

The effect of depth on the feeding rates of arrow worms is also mentioned in some studies [24, 25]. In most cases, feeding activity decreases with increasing depth, which can be explained by food scarcity and low temperatures. Among the three studied species, only *E. hamata* shows a clear depth-related decline in δ^{15}N values. *Parasagitta elegans* and *P. maxima* seem to have no depth-related trends, but this may also be due to sampling artefacts. In this study, chaetognaths were collected by a trawl net close to the bottom, but we cannot exclude the possibility of catching them from the upper layers of the water column.

5 Conclusion

The data presented in this study provide new insights into how ecological niche partition occurs in the three coexisting species of chaetognaths from the Barents Sea. Stable isotope analysis revealed significant differences in their δ^{15}N values and TLs, as well as different size-related and depth-related trends among the species. According to the results obtained, *Parasagitta elegans* has the highest TL, *Pseudosagitta maxima* has intermediate TL, and *Eukrohnia hamata* has the lowest TL. It appears that the observed differences are mostly related to their metabolic activities rather than their body size. Based on these data, we assume different feeding strategies for the three species: *P. elegans* is an active predator, *P. maxima* is an ambush predator, and *E. hamata* is predominantly an omnivore. The results of this study should be considered when estimating predation rates of arrow worms in the polar ecosystems, and further research is required to more accurately establish their trophic role in the pelagic food webs.

Acknowledgements. We are grateful to the researchers and crews onboard the vessels collecting the samples. We also acknowledge K. N. Kosobokova and A. K. Zalota (Shirshov Institute of Oceanology, RAS), K. M. Gorbatenko (Pacific branch of the "VNIRO") and E. A. Pakhomov (University of British Columbia) for their constructive comments on the study. This work was supported by the Russian Science Foundation (grant No. 23-17-00121).

References:

1. Hopcroft, R.R., Clarke, C., Nelson, R.J., Raskoff, K.A.: Zooplankton communities of the Arctic's Canada Basin: the contribution by smaller taxa. Polar Biol. **28**, 198–206 (2005)
2. Feigenbaum, D.L., Maris, R.C.: Feeding in the Chaetognatha. Oceanogr. Mar. Biol. Annu. Rev. **22**, 343–392 (1984)
3. Grigor, J.J., et al.: Non-carnivorous feeding in Arctic chaetognaths. Prog. Oceanogr. **186**, 102388 (2020)
4. Post, D.M.: Using stable isotopes to estimate trophic position: models, methods, and assumptions. Ecology **83**(3), 703–718 (2002)
5. Søreide, J.E., Hop, H., Carroll, M.L., Falk-Petersen, S., Hegseth, E.N.: Seasonal food web structures and sympagic–pelagic coupling in the European Arctic revealed by stable isotopes and a two-source food web model. Prog. Oceanogr. **71**(1), 59–87 (2006)
6. Søreide, J.E., Carroll, M.L., Hop, H., Ambrose Jr, W.G., Hegseth, E.N., Falk-Petersen, S.: Sympagic-pelagic-benthic coupling in Arctic and Atlantic waters around Svalbard revealed by stable isotopic and fatty acid tracers. Mar. Biol. Res. **9**(9), 831–850 (2013)

7. Shevelev, M.S.: Instructions and Guidelines for the Collecting and Processing of Biological Information in PINRO Research Areas. Publishing House PINRO, Murmansk (2001). (in Russian)
8. Sarakinos, H.C., Johnson, M.L., Zanden, M.J.V.: A synthesis of tissue-preservation effects on carbon and nitrogen stable isotope signatures. Can. J. Zool. **80**(2), 381–387 (2002)
9. Rau, G.H., Ohman, M.D., Pierrot-Bults, A.: Linking nitrogen dynamics to climate variability off central California: a 51 year record based on $^{15}N/^{14}N$ in CalCOFI zooplankton. Deep Sea Res. Part II: Topical Stud. Oceanogr. **50**(14–16), 2431–2447 (2003)
10. Bohata, K., Koppelmann, R.: Chaetognatha of the Namibian upwelling region: taxonomy, distribution and trophic position. PloS One. **8**(1), e53839 (2013)
11. Berge, J., Daase, M., Hobbs, L., Falk-Petersen, S., Darnis, G., Søreide, J.E.: Zooplankton in the polar night. In: Berge, J., Johnsen, G., Cohen, J.H. (eds.) Polar Night Marine Ecology. APE, vol. 4, pp. 113–159. Springer, Cham (2020). https://doi.org/10.1007/978-3-030-33208-2_5
12. Grigor, J.J., et al.: Polar night ecology of a pelagic predator, the chaetognath *Parasagitta elegans*. Polar Biol. **38**, 87–98 (2015)
13. Cleary, A.C., Søreide, J.E., Freese, D., Niehoff, B., Gabrielsen, T.M.: Feeding by *Calanus glacialis* in a high arctic fjord: potential seasonal importance of alternative prey. ICES J. Mar. Sci. **74**(7), 1937–1946 (2017)
14. Sullivan, B.K.: In situ feeding behavior of *Sagitta elegans* and *Eukrohnia hamata* (Chaetognatha) in relation to the vertical distribution and abundance of prey at Ocean Station "P" 1. Limnol. Oceanogr. **25**(2), 317–326 (1980)
15. Falkenhaug, T.: Prey composition and feeding rate of *Sagitta elegans* var. *arctica* (Chaetognatha) in the Barents Sea in early summer. Polar Res. **10**(2), 487–506 (1991)
16. Samemoto, D.D.: Vertical distribution and ecological significance of chaetognaths in the Arctic environment of Baffin Bay. Polar Biol. **7**, 317–328 (1987)
17. Øresland, V.: Feeding and predation impact of the chaetognath *Eukrohnia hamata* in Gerlache Strait, Antarctic Peninsula. Mar. Ecol. Prog. Ser. Oldendorf. **63**(2), 201–209 (1990)
18. Wada, E., Terazaki, M., Kabaya, Y., Nemoto, T.: ^{15}N and ^{13}C abundances in the Antarctic Ocean with emphasis on the biogeochemical structure of the food web. Deep Sea Res. Part A Oceanogr. Res. Pap. **34**(5–6), 829–841 (1987)
19. Pomerleau, C., et al.: Spatial patterns in zooplankton communities across the eastern Canadian sub-Arctic and Arctic waters: insights from stable carbon ($\delta^{13}C$) and nitrogen ($\delta^{15}N$) isotope ratios. J. Plankton Res. **33**(12), 1779–1792 (2011)
20. Thuesen, E.V., Childress, J.J.: Enzymatic activities and metabolic rates of pelagic chaetognaths: lack of depth-related declines. Limnol. Oceanogr. **38**(5), 935–948 (1993)
21. Perez, Y., Casanova, J.P., Mazza, J.: Degrees of vacuolation of the absorptive intestinal cells of five *Sagitta* (Chaetognatha) species: possible ecophysiological implications. Mar. Biol. **138**(1), 125–133 (2001)
22. Pearre Jr, S.: Feeding by Chaetognatha: the relation of prey size to predator size in several species. Mar. Ecol. Prog. Ser. **3**, 125–134 (1980)
23. Saito, H., Kiørboe, T.: Feeding rates in the chaetognath *Sagitta elegans*: effects of prey size, prey swimming behaviour and small-scale turbulence. J. Plankton Res. **23**(12), 1385–1398 (2001)
24. Terazaki, M.: Deep-sea adaptation of the epipelagic chaetognath *Sagitta elegans* in the Japan Sea. Mar. Ecol. Prog. Ser. Oldendorf. **98**(1), 79–88 (1993)
25. Terazaki, M., Marumo, R.: Feedings habits of mesopelagic and bathypelagic chaetognatha, *Sagitta zetesios* Fowler. Oceanologica Acta **5**(4), 461–464 (1982)

Vertical Distribution of Viral Particles in the Deep Part of the Black Sea

M. A. Ufimtseva(✉) and V. S. Mukhanov

A.O. Kovalevsky Institute of Biology of the Southern Seas, Russian Academy of Sciences, Sevastopol, Russia
ritica011@bk.ru

Abstract. The study was carried out in the deep-water part of the Black Sea in October 2022. Using flow cytometry and epifluorescence microscopy, the first data for the Black Sea on the abundance and vertical distribution of virioplankton were obtained, as well as the abundance of bacterioplankton and the viruses-to-bacteria ratio (VBR) were estimated. The number of viruses varied in the range from 0,4 to 9,4 × 10^6 cells ml^{-1} with maxima in the aerobic zone. A decrease in the number of viral particles from the oxycline to the suboxic layer by more than an order of magnitude was revealed. Within the hydrogen sulfide zone, the number of viruses was found to increase insignificantly with increasing depth.

Keywords: viral particles · virioplankton · vertical distribution · abundance · Black Sea · flow cytometry · epifluorescence microscopy

1 Introduction

Viruses are the most numerous components of the biota in aquatic ecosystems: their numbers reach 10^6 to 10^8 particles ml^{-1} [1]. As one of the most important factors in the death of microorganisms [2], they play a key role in regulating the abundance and diversity of microbial communities in the World Ocean, take an active part in global biogeochemical cycles, control the flows of matter and energy in marine ecosystems through the so-called viral 'shunt', the essence of which consists in the death of virus-infected hydrobionts and, as a result, the replenishment of dissolved and suspended organic matter. Essentially, the 'viral shunt' counteracts the microbial 'loop' by preventing organic matter from returning back into the food chain [3]. Despite the constant increase in the number of studies of marine virioplankton, the Black Sea viruses are practically not studied. Nothing is known about their abundance and vertical distribution in the deep part of the Black Sea including the hydrogen sulfide zone. Information about their abundance is contained in rare microbiological studies of coastal waters and bottom sediments of the Black Sea [4–6], however, the reliability of some results is questionable due to the use of incorrect methods and a significant underestimation of viral particles in water samples. Thus, the Black Sea is a white spot on the map of studies of the virioplankton of the World Ocean. In the framework of this work, the task was to fill this gap, obtaining the first data on the Black Sea, virioplankton using flow cytometry and epifluorescence microscopy.

T. Chaplina (Ed.): CIWO 2023, SPEES, pp. 217–222, 2023.
https://doi.org/10.1007/978-3-031-47851-2_25

2 Material and Methods

The material was collected on the 124 cruise of the R/V Professor Vodyanitsky in October 2022 on the shelf south of the Crimean Peninsula and on the section from Cape Chersonese to the center of the western gyre (the sampling areas are shown in Fig. 1).

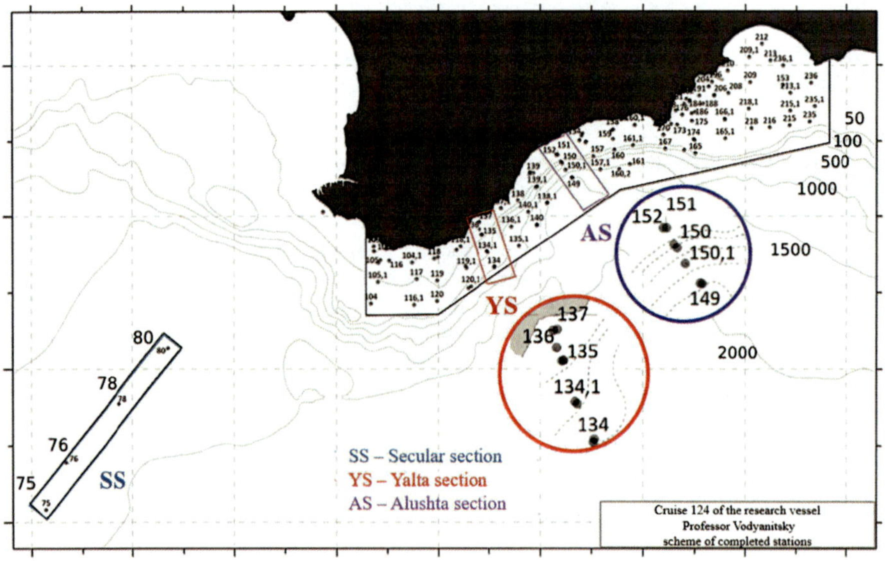

Fig. 1. Water sampling stations in the Black Sea (124th cruise of the R/V Professor Vodyanitsky) on the secular section (SS) from the center of the western cyclonic gyre (station 75) towards Cape Chersones, Yalta (YS) and Alushta (AS) sections on shelf south of the Crimean peninsula

Samples were taken with a Niskin bottle at several horizons from the surface to the bottom layer (about 2000 m), taking into account the CTD profiling data on the hydrological and hydrochemical structure of the water column. Seawater samples were poured in 5-ml cryovials in 3 replicates, fixed with formalin (2% final conc.), and frozen in in liquid nitrogen to store them until they were processed at a stationary laboratory. After thawing, a 1 ml aliquot of the sample was filtered onto a membrane with a pore size of 0.02 μm (Whatman Anodisc 25), stained on a drop of SYBR Gold, and washed according to [7]. SYBR Gold is the brightest fluorescent stain in the SYBR family of stains for nucleic acid detection and imaging. Viral particles were visualized by epifluorescence microscopy [8] using a Jenalumar-a/d microscope equipped with an HBO-202 mercury lamp. Micrographs of the visual fields were processed using the ImageJ program (National Institutes of Health) for automatic particle counting. This is the standard method for quantifying viruses in seawater. Figure 2 illustrates the "starry sky" of virus particles and bacteria from the Black Sea.

The abundances of viral particles and bacteria were also determined using the flow cytometry method. Before flow cytometry, the samples were diluted tenfold with TE buffer (10 mM Tris–HCl, 1 mM EDTA pH 8.0) and stained with SYBR Green I at +

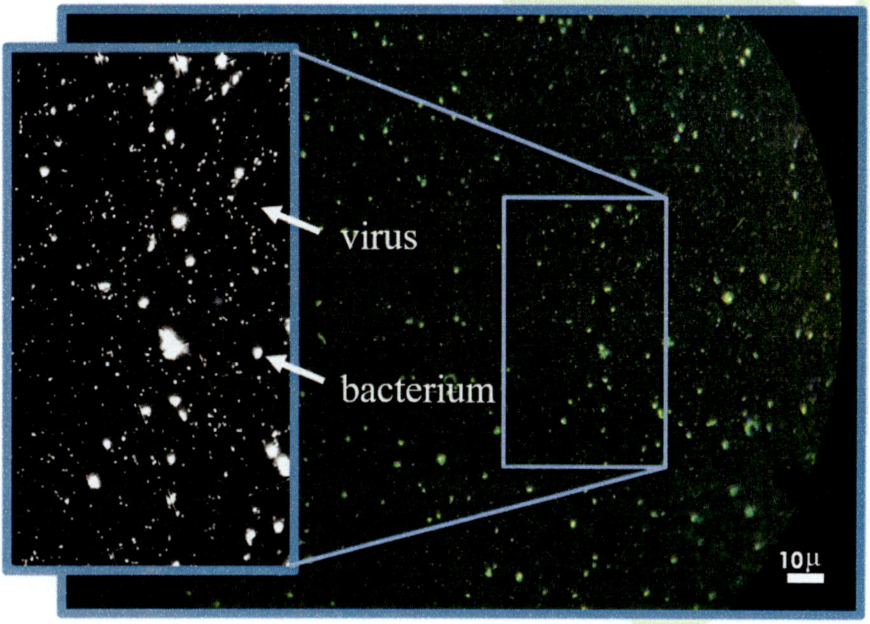

Fig. 2. Marine viruses and bacteria in the field of view of an epifluorescence microscope

80 °C according to [9]. Comparison of these two methods showed a significant underestimation of the number of viruses using flow cytometry, however, the correlation between the results of cytometry and microscopy was high (>0.7), which made it possible to intercalibrate these methods for the Black Sea virioplankton samples. The abundance of bacterioplankton was also determined by flow cytometry according to [9].

Virioplankton abundance was determined by gating the cell population on 2-parameter side-scattering (SS) cytograms and SYBR Green I fluorescence in the green region of the spectrum (FL1 channel, 525 nm) on dimensionless logarithmic scales, where populations of bacterioplankton (BAC) and viral particles (VIR). The abundance of bacterioplankton was also determined by flow cytometry [10] by staining a seawater sample with the same dye SYBR Green I, but without dilution.

The concentration of bacterio- and virioplankton cells was calculated from the sample flow rate (60 μl min^{-1}), counting time (180 s), and the number of cells registered during this period of time (at least 4000 particles in virioplankton samples). The quality control of the measurements was performed using Flow-Check™ calibration fluorospheres (Beckman Coulter) with a known concentration in the sample.

3 Results and Discussion

In accordance with the results obtained, the abundance of virioplankton in the Black Sea varied in the range from 10^6 to 10^7 cells ml^{-1}. The vertical distribution of viral particles turned out to be heterogeneous: in the aerobic zone, their number varied in the range from

0,74 to 9,36 × 10^6 cells ml^{-1} and, on average, was 4,01 ± 1,94 × 10^6 cells ml^{-1} (n = 44, standard deviation indicated here and further). At the border of the hydrogen sulfide zone, the number of virioplankton sharply decreased, and up to the maximum depths, its values changed in a narrow range (between 0,43 and 1,62 × 10^6 particles ml^{-1}, n = 15), and the average value (0,97 ± 0,34 × 10^6 particles ml^{-1}) was significantly lower than in the aerobic layer. From a depth of about 800 m to the bottom layer, a gradual increase in the number of viruses up to 2 × 10^6 cells ml^{-1} was observed, the reasons for which are not yet clear. Figure 3 summarizes the data obtained at different stations in order to demonstrate the general patterns of the vertical distribution of virioplankton in the deep part of the Black Sea.

The ratio of the numbers of viruses and bacteria (VBR) varied in a fairly wide range - from 3 to 90. The maximum and minimum were identified, respectively, at the depths of about 20 m and 160 m. The nature of the vertical distribution of this indicator was similar, with high values in the aerobic layer (35 ± 19) and low in the hydrogen sulfide zone (13 ± 8).

The ranges of changes in virioplankton abundance and VBR values in the deep part of the Black Sea, as well as their decrease with depth, corresponded to observations in other areas of the World Ocean (REFS). At the same time, we failed to reveal similarities with the vertical structure of the virioplankton of the Baltic Sea, in the anaerobic zone of

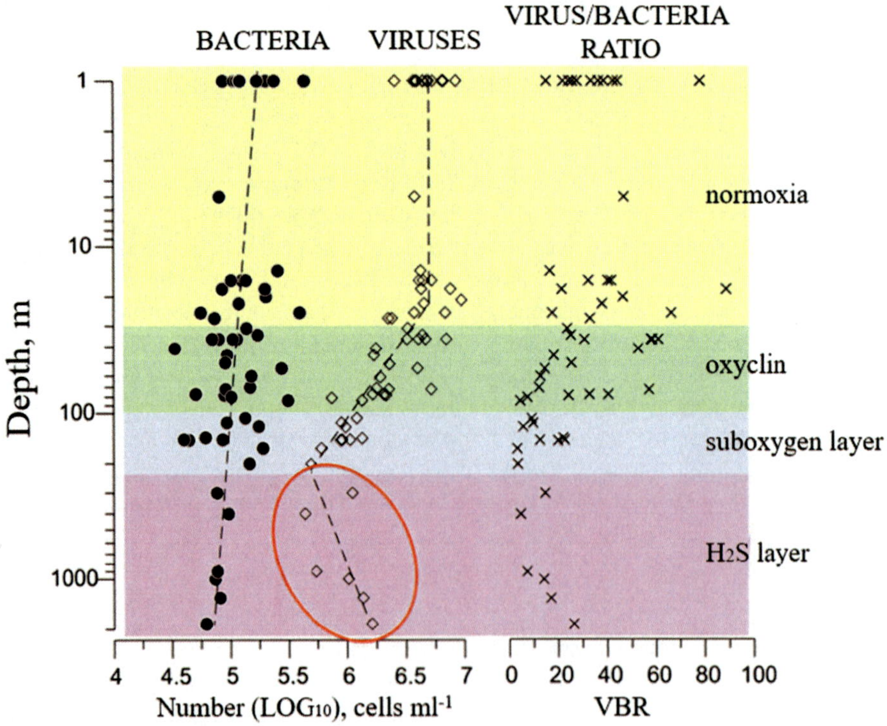

Fig. 3. Vertical distribution of bacteria, viruses, virus/bacteria ratios in the deep Black Sea

which the high numbers of viruses (up to 2×10^7 cells ml^{-1}) were due to their extremely low rates of decomposition [10].

Thus, the number of bacteria had a trend of monotonous decrease with increasing depth and, apparently, did not depend on the stratification of the water column.

Viruses, on the contrary, retained relatively high concentrations in the aerobic layer. From the oxycline to the suboxic layer, their number decreased rapidly, by more than an order of magnitude. And in the hydrogen sulfide zone, it gradually increased with depth, which looks quite intriguing. The nature of such a vertical structure of virioplankton in the Black Sea is not yet clear.

Our first data on the virioplankton of the Black Sea do not yet allow us to put forward any hypotheses in this regard, but, nevertheless, we can make assumptions based on a recent study of viruses in the Baltic Sea [10], which resembles the Black Sea in terms of its conditions. Its authors have shown that (i) redoxcline viruses of the Baltic Sea direct more resources to the formation of defense mechanisms (capsid is more durable and resistant to aggressive environment); (ii) in the redoxcline of the Baltic Sea, the low rate of elimination (destruction) of viral particles leads to an increase in their abundance. It is possible that similar mechanisms also underlie the increase in the number of virioplankton in the anaerobic depths of the Black Sea. This remains to be seen in future studies.

4 Conclusion

The estimates of virioplankton abundance (VIR) and virus/bacteria ratio (VBR) have been obtained first for the Black Sea. In autumn, on the Crimean shelf and in the deep part of the Black Sea, the number of viruses varied in the range from 0,4 to $9{,}4 \times 10^6$ cells ml^{-1} with maxima in the aerobic zone; VBR values ranged from 3 to 90.

A significant (more than an order of magnitude) decrease in the number of viral particles from the Black Sea oxycline to the suboxic layer was revealed. Within the hydrogen sulfide zone, an insignificant increase in the number of viruses with depth was observed.

Acknowledgements. The study was carried out within the framework of state assignment No. 121040600178-6 and with the financial support of the Russian Foundation for Basic Research (No. 21-55-52001).

References:

1. Suttle, C.A.: Viruses in the sea. Nature **437**(7057), 356–361 (2005)
2. Weitz, J.S., Wilhelm, S.W.: Ocean viruses and their effects on microbial communities and biogeochemical cycles. F1000 Biol. Rep. **8**, 2–9 (2012)
3. Breitbart, M., Bonnain, C., Malki, K., Sawaya, N.A.: Phage puppet masters of the marine microbial realm. Nat. Microbiol. **3**, 754–766 (2018)
4. Rylkova, O.A., Gulin, S.B., Pimenov, N.V.: Determination of the total number of microorganisms in the bottom sediments of the Black Sea by flow cytometry. Microbiology **88**(6), 685–694 (2019). (in Russian)

5. Stepanova, O.A.: Marine bacteria and viruses in water and bottom sediments near the Crimean coast. Ecol. Saf. Coast. Shelf Zones **9**, 255–261 (2003). (in Russian)
6. Mironov, O.G., Poltaruha, O.P., Kovalchuk, Yu.L.: Seasonal dynamics of virus-like particles and nucleic acids in water and bottom sediments of the Sevastopol Bay of the Black Sea. Water Chem. Ecol. **2**, 27–30 (2010). (in Russian)
7. Chen, F., Lu, J.R., Binder, B., Hodson, R.E.: Enumeration of viruses in aquatic environments using SYBR Glod stain: application of digital image analysis and flow cytometer. Appl. Environ. Microbiol. **67**, 539–545 (2001)
8. Hara, S., Terauchi, K., Koike, I.: Abundance of viruses in marine waters: assessment by epifluorescence and transmission electron microscopy. Appl. Environ. Microbiol. **57**(9), 2731–2734 (1991)
9. Brussaard, C.P.D., Payet, J.P., Winter, C., Weinbauer, M.: Quantification of aquatic viruses by flow cytometry. Man. Aquat. Viral Ecol. **11**(2004), 102–109 (2010)
10. Köstner, N., Scharnreitner, L., Jürgens, K., Labrenz, M., Herndl, G.J., Winter, C.: High viral abundance as a consequence of low viral decay in the Baltic Sea redoxcline. PLoS One **12**(6), e0178467 (2017)

MtDNA Polymorphism of *Macrohectopus branickii* Dybowsky, 1974 (Amphipoda) – An Endemic Pelagic Key Species of Lake Baikal

I. Y. Zaidykov(✉), E. Y. Naumova, and L. V. Sukhanova

Limnological Institute, Siberian Branch of the Russian Academy of Sciences, Ulan-Batorskaya Str., 3, Irkutsk 664033, Russia
igorrock11@mail.ru

Abstract. *Macrohectopus branickii* is species endemic to Lake Baikal. The number of amphipod species in Lake Baikal is over 350, but *M. branickii* is the only species with an exclusively pelagic lifestyle. This work we are study of the population structure *M. branickii* based on the molecular genetic data. In our study, we used the COI marker - fragment of the cytochrome oxidase subunit 1 of mitochondrial DNA (mtDNA). Based on these sequences, we constructed a haplotype tree using the Network software as well as a phylogenetic tree using the BEAST software packages and molecular clock calibrations. The network showed the presence of separate haplotype groups in the population structure. The dates in the nodes of the phylogenetic tree of *M. branickii* were near with those of other Baikal endemic amphipod *Gmelinoides fasciatus* Stebbing, 1899. This may indicate a commonality of events that influenced their population structure in the past, such as the influence of past geological and climatic changes.

Keywords: Pelagic species · *Macrohectopus branickii* · endemic to Lake Baikal · amphipods · COI · climate change · molecular clock

1 Introduction

Baikal is a unique freshwater lake, is the deepest (1637 m) lake of the world. It is located in the southern part of Eastern Siberia of the Russian Federation.

The only species of zooplankton in Lake Baikal is *Macrohectopus branickii*, which is endemic to the lake and has a female maximum body length of up to 38 mm, while the dwarf males have a maximum length of only 6 mm [1]. The number of amphipod species in Lake Baikal is over 350, but *M. branickii* is the only species with an exclusively pelagic lifestyle [2, 3]. This species is one of the major consumer of zooplankton and have a key role in the pelagic fish nutrition of Lake Baikal. It is occupies the same ecological niche as mysids in large European or North American lakes and oceanic krill [3, 4]. *M. branickii* inhabitants of 100–700 depths and is distributed throughout Baikal [5]. Due to its mobility, large aggregations of this species are formed in shallow water (150–250 m

T. Chaplina (Ed.): CIWO 2023, SPEES, pp. 223–229, 2023.
https://doi.org/10.1007/978-3-031-47851-2_26

depth) and in the most productive bays (Barguzin Bay and Chivyrkuy Bay). During the day, *M. branickii* may form nearbottom concentrations [1]. Vertical migration is performed by all age groups throughout the year; at night, 50–60% of individuals move from deeper layers to the surface and at dawn it migrates back to the depths [2, 3, 6].

The distribution and biology of *M. branickii* in Baikal have a long history of the study [1–10]. However, many aspects of this species ecology are still unknown or disputable. Molecular genetic studies of this species were mainly associated with phylogenetic reconstructions of amphipods [11–13]. The structure of mitochondrial DNA was also investigated [13], which showed that *M. branickii* is distinguished by the large size of the mitochondrial genome (42 Kbp). However, the molecular genetic data has not previously been used to study the population structure of the species.

2 Material and Methods

M. branickii were collected from 2010 to 2013 from a depth of 400 to 0 m using a DJOM net (modified Juday net with mesh size 160 μm). The samples were collected from 5 point of Lake Baikal (Fig. 1.A). The animals were fixed in 96% ethanol. DNA isolation from individuals was carried out using the AmpliSens ® DNA-sorb-D kit. The universal primers LCO1490 and HCO2198 are used to amplify the COI gene fragment [14].

A fragment of the nucleotide sequence with the size of 480 bp was determined for 58 samples. The data were placed in the GenBank - genetic sequence database (https://www.ncbi.nlm.nih.gov/; OR272120-OR272177).

The MEGA 7.0 program's ClustalW [15] software was used to align the sequences and to search for the best evolutionary model. Xia-test [16] in the Dambe v.6.4. [17] program was determined low saturation level of the substitutions (Table 1).

In the NETWORK 10 software [18], a median-joining algorithm was utilized to evaluate the genealogy of the haplotypes. We built a phylogenetic tree using the BEAST software using a molecular clock with calibrations developed for another Baikal amphipod *Gmelinoides fasciatus* (1.7% per million years) [19]. The sequence of amphipods *Micruropus fixseni* from GenBank (AY926681) was used as an outgroup. In a number of phylogenetic studies [18, 19] of amphipods, this species was identified as one of the genetically closest to *M. branickii*.

We were running 100 million generations Markov chain Monte Carlo (MCMC) with sampling every 10000th tree. Speciation models: yule process and birth-death process [21]; molecular clock models: strict clock and uncorrelated exponential relaxed clock [22] (Table 2) were used for phylogenetic analysis. By comparing log-likelihood values in the Tracer v1.5 program, the best evolutionary models were selected. The ESS statistics in the Tracer v. 1.5 program from the BEAST software package were used to determine the convergence of phylogenetic reconstruction.

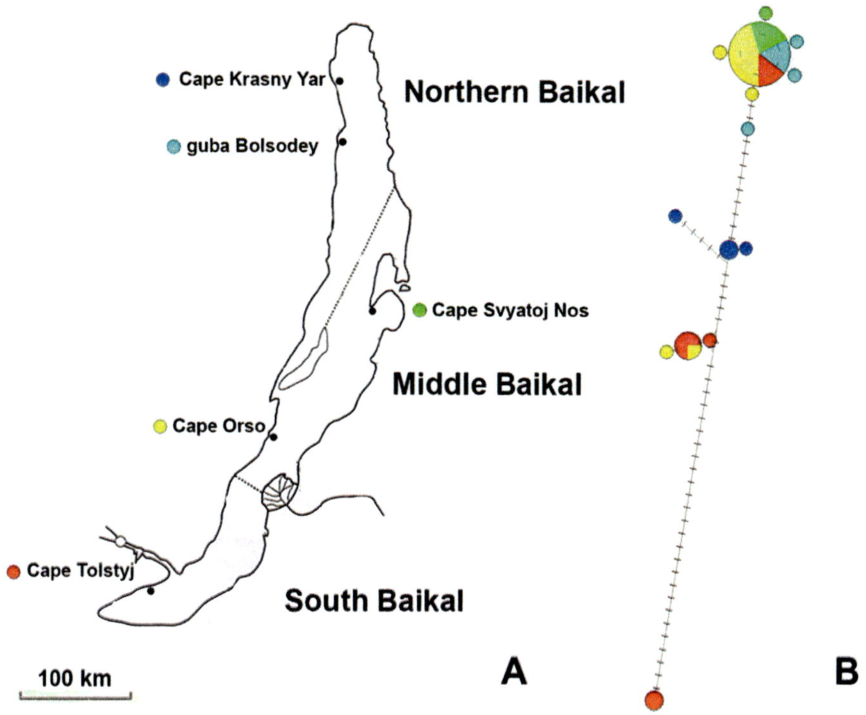

Fig. 1. A – Map of the sampling sites in Lake Baikal. B – COI haplotypes of *M. branickii* are used to construct median-joining networks. The size of the haplotype circle is proportional to the number of individuals. Node haplotypes that are missing are represented by small grey dashes. The number of mutational steps is proportional to the length of the branch. The different sampling sites are represented by colors.

Table 1. Results of Xia-test [16] on saturation of the substitution. Testing whether the observed "Iss" is significantly lower than "Iss.c". "Iss" is index of substitution saturation; "Iss.c" is critical Iss value at which the sequences will begin to fail to recover the true tree; "Iss.cSym" is "Iss.c" assuming a symmetrical topology; "Iss.cAsym" is "Iss.c" assuming an asymmetrical topology.

	M. branickii	*M. branickii* + *M. fixseni*
Iss	0,0346	0,0825
Iss.cSym	0,7128	0,7153
Iss.cAsym	0,5245	0,5375

Table 2. Evolutionary models were used for phylogenetic analysis.

Speciation model	Clock model	Log-likelihood
Birth-death	Uncorrelated exponential relaxed	−1310,537
Yule	Uncorrelated exponential relaxed	−1310,588
Birth-death	Strict	−1325,548
Yule	Strict	−1328,263

3 Results and Discussion

Analysis of the COI sequences of *M. branickii* revealed 47 variable positions. 9 nonsynonymous substitution was present. A total of 42 out of 49 positions were phylogenetically informative. Haplotype and nucleotide diversity of *Macrohectopus* were estimated as 0,523 and 0,006.

The Network program was used to construct a CO1 fragment haplotype network (Fig. 1 B). The network showed the presence of separate haplotype groups in the population structure. The largest group of haplotypes (40 out of 58 nucleotide sequences) had a geographically mixed composition it included sequences of 4 out of 5 sampling sites. Other haplotype groups were more or less geographically isolated, but consisted of a small number of sequences.

To realize the reasons for the presence of separate haplotype groups, phylogenetic trees were built (Fig. 2 B) Based on the AICc information criterion, GTR+G was chosen as the best model of nucleotide substitution. The speciation model birth-death process with uncorrelated exponential relaxed clock had the best likelihood indicators (Table 2).

Interestingly, some dates in the nodes of the phylogenetic tree of *M. branickii* were near with those of *G. fasciatus* (Fig. 2 A). This may indicate a commonality of events that influenced their population structure in the past. The population structure of *G. fasciatus* was formation by influenced of past geological and climate change [19]. During 0–1.8 million years ago, Lake Baikal experienced the most abrupt climate changes. The river influx into the lake shrinks on a periodic long-term basis due to climate changes, resulting in a decrease in the amount of biogenic elements and the lake's primary production [19, 23]. Depending on the plankton composition and age, *M. branickii* can feed phytoplankton or zooplankton and possibly golomyanka larvae [1, 2]. The decrease of the lake's primary production is not only a decrease in the number of phytoplankton, but also zooplankton, which feeds on it.

The authors of the article on *G. fasciatus* [19], arguing about the reasons for the formation of population subdivision, note: «Limited water inflow during the summer period in local sites of the Lake Baikal littoral supported the inflow of biogenic elements, providing vital activity of primary producers. In these local places of the Baikal littoral, the existence of geographically isolated groups of *G. fasciatus* individuals was maintained». Although *G. fasciatus* is a littoral species, and *M. branickii* is pelagic, this statement may also be relevant for *M. branickii* too. Perhaps it was the areas of river mouths that served as places where isolated populations of *M. branickii* were preserved

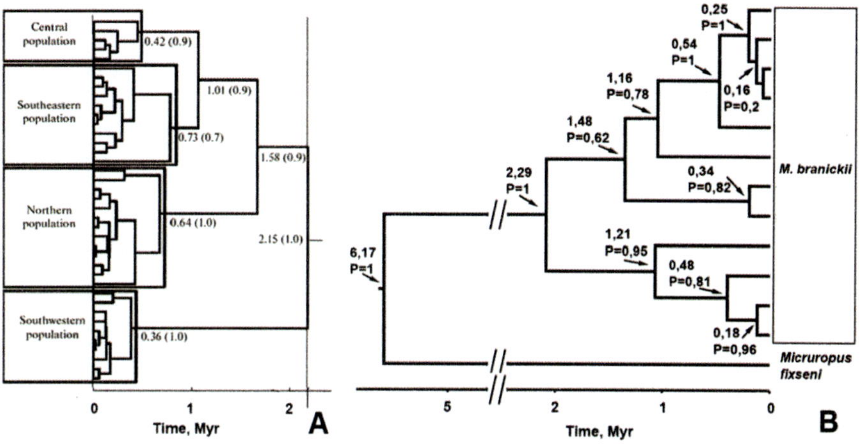

Fig. 2. **A** - from the article [19]. Phylogenetic tree for unique haplotypes of the mitochondrial COI gene from different populations of *G. fasciatus* reconstructed with the use of Bayesian analysis with molecular clock assumption. The tree nodes show dating in millions of years; in brackets are the posteriori probabilities (statistical support). **B** - A phylogenetic tree based on the unique haplotypes of the *M. branickii* COI fragment. The tree is constructed using Bayesian analysis in the BEAST software package. The tree nodes contain dates and statistical supports (P).

during periods of decline in the total primary production of the lake. It is possible that the result of such isolation we observe on the network of haplotypes.

It is not yet clear whether the structure of the COI network reflects only demographic events of the past in the now unified population of *M. branickii*. Or this structure reflects the currently existing subdivision of the population. The possibility of the existence of a group of cryptic species or subspecies of *Macrohectopus* cannot be excluded. The resolution of this issue needs further research.

4 Conclusion

Molecular genetic analysis of the population structure of *M. branickii* based on the nucleotide sequence of the mtDNA COI gene revealed separate groups of haplotypes. The most obvious reason for this haplotype subdivision could be the climatic and geological changes of the past that took place in Lake Baikal. But whether such subdivision is a reflection of the current state of the *M. branickii* population is a matter for further research.

Acknowledgements. This research was funded by the Russian governmental projects nos. 0279–2021–0005.

References

1. Naumova, E.Y., Zaidykov, I.Y., Makarov, M.M.: Recent quantitative values of Macrohectopus branickii (Dyb.) (amphipoda) from Lake Baikal. J. Gt. Lakes Res. **46**, 48–52 (2020)

2. Kozhov, M.M.: Lake Baikal and Its Life, p. 344. DR. W. Junk Publishers, The Hague (1963)
3. Melnik, N.G., Timoshkin, O.A., Sideleva, V.G.: Distribution and some characteristics of Macrohectopus branickii (Dyb.) ecology. In: Guide and Key to Pelagic Animals of Baikal, pp. 511–522 (1995). (in Russian)
4. Rudstam, L.G., Melnik, N.G., Timoshkin, O.A., Hansson, S., Pushkin, S.V., Nemov, V.: Diel dinamics of an aggregation of Macrohectopus branickii (Dyb.) (Amphipoda, Gammaridae) in the Barguzin Bay, Lake Baikal, Russia. J. Great Lakes Res. **18**(2), 286–297 (1992)
5. Bekman, M.Yu., Afanasyeva, E.L.: Distribution and production of *Macrohectopus*. In: Biological Productivity of Baikal Pelagic Region and Its Dynamics, pp. 76–98. Nauka, Novosibirsk (1977). (in Russian)
6. Karnaukhov, D.Yu., Dolinskaya, E.A., Biritskaya, S.A., Teplykh, M.A., Khomich, A.S., Silow, E.A.: Effect of artificial light on the migratory activity of the pelagic amphipod Macrohectopus branickii during daily vertical migration in Lake Baikal. Ecol. Environ. Conserv. **25**(4), 208–210 (2019)
7. Vilisova, I.K.: To the ecology of the Baikal pelagic amphipod Macrohectopus branickii Dyb. In: Systematics and Ecology of Crustaceans of Lake Baikal, pp. 156–171. Nauka, Novosibirsk (1962). (in Russian)
8. Melnik, N.G., Timoshkin, O.A., Sideleva, V.G., Pushkin, S.V., Mamylov, V.S.: Hydroacoustic measurement of the density of Baikal macrozooplankter Macrohectopus branickii. Limnol. Oceanogr. **38**, 425–434 (1993)
9. Tereza, E.P., et al.: Vertical distribution and feeding activity of *Epischura baicalensis* Sars (Copepoda) nauplii in response to two predators in Lake Baikal in winter. Fundam. Appl. Limnol. **169**(3), 211–216 (2007)
10. Karnaukhov, D.Y., Biritskaya, S.A., Dolinskaya, E.M., Silow, E.A.: Some traits of the pelagic amphipod *Macrohectopus branickii* (Dyb.) distribution in Lake Baikal. UPI J. Chem. Life Sci. **1**(2), JCLS11 (2018)
11. Macdonald III, K.S., Yampolsky, L., Duffy, J.E.: Molecular and morphological evolution of the amphipod radiation of Lake Baikal. Mol. Phylogenet. Evol. **35**, 323–343 (2005)
12. Mats, V.D., Shcherbakov, D.Y., Efimova, I.M.: Late Cretaceous–Cenozoic history of the Lake Baikal depression and formation of its unique biodiversity. Stratigr. Geol. Correl. **19**(4), 404–423 (2011)
13. Romanova, E.V., Bukin, Y.S., Mikhailov, K.V., Logacheva, M.D., Aleoshin, V.V., Sherbakov, D.Y.: The Mitochondrial Genome of a Freshwater Pelagic Amphipod Macrohectopus branickii is among the longest in Metazoa. Genes **12**, 2030 (2021)
14. Folmer, O., Black, M., Hoen, W., et al.: DNA primers for amplification of mitochondrial cytochrome c oxidase subunit I from diverse metazoan invertebrates. Mol. Mar. Biotechnol. **3**, 294–299 (1994)
15. Kumar, S., Stecher, G., Tamura, K.: MEGA7: molecular evolutionary genetics analysis Version 7.0 for bigger datasets. Mol. Biol. Evol. **33**, 1870–1874 (2016)
16. Xia, X., Xie, Z., Salemi, M., et al.: An index of substitution saturation and its application. Mol. Phylogenet. Evol. **26**(1), 1–7 (2003)
17. Xia, X., Lemey, P.: Assessing Substitution Saturation with DAMBE. The Phylogenetic Handbook, pp. 615–630. Cambridge University Press, Cambridge (2009)
18. Bandelt, H.-J., Forster, P., Rohl, A.: Median-joining networks for inferring intraspecific phylogenies. Mol. Biol. Evol. **16**, 37–48 (1999)
19. Bukin, Yu.S., Petunina, J.V., Sherbakov, D.Yu.: The mechanisms for genetic diversity of Baikal endemic amphipod *Gmelinoides fasciatus*: relationships between the population processes and paleoclimatic history of the lake. Russ. J. Genet. **54**, 1059–1068 (2018)
20. Drummond, A.J., Rambaut, A.: BEAST: Bayesian evolutionary analysis by sampling trees. BMC Evol. Biol. **7**(1), 214 (2007)

21. Gernhard, T.: The conditioned reconstructed process. J. Theor. Biol. **253**(4), 769–778 (2008)
22. Drummond, A.J., Ho, S.Y.W., Phillips, M.J., Rambaut, A.: Relaxed phylogenetics and dating with confidence. PLoS Biol. **4**, e88 (2006)
23. Mats, V.D., Ufimtsev, G.F., Mandelbaum, M.M., et al.: The Baikal Basin in the Cenozoic: Structure and Geologic History, p. 252 Publishing House of SB RAS, Novosibirsk (2001)

Ocean Chemistry

Marine Carbonate System Parameters of the West Spitsbergen Fjords in Late Summer 2022

N. K. Alekseeva[1,2](✉), A. L. Nikulina[1], I. V. Ryzhov[1], A. E. Novikhin[1], R. V. Kornilova[1], N. A. Smirnov[1], and A. A. Fedorova[1]

[1] Arctic and Antarctic Research Institute, Saint Petersburg 199397, Russia
nkalekseeva@aari.ru

[2] Saint-Petersburg State University, Saint Petersburg 199034, Russia

Abstract. The focus of research was to assess the current state of carbonate system parameters and CO_2 air-sea flux also. We used the data from expeditions to the Isfjorden and Grønfjorden, West Spitsbergen, in September 2022. We carried out an investigation of surface distribution of total alkalinity, pH, salinity, temperature and meteorological observations of CO_2 air concentration and wind speed. The obtained data were applied for carbonate parameters calculation. Our observations identify a minor change of distribution of carbonate system parameters. The results showed that both fjords are characterized by carbon dioxide invasion and sufficiently saturated with aragonite and calcite.

Keywords: Carbonate system · CO_2 · alkalinity · pH · Isfjorden · Grønfjorden

1 Introduction

The Arctic Ocean is particularly vulnerable to ocean acidification, a process that is mainly driven by the uptake of anthropogenic carbon from the atmosphere [1, 2]. The dissolving of carbon dioxide (CO_2) acidifies the seawater (lowers pH) and shifts the equilibrium of carbonate species, decreasing carbonate ion and increasing bicarbonate concentration [3, 4]. The carbonate system is very important since it regulates the pH of seawater, and controls the circulation of carbon dioxide between the biosphere, the lithosphere, the atmosphere and the oceans [5]. Its main parameters are total alkalinity (Alk), pH, partial pressure of carbon dioxide (pCO_2), hydrocarbonate (HCO_3^-) and carbonate (CO_3^{2-}) ions, dissolved carbon dioxide (CO_2), total dissolved inorganic carbon (TIC), saturation state for aragonite (ΩAr) and calcite (ΩCa) [6, 7]. Therefore, knowing these parameters we can make a quantitative assessment of whether the ocean absorbs or releases carbon dioxide into the atmosphere. This observation focused on assessing the current state of carbonate system parameters. We carried out our study in Isfjorden and Grønfjorden in the late summer period. Important to note that in the study area, there is not much field data of total alkalinity and calculated parameters of the carbonate system. Thus, our work makes some contribution to replenishing the database for marine carbonate system of the fjords.

T. Chaplina (Ed.): CIWO 2023, SPEES, pp. 233–238, 2023.
https://doi.org/10.1007/978-3-031-47851-2_27

2 Material and Methods

The Western Spitsbergen fjords are located in the active interaction zone of warm and saline Atlantic waters inflow and entering the Arctic waters from the Barents Sea. Study areas include Isfjorden is one of the largest fjords in the Arctic and second longest in the Svalbard archipelago. Grønfjorden is a relatively small fjord located on the south side of Isfjorden, near its mouth. Atlantic waters spread along the archipelago in the form of the West Spitsbergen Current, which is a continuation of the Norwegian Current. Arctic waters with lower salinity and temperature values compared to Atlantic Water enter the area of the continental shelf of Western Spitsbergen with the waters of the East Spitsbergen Current, making a border for the Western Spitsbergen fjords from the direct constant influence of Atlantic Water. Thus, frontal zone is formed between the Atlantic and Arctic waters. Continental glaciers and river runoff are the main source of fresh water entering the fjords. For both fjords, there are four types of water masses: surface water, intermediate water, transformed Atlantic water, and Atlantic water [8].

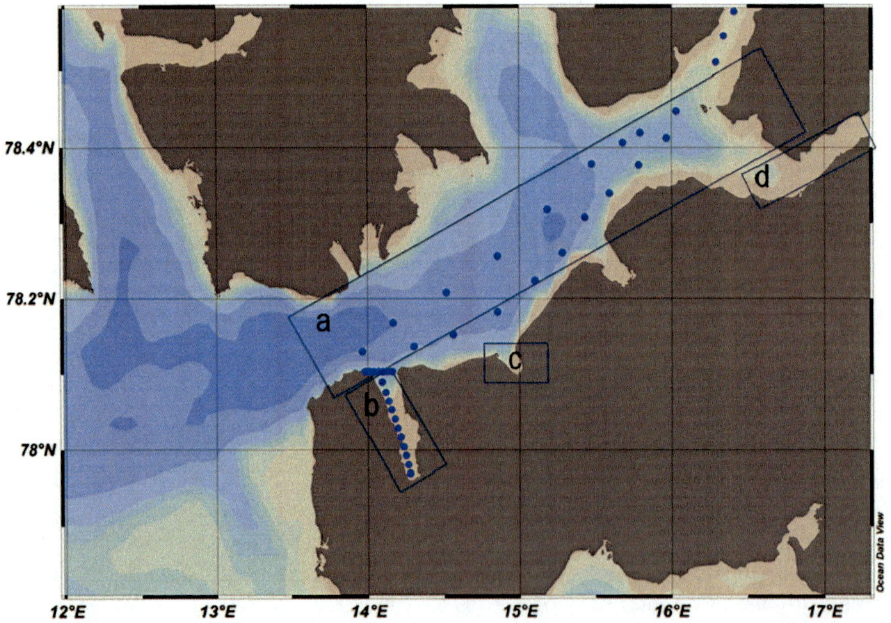

Fig.1. Study area in September 2022, where a - Isfjorden, b - Grønfjorden, c - Colesbuckta, d - Temlefjord.

We used data from the expedition to Spitsbergen archipelago in September 2022. Study area shown in Fig. 1. The oceanological and hydrochemical works carried out in Isfjorden and Grønfjorden. We collected surface water samples and analyzed them in an analytical laboratory in Barentsburg. We used a CTD profiler SeaBird SBE-19plusV2 SeaCat for temperature, salinity and pressure measuring. First of all, samples were analyzed for pH and total alkalinity. The pH measurement analyzed using with pH meter

"Mettler Toledo Seven Compact S220". Calibration was carried out in three points using buffer solutions 4.01, 7.0, 9.21 Mettler Toledo. To calculate the parameters of the carbonate system, the NBS pH scale was used. Determination of total alkalinity involved sample titration with a hydrochloric acid solution with simultaneously passing a stream of CO_2-free air through the titrated sample to an equivalence point of pH 5.5. Titrimetric analysis was performed with a Biohit Biotrate 50 ml digital burette. The equilibrium carbon dioxide partial pressure (pCO_2), hydrocarbonate (HCO_3^-) and carbonate (CO_3^{2-}) ions, total dissolved carbon dioxide (TIC), total inorganic carbon and saturation state for aragonite and calcite calculated using the Program Developed for CO_2 System calculations [9]. Input data included NBS scale for pH, total alkalinity (TA), pressure, salinity and the carbonic acid dissociation constants (K_1 and K_2) of Mehrbach, as refitted by Dickson and Millero. The carbon dioxide air-sea flux was estimated using cubic parameterization of Wanninkov. The carbon dioxide air concentration was measured using with CO12M analyzer of carbon monoxide and carbon dioxide by Environment S.A. The wind speed in the surface layer of the atmosphere (up to 2 m) was determined using with the Campbell scientific meteorological complex.

3 Results and Discussion

During the late summer expedition of 2022 we obtained an insignificant spatial spread of hydrophysical and carbonate parameters. The distribution of temperature in the surface layer ranged from 5.4 °C in Grønfjorden to 6.1 °C in Isfjorden. The surface salinity changes were not significant from Grønfjorden (31.4 psu) to Isfjorden (30.9 psu). The concentration of air carbon dioxide measured about 412 ppm for 08 September 2022 and 424 ppm for 13 September 2022 during fieldwork in Isfjorden and Grønfjorden respectively. Wind speed registered 1.4 m/sec 08 September 2022 and 3, 4 m/s (13 September 2022).

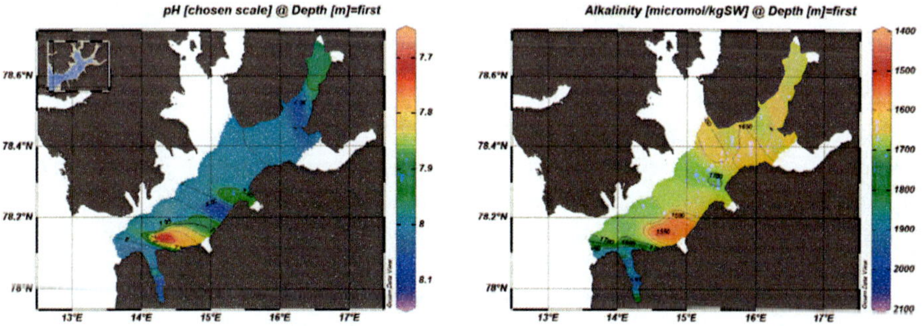

Fig. 2. The surface distribution of pH and Total Alkalinity in the Isfjorden and Grønfjorden.

Figures 2, 3 and 4 present the results of chemical compounds distribution in surface layer. The concentration of total alkalinity (TA) ranged from 1436 to 2099 μmol/kg seawater (SW) for both fjords (see Fig. 2). These values differed from the open oceans and seas values (2000–2500 μmol/kg SW). Thus, 1540–1711 μmol/kg SW for alkalinity

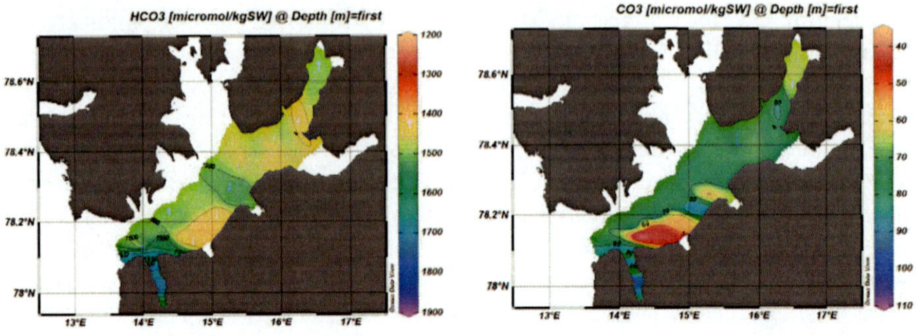

Fig. 3. The surface distribution of HCO_3^- and CO_3^{2-} in the Isfjorden and Grønfjorden.

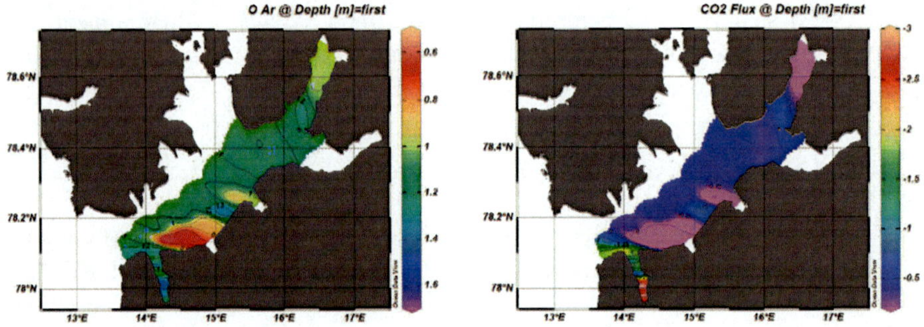

Fig. 4. The surface distribution of saturation state for Aragonite (ΩAr) and CO_2 flux in the Isfjorden and Grønfjorden.

in Isfjorden and 1436–2099 μmol/kg SW in Grønfjorden. The pH scale ranged from 7.7–8.12. An average value of pH 7.97 in Isfjorden and 8.02 in Grønfjorden (see Fig. 2). Carbonate system parameters spatial distribution are defined by following changes. The partial pressure of carbon dioxide in surface water varied from 219 to 693 ppm. The dissociation products of carbonic acid were in the form of a hydrocarbonate (HCO_3^-) ion and a carbonate (CO_3^{2-}) ion in Grønfjorden. The average values are 1667 μmol/kg SW for HCO_3^- and 83.5 μmol/kg SW for CO_3^{2-}. The average values are 1472 μmol/kg SW for HCO_3^- and 68 μmol/kg SW for CO_3^{2-} in Isfjorden (see Fig. 3). The total amount of inorganic carbon registered about 1559 μmol/kg SW and about 1769 μmol/kg SW in Isfjorden and in Grønfjorden respectively. The saturation state for aragonite indicated 1.04 in Isfjorden and 1.27 in Grønfjorden (see Fig. 4), which provides favorable environmental conditions for marine organisms. The determined concentrations of dissolved carbon dioxide changed from 11.5 μmol/kg SW to 36 μmol/kg SW. The air-sea CO_2 flux estimation showed negative values −0.2 mmol/m^2 *day in Isfjorden and −1.9 mmol/m^2 *day in Grønfjorden (see Fig. 4).

We detected a plume in the part of Isfjorden near Colesbuckta (see Fig. 1c). The measured parameters differenced from the open water fjords and detected a more low value. The alkalinity concentration was about 1400 μmol/kg SW. The pH value was in the range of 7.7–7.9. The saturation state for aragonite was 0.8. The plume appearance can be explained by the transformation of the terrigenous runoff into the open part of the fjord. We noted that main characteristics are typical for river and glacial runoff. The surface layer of Grønfjorden southern part has relatively higher silicate concentrations. It could be suggested that it locally are more influenced by fresh river and glacier runoff while the northern open part reveal more marine conditions than Isfjorden in general.

Both regions have quite low surface nutrient concentrations with no significant evidence of active organic matter destruction. From other hand, the results shows that relatively small Grönfjorden demonstrates about 10 times higher CO_2 flux from the atmosphere into the water column. At the same time average pH values in both fjords are quite similar 8.02 for Grønfjorden and 7.97 in Isfjorden. One of suggestions could be more active photosynthetic process in Grønfjorden as a region in the frontal zone between marine and terrigenous waters, which became more evident in small scales of Grønfjorden.

Similar research carried out at Templefjord [10]. The research group performed a complex investigation of the seasonal changes of the biogeochemical regime, including the parameters of the carbonate system. They used data from five expeditions to the Templefjord (see Fig. 1d), West Spitsbergen in different seasons (February 2011, September 2011, March 2014, June 2015 and July 2017). We have chosen to compare September 2011 with our data for September 2022. The Templefjord surface layer characterized by 1988 μmol/kg for alkalinity, where alkalinity decreased from the glacier to the central fjord, from 1756 μmol/kg to 1988 μmol/kg. TIC increased from 1635 μmol/kg to 1828 μmol/kg and HCO_3 from 1531 μmol/kg to 1700 μmol/kg. Our results are looking similar with ones obtained in Templfjord [10]. For example, the authors also revealed in the fjord the river plume with low alkalinity and high silicate concentrations.

4 Conclusion

During late summer expedition of 2022, we found some difference of spatial carbonate parameters distribution in the fjords surface layer. Presumably, the more actively biogeochemical processes observed in the relatively small Grønfjorden due to a more closed area location in the frontal zone between marine and terrigenous waters.

The air-sea CO_2 flux depends on the difference between the partial pressure of carbon dioxide in air (pCO2) and pCO_2 in the water. When the pCO_2 in the air is higher, then carbon dioxide invades water, if the pCO_2 (air) is lower, than carbon dioxide evasion [11]. We detected carbon dioxide invasion in both fjords. The surface layer of the sea is sufficiently saturated with aragonite ($\Omega Ar > 1$). The saturation for aragonite of fjords surface layer provides a favorable environment for marine organisms: corals, mollusks etc.

During the study, a plume recorded near Colesbukhta and its presence is most likely due to terrigenous runoff.

The obtained results based on a short time scale and local observations. To understand in more details the processes in West Spitsbergen fjords it is necessary to get more data from the study region.

Acknowledgements. Our research group would like to thank Russian scientific center on Spitsbergen (Svalbard) archipelago for logistical support of expedition. This study was carried out within the project "United National System for Monitoring Climate-Active Substances" (Agreement 169-15-2023-002).

References

1. Terhaar, J., Tanhua, T., Stöven, T., Orr, J.C., Bopp, L.: Evaluation of data-based estimates of anthropogenic carbon in the Arctic Ocean. J. Geophys. Res. Oceans **125**(6), 3–15 (2020)
2. Zeebe, R.E.: History of seawater carbonate chemistry, atmospheric CO2, and ocean acidification. Annu. Rev. Earth Planet Sci. **40**, 141–165 (2012)
3. Friedlingstein, P., Jones, M.W., O'Sullivan, M., Andrew, R.M., Bakker, D.C.E., Hauck, J., et al.: Global carbon budget 2021. Earth Syst. Sci. Data **14**, 1917–2005 (2018)
4. Sabine, C.L., Tanhua, T.: Estimation of anthropogenic CO_2 inventories in the ocean. Annu. Rev. Mar. Sci. **2**, 175–198 (2010)
5. Millero E.J.: The Carbonate System in Marine Environments. Chemical Processes in Marine Environments. Springer, Berlin (2000). 446 p.
6. Dickson, A.G.: The carbon dioxide system in seawater: equilibrium chemistry and measurements. In: Guide to Best Practices for Ocean Acidification Research and Data Reporting, pp. 17–40 (2010)
7. Millero, F.J.: The marine inorganic carbon cycle. Chem. Rev. **107**, 308–341 (2007)
8. Bloshkina, E.V., Filchuk, K.V.: The present water masses conditions of West Spitsbergen fjords. Problemy Arktiki i Antarktiki **64**(2), 125–140 (2018). (in Russian)
9. Robbins, L.L., Hansen, M.E., Kleypas, J.A., Meylan, S.C.: CO2calc- a user-friendly seawater carbon calculator for Windows, Max OS X, and iOS (iPhone). U.S. Geological Survey Open-File Report 2010, vol. 1280, p. 17 (2010)
10. Pogojeva, M., Polukhin, A., Makkaveev, P., Staalstrøm, A., Berezina, A., Yakushev, E.: Arctic inshore biogeochemical regime influenced by coastal runoff and glacial melting (case study for the Templefjord, Spitsbergen). Geosciences **12**, 44 (2022)
11. Malinin, V.N., Obratsova, A.A.: Variability of carbon dioxide exchange in the ocean-atmosphere system. Society Environ. Dev **4**, 220–226 (2011). (in Russian)

The Hydrochemical Composition of the Watercourses of the Bay Oga, Tzivolki, Sedova (The Coast of the Kara Sea, Novaya Zemlya)

G. V. Borisenko(✉), A. A. Polukhin, and D. V. Sharmar

Federal State Budgetary Institution of Science Shirshov Institute of Oceanology of the Russian Academy of Sciences, Moscow 117997, Russia
gennady.val.borisenko@gmail.com

Abstract. Hydrochemical studies of watercourses and the water area of Tzivolki, Oga, Sedova bays (Novaya Zemlya, Arctic, Russia) were carried out. The concentrations of nutrients in rivers and streams are higher than the concentrations in the water area. It is shown that the concentration of silicon in constantly flowing rivers is 1–13 μM, the concentration of NO^{3-} 0.5–8, for watercourses flowing directly from the glacier these values are in the range of 7–10 μM Si, 3 μM NO^{3-}. The influence of streams and rivers flowing into the Bays on the water area of the bay is local and extends to 1–5 km from the mouth, and does not influence on Kara Sea nutrient content.

Keywords: Kara Sea · Novaya Zemlya · nutrients · river flow · streams

1 Introduction

From 1971 to 2019, an increase in the average temperature of the surface layer in the Arctic by 3.1 °C was recorded; the average summer temperature reached 10.6 °C. Marine-type glaciers in direct contact with the sea have shrunk and continue to retreat. Since the end of the 20th century on Novaya Zemlya (an archipelago between the Kara and Barents Seas), the area of glaciers has decreased by 1000 km^2, the volume of ice has decreased by 380 km^3 [1]. Various studies have shown that the reduction in the area of ice leads to changes in the hydrochemical parameters of the adjacent water area. Thus, using icelandic rivers as an example, it is shown that an increase in river runoff leads to an increase in mechanical and chemical weathering of rocks [2]. Weathering products enter the fjords and enrich the waters of the fjords with carbonate ions and metal ions. Thus, there is a relationship between changes in river runoff and the intensity of rock weathering [3].

Novaya Zemlya - an archipelago located between the Kara and Barents Seas, is a flooded mountain range and consists of 2 large islands - Severny and Yuzhny. The coastline is indented by narrow bays - fjords. There is very little research on the bays of Novaya Zemlya; the Blagopoluchia Bay on the Severny Island is the best studied. It is

T. Chaplina (Ed.): CIWO 2023, SPEES, pp. 239–244, 2023.
https://doi.org/10.1007/978-3-031-47851-2_28

known that the main source of desalination in it is glacial runoff, and one of the main sources of biogenic elements (N, P, Si) in the water area of the bay are small streams and rivers flowing into the bay. In addition, it has been experimentally shown that rocks can release inorganic nitrogen and silicon. However, the influence of watercourses is limited to the water area of the bay and has little effect on the Kara Sea adjacent to Novaya Zemlya [4].

In the current work, we consider the hydrochemical parameters (concentrations of nitrate nitrogen, dissolved silicon) of the Oga, Tsivolki, and Sedova bays. The study of the concentrations of nutrients in the Kara Sea is especially important in the aspect of the oligotrophy of the Kara Sea. Climate change can change the hydrochemical regime of Novaya Zemlya fjords and, possibly, the Kara Sea [5].

Sedova Bay is located in the central part of the North Island. The length of Sedova Bay is $\approx$ 20 km. A large number of small streams-watercourses flow into the bay **The Bay of Oga** is 24 km long. The Goluboy glacier is unloading into the bay. **The Tzivolki Bay** protrudes 30 km deep into the Severny Island in the NW direction to the frontal part of the Serp i Molot glacier. Watercourses of 2 types flow into the Tsivolki Bay: those with a catchment area and those flowing directly from the glacier. The location of the bays on the map is shown in Fig. 1 [6, 7].

2 Materials and Methods

The work is based on hydrochemical data obtained during the oceanographic cruises of the vessels "Akademik Mstislav Keldysh" and "Professor Shtokman" in 2014–2016. In all years, the time of sampling coincided and fell on the end of August - beginning of September. Water samples were taken both from the vessel and during landings on the shores of the Tzivolki, Oga, and Sedova bays.

Water samples in rivers and in marine part of Bay were sampled with a plastic Niskin and then transferred into specially prepared glass and plastic bottles with screw caps, if necessary, canned and stored at a low temperature without access to light or in a freezer. Water samples from rivers were preliminarily filtered through Millipore filters with a pore size of 0.45 μm. Water temperature and salinity were measured using a CTD probe SeaBird SBE19plus. Water temperature and salinity during the landings were determined using a portable salt thermometer [8].

Determination of dissolved silicon was performed by blue molybdenum complex. Traces of silica are determinable down to 0.28 μM $SiO_3{}^{2-}$ liter-1 with an error of $\pm 3\%$. The determination of dissolved forms of nitrogen (nitrates) was carried out colorimetrically using the Griess reagent after reduction of nitrates to nitrites in cadmium reducers. The precision of the method is 0.02 μM. ERA5 reanalysis data were used to identify the wind speed and direction [9].

3 Results and Discussion

The results of measurements of the concentrations of biogenic elements (N, Si) in watercourses flowing into the Tzivolki Bay are shown in Figs. 2. Measurements 2014–2016 showed that the highest concentrations of nitrate nitrogen in watercourses are observed

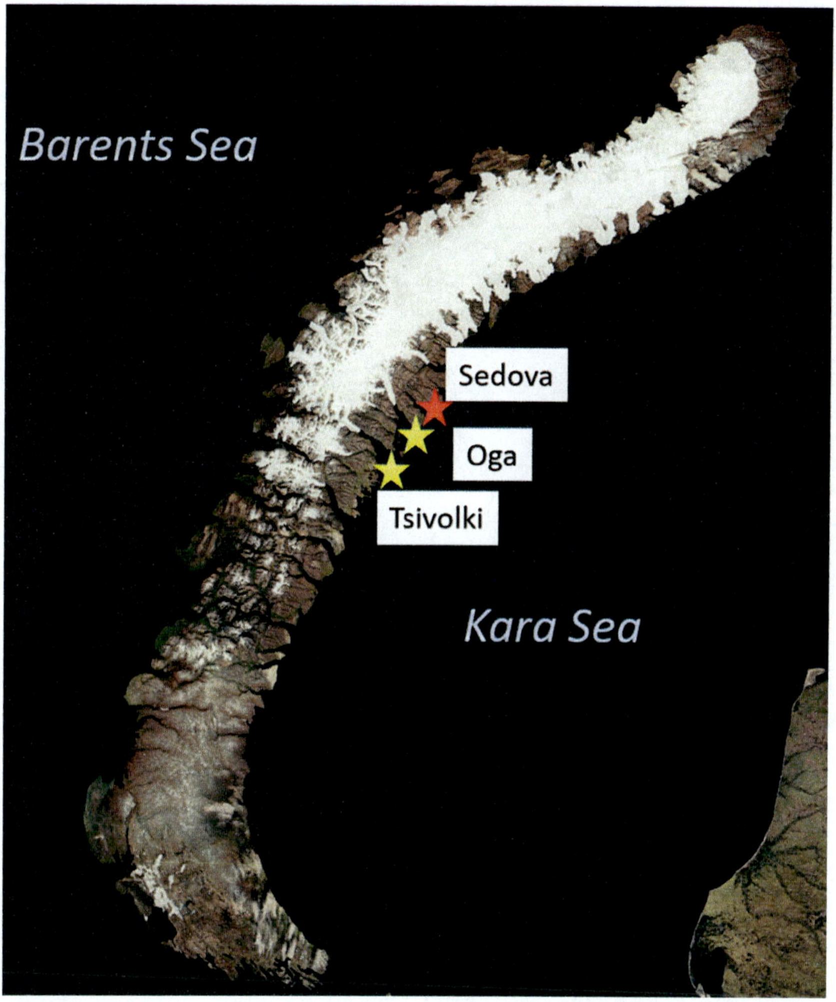

Fig. 1. Location of the Tzivolki, Oga, Sedova bays on the Novaya Zemlya archipelago

in watercourses with a certain catchment area, the average value is 2.5 μM, at the same time, watercourses flowing directly from the Serp i Molot glacier contain an average of 0.5 μM NO^{3-}. The concentrations of dissolved silicon in the watercourses of both types are comparable.

Differences in the content of nitrate nitrogen are connected with the different weathering ability of watercourses flowing directly from the glacier, and watercourses with a catchment area. On the example of the fjords of Svalbard and Greenland, it is shown that the type of watercourse is important for the concentration of dissolved substances. The largest amount of dissolved silicon (up to 200 μM in Greenland and up to 60 μM in Svalbard) was found in streams flowing from recently deglaciated slopes.

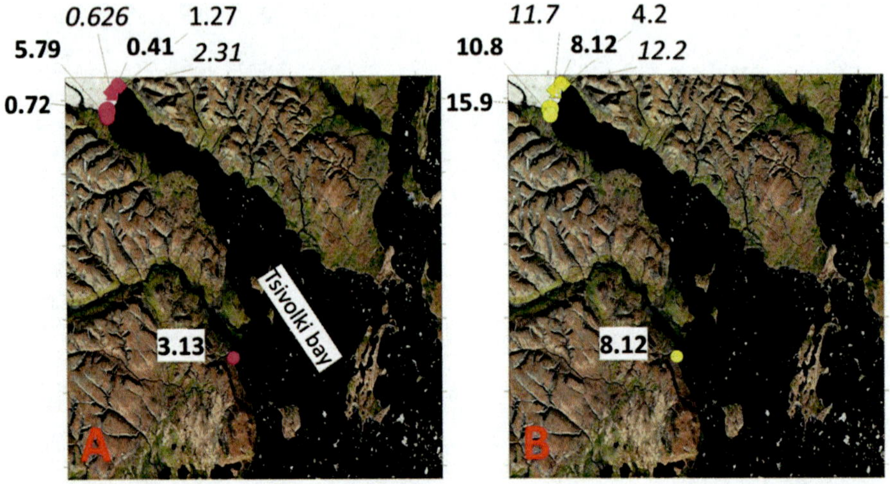

Fig. 2. The concentrations of nitrate nitrogen (A) and dissolved silicon in watercourses flowing into the Tsivolki. Different font types reflect different years: 2014 bold, 2015 italic, 2016 unchanged.

During landings to the shores of Oga and Sedov bays in 2016, water samples were collected from streams and rivers. The results of measurements of the concentrations of biogenic elements are presented in Fig. 3.

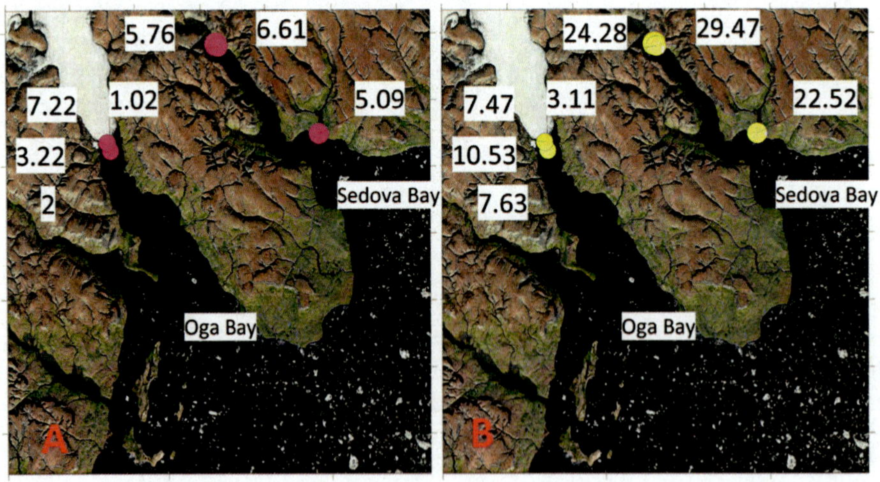

Fig. 3. The concentrations of nitrate nitrogen (A) and dissolved silicon (B) in watercourses flowing into the Sedova Bay and Oga Bay in 2016.

The average concentration of nitrate nitrogen in the watercourses flowing down from the Goluboy glacier (Oga bay) was 3.4 μM; Sedova, this value was about 5.5 μM. The highest concentrations of dissolved silicon were also observed in watercourses Sedova bay (25 μM on average), while in Oga, this value averaged 6 μM. Field measurements were obtained with a difference of 3 days. In Oga and Sedova bays, the type of watercourses also differs: in Oga Bay - watercourses flowing directly from the glacier, in Sedov Bay - watercourses with a certain catchment area.

The influence of watercourses on the adjacent water area of the bay is local, nutrients are distributed at a distance of 5 km from the source of supply. According to the literature data available for the fjords of the Greenland Ice Sheet, the distance of influence of watercourses determined by the fjord boundary. For the Godthabsfjord fjord, the nutrient signal is visible at a distance of 20 km, which is primarily determined by the difference in the location of the mouths of the watercourses. For Young Sound Fjord, the distance of source influence (increased silicon concentrations) is also 20 km [10]. The distribution of nutrients in the water area of the bay, in our opinion, is influenced by the morphology of the bay (whether it is open or closed from the waters of the Kara Sea), wind conditions during the observation period. According to the wind speed and direction data, which were obtained from the ERA5 reanalysis, it was found that the wind situation during the period of changes affects the advection of waters from the Kara Sea. Thus, the bays of Novaya Zemlya are the interface between the sea and land in the area of the western part of the Kara Sea.

4 Conclusion

On the example of watercourses b. Tzivolki, Oga, Sedov showed that the total concentration of nitrate nitrogen and dissolved silicon is influenced by the type of watercourse - flowing directly from the glacier, or having a catchment area. Earlier, in the course of laboratory experiments, it was shown that the ankeritolite shales that make up the shores of Novaya Zemlya can be a source of nutrient in water. Thus, the presence of high concentrations of biogenic elements for the Kara Sea, we associate with the leaching of nutrients from rocks. We observe this process in the current study.

Acknowledgements. The author thanks the staff of the Laboratory of Biohydrochemistry of the IO RAS and personally Academician M.V. Flint for their participation in the work.

References:

1. AMAP Climate Change Update 2019: An Update to Key Findings of Snow. Water, Ice and Permafrost in the Arctic (SWIPA), C. 12 (2017)
2. Gislason, S.R., et al.: Direct evidence of the feedback between climate and weathering. Earth Planet. Sci. Lett. **277**(1–2), 213–222 (2009)
3. Polukhin, A., et al.: Leaching of inorganic carbon and nutrients from rocks of the Arctic archipelagos (Novaya Zemlya and Svalbard). Russ. J. Earth Sci. **21**(4), ES4002 (2021)
4. Dubinina, E.O., Kossova, S.A., Miroshnikov, A.Y.: Sources and mechanisms of seawater freshening in Tsivolky and Sedov Bays (Novaya Zemlya Archipelago) based on isotope data (δD and δ 18 O). Oceanology **59**, 836–847 (2019)
5. Udalov, A.A., et al.: Bottom communities of Sedova Bay (Novaya Zemlya, the Kara Sea). Oceanology **60**, 617–624 (2020)

6. Udalov, A.A., et al.: Benthic fauna of Oga Bay (Novaya Zemlya, Kara Sea). Oceanology **59**, 931–940 (2019)
7. Chava, A.I., et al.: Benthic fauna of Tsivolki Bay (Novaya Zemlya Archipelago, Kara Sea). Oceanology **57**, 144–153 (2017)
8. Grasshoff, K., Kremling, K., Ehrhardt, M. (eds.): Methods of Seawater Analysis. Wiley, Hoboken (2009)
9. Hersbach, H., et al.: The ERA5 global reanalysis. Q J. Royal Meteorol. Soc. **146**(730), 1999–2049 (2020)
10. Meire, L., et al.: High export of dissolved silica from the Greenland Ice Sheet. Geophys. Res. Lett. **43**(17), 9173–9182 (2016)

Characteristics of Bottom Sediments of the North-Eastern Part of the Black Sea Coastal Zone

Y. S. Gurova(✉), K. I. Gurov, and N. A. Orekhova

Marine Hydrophysical Institute of RAS, Sevastopol 299011, Russia
kurinnaya-jul@yandex.ru

Abstract. The features of the spatial distribution of geochemical characteristics and the chemical profile of the pore waters of bottom sediments of the north-eastern coastal zone of Black Sea are studied. It has been established that the surface (0–5 cm) sediment layer is formed mainly by silty material, the proportion of which increases with increasing sampling depth, as well as in the areas where large rivers of the Caucasus flow in. The results of the analysis of the chemical composition of pore water showed that in most of the studied areas, oxic conditions prevail in the upper layer of sediments and an oxygen fluxes are formed at the boundary with the bottom water layer. The oxygen flux values are determined by the oxygen concentration gradient between the surface layer of sediments and the bottom layer of water, as well as the granulometric composition of the sediment.

Keywords: Bottom sediments · Pore water · Oxygen · Granulometric composition · Organic carbon · Black Sea

1 Introduction

The criteria for assessing the ecological state of marine ecosystems include the concentration of oxygen in the near-bottom water layer and the upper layer of bottom sediments, the content of organic matter, and the granulometric composition of sediments. Also, the ecological situation is influenced by water dynamics - stratification of the water column and difficult water exchange limit the flow of oxygen into sediments, which contributes to the development of its deficiency and the emergence of environmental risk zones.

The region of the north-eastern part of the Black Sea is the area from the Taman Peninsula to the city of Adler, characterized by intense water dynamics and variability of coastal currents [1, 2]. The area is also characterized by a narrow shelf [3].

The north-eastern coast is characterized by recreational and sanitary-resort activities, as a result of which the pace of development of the coastal strip is constantly growing [4]. Active anthropogenic development of coastal waters contributes to the entry of an additional amount of organic matter and biogenic elements into the ecosystem. This leads to an increase in the rate of sedimentation and determines the high rate of change of the coastal waters characteristics.

T. Chaplina (Ed.): CIWO 2023, SPEES, pp. 245–253, 2023.
https://doi.org/10.1007/978-3-031-47851-2_29

The basis of redox processes is predominantly reactions involving organic matter and available electron acceptors, most often dissolved oxygen or oxygen-containing compounds.

The sequence of these reactions (Table 1) is determined by the thermodynamic characteristics [5, 6]. Aerobic oxidation of organic matter is the most energetically favorable (Table 1, Eq. 1), after oxygen is depleted, reactions with the participation of other oxidizing agents proceed (Table 1, Eqs. 2–5), resulting in the formation of suboxic conditions and anoxic conditions (lack of oxygen and the appearance of reduced forms of sulfur) [5–7]. After the exhaustion of oxygen in the bottom sediments, and then in the water column, the oxygen deficiency develops. Ultimately, this leads to the development of hypoxia/anoxia and the emergence of environmental risk zones [8].

The aim of the work was to identify the features of the spatial distribution of geochemical parameters, and their contribution to the formation of the vertical profile of pore water and the redox conditions of bottom sediments in the north-eastern part of the Black Sea under anthropogenic load.

Table 1. Sequence of reactions of organic matter oxidation in bottom sediments

Eq. No.	Process name	Reaction scheme	Redox conditions
1	Aerobic oxidation	$C_{106}H_{175}O_{42}N_{16}P + 150\ O_2 \rightarrow 106\ CO_2 + + 16\ HNO_3 + H_3PO_4 + 78\ H_2O$	Oxic
2	Denitrification	$C_{106}H_{175}O_{42}N_{16}P + 104\ HNO_3 \rightarrow 106\ CO_2 + + 60\ N_2 + H_3PO_4 + 138\ H_2O$	Suboxic
3	Manganese reduction	$C_{106}H_{175}O_{42}N_{16}P + 260\ MnO_2 + 174\ H_2O \rightarrow 106\ CO_2 + 8\ N_2 + H_3PO_4 + 260\ Mn(OH)_2$	Suboxic
4	Iron reduction	$C_{106}H_{175}O_{42}N_{16}P + 236\ Fe_2O_3 + 410\ H_2O \rightarrow 106\ CO_2 + 16\ NH_3 + H_3PO_4 + 472\ Fe(OH)_2$	Suboxic
5	Sulfate reduction	$C_{106}H_{175}O_{42}N_{16}P + 59\ H_2SO_4 \rightarrow 106\ CO_2 + + 16\ NH_3 + H_3PO_4 + 59\ H_2S + 62\ H_2O$	Anoxic

2 Material and Methods

Data were obtained in August 2021. A total of 18 samples of the surface bottom sediments (0–5 cm) and 5 bottom columns were taken (Fig. 1). Sampling and preparation of bottom sediment samples were carried out according with regulatory documents (GOST 17.1.5.01-80; ISO 5667-19:2004).

The content of dissolved oxygen in water samples was determined by the Winkler volumetric titration method modified by Carpenter [9]. The technique allows obtaining

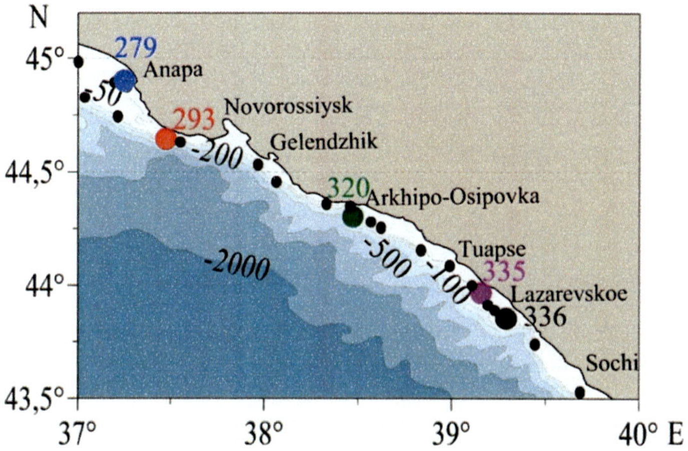

Fig. 1. Scheme of sampling stations (small black dots - surface layer of sediments (0–5 cm), large colored dots - columns) (Color figure online)

results with an accuracy of ± 0.010 ml/l (±0.4 μM). The degree of oxygen saturation of the near-bottom water layer (%) was calculated in accordance with [10]:

$$[O_2]sat = (C[O_2] \times 100)/\ln C,$$

where $[O_2]$sat – degree of the water saturation with oxygen, %;

$C[O_2]$ – oxygen concentration, ml/l.

$$\ln C = A_1 + A_2(100/T) + A_3 \ln (T/100) + A_4(T/100)$$
$$+S\left[B_1 + B_2(T/100) + B_3(T/100)^2\right],$$

where C – oxygen solubility at a total pressure of 1 atm, taking into account the pressure of saturated water vapor, ml/l;

$A_{(1,2,3,4)}$ and $B_{(1,2,3)}$ – constants (A_1 = –173,4292; A_2 = 249,6339; A_3 = 143,3483; A_4 = –21,8492; B_1 = –0,033096; B_2 = 0,014259; B_3 = –0,0017);

T – absolute temperature, K;

S – salinity, ‰.

To obtain the chemical profile of pore water, the polarographic method of analysis with a glass Au-Hg microelectrode was used [11, 12]. The main advantage of the method is the ability to analyze the composition of pore water under conditions as close as possible to natural ones, without sample destruction and additional sample preparation. The detection limit for O_2 was 5 μM, H_2S – 0.5 μM, Fe(II) – 10 μM, Mn(II) – 5 μM, respectively. For all measurements, the determination error did not exceed 10%. As a result of the formation of complexes in the case of Fe(III), their concentration cannot be determined [13]; therefore, their semiquantitative content in microamp (μA) is considered in this work.

The granulometric composition of bottom sediments was determined by the combined method of decantation and scatter (GOST 12536-2014).

The content of organic carbon (C_{org}) was determined coulonometrically on an AN 7529 express analyzer according to a procedure adapted for marine bottom sediments [14, 15].

The oxygen flux in the pore waters of bottom sediments were calculated using Fick's first diffusion law by calculating the diffusion gradient through the diffusion boundary layer and using the oxygen diffusion coefficient [16, 17]:

$$J = -\varphi D_s dC/dZ,$$

where J – flux, M/year · m^2;

φ – porosity;

dC/dZ – concentration gradient, M/m^3 · m;

D_s – molecular diffusion coefficient with regard to viscosity, m^2/ year.

To obtain a smoothed vertical profile of the oxygen concentration gradient, a polynomial approximation was used using the Microsoft Excel program. The values of the oxygen diffusion coefficients for the water column, at different temperatures and salinity of the water, are taken from the table [18]. The temperature of the near-bottom water layer, measured at the time of sampling, was taken into account. The equation:

$$D_s = D_0/\theta^2$$

where D_0 – water molecular diffusion coefficient, m^2/year;

θ – viscosity.

The dependence of viscosity (θ) on porosity (φ) is described by the equation:

$$\theta^2 = 1 - \ln(\varphi^2)$$

The porosity values of bottom sediments were calculated using the formula from the work [19].

3 Results and Discussion

According to the obtained data, the upper (0–5 cm) layer of bottom sediments in the study area is formed by material of different geochemical composition.

In some shallow areas, in the region of the Kerch fore-strait area and the city of Anapa, coarse-grained gravel-sand material prevails, the C_{org} content in the sediments varied from 0.2 to 0.5% (Fig. 2, b). An increase in the share of coarse-grained fraction and a decrease in the content of the clay fraction to less than 30% and the content of C_{org} with a decrease in the depth of sampling (Fig. 2, a) in this region corresponds to the literature data [20].

In the areas of large rivers (Pshada, Tuapse, Sochi, Mzymta) inflow from the city of Tuapse to the city of Adler, the predominance of silt material (97%) in bottom sediments was noted (Fig. 2, a). In these areas an increased content of C_{org} (up to 1.9%) was noted, which is most likely a consequence of the influx of organic matter and nutrients with river runoff [21], as well as the accumulation of a fine-grained fraction in sediments.

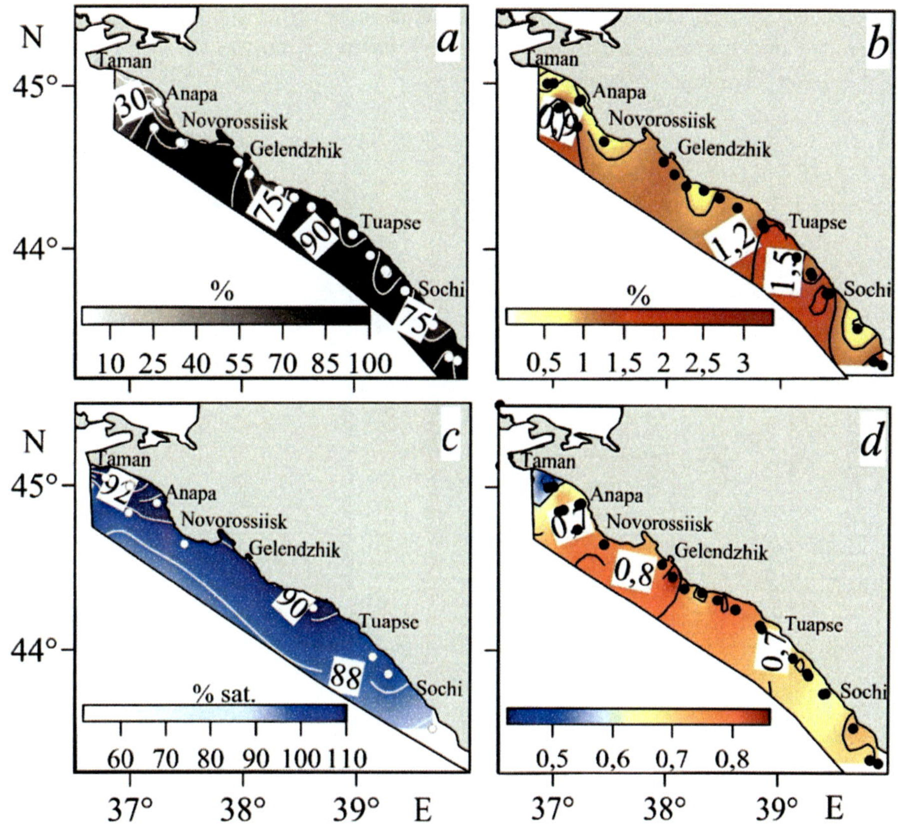

Fig. 2. Spatial distribution of physical and chemical characteristics: a) clay fraction, b) C_{org} in the surface layer of sediments, c) degree of oxygen saturation of the near-bottom water layer, d) porosity of bottom sediments of the north-eastern Black Sea

The influx of an additional amount of suspended matter and organic matter with river runoff and their accumulation in bottom sediments leads to oxygen consumption in the near-bottom water layer.

The concentration of oxygen and the degree of saturation of the near-bottom water layer with it are important factors that determine the chemical composition of the pore waters of bottom sediments and the occurrence of redox processes in them. On the one hand, the high dynamics of water in the study area contributes to the saturation of near-bottom waters and the upper layer of bottom sediments with oxygen. On the other hand, the supply of organic matter with the waters of the Sea of Azov and Caucasus river runoff contributes to the consumption of oxygen for its oxidation. The analysis of the obtained results showed that the saturation of the near-bottom water layer with oxygen corresponded to 88–91% sat., which is typical for the studied depth range (Fig. 2, c).

Oxygen was detected in bottom sediments at most of the stations studied; it was penetrated into the sediment to the maximum (up to 13 mm) at stations in the area of the village Arkhipo-Osipovka (station 320; Fig. 3, a), which can be explained by

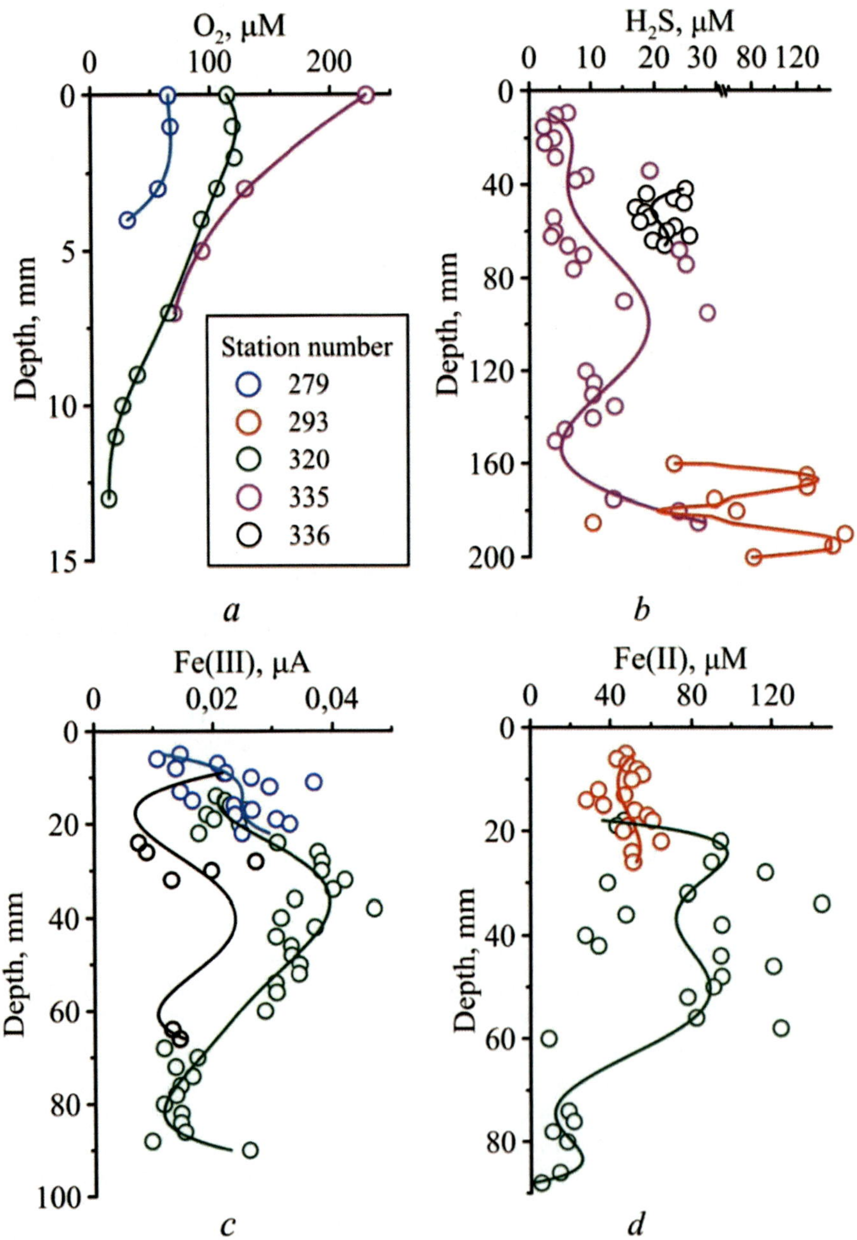

Fig. 3. Vertical distribution of the main components of the pore waters of the bottom sediments of the north-eastern part of the Black Sea

the saturation of the near-bottom water layer with oxygen and a low content of C_{org} in the sediments, despite the high proportion of clay fraction in the sediment. At the station near the city of Tuapse, a high, relative to other parts of the north-eastern region, oxygen concentration was noted on the sediment surface (231 μM; Fig. 3, a). However, its concentration decreased sharply with depth, which is a consequence of the active consumption of oxygen for oxidation C_{org} (Table 1, Eq. 1). In the area of the city of Anapa, a low oxygen content was noted on the surface of the sediment (65 μM, Fig. 3, a), which was explained by the low, for a given depth, saturation of the near-bottom water layer with oxygen (90%, Fig. 2, c).

It can be concluded that at most stations (stations 279, 320, 335) aerobic conditions in the upper layer of sediments were formed, which changed to suboxic (stations 279, 320) and anoxic (st. 335) in deeper layers. At stations 293 and 336, suboxic conditions were already observed in the upper layer of sediments.

In the anoxic zone, high oxygen fluxes in the upper layer of sediment lead to intensive oxygen consumption from the near-bottom layer of waters and the development of its deficiency in the future. In turn, high concentrations of hydrogen sulfide in bottom sediments determine its flow from bottom sediments to the near-bottom water layer. These features play an important role in the formation of the biogeochemical structure of bottom sediments and the near-bottom water layer, and are also extremely important in assessing and predicting the formation of ecological risk zones in marine ecosystems.

The oxygen flux values vary with depth and are determined by the oxygen concentration gradient, as well as the particle size distribution of the sediment (Fig. 4). On average, the oxygen flux increases at stations from the city of Anapa to the city of Tuapse, which

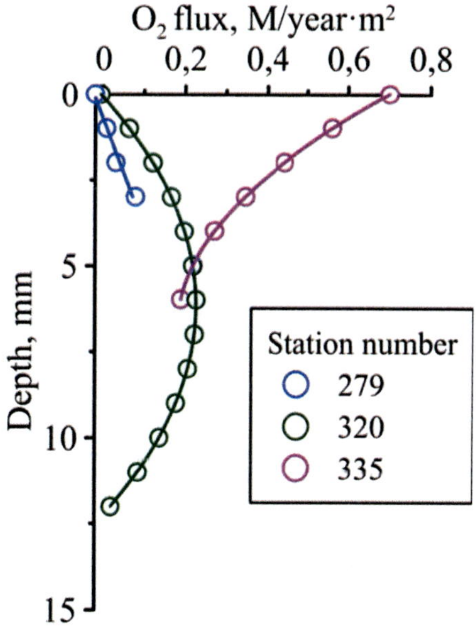

Fig. 4. Oxygen flux in the upper layer of bottom sediments

corresponds to an increase in the oxygen concentration in the bottom water layer, the proportion of the clay fraction, and the C_{org} content in the sediment.

4 Conclusion

The data obtained made it possible to identify the features of the geochemical characteristics, as well as the chemical profile of the pore waters of the bottom sediments of the coastal zone of the north-eastern part of the Black Sea. It has been established that, on average, the proportion of the clay fraction in the sediment increases in the direction from the city of Anapa to the city of Sochi. The content of C_{org} in the sediments varied from 0.2 to 1.9% and increased with an increase in the depth of sampling and the proportion of the fine-grained fraction in the sediment.

It was found that at most stations, aerobic conditions were formed in the upper layer of sediments, which changed to suboxic and anoxic conditions in deeper layers.

The oxygen flux values varied with depth and were determined by the oxygen concentration gradient, as well as the granulometric composition of the sediment.

Acknowledgements. The investigation was carried out within the framework of the state assignment of the MHI RAS on themes FNNN-2021–0005 "Complex interdisciplinary studies of oceanologic processes which determine functioning and evolution of ecosystems in the coastal zones of the Black Sea and the Sea of Azov" and FNNN-2022–0002 «Monitoring of the carbonate system, CO_2 and its fluxes in the marine environment of the Black and Azov Seas», and with the support of the RFBR project No. 20–35-90103 "Factors determining oxygen and sulfides fluxes at bottom water-sediments boundary in marine ecosystems".

References:

1. Artamonov, Yu.V., et al.: Water circulation in the northern part of the Black Sea in summer - winter 2018. Ecol. Safety Coastal Shelf Zones Sea **1**, 69–90 (2020). https://doi.org/10.22449/2413-5577-2020-1-69-90. (in Russian)
2. Lavrova, O.Yu., Mityagina, M.I., Sabinin, K.D., Serebryany, A.N.: Study of hydrodynamic processes in the shelf zone based on satellite information and sub-satellite measurement data. Modern Probl. Remote Sens. Earth Space **12**(5), 98–129 (2015). (in Russian)
3. Korshenko, A.N. (ed.): The Quality of Sea Waters by Hydrochemical Indicators. Yearbook 2020. Science, Moscow (2020). 281 p. (in Russian)
4. Investigation of the state of the ecosystem of the coastal part of the Black Sea in the Greater Sochi area. Water Ind. Russia Probl. Technol. Manage. **5**, 6–25 (2008). (in Russian)
5. Sarmiento, J.L., Gruber, N.: Ocean Biogeochemical Dynamics. Princeton University Press, Princeton (2006). 503 p.
6. Rickard, D., Luther, G.W.: Chemistry of iron sulfides. Chem. Rev. **107**(2), 514–562 (2007). https://doi.org/10.1021/cr0503658
7. Volkov, I.I. (ed.): Ocean Chemistry. Geochemistry of Bottom Sediments. "Science", vol. 2 (1979). 536 p. (in Russian)
8. Zhang, J., Gilbert, D., Gooday, A.J., et al.: Natural and human-induced hypoxia and consequences for coastal areas: synthesis and future development. Biogeosciences **7**(5), 1443–1467 (2010). https://doi.org/10.5194/bg-7-1443-2010

9. Eremeev, V.N., Konovalov, S.K., Romanov, A.S.: The distribution of oxygen and hydrogen sulfide in Black Sea waters during winter-spring period. Phys. Oceanogr. **9**(4), 259–272 (1998). https://doi.org/10.1007/BF02522712
10. Weiss, R.F.: The solubility of nitrogen, oxygen and argon in water and seawater. Deep Sea Res. Oceanogr. Abstr. **17**(4), 721–735 (1970). https://doi.org/10.1016/0011-7471(70)90037-9
11. Luther, G.W., III.: Use of voltammetric solid-state (micro)electrodes for studying biogeochemical processes: laboratory measurements to real time measurements with an in situ electrochemical analyzer (ISEA). Marine Chem. **108**(3–4), 221–235 (2008)
12. Orekhova, N.A., Konovalov, S.K.: Polarography of bottom sediments of the Sevastopol Bay. Marine Hydrophys. J. **2**, 52–66 (2009). (in Russian)
13. Taillefert, M., et al.: Reactivity of freshly formed Fe (III) in synthetic solutions and (pore)waters: voltammetric evidence of an aging process. Environ. Sci. Technol. **34**, 2169–2177 (2000). https://doi.org/10.1021/ES990120A
14. Lyutsarev, S.V.: Determination of organic carbon in sea bottom sediments by dry burning. Oceanology **26**(4), 704–708 (1986). (in Russian)
15. Zabegaev, I.A., Shulgin, V.F., Orekhova, N.A.: Application of instrumental methods of analysis of bottom sediments for ecological monitoring of marine ecosystems. Uchenye zapiski V. I. Vernadsky Crimean Federal University. Biol. Chem. **7**(73)(4), 242–254 (2021). (in Russian)
16. Hyacinthe, C., Anschutz, P., Carbonel, P., Jouanneau, J.-M., Jorissen, F.J.: Early diagenetic processes in the muddy sediments of the Bay of Biscay. Marine Geol. **177**(1–2), 111–128 (2001). https://doi.org/10.1016/S0025-3227(01)00127-X
17. Ullman, W.J., Aller, R.C.: Diffusion coefficients in nearshore marine sediments. Limnol. Ocenogr. **27**(3), 552–556 (1982)
18. Ramsing, N., Gundersen, J.: 2014 Seawater and Gases. Tabulated physical parameters of interest to people working with microsensors in marine systems. Unisense (2014)
19. Avnimelech, Y., Ritvo, G., Meijer, L.E., Kochba, M.: Water content, organic carbon and dry bulk density in flooded sediments. Aquacult. Eng. **25**(1), 25–33 (2001). https://doi.org/10.1016/S0144-8609(01)00068-1
20. Rukhin, L.B.: Fundamentals of Lithology, 3rd edn. Nedra (1969). 703 p. (in Russian)
21. Kostyleva, A.V.: Distribution of dissolved organic carbon in the estuarine areas of Greater Sochi (north-eastern part of the Black Sea). Oceanology **5**(2), 224–230 (2015). (in Russian)

Marine-Freshwater Ratio by Hydrochemical Proxies in the Surface Desalinated Layer of the Kara Sea

U. A. Kazakova(✉) and A. A. Polukhin

Shirshov Institute of Oceanology, Russian Academy of Sciences, Moscow 117997, Russia
ulya_kazakova_2910@mail.ru

Abstract. Freshwater in the Arctic seas is an important factor in the circulation of water masses and the functioning of the marine ecosystem. The Kara Sea is a unique object of interaction between sea waters and freshwaters of different genesis. The paper presents the results of calculating the percentage contribution of freshwater of different origin to the surface desalinated layer using hydrochemical parameters. The structure of the surface desalinated layer in different seasons is considered. The predominance of meltwater and river waters of the Ob' and Yenisei was noted, depending on the observation period. The variability of the structure of the surface layer and the redistribution of freshwater during the ice-free period are shown.

Keywords: freshwater · surface desalinated layer · Kara Sea · alkalinity · dissolved silicates

1 Introduction

The Kara Sea receives more than half of the total volume of river runoff of rivers flowing into the seas of the Arctic Ocean. Incoming river waters form the so-called surface desalinated layer (SDL). Its spreads over the sea area depends on the topography of the seabed [1] and various hydrometeorological factors [2]. In general, the waters of the Kara Sea are the result of a mixture of Atlantic waters that flow through deep-water troughs and riverine waters [3]. The waters of large rivers flowing into the sea, such as the Ob and Yenisei, which account for more than 80% of the river flow of the Kara Sea, interact both with sea waters and with each other, forming a complex SDL structure. The Ob and Yenisei waters differ in the magnitude of various hydrochemical parameters [4], in particular, total alkalinity, dissolved inorganic carbon and dissolved silicates [5]. In addition to river runoff, the waters from the melting sea ice contribute to the formation of the SDL [6]. Incoming fresh water plays an important role both on water circulation and on marine ecosystems not only in the Kara Sea, but also in the entire Arctic Ocean [7, 8].

According to the results in [5], the values of alkalinity and silicates have a linear dependence on salinity. At the same time, due to the fact that the catchment areas of

T. Chaplina (Ed.): CIWO 2023, SPEES, pp. 254–261, 2023.
https://doi.org/10.1007/978-3-031-47851-2_30

rivers are located in different natural zones, the waters of rivers differ from each other in hydrochemical parameters.

The main purpose of the work is to determine the ratio of marine and fresh waters of different origin in the surface desalinated layer to pycnocline using hydrochemical parameters.

2 Material and Methods

The work is based on field data obtained during the expeditions of the Institute of Oceanology of the Russian Academy of Sciences to the Kara Sea in 2015–2021. All hydrochemical parameters were processed in accordance with standard methods [9].

All the data were divided into several groups in order to trace the seasonal variability of the ratio of fresh and sea waters in the surface layer. The first group of data includes expeditions that were conducted in July (2016 and 2019). The second group included data from expeditions at the end of the summer period (August-September): 2015, 2017, 2018 and 2020. A separate group considered the data obtained in 2021 in the months of June, August and October for the analysis of seasonal variability within one year.

All available sampling data were interpolated to standard horizons: 0, 5, 10, 15 and 20 m. Since the maximum thickness of the SDL rarely exceeds 20 m and outside the zone of influence of river runoff, the concentration of silicates increases with depth, which may make an error in estimating the percentage contribution of freshwater.

The calculation of the percentage contribution of waters of different origin was carried out for each horizon when solving a system of Eqs. (1). The system of equations includes linear mixing equations and a balance equation [6, 10, 11].

$$\begin{aligned} XS_1 + X_2S_2 + \ldots + X_nS_n &= S \\ X_1TA_1 + X_2TA_2 + \ldots + X_nTA_n &= TA \\ X_1Si_1 + X_2Si_2 + \ldots + X_nSi_n &= Si \\ X_1 + X_2 + \ldots + X_n &= 1 \end{aligned} \tag{1}$$

X_i – percentage contribution of a certain water;
n – the number of waters involved in mixing;
S_n, S – salinity in water sources and observed in the water sample;
TA_n, TA – total alkalinity in water sources and observed in the water sample;
Si_n, Si – dissolved silicates in water sources and observed in the water sample.

The system of equations was adapted for the case of 4-component mixing of waters of different genesis, three of which are fresh waters: river waters of the Ob' and Yenisei and freshwater of a different genesis. They include waters of the melting sea ice and precipitation.

3 Source of the Freshwater Types and Their Parameters

Determining the parameters of each source of freshwater which takes part in SDL formation is an important task in calculating the percentage contribution of freshwaters of different genesis. The interannual and seasonal variability of these characteristics is the main difficulty in determining the value of the parameter used.

Due to the lack of field observations in the estuaries of rivers, the parameters of river waters were found using the regression Eq. (2).

$$y = a \times S + b \tag{2}$$

y – total alkalinity or dissolved silicates;
S – salinity.
a,b – empirical coefficients.

The parameters of river waters (alkalinity, dissolved silicates) are presented in Table 1.

Table 1. Parameters of various freshwater sources (according to the regression equation) and confidence interval

Year	Freshwater source	*TA, μM*	*ΔTA*	*Si, μM*	*ΔSi*	*n*
2015	Ob'	801	±38.86	54.17	±1.96	10
	Yenisei	910	±52.45	77.73	±6.22	16
2016	Ob'	1250	±69.8	111.28	±11.35	31
	Yenisei	459	±40.36	34.63	±2.48	27
2017	Ob'	1211	±47.6	93.37	±3.28	10
	Yenisei	918	±30.68	72.05	±2.6	10
2018	Ob'	823	±73.55	63.21	±6.22	17
	Yenisei	1029	±46.33	80.18	±5.74	41
2019	Ob'	1250	±53.31	94.99	±9.91	12
	Yenisei	1131	±27.46	35.02	±1.65	18
2020	Ob'	807	±25.71	39.69	±4.95	14
	Yenisei	888	±17.44	69.39	±1.89	24
2021*	Ob'	861 1167 1131	±42.36 ±46.78 ±51.66	61.2 121.94 74.57	±6.12 ±4.42 ±1.703	14 6 7
	Yenisei	699 798 465	±6.89 ±31.28 ±30.36	32.71 55.44 20.51	±3.27 ±3.75 ±1.86	14 6 7

* The parameters are listed in chronological order: June, August, October

The variability of the values of alkalinity and dissolved silicates for each year and season (for 2021) confirms the chosen approach to determining the value of the final element of mixing for each season/year, rather than using an integral value for each source for the entire period. Changes in the concentration of alkalinity and silicon cannot be described by any regularities regarding the flow of rivers.

4 Share of Freshwater in the Surface Layer

Several stations were selected for each group for a detailed analysis of the percentage of marine water and freshwater (Fig. 1).

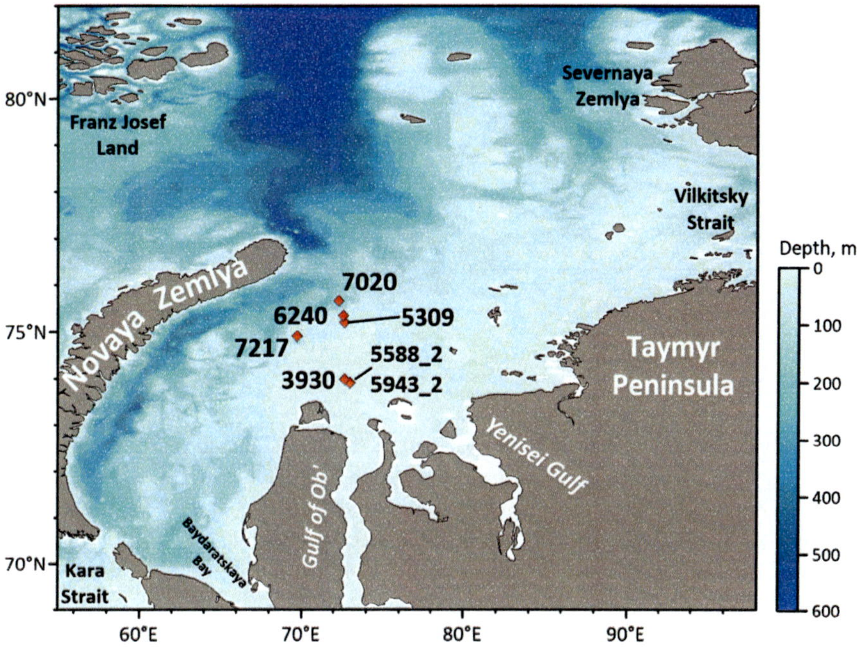

Fig. 1. Station map

The ratio of freshwater in the SDL at the beginning of the summer period (July 2016 and 2019).

At the station closest to the estuary area of the rivers (5309), the total contribution of freshwater to the SDL varies from 32% in the surface layer to 6% at a depth of 20 m (Fig. 2). The predominant source of freshwater on all horizons is water formed as a result of melting ice and precipitation (other freshwater) since the observation period was close to the period of ice melting. The contribution of the Yenisei and Ob' waters is slightly less in the surface layer. The contribution of the Ob' waters decreases greatly with depth. Station 6240 was located near station 5309 (2016), but the vertical structure of the SDL on them is very different. In 2016, waters from other sources prevailed on all horizons, and in 2019, the largest contribution to the SDL was made by waters of Yenisei origin, and the contribution of meltwater was minimal. There is a significant contribution of both Yenisei and Ob' waters at the horizons of 0 and 5 m. The contribution of waters of Ob' origin is almost zero at a depth of 10 m. Perhaps this distribution is due to the process of uneven dilution of the mixture of freshwater with marine. The total contribution of fresh water to the SDL does not differ much from that at station 5309 (about 30%).

The ratio of freshwater in the SDL at the end of the summer period (September 2017 and 2018).

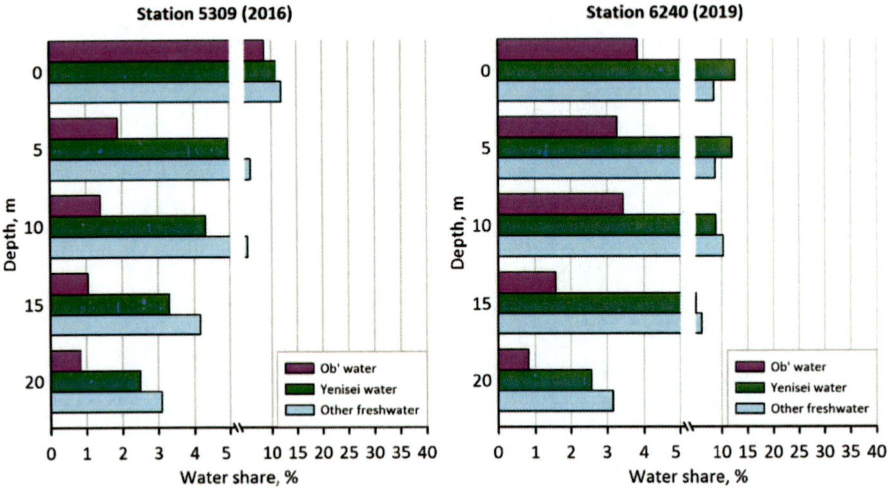

Fig. 2. Percentage contribution of freshwater to the SDL in July

Over time, the largest contribution of fresh water to the SDL was made by the riverine waters of the Yenisei and the Ob'. At the same time, the Yenisei waters prevailed in the surface layer, and the Ob' waters prevailed at a depth of 10 m. The predominant source of freshwater was waters of a different genesis (other freshwater) on the underlying horizons. The total contribution of freshwater varied from 48% on the surface to 11% on the horizon of 20 m.

Station 5943_2 (2018) was located at the same point as station 5588_2 (2017). In the distribution of fresh water in the 20-m thickness, the main tendency to decrease the contribution of fresh water with depth is also determined. But, unlike the case in 2017, the distribution of 2018 has its own peculiarities, which are shown in Fig. 3. Firstly, the predominance of Ob waters over other freshwater sources in the surface 5-m layer at the end of August and 10-m layer at the end of September. Secondly, the change in the ratio of the waters of the Yenisei and other fresh springs over time. At the end of September, after the waters of the Ob, other fresh sources are the predominant source of water. Such a change can be explained by the seasonal variability of the parameters of the final mixing components (alkalinity and silicates), which cannot be taken into account as a result of determining the parameter value at zero salinity during regression analysis.

The ratio of freshwater in the SDL during ice-free period in 2021.

The total contribution of freshwater in the entire 20-m surface layer of the Kara Sea does not exceed 7% in June 2021 (Fig. 4). The largest contribution among all sources is presented by the waters of non-river genesis, which is associated with the observation period during the intense melting of seasonal ice in the western part of the sea.

In August 2021, the distribution of freshwaters is characterized by their large total contributions and heterogeneity of distribution compared to that in June. More than 70% of continental runoff enters the sea by August. The most pronounced transformation of the SDL structure can be traced at stations located at the estuarine area. The total

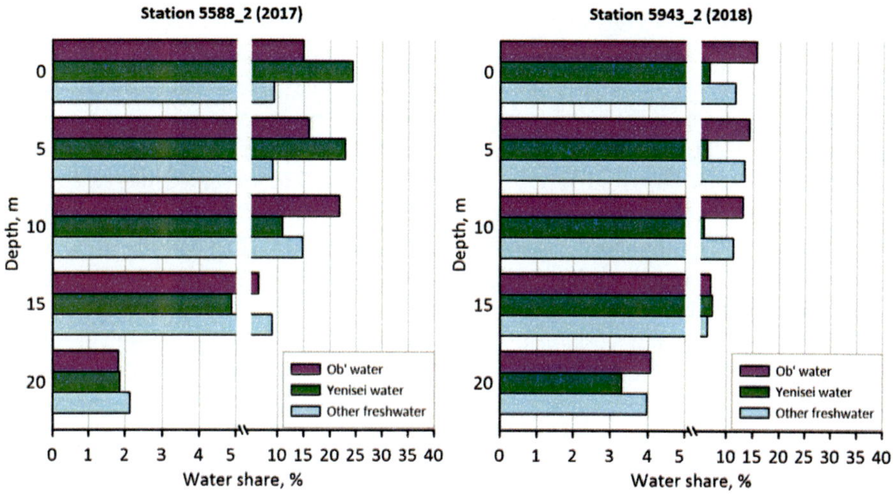

Fig. 3. Percentage contribution of freshwater to the SDL in September

percentage contribution of freshwater to the surface at station 3930 is more than 60%, with depth it decreases to 14%.

The predominant sources of freshwater on the surface were the waters of the Yenisei, which at 15-m depth were replaced by freshwater from other sources. The total contribution of freshwater varied from 25% at the surface to 10% at 20 m depth. The contribution of the Ob' waters did not exceed 5%.

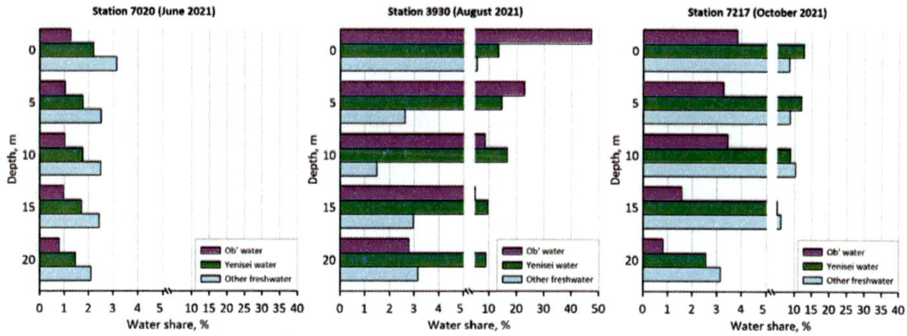

Fig. 4. Percentage contribution of freshwater to the SDL in different seasons in 2021

5 Conclusion

The predominance of the contribution of meltwater and precipitation to the SDL structure is noted in almost the entire water area under consideration in June. During this period, seasonal ice cover was observed at most of the Kara Sea area, and the river runoff of the high water period on both Ob' and Yenisei presumably either did not have time to spread across the water area, or spread to the east under the influence of the Coriolis force.

Obviously, over the time, the contribution of meltwater should decrease due to the dilution of fresh water of ice melting origin with ambient marine water. However, according to the results obtained, there is an increase in the contribution of other sources (meltwater and precipitation) from July to October in 2021. The observed increase in the percentage contribution of other freshwater sources (mostly of the melted sea ice) can be explained by several factors: the seasonal ice cover in 2021, that was observed in the northeastern part of the Kara Sea during the entire period, as well as an increase in the amount of precipitation in the region under consideration associated with the observed climatic changes [12]. It is possible that the observed (rather high) proportion of melted ice and atmospheric sources of fresh water retains a seasonal value annually, however, except for 2021, we do not have additional data to support or refute this assumption.

Spatial variability and relative values of the Ob and Yenisei waters in the SDL structure are quite characteristic of different types of SDL dynamics depending on wind action.

The marked predominance of fresh waters of Yenisei origin in the surface layer is confirmed by previously published results [6, 13]. However, it is shown that this trend is not observed in all sea areas. At the end of the summer period near the Gulf of Ob, the greatest contribution to the SDL was made by waters of Ob origin, and in the southwestern part of the sea in June-July, meltwater prevailed. Also, the predominance of Yenisei waters in the SDL structure in October 2021 can be justified by the situation observed in 2021: the average monthly discharge of Yenisei River exceeded the average monthly discharge of the Ob River throughout the year. In general, the heterogeneity of the SDL structure due to the inflow of riverine waters of different genesis, obtained as a result of calculating the percentage contribution by hydrochemical parameters, confirms the complex structure of the riverine plume formed by the interaction of the Ob and Yenisei waters.

Acknowledgements. The work was carried out within the framework of the state assignment of the Shirshov Institute of Oceanology (theme № FMWE-2021-0007).

References:

1. Johnson, D.R., McClimans, T.A., King, S., Grenness, Ø.: Fresh water masses in the Kara sea during summer. J. Marine Syst. **12**, 127–145 (1997)
2. Zatsepin, A.G., Zavialov, P.O., Kremenetskiy, V.V., Poyarkov, S.G., Soloviev, D.M.: The upper desalinated layer in the Kara Sea. Oceanology **50**(5), 698–708 (2010)
3. Hanzlick, D., Aagard, K.: Freshwater and Atlantic water in the Kara sea. J. Geophys. Res. **85**, 4937–4942 (1980). https://doi.org/10.1029/JC085IC09P04937
4. Pivovarov, S.: Chemical oceanography of the Arctic seas of Russia (study guide). Gidrometeoizdat, SPb (2001)
5. Stunzhas, P.: Separation of the waters of the Yenisei and the Ob in the Kara Sea by alkalinity and silicon. Oceanology **35**(2), 215–219 (1995)
6. Polukhin, A.A., Makkaveev, P.N.: Features of the continental runoff distribution over the Kara Sea. Oceanology **57**(1), 25–37 (2017). https://doi.org/10.7868/S0030157417010142

7. Aagard, K., Carmack, E.C.: The role of sea ice and other freshwater on the Arctic circulation. J. Geophys. Res. **94**, 485–498 (1989)
8. Carmack, E.C., et al.: Freshwater and its role in the Arctic Marine System: sources, disposition, storage, export, and physical and biogeochemical consequences in the Arctic and global oceans. J. Geophys. Res. Biogeosci. **121**, 675–717 (2016). https://doi.org/10.1002/2015JG003140
9. Parsons, T.R.: A Manual of Chemical & Biological Methods for Seawater Analysis. Elsevier (2013)
10. Yamamoto-Kawai, M., Tanaka, N., Pivovarov, S.: Freshwater and brine behaviors in the Arctic Ocean deduced from historical data of δ18O and alkalinity (1929–2002 AD). J. Geophys. Res. Oceans **110**(C10003), 1–16 (2005). https://doi.org/10.1029/2004JC002793
11. Nedashkovsky, A.P.: Hydrochemical Variability in the Surface Mixed Layer along the Trajectory of the Drifting Station "North Pole 35." Oceanology **52**(4), 498–508 (2012)
12. Zhang, P., et al.: A stratospheric pathway linking a colder Siberia to Barents-Kara Sea sea ice loss. Sci. Adv. **4**(7), 1–8 (2018). https://doi.org/10.1126/sciadv.aat6025
13. Zavialov, P.O., Izhitskiy, A.S., Osadchiev, A.A., Pelevin, V.V., Grabovskiy, A.B.: The structure of thermohaline and bio-optical fields in the upper layer of the Kara Sea in September 2011. Oceanology **55**(4), 461–471 (2015). https://doi.org/10.1134/s0001437015040

Evidence of Atlantic-Origin Water Shoaling in the Western Chukchi Sea

K. V. Kodryan(✉) and K. K. Kivva

Russian Federal Research Institute of Fisheries and Oceanography, 105187 Moscow, Russia
angelloka@yandex.ru

Abstract. In August 2019, VNIRO Transarctic expedition aboard the R/V Professor Levanidov accomplished hydrological and hydrochemical studies in the western part of the Chukchi Sea. We observed transformed Atlantic water (AW) at depth of 117–137 m, which is not typical for this water mass. Salinity values within transformed AW were in range from 34.2 to 35.7. This water had unusually high content of dissolved Si (93–97 μM) and low apparent oxygen utilization (240–270 μM). Similar situations (but with lower Si concentration) have already been observed in approximately the same region by scientists from Japan, Sweden, and Korea. Combination of our results and previously published data allow to conclude that, at least since 2004, there is a tendency towards increase of mineral Si concentration and decrease of dissolved oxygen concentration in the waters of the upper halocline in the study area.

Keywords: Atlantic water · Arctic Ocean · nitrate nitrogen · phosphorus · dissolved oxygen · silicon · spatial distribution · water masses

1 Introduction

The Chukchi Sea represents an important transition zone between waters of the Pacific and Arctic Oceans [1]. It is characterized by high primary productivity [2] related to distribution of nutrient-rich water of Pacific origin. Recent studies have documented the substantial changes in the Chukchi Sea ecosystems, including changes in sea ice concentration, water temperature and seasonality [3], as well as shifts in species distribution [4]. Those changes are at least partially related to increase of Pacific water inflow into the sea [5] and global climate change patterns. The understanding and forecasting of those changes is crucial for nature conservation and fisheries in the region. Despite several international projects aimed at investigation of the sea, its western part is less studied comparing to the eastern part [6]. In particular, there is a lack of information on spatiotemporal variability of water masses (WM) in the western Chukchi Sea (WCS) and particularly on bottom WMs.

Several WMs are usually distinguished in the WCS. There is one WM of Atlantic origin which is often referenced as transformed AW. AW is characterized by temperature values above − 1.26 °C and salinity values above 33.6 [1]. The AW typically resides below 150 m and constitutes the upper part of halocline in the Pacific and Central regions

T. Chaplina (Ed.): CIWO 2023, SPEES, pp. 262–267, 2023.
https://doi.org/10.1007/978-3-031-47851-2_31

of the Arctic Ocean. In the WCS, the AW may be found in the bottom layer. The other WMs of WCS include waters of Pacific origin: Alaskan Coastal Water (ACW), Bering Summer Water (BSW), and Remnant Winter Water (RWW). The WMs of local Arctic origin in the WCS are Siberian Coastal Water (SCW), and Melt Water (MW) [1]. Only RWW, BSW and AW are usually found in the bottom layer. Between them, the AW is the saltiest one. AW is advected along the Arctic shelf break where primary production is relatively low and biological pump is relatively weak, therefore it is usually reach in oxygen and deplete in nutrients comparing to RWW and BSW. The latter two WMs are exposed to the bottom sediments of the Chukchi Sea. The high productivity along the relatively shallow Chukchi Sea, and especially the WCS, results in detritus sedimentation and destruction. This leads to suboxic or anoxic conditions within bottom sediments and in near-bottom layer of water column. Thus, RWW and BSW may keep footprints of chemical exchange with the bottom sediments.

Water mass investigation was carried out in the WCS in September 2019 [6]. It has been shown, along with other findings, that the AW had unusual characteristics and location. Considering the lack of information on AW variability in the WCS, the main aim of this study is to describe recent findings related to the AW in the WCS in details and to discuss possible reasons of changes in the state of AW.

2 Material and Methods

Samples of water were collected during VNIRO Transarctic expedition (R/V Professor Levanidov) in August 2019. In the Western Chukchi Sea, 55 stations were performed. Hydrological and hydrochemical studies were carried out at every station (Fig. 1).

Temperature and salinity data was collected with SBE 19plus V2 SeaCAT Profiler. The CTD was calibrated prior to the expedition. The content of nutrients (NO_2, NO_3, NH_4, PO_4 and SiO_3) and dissolved O_2 in samples was determined immediately after sampling in the onboard hydrochemical laboratory. Chemical analyses were accomplished in accordance to standard procedures [7] using Shimadzu UV-1601PC spectrophotometer.

Calculated parameters such as N* and apparent oxygen utilization (AOU) were used as additional water mass tracers. N* was calculated according to [8], Eq. 1:

$$N*(\mu M) = \left(\left[NO_3^-\right] + \left[NO_2^-\right] + \left[NH_4^+\right]\right) - 16[PO_4] + 2.9, \quad (1)$$

where $[NO_3^-]$, $[NO_2^-]$, $[NH_4^+]$ and $[PO_4]$ are the concentration of nitrate, nitrite, ammonium nitrogen, and dissolved inorganic phosphorus in μM, respectively [8]. This parameter reflects deficiency of fixed dissolved inorganic nitrogen relative to dissolved inorganic phosphorus as predicted stoichiometrically in accordance to the average N:P ratio for the World Ocean [6, 8]. Negative N* values indicate denitrification processes occurring in water column or in the pore water within bottom sediments.

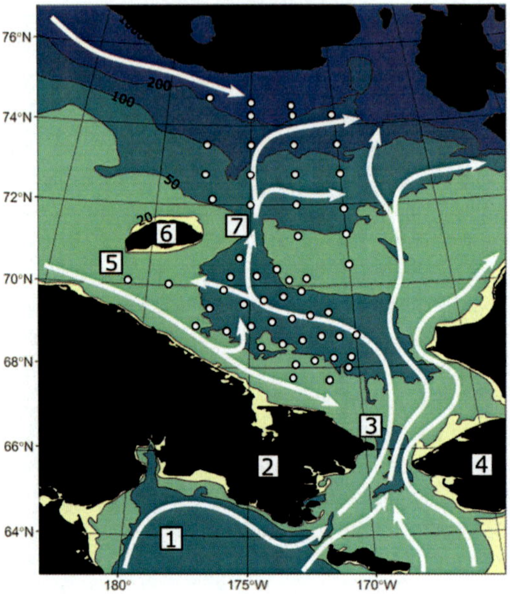

Fig. 1. Sketch map of study area. Stations of R/V Professor Levanidov are marked with dots. White arrows represent main currents ([1] with additions). 1, Anadyr Bay; 2, Chukchi Peninsula; 3, Bering Strait; 4, Seward Peninsula; 5, Long Strait; 6, Wrangel Island; 7, Herald Canyon (picture from [6])

3 Results and Discussion

A comprehensive survey was carried out over the western part of the sea; however the most interesting data were obtained in the northern deeper part. A water mass of Atlantic origin was observed in the near-bottom layer at 117–137 m at three stations to the north of Wrangel Island. This vertical residence of AW is relatively shallow comparing to number of previous observations. The upper margin of AW is typically found deeper than 150 m in the study area. Besides unusual location, AW observed in WCS in 2019 also had uncharacteristically high concentration of dissolved Si (93–97 μM) (Fig. 2), low concentration of dissolved O_2 (2.0–2.8 ml/l), low relative oxygen content (25–34%), as well as high values of AOU (240–270 μM), and low N* (down to − 24.0 μM) [6]. In addition, this WM had high content of nitrate nitrogen (up to 19 μM) and phosphates (up to 3.0 μM) [6]. It's worth noting that AW at locations deeper than 140 m had considerably lower concentration of dissolved Si (less than 53 μM) as well as higher concentration of dissolved O_2 (3.3–5.5 ml/l with 42–68% of oxygen saturation).

High concentration of dissolved Si was also measured in the bottom layer to the north and northwest of Wrangel Island at stations whose depth did not exceed 70 m (Fig. 3), as well as in the northern East Siberian Sea. The concentration of silicon in the bottom layer in this area ranged from 83 to 108 μM and was associated with water with salinity between 32.4 and 32.7. In addition to the high content of silicon, this water was characterized by a high content of nitrate nitrogen (12.33–14.39 μM) and phosphates (from 2.33 to 3.10 μM), as well as high AOU values (up to 213 μM) and low values of

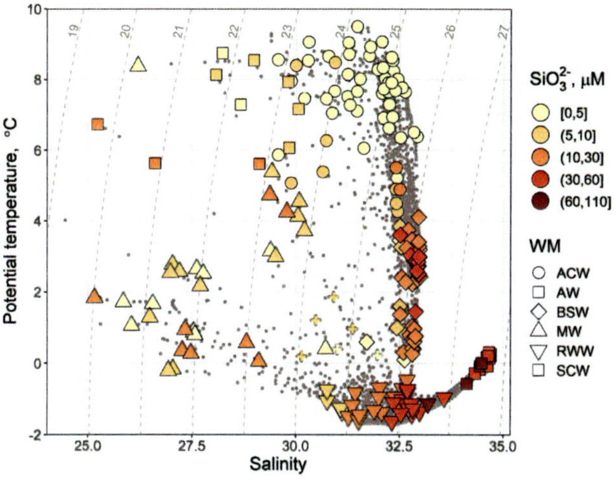

Fig. 2. θ–S diagram based on R/V Professor Levanidov data. Larger symbols indicate water samples with hydrochemical data available. Gray dots mark data of CTD casts. Color represents concentration of dissolved Si, shape of symbol corresponds to water mass (WM) (picture from [6])

N * (down to – 28 μM). These waters, according to hydrological parameters (S≈32.6, T≈ – 1.5 °C), had the characteristics of the RWW, which means they are of Pacific Ocean origin.

High nutrient content and low dissolved oxygen content usually indicate the mineralization of organic matter. The combination of negative N* and high AOU in the bottom layer within study area indicates that the corresponding WM had contacted bottom sediments with hypoxic conditions for relatively long time. Thus, silicon maxima observed in WCS in 2019 were most likely formed in contact with sediments.

It is important to note that similar situations have already been observed in approximately the same region. For example, in September–October 2004, Japanese researchers observed a maximum of mineral dissolved silicon (up to ≈35 μM) in AW (salinity 34.1–34.5) in the same area at 120–130 m [9]. The parameter N** estimated by them (rather than N* as in our case) was about – 5 μM. In July–September 2014, SWERUS-C3 expedition observed water with thermohaline characteristics of AW and relatively high concentration of mineral dissolved silicon along the East Siberian Sea continental slope. However, the maximum concentration of silicon in the AW accompanied by a minimum of dissolved oxygen was about 50 μM [10]. In both mentioned cases, the silicon maximum was explained by the accumulation and dissolution of diatom valves in the halocline near the continental slope. However, such processes could unlikely lead to such a noticeable decrease in the concentration of dissolved oxygen. In 2015–2017, several hydrographic studies were carried out in the northwestern part of the Chukchi Sea [11]. The authors do not report the concentration of dissolved Si, but show that in the area north of Wrangel Island in the 80–100 m layer, a significant change in N* was observed from about – 8 μM in 2015 to – 10 μM in 2016 and – 15 μM in 2017.

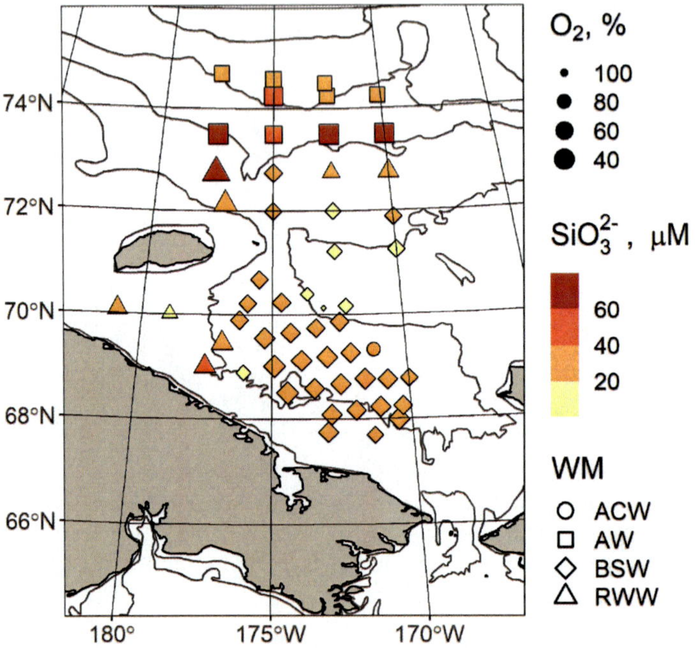

Fig. 3. Distribution of dissolved Si (μM, color) and oxygen saturation (%, symbol size) in the bottom layer. Water masses are indicated by the shape of the symbol (picture from [6])

In August 2018, during the expedition of the Pacific Branch of VNIRO to the northern Chukchi Sea onboard the R/V TINRO, approximately in the same area, high concentrations of mineral dissolved silicon (up to 72.8–82.9 μM) were also observed in waters with a salinity of 33.40–34.59 at depths of 136–185 m [A.S. Vazhova, personal communication].

4 Conclusion

Based on the information above, it can be concluded that, at least since 2004, there is a tendency towards an increase of concentration of mineral silicon, a decrease of concentration of dissolved oxygen, and an increase of deficiency of mineral nitrogen (decrease of N*) in the waters of the upper halocline in the study area. Those changes in the AW are most likely associated with changes in location of AW. Combination of our results and published information allowed to conclude that the AW tend to move inshore. In other words, this means at least episodic appearance of AW in the shelf areas with high vertical flux of organic matter to the bottom layer and sediments.

Acknowledgements. The work was carried out in accordance with the Plan of resource research and state monitoring of aquatic biological resources for 2019 (order of the Federal Agency for Fishery of the Russian Federation, 30th of November 2018, No. 701, paragraphs. 22-25, 38, 91, 92).

References

1. Pisareva, M.N., Pickart, R.S., Spall, M.A., et al.: Flow of pacific water in the western Chukchi Sea: results from the 2009 RUSALCA expedition. Deep-Sea Res. **105**, 53–73 (2015)
2. Sakshaug, E.: Primary and secondary production in the Arctic Seas. In: Stein, R., MacDonald, R.W. (eds.) The Organic Carbon Cycle in the Arctic Ocean, pp. 57–81. Springer, Heidelberg (2004). https://doi.org/10.1007/978-3-642-18912-8_3
3. Arrigo, K.R., van Dijken, G.L. Secular trends in Arctic Ocean net primary production. J. Geophys. Res. **116**(C9), 011 (2011)
4. Woodgate, R.A.: Increases in the Pacific inflow to the Arctic from 1990 to 2015, and insights into seasonal trends and driving mechanisms from year round Bering Strait mooring data. Prog. Oceanogr. **160**, 124–154 (2018)
5. Polyakov, I.V., Alkire, M.B., Bluhm, B.A., et al.: Borealization of the arctic ocean in response to anomalous advection from sub-arctic seas. Front. Marine Sci. **7**, 491 (2020)
6. Kodryan, K.V., Kivva, K.K., Zubarevich, V.L., Pedchenko, A.P.: Water masses in the Western Chukchi Sea in august 2019 and their hydrochemical features. Oceanology **63**(3), 314–324 (2023)
7. Guidance on the Chemical Analysis of Marine and Fresh Waters in the Ecological Monitoring of Fishery Reservoirs and Areas of the World Ocean That Are Promising for Fishing. VNIRO, Moscow (2003). [in Russian]
8. Gruber, N., Sarmiento, J.: Global patterns of marine nitrogen fixation and denitrification. Global Biogeochem. Cycl. **11**, 235–266 (1977)
9. Nishino, S., Shimada, K., Itoh, M., Chiba, S.: Vertical double silicate maxima in the sea-ice reduction region of the western arctic ocean: implications for an enhanced biological pump due to sea-ice reduction. J. Oceanogr. **65**, 871–883 (2009)
10. Anderson, L.G., Björk, G., Holby, O., et al.: Shelf-Basin interaction along the East Siberian Sea. Ocean Sci. **13**(2), 349–363 (2017)
11. Jung, J., Cho, K.-H., Park, T., et al.: Atlantic-origin cold saline water intrusion and shoaling of the nutricline in the Pacific Arctic. Geophys. Res. Lett. **48**(6), e2020GL090907 (2021)

Carbonate System and CO_2 Fluxes in the Partizanskaya River Estuary (Nakhodka Bay, Sea of Japan)

K. A. Kryzhova[1] (✉) and P. Yu. Semkin[2]

[1] Saint Petersburg State University, St. Petersburg 199034, Russia
2002kristin@gmail.com
[2] POI FEB RAS, Vladivostok 690041, Russia

Abstract. Coastal waters connect land, oceans, atmosphere, biota, and sedimentary deposits. Although estuarine areas represent only a small fraction of the total ocean surface area, they play a significant role in the transport of carbon and biogenic elements from continents to the open sea. In this study, a biogeochemical investigation was conducted, focusing on the carbonate system and Carbon dioxide (CO_2) fluxes in the estuary of the Partizanskaya River (Nakhodka Bay, Sea of Japan). The vertical variability of temperature, salinity, and chlorophyll-a was presented based on CTD (Conductivity-Temperature-Depth) profiler data. The main parameters of the carbonate system, including pH, pCO_2 (partial pressure of CO_2), DIC (dissolved inorganic carbon), and TA (total alkalinity), were described for the four seasons of the year. The CO_2 flux (FCO_2) at the water-air interface was calculated. The need for further hydrochemical research in coastal areas to identify the drivers of changes in the directionality of one of the major greenhouse gases' fluxes was justified. Intensive bloom of photosynthetic algae was detected during the winter period, with chlorophyll-a concentrations reaching 49,3 mg/l under the ice cover. In samples collected at stations located in the downstream part of the river and in the open section of the river estuary, pCO_2 varied significantly, exceeding equilibrium values (taken as 390 μatm) in some cases (ranging from 1000 to more than 2000 μatm), while being lower in others. Overall, during the autumn-winter period, the Partizanskaya River estuary acted as a source of CO_2 to the atmosphere due to high organic matter decomposition rates, as evidenced by the positive values of FCO_2, with an average value of 83.05 mmol·m-2 · day-1. In the spring-summer period, the fluxes changed direction at some sampling points, ranging from −74.01 mmol·m-2 · day-1 (gas influx) to 198.41 mmol · m-2·day-1 (gas efflux). In conclusion, the carbonate system explains the processes related to organic matter production and decomposition in river estuaries, providing a better understanding of the intensity and directionality of carbon flux fluctuations in coastal areas, enabling the control of eutrophication even in oligotrophic water bodies, and overall buffering capacity of ecosystems.

Consequently, the relevance of conducting hydrochemical studies of both small and large rivers around the world is increasing regularly.

Keywords: hydrochemistry · carbonate system · carbon dioxide fluxes · production-destruction processes of organic matter · estuary

T. Chaplina (Ed.): CIWO 2023, SPEES, pp. 268–279, 2023.
https://doi.org/10.1007/978-3-031-47851-2_32

1 Introduction

Investigating the gas exchange processes at the ocean-atmosphere interface represents a crucial undertaking within the field of ocean chemistry.

The oceans have played a significant role during the industrial age, absorbing roughly 40% of the CO_2 emissions, which has consequently led to a decrease in the pH of surface seawater by around 0.1 units. Anticipations suggest that the pH might further descend by 0.2 to 0.3 units by the end of this century, potentially resulting in serious impacts on vulnerable organisms and ecosystem [17].

Global estimates of estuary CO_2 emissions are approximately 21% of the total CO_2 uptake by the continental shelf [3]. The general increase in the concentration of CO_2 in the atmosphere, together with a significant input of biogenic elements into marine ecosystems as a result of anthropogenic activities, resulted in the transformation of carbon cycle in coastal waters [16].

Estuaries act as the only thruway for the terrestrial carbon to access the open sea through the water. Yet, inner estuaries provide a by-pass of carbon towards the atmosphere and tend to be supersaturated with CO_2 due to a higher CO_2 partial pressure than atmospheric equilibrium. Near-shore areas suffer from acidification, nutrient anthropogenic load, physical and biochemical processes much more often that the open sea.

The timeliness of this study is associated with the importance of studying ecosystems at the river-sea interface, particularly in underoccupied secluded coastal areas. Publicly available databases are lacking due to the challenges in accurately measuring various hydrochemical indicators using non-contact methods. Therefore, a representative database for a specific water body from different parts of the world is crucial in forming an imagery of estuaries' contribution to the global carbon budget.

The aim of this study is to identify the intra-annual variability of inorganic carbon content within the mixing zone of the Partizanskaya River using in-situ data and to quantitatively assess the fluxes of CO_2 and biogenic elements in order to understand how an estuary responds to the dual stresses of eutrophication and acidification.

2 Material and Methods

The Partizanskaya River (before 1972, the Suchan River) originates on the southern spurs of the Sikhote-Alin Ridge (Primorsky Kray) and falls near the city of Nakhodka into a profitable bay that falls in the eastern part of the hall. Peter the Great and limited by Cape Povorotny. This is one of the most important rivers of southern Primorye. The river catchment area is 3120 km^2 in the section near Partizansk (43 km from the mouth) and 4140 km^2 in the mouth section, the length is 142 km. The river and its tributaries are mountainous in the upper reaches and semi-mountainous in the lower reaches. In the water regime Partizanskaya is distinguished by winter low water, spring flood and summer-autumn floods, as well as periods of low water in the summer and autumn seasons.

On May 18–19, July 18–19 and September 16, 2012, water samples from surface and bottom layers were collected at 16 stations each and at 8 stations 25–26 January, along the profile of the maximum depths using Niskin bottles. Additionally, hydrological sounding of the water column was conducted, using RBR-XRX620 profiler (Canada), equipped with a CTD-sensor. Water sampling, analysis of hydrochemical characteristics (pH, alkalinity, dissolved oxygen concentration was determined on the day of sampling, and the partial pressure of carbon dioxide (pCO2) and dissolved inorganic carbon (DIC) were analytically calculated. The parameters of the carbonate system were calculated according to the standard and original methods described in detail earlier in [14].

The calculation of parameters for the carbonate system involves knowing the salinity values, in-situ temperature of the water sample, as well as the depth from which the sample was taken. An essential aspect of the calculation is the selection of acid dissociation constants, which, in turn, depends on the pH scale used. In our study of the carbonate system, we use two scales: the Pitzer scale and the "total hydrogen concentration" scale. A detailed description of these scales is given in [15] and a more detailed explanation on the methods of measuring pH values as well as description of is mentioned in paper [12].

The calculation was based on pH and total alkalinity (TA) measurements, taking into account the concentration of humic substances (HS) and correcting for organic alkalinity (OA). pH measurements were performed at a temperature of 15 ± 0.05 °C in a flow-through cell with a volume of approximately 80 cm^3 without liquid junction [13].

DIC was determined as follows (1):

$$\mathbf{DIC} = [\mathbf{CO_2}] + \left[\mathbf{HCO^{3-}}\right] + \left[\mathbf{CO_3^{2-}}\right], \tag{1}$$

$[CO_2]$ dissolved gas = $[H_2CO_3]$ - dissolved gas + $[H_2CO_3]$ molecular carbonic acid.

The concentrations of dissolved inorganic carbon forms (CO_2, HCO_3^-, and CO_3^{2-}) and the values of pCO_2 partial pressure of CO_2 were calculated based on the measured pH and TA values using an algorithm for calculating carbonate system parameters in freshwater.

Apparent Oxygen Utilization (AOU) was calculated, representing the difference between the observed oxygen concentration and the oxygen concentration at 100% water saturation.

$$\mathbf{AOU} = [\mathbf{O_2}]\,\mathbf{equilibrium} - [\mathbf{O_2}], \tag{2}$$

$[O_2]$ equilibrium is the concentration of dissolved oxygen (mmol/kg) calculated as a function of temperature, salinity at the site, and total pressure at one atmosphere.

FCO_2 was calculated into account the gas transfer coefficient, wind speed, intensity of mixing of various water masses and water temperature (3).

$$F\mathbf{CO_2} = \mathbf{kK_0}\left(\mathbf{pCO_{2water}} - \mathbf{pCO_{2air}}\right), \tag{3}$$

k – transfer CO_2 speed; K_0 – solubility of CO_2 at a certain temperature and salinity.

In this work, the average wind speed is for the area according to the reanalysis data. This decision is due to the frequent variability of wind speed and direction during the day, so we neglect daily fluctuations. Still, it is important noting that it is the wind speed that primary affects the increase in the value of the CO_2 flux, exponentially. In addition, the gas exchange coefficient affects the flow rate.

3 Results

The range of values for physical, chemical and biological variables is shown in Table 1. The river discharge significantly depends on the season, since it is determined by the monsoonal nature of the climate. In 2012, the river discharge was quite low in winter (4.45 ± 0.02 m^3/s). In the spring, the flood peak was observed in May (river discharge 60.87 ± 6.64 m^3/s). In July, the observations were carried out when the flow of the river dropped to 20.10 m^3/s. The absolute maximum was observed in August-September (112.24 ± 18.95 m^3/s), which is associated with the consequences of the monsoon season.

The mixing of sea waters with river waters is accompanied by a decrease in the buffer capacity and, as a result, a more intensive increase in pCO_2 during the oxidation of organic matter, which can also be the reason for a significant difference in this value for bottom waters in the hypoxic area at a salinity of more than 30 ‰ and in the reaches at the beginning of the mixing zone in brackish waters.

Table 1. The Partizanskaya river estuary seasonal values

Sampling characteristics for seasons	Winter	Spring	Summer	Autumn
Wind speed ($m\ s^{-1}$)	3	10	12	3
T (°C)	0 ± 0,5	10,3 ± 2,1	20,1 ± 0,6	17,8 ± 0,9
S, (‰)	0,0–8,7	0,0–31,9	0,0–21,8	0,0–32,8
Q a ($m^3\ s^{-1}$)	5,12	73,67	21,47	19,5
Water-air CO2 flux (mmol C $m^{-1}\ d^{-1}$)	78,32 ± 34,30	367,2 ± 622,3	347,18 ± 465,23	84,27 ± 71,82
Chl *a* ($mg\ m^{-1}$)	0,3–49,3	0,3–2,5	0,7–29,7	0,3–40,6
pH	7,3–7,6	7,2–8,2	7,4–8,1	7,2–8,0
pCO_2 (μatm)	919,0–2020,0	175,1–2043,0	233,5–1612, 0	430,9–2562,6
DIC (mmol kg^{-1})	1,3–1,5	0,6–1,4	0,8–1,6	0,9–1,9
TA (mmol kg^{-1})	1,31–1,53	0,61–1,44	0,80–1,74	0,84–1,03
AOU (μmol/kg)	−106,7–(−6,8)	−35,0–22,2	−40, 0–44,7	– 34 –3,4
DO (ml/l)	9,9–12,1	7,4–8,4	5,5–6,5	5,6–7,2

Oxidation of organic matter, which provokes production or destruction, limits the flow direction. Due to the high intensity of biogeochemical processes in coastal ecosystems, the prevalence of a biogenic or abiogenic component at a certain point in time can lead to a shift in equilibrium. As a result, there is a rapid transition from the invasion of CO_2 in the atmosphere to its evasion.

Positive values of the flux characterize the active absorption of CO_2 by water from the atmosphere, and negative values correspond to the flow of CO_2 from the atmosphere into water. The calculated fluxes of carbon dioxide (FCO_2) show that in winter the waters were in the state of evasion 78.32 ± 34.30 (mmol m-2 day-1).

The spring period is characterized by a clear change in the sign of the flow at the boundary of two water masses. Thus, river water was a source of CO_2 (783.84 ± 528.39 (mmol m-2 day-1)), which was determined by the supersaturation of river waters with CO_2 relative to its content in the atmosphere, and the estuary from the beginning of the mixing zone was a sink of CO_2 - the average value −168.38 ± 29.46 (mmol m-2 day-1). Large amplitudes of CO2 flux intensity could be related to wind activity during these surveys. And the highest value of the flux during the season could be associated with the high temperature of the water, corresponding to the warm season (23 °C).

The occurrence of redox reactions contributed to the formation of a sufficient amount of carbon dioxide for consumption by autotrophic organisms. In autumn, the entire study area was a source of CO_2: 84.27 ± 71.82 (mmol m-2 day-1).

In general, in the autumn-winter period, the estuary of the river Partizanskaya was a source of CO_2, due to the high rate of destruction of organic matter, the average value for the region was positive: 83.05 (mmol m-2 day-1), and in the spring-summer period the flows changed direction and, at different parts of the estuary changed from −74.01 (mmol m-2 day-1) - gas invasion, to 198.41 (mmol m-2 day-1) - gas evasion (Fig. 1).

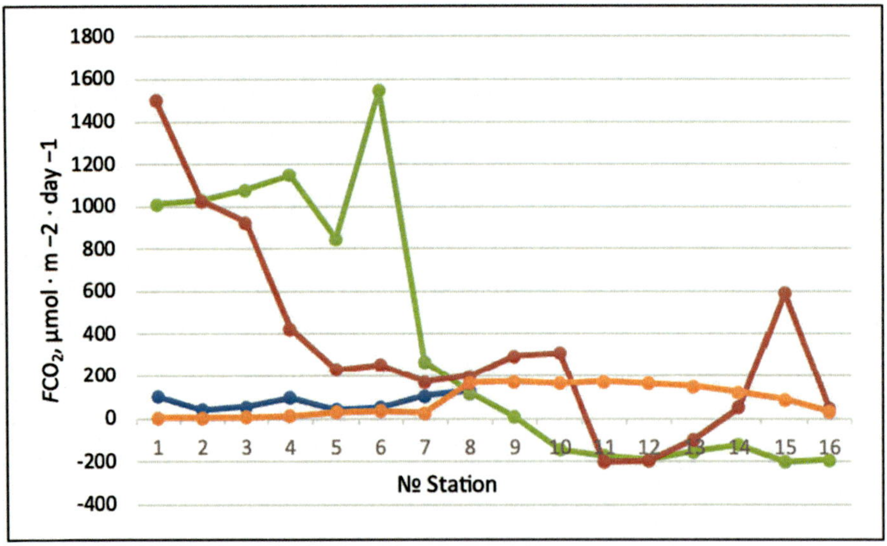

Fig. 1. Carbon Dioxide flux in the mixing zone of Partizanskaya River Estuary. On the left is the river, on the right is the sea. On the X axis - the distance from the crest of the mouth bar of the river. Colors outline the seasons of the year: blue – winter, green – spring, red – summer, orange – autumn.

The boxplots in Appendix provide a better view of the intra-annual variability and the range of chemical parameters' concentrations on intra-annual variability.

The surface layer as a whole was better saturated with oxygen throughout the year, the maximum was in winter, when the low water temperature contributed to the high solubility of gases (~11.5 mg/l) and active photosynthesis occurred, and the minimum values (~6.75 mg/l) were typical for summer, when the processes of destruction and oxygen consumption predominated. For the bottom layer, the general seasonal trend persists, but the concentrations shift to a lower side. Remarkably, that hypoxic features (1.5 mg/l) were indicated in summer, and a large range of values is associated with a large difference between the characteristics of the near-bottom horizon in the river and the outer estuary. Common to all seasons is the inversely-proportional dependence of oxygen concentration on salinity: less saline river waters were better at dissolving oxygen than salty sea waters.

The variability of TA and DIC is directly related to salinity changes. Overall, salinity increased downstream. In general, the surface can be considered fully freshwater due to the high river discharge, which forms a plume 0–6‰ above the denser saline waters. Consequently, the TA (Fig. 18) is not constant, and we observe a systematic increase in concentrations from spring (0.6 mmol/kg) to autumn (0.9 mmol/kg). The spring minimum (0,66 mmol/kg) is associated with spring flooding and the influx of a large amounts of carbonates from the river discharge, while the winter maximum (2,23 mmol/kg) is linked with increased salinity under ice cover. The bottom waters, as a whole, are characterized by higher TA values.

The general trend that persists for all seasons is that fresh river waters contain less total alkalinity, and its value increases as it approaches an open estuary or deepens (where the layer of salty sea water is located).

The same factors influence the variation in dissolved inorganic carbon (DIC), ranging from 0.6 (mmol/kg) in the surface layer during spring to 1.45 (mmol/kg) in winter, with the maximum amplitude observed in the bottom horizon ranging from 0.6 to 2.0 μmol/kg. Alkalinity is a linear function of salinity in the marine portion of an estuary. The concentration of total inorganic carbon (DIC) deviates to a greater extent from the linear relationship, as DIC includes the concentration of carbon dioxide, which is highly variable in seawater and primarily depends on the intensity of organic matter formation/decomposition processes.

Considering pCO_2, we can outline that the intra-annual variation of the value is different on the surface layer and near the bottom (we also study the near-bottom layer, since the frequency of oscillations of the sections does not exceed 10 m). Median value for each season of the year exceeded the equilibrium pCO_2 and amount to 1400 ppm in winter, 450 μatm in spring, 590 μatm in summer and 1150 μatm in autumn. There is an inverse relationship with pH, because in the autumn, the pH of the water was 7.2, while the absolute maximum of pCO_2 also fell in autumn and was 2380 μatm.

The pH value in winter varies depending on the depth. So, for most stations, the bottom waters were more shifted to the alkaline environment (7.9–8.3), while the pH at the surface was (7.3–7.7). This can be attributed to winter convection and the predominance of cold water on the surface. In autumn, the river runoff increases, the river plume propagates deep into the bay; therefore, the surface layer is characterized by low salinity values. A weak direct dependence of salinity and pH can be noted: thus, with an increase in salinity, pH shifted to an alkali environment (7,8–8,1). The entire study area was a source of CO2 to the atmosphere (Fig. 2).

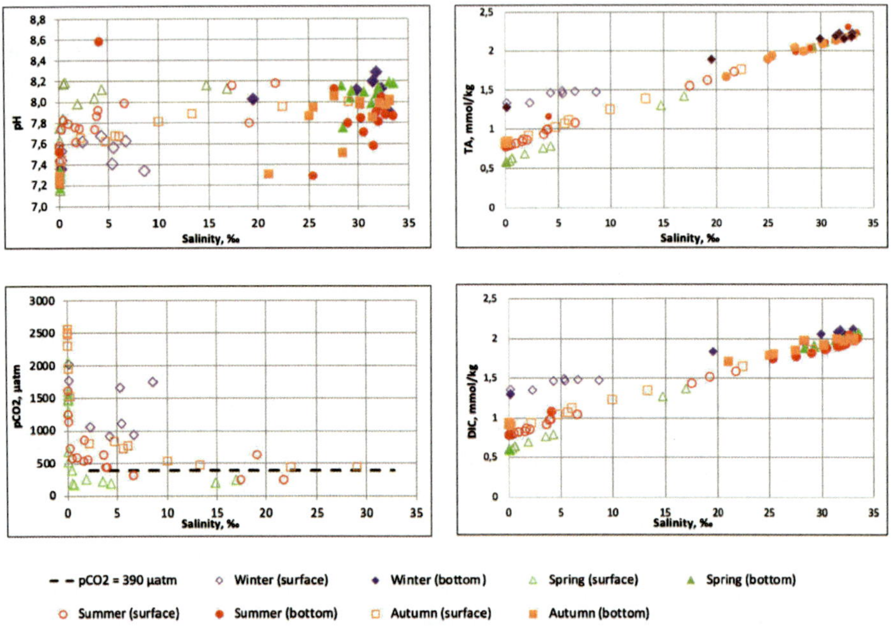

Fig. 2. Carbonate system parameters of the Partizanskaya River Estuary

4 Discussion

As a result of intense estuary circulation, high stratification of layers of water with an increase is determined in the pycnocline, which is typical for microtidal estuaries [7]. The emerging pycnocline can be configured as a liquid bottom where phytoplankton accumulates. Besides river discharge, temperature fluctuations, turbidity, intensity and direction of flow turbulence can also cause stratification, but it is the characteristic salinity in estuarine environments that makes the main contribution. At the same time, the inflow of fresh water into a saline water mass is not a sufficient condition for the formation of a highly stratified estuary; the freshwater influx rate must exceed the mixing rate, which is usually a function of tidal action.

The considerable influx of terrestrial carbon can be attributed to heterotrophic metabolism and the mineralization of organic matter, thereby leading to an excessive saturation of CO_2 compared to its atmospheric concentration.

Additionally, scatterplots were presented to represent the main parameters of the carbonate system. The abscissa shows salinity as the main significant limiting factor in the mixing zone. Low salinity in the coastal zone is associated with an increase in the flow of rivers or CO_2-rich pore water, which contributes to CO_2 efflux. Surface waters may have been transformed by the influence of ice formation due to slow turbulent exchange and physical processes in the river.

At the same time, the mixing of estuary waters with river plumes is especially important in winter, when snowmelt increases river runoff, thus the plume plays a dominant role in the formation of certain alkalinity values. In winter, production processes predominate over destruction processes, which is indicated by oxygen oversaturation and carbon dioxide undersaturation.

As mentioned above, DIC and TA parameters behaved conservatively. But formation/degradation of organic matter in the estuary, since one of the DIC components, dissolved CO_2, is not a component of TA made the slight shift in the linearly. As already noted, photosynthesis usually leads to a decrease in the partial pressure of CO_2, an increase in pH and the concentration of dissolved O_2, which is clearly seen in the spring-summer period.

In addition to photosynthesis and destruction of organic matter, FCO_2 is subjected to a wind-dependent gas transfer coefficient, as well as CO_2 solubility. The Fig. 3 shows pCO_2 versus AOU. Conventionally, this can be interpreted as a conservative function of temperature, because AOU increases with decreasing temperature and shifts to positive values, since DO equilibrium increases. Likewise, pCO_2 will decrease due to the solubility of CO_2 with increasing temperature. Therefore, the relationship between the two parameters is direct, and is statistically significant only for the winter ($r^2 = 0.80$) and summer periods ($r^2 = 0.75$). We assume that the presence of such a relationship is primarily determined with a small runoff in winter and summer. The lack of linkage between spring and autumn may also indicate the influence of biotic processes on the dynamics and transformation of these components, or that the formation of water occurred too quickly to balance the chemical parameters under changing conditions.

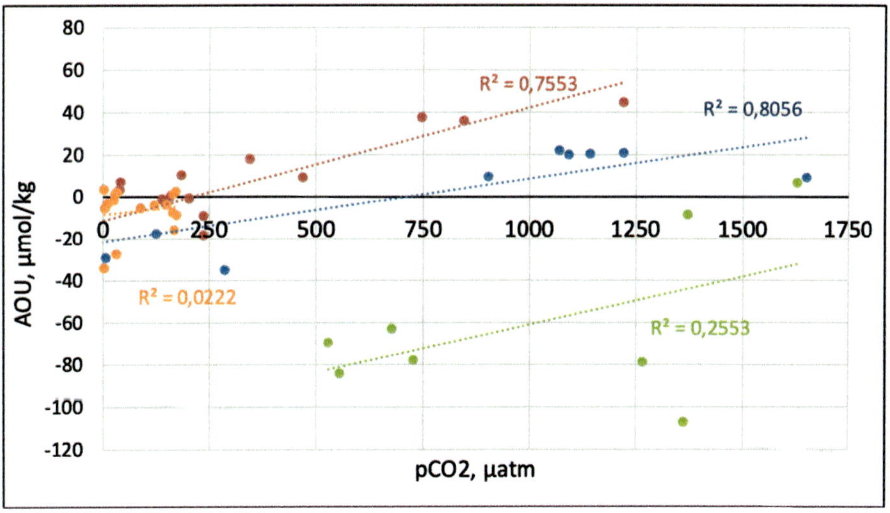

Fig. 3. Linear connection of pCO_2 (partial pressure of Carbon Dioxide) on AOU (Apparent Oxygen Utilization). Blue color corresponds to the winter season, green – to spring, red – to summer, orange – to autumn.

It has been confirmed that coastal ecosystems have the capacity for self-restoration through complex and sometimes unpredictable mechanisms by ecosystem management, even in the presence of anthropogenic impacts. This positive response to restoration in coastal areas may contribute to the continuation of research on eutrophication and acidification in coastal systems that show signs of recovery.

5 Conclusion

This study presents, the visualization of field data characterizing the magnitude and direction of CO_2 fluxes at the water-atmosphere interface in the Partizanskaya River estuary for the four seasons of 2012. In some sections of the river, pCO_2 is nearly five times higher than in the atmosphere. The study revealed an oversaturation of pCO_2 and undersaturation of O_2, indicating the decomposition of labile organic carbon as the primary controlling process.

Our study suggests that the changes in the carbonate system parameters of the Partizanskaya River are likely the result of much more complex processes than simple mixing of freshwater and seawater that undoubtfully plays a key role in the distribution of chemical and biological water components.

Thus, the predominantly studied waterbody was in a state of evasion throughout the year, i.e., CO_2 came from the water into the atmosphere, and the highest flux values were typical for spring and summer. This can be explained by physicochemical processes (deceleration of gas solubility in the warm season), as well as by the influence of hydrodynamic factors. At the same time, only in the spring-summer period, sea waters propagating in the upper layer from the mixing zone were in a state of invasion, i.e., they absorbed CO_2 from the atmosphere.

Furthermore, this study emphasizes the significance of small estuaries in emitting a substantial flux of CO_2 into the atmosphere. The findings contribute to expanding the dataset on pCO_2 and CO_2 emissions from river estuaries worldwide, aiming to accurately assess the role of estuaries in the global carbon balance.

Acknowledgments. The work was carried out within the framework of the Russian Science Foundation grant № 23-77-10001.

Appendix

See Fig. 4.

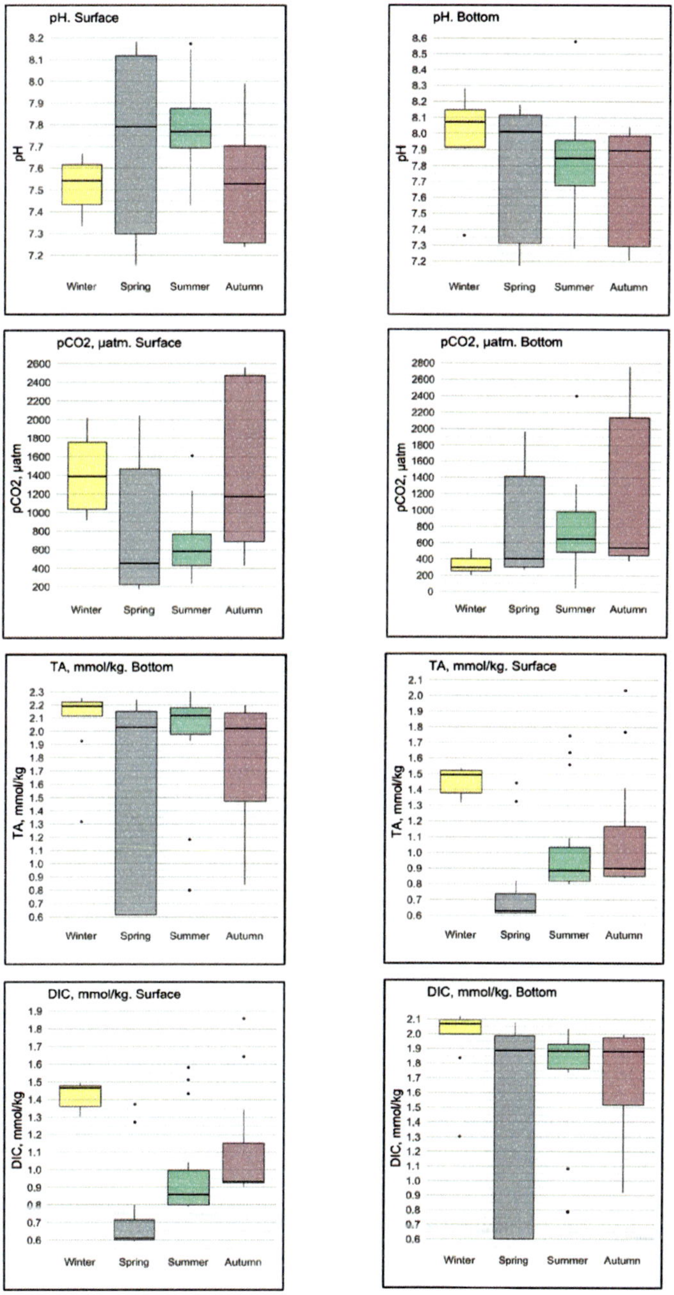

Fig. 4. Box-whisker plot of DIC, pH, TA pCO2, Salinity, DO in the Partizanskaya River Estuary. The x-axis shows aggregated data for each of the four seasons of the year, and the y-axis refers to the corresponding parameter's value. The colored block represents the spread of the second (25–50%) and third (50–75%) quartiles, and the horizontal line in it represents the median of the entire distribution. The block tails show, respectively, the first and fourth quartiles of the magnitude distribution - the minimum and maximum values.

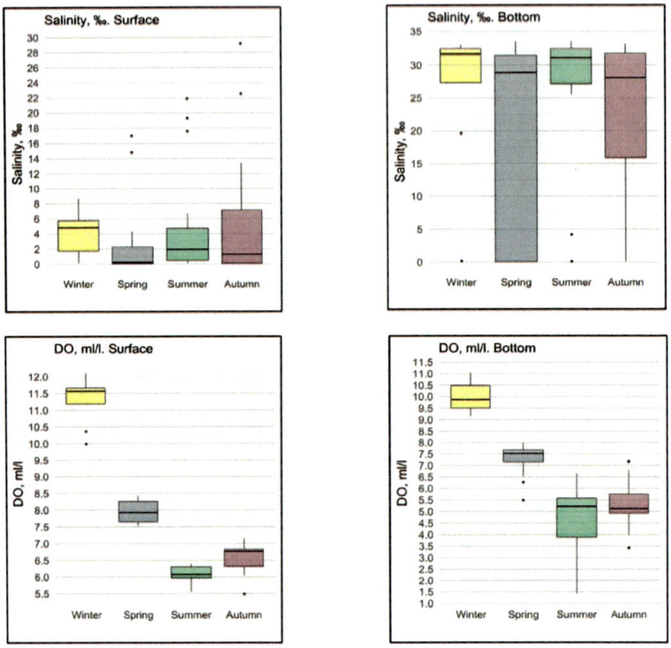

Fig. 4. (*continued*)

References

1. Borges, A.V.: Present day carbon dioxide fluxes in the coastal ocean and possible feedbacks under global change. In: Duarte, P., Santana-Casiano, J. (eds.) Oceans and the Atmospheric Carbon Content, pp. 47–79. Springer, Dordrecht (2011). https://doi.org/10.1007/978-90-481-9821-4_3
2. Cameron, W.M., Pritchard, D.W.: Estuaries. The Sea, vol. 2, pp. 306–324. Wiley Interscience Publ., Hoboken (1963)
3. Chen, C..-T..A.., Borges, A..V..: Reconciling opposing views on carbon cycling in the coastal ocean: continental shelves as sinks and near-shore ecosystems as sources of atmospheric CO2. Deep Sea Res. II Top. Stud. Oceanogr. **56**(8), 578–590 (2009)
4. D'Adamo, N., Lukatelich, R.: Water quality of the Murray River Estuary: summary report. "Environmental dynamics report ED-85-114." Bibliography: p. 17
5. Elliott, M., Whitfield, A.K.: Challenging paradigms in estuarine ecology and management. Estuarine Coastal Shelf Sci. **94**, 306–314 (2011)
6. Hunt, G.L., et al.: Climate impacts on eastern Bering Sea foodwebs: a synthesis of new data and an assessment of the Oscillating Control Hypothesis. ICES J. Marine Sci. **68**(6), 1230–1243 (2011). https://doi.org/10.1093/icesjms/fsr036
7. Mikhailov, V.N., Dobrovolsky, A.D.: General Hydrology: Textbook. Higher School, Moscow (1991). 368 p.
8. Oliveira, A.P., Cabeçadas, G., Mateus, M.D.: Inorganic carbon distribution and CO2 fluxes in a large European estuary (Tagus, Portugal), Scientific reports (2017)
9. Orekhova, N.A., Konovalov, S.K., Polukhin, A.A., Seliverstova, A.M.: 2023 oxygen and pCO2 in the surface waters of the Atlantic Southern Ocean in 2021–2022. Water **15**(9),1642 (2023)

10. Semkin, P.Yu., Tishchenko, P.Ya.1, Lobanov, V.B., et al.: Dynamics and structure of waters in the estuary of the Partizanskaya River (Nakhodka Bay, Sea of Japan). Water Res. **46**(1), 24–34 (2019). (in Russian)
11. Semkin, P.Yu., Tishchenko, P.Ya., Mikhailik, T.A., et al.: Hydrochemical studies of the estuary of the Partizanskaya River (Nakhodka Bay, Sea of Japan) during the summer low water period. Izvestiya TINRO **193**, 143–152 (2018). (in Russian)
12. Tishchenko, P.Ya.: Standardization of pH measurements based on ion approach. Russ. Chem. Bull. **49**(4) (2000)
13. Tishchenko, P.Ya., Chichkin, R.V., Ilyina, E.M. et al.: Measurement of pH in estuaries using a liquid-free compound cell. Oceanology **42**(1), 32–41 (2002)
14. Tishchenko, P.Ya., Wong, Ch.Sh., Volkova, T.I., et al.: Carbonate system of the estuary of the Razdolnaya River (Amur Bay of the Sea of Japan). Biol. Sea **31**(1), 51–60 (2005)
15. Tishchenko, P.Ya., Wahlmann, K., Vasilevskaya, N.A.: Contribution of organic matter to the alkaline reserve of natural waters. Oceanology **46**(2), 211–219 (2006). (in Russian)
16. Orekhova, N.A., et al.: Characteristics of the carbonate system of the waters of the Sevastopol Bay in 2009–2015. Experimental and expeditionary research (2016). (in Russian)
17. Zhou, J., Zheng, Y., Hou, L., et al.: Effects of acidification on nitrification and associated nitrous oxide emission in estuarine and coastal waters. Nat. Commun. **14**, 1380 (2023)

Results of the Carbonate System Features Study of the South-Eastern Part of the Baltic Sea in 2022

A. A. Muratova[1,2](✉), E. S. Bubnova[1,2], and A. A. Polukhin[1]

[1] Shirshov Institute of Oceanology RAS, Moscow, Russia
alexmurashkina24@gmail.com
[2] Immanuel Kant Baltic Federal University, Kaliningrad, Russia

Abstract. Measurements of total alkalinity and pH were carried out in two expeditions in autumn-winter season of year 2022 in the Russian exclusive economic zone in the south-eastern part of the Baltic Sea. The main distribution patterns of the total alkalinity and pH in were described, and the contribution of the sulphide component to the distribution of these parameters was also noted.

Keywords: carbon dioxide · total alkalinity · carbonates · pH · Baltic Sea · Gdansk Basin · eutrophication

1 Introduction

One of the most important factors of human activity at the present moment is the increase of CO_2 concentration in the atmosphere, which not only affects the climate but also the oceans and lithosphere. Therefore, there is increasing research interest in the ocean as a major sink and regulator of atmospheric CO_2 concentration, mitigating the unfavourable climatic and environmental effects of this increase. The carbonate system of marine waters plays an important role in ocean chemistry and is the main buffer system of waters. This buffer regulates the hydrogen index (pH) and thus prevents the acidification of water, affecting many biogeochemical and geochemical processes. It is known [1] that coastal and estuarine zones of shelf seas play crucial role in the carbon dioxide cycle between the atmosphere and the ocean. The heavily eutrophied inland Baltic Sea looks particularly interesting to study in this regard. Changes in concentrations as well as the distribution of total alkalinity in the Baltic Sea depend both on Major Baltic inflows (MBI) and river discharge as well as on internal mixing between the sea basins. According to [2] rivers flowing into the south-eastern part of the Baltic Sea tend to increase the total alkalinity of the water area as they flow through limestone-rich catchments. At the same time, MBIs also contribute into total alkalinity rise due to rise of salinity [3, 4], so the Baltic Sea may potentially be the area for acidification process mitigation. Moreover, study area includes one of the first in Russia marine carbon polygones (CP), where fluxes of greenhouse gases are studied. This CP in particular is located at the CH_4 anomaly [5]. Therefore, the aim of this study was to measure and calculate components of the carbonate system in the south-eastern part of the Baltic Sea in comparison with hydrological conditions, demonstration MBI and river runoff influence.

T. Chaplina (Ed.): CIWO 2023, SPEES, pp. 280–286, 2023.
https://doi.org/10.1007/978-3-031-47851-2_33

2 Material and Methods

The research was based on data from 54th cruise of the R/V "Akademik Sergey Vavilov" (04–10.11.2022) (e.g. "autumn cruise") and 91st cruise of the R/V "Akademik Mstislav Keldysh" (10–28.12.2022) (e.g. "winter cruise") (Fig. 1), so data represent autumn-winter season. Water samples for determination of pH and total alkalinity (TA) were taken by Hydrobios MWS 12 Slimline hydrological complex equipped with Niskin bathometers. Salinity was measured with Sea&Sun Tech CTD 90 and Idronaut Ocean 316 multi-channel hydrophysical probes. Sampling and analyses were carried out according to standard methods accepted in Russian practice [6]. pH in situ was calculated via CO2SYS software [7]. pH is mentioned in NBS scale [8].

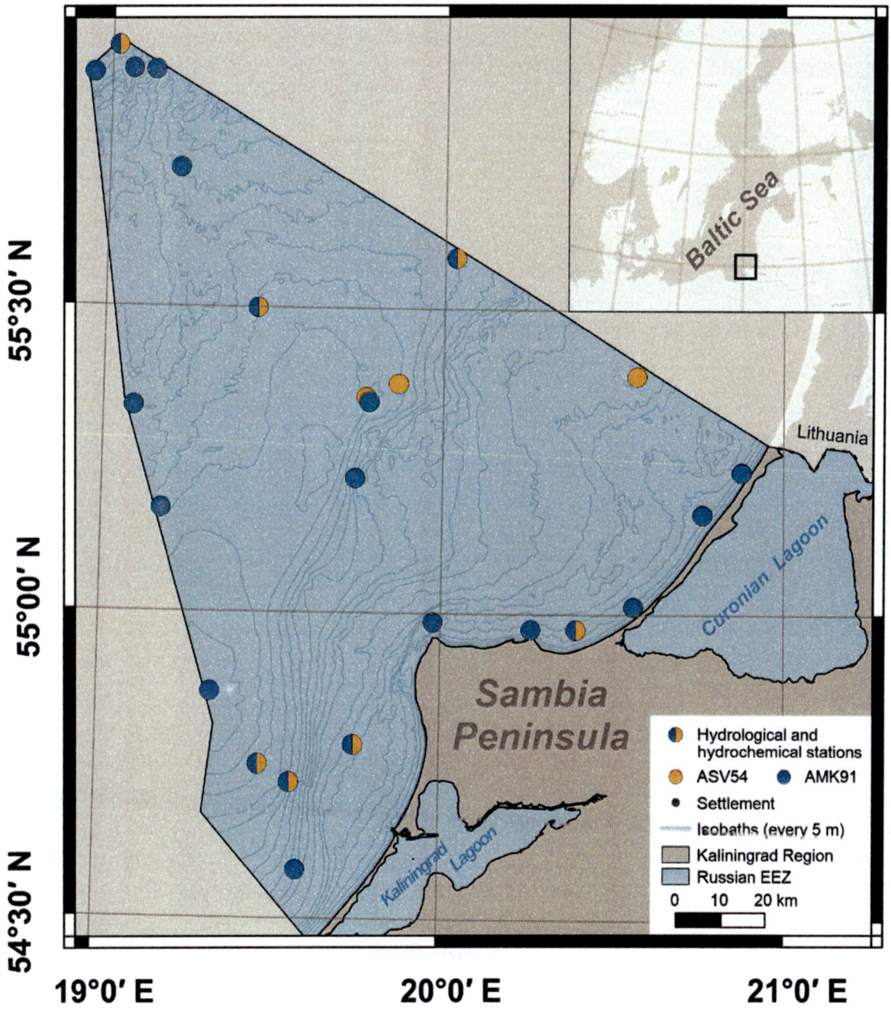

Fig. 1. The study area—south-eastern part of the Baltic Sea (Russian sector)

3 Results and Discussion

Measured TA showed little variability in surface layer during the autumn expedition (Fig. 2a), as well as salinity, which was almost homogeneous for all stations and varied from 7.08 to 7.34 psu; TA at the same time varied from 1630 to 1740 μmol/l. Changes in salinity and TA in the bottom layer at shallow stations are negligible due to intense mixing and desalination by river discharge: salinity varied from 7.08 to 7.34 psu, while TA was measured in a range from 1680 to 1780 μmol/l. The spatial distribution of TA for deep-water stations had the widest range in values: salinity varies from 10.86 to 11.35 psu, TA varies subsequently from 1780 to 1930 μmol/l.

pH level during the autumn expedition in the surface layer did not show strong spatial differences as well due to intensive water mixing (Fig. 2b), average pH was 8.05. The average hydrogen ion concentration in the deep-water layer, due to its isolation, the presence of hydrogen sulphide, the accumulation of CO_2 in it, and the decomposition process of organic matter, reaches pH 7.44.

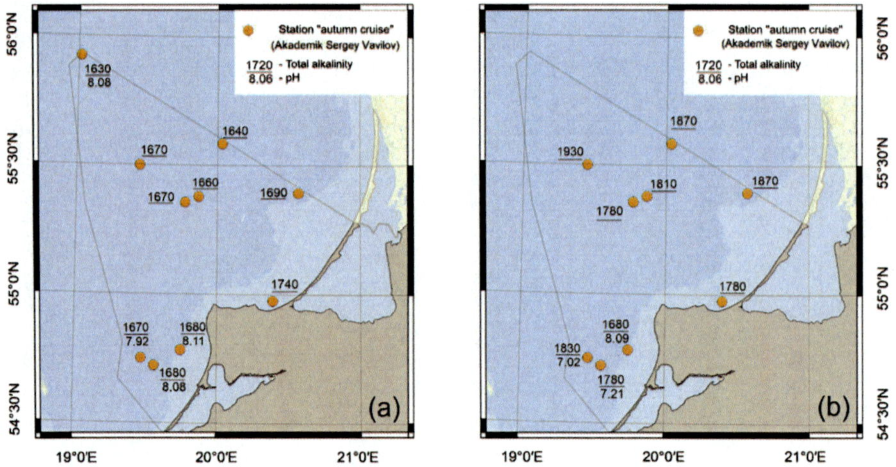

Fig. 2. Spatial distribution of total alkalinity and pH in the surface (a) and bottom (b) layer in autumn

The winter expedition also showed the insignificant TA and salinity fluctuations in the surface layer (Fig. 3a): from 7.09 to 7.57 psu and from 1620 to 1720 μmol/l, respectively. The coastal zone (depth less than 30 m) experienced low variability in salinity and TA as well: from 7.10 to 7.44 psu and from 1660 to 1740 μmol/l, respectively. All stations, that were located outside the coastal zone, showed wider range for both salinity and TA: from 10.45 to 12.33 psu and from 1470 to 1860 μmol/l. Bottom layer salinity has risen from autumn to winter.

The spatial distribution of pH in the surface layer in winter changed insignificantly (from 7.89 to 8.11). At the same time, pH in bottom layer varied from 7.15 in the deep-water part to 8.02 in the coastal zone (Fig. 3b).

Peculiar anomaly was noticed alongside the Curonian Spit during both cruises: TA decreased with depth while salinity increased. This inversion of vertical distribution may be related to the oxidation of organic matter supplied by the Curonian Lagoon runoff.

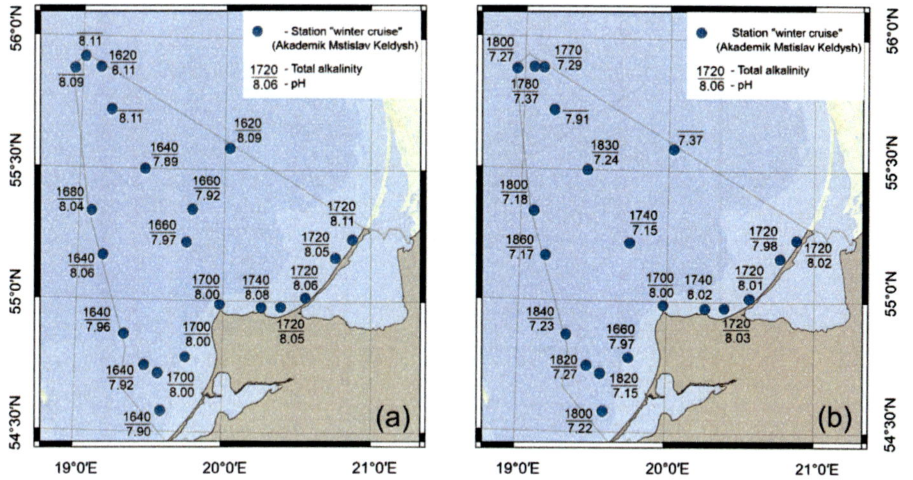

Fig. 3. Spatial distribution of total alkalinity and pH in the surface (a) and bottom (b) layer in winter

All stations, that were carried out in autumn, showed the same trend: salinity and TA values increase proportionally with depth, which corresponds to the classical distribution of TA [9, 10]. Yet there is one specific for the Baltic Sea factor, that may change the regular distribution of TA—a sulphide component, which leads to a linear direct dependence of TA from hydrogen sulphide concentration [11, 12] Almost all stations in the autumn expedition showed the presence or trace amounts of dissolved hydrogen sulfide. The very maximum of TA (1930 μmol/l) was measured in the bottom layer of the northern part of the Gdansk Deep alongside with the absolute hydrogen sulfide maximum (2.26 mg/l).

The result of winter expedition followed the same patter as autumn results: direct dependance TA from salinity with anomaly in the layer, rich in hydrogen sulfide. A jump of TA values occurs at the upper boundary of this layer (Fig. 4). The hydrogen sulphide maximum (1.2 mg/l) coincides with TA maximum (1860 μmol/l).

The pH distribution is characterised by decreasing pH values with depth from 8.11 to 7.15. Deviations from this pattern were observed in the central part of the Gdańsk Deep: pH decreased uniformly from the surface to the horizon with the maximum hydrogen sulphide concentration (80 m) from 7.96 to 7.11, while deeper pH increased uniformly towards the bottom up to 7.27 (Fig. 4).

The area, coinciding with the CP location, was studied more precisely due to potential anomalies, that may occur here. Hydrological conditions were surveyed at the CP itself and two stations nearby. The CP area did not experience significant salinity fluctuations from autumn to winter in the surface layer (just a slight decrease), while the bottom layer salinity has risen from 10.5 psu to 11.5 psu. Temperature has shown clear seasonal dynamic. The main difference is visible presence of the upper mixed layer along with the

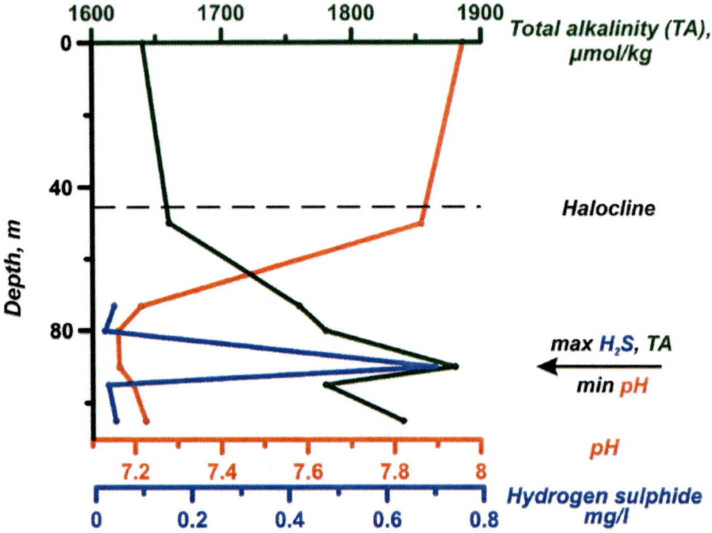

Fig. 4. Vertical distribution of TA and pH in the Gdansk Deep with hydrogen sulfide presence

thermocline and the intermediate cold layer in autumn, while winter vertical structure shows only deepened cold uniform layer from a surface to permanent halocline at 60–80 m deep (Fig. 5).

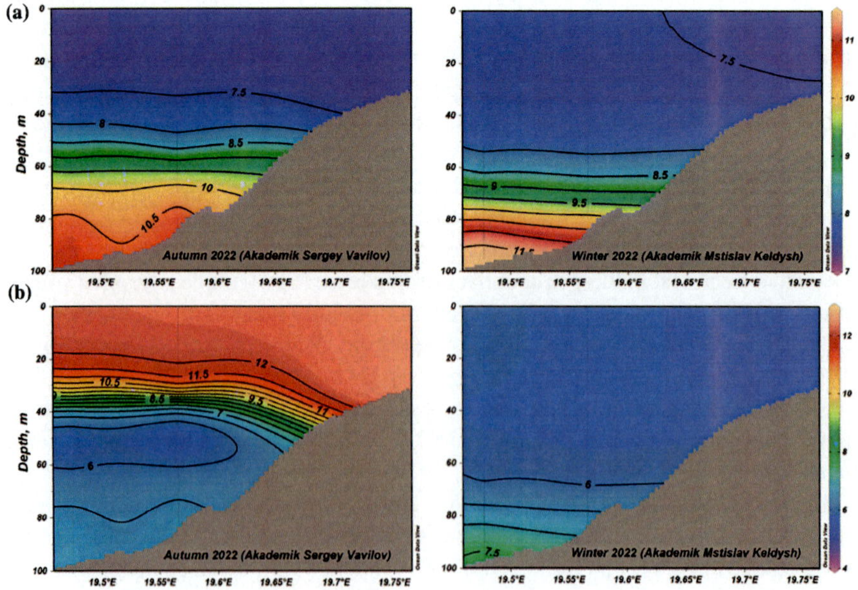

Fig. 5. The distribution salinity (a) and temperature (b) at the carbon polygon

The transition from autumn to winter conditions in a matter of vertical TA distribution was shown as slight decrease in surface TA level (as well as salinity). The TA level of 1700 μmol/l has mover deeper from 40 m to 60 m alongside with the winter convection of cold water. It is noticeable that bottom TA level hasn't risen along with salinity rise.

The change in pH vertical distribution was more noticeable: the homogenous in pH upper layer coincided with upper cold layer in winter. It seemed that that fact that bottom water in winter was a bit warmer caused the slightest pH rise as well (Fig. 6).

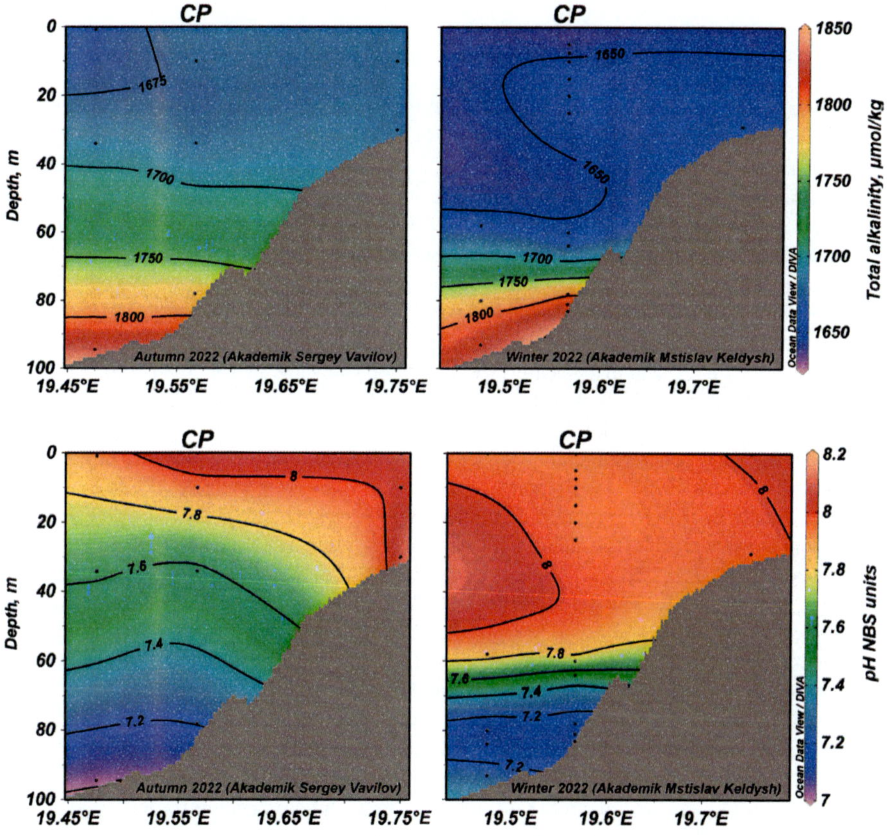

Fig. 6. The distribution of total alkalinity (a) and pH (b) at the carbon polygon (CP)

4 Conclusion

The data, collected during two expeditions, allowed us to notice that distribution of a carbonate system components in the south-eastern part of the Baltic Sea simultaneously follows natural patterns, but also shows some anomalies.

TA concentrations increase with depth in proportion to salinity, yet strong influence of the sulphide component on the carbonate system of the Baltic Sea was confirmed. TA increases in the presence of hydrogen sulphide, while pH decreases.

Anomalies in standard salinity-TA relationship were noticed alongside the Curonian Spit: TA has decreasing in a water with risen salinity. Presumably, this may be related to denitrification process and organic matter removal.

Acknowledgements. The analysis of the hydrological data was carried out within the framework of the IO RAS state assignment №FMWE-2021-0012, the characteristics of the carbonate system were studied within the framework of IKBFU state assignment №FZWM-2023-0002.

References

1. Borges, A.V., Schiettecatte, L.S., Abril, G., Delille, B., Gazeau, F.: Carbon dioxide in European coastal waters. Estuarine Coastal Shelf Sci. **70**(3), 375–387 (2006)
2. Hjalmarsson, S., Wesslander, K., Anderson, L.G., Omstedt, A., Perttilä, M., Mintrop, L.: Distribution, long-term development and mass balance calculation of total alkalinity in the Baltic Sea. Continental Shelf Res. **28**(4–5), 593–601 (2008)
3. Beldowski, J., Löffler, A., Schneider, B., Joensuu, L.: Distribution and biogeochemical control of total CO2 and total alkalinity in the Baltic Sea. J. Marine Syst. **81**(3), 252–259 (2010). https://doi.org/10.1016/j.jmarsys.2009.12.020
4. Müller, J.D., Schneider, B., Rehder, G.: Long-term alkalinity trends in the Baltic Sea and their implications for CO2-induced acidification. Limnol. Oceanogr. **61**(6), 1984–2002 (2016)
5. Ulyanova, M., Sivkov, V., Kanapatskij, T., Sigalevich, P., Pimenov, N.: Methane fluxes in the Southeastern Baltic Sea. Geo-Mar. Lett. **5**(32), 535–544 (2012)
6. Bordovsky, O.K., Ivanenkov, V.N.: Modern methods of hydrochemical studies of the ocean. IORAS, Moscow (1992). (in Russian)
7. , E.R., Wallace, D.W.R.: Program developed for CO2 system calculations. – Environmental System Science Data Infrastructure for a Virtual Ecosystem (ESS-DIVE) (United States), № cdiac: CDIAC-105 (1998)
8. Perez, F.F., Fraga, F.: The pH measurements in seawater on the NBS scale. Marine Chem. **21**(4), 315–327 (1987)
9. Kondratev, S.I., Medvedev, E.V., Konovalov, S.K.: Total alkalinity and pH in the Black Sea waters in 2010–2011. Phys. Oceanogr. **4**, 35–45 (2017)
10. Wolf-Gladrow, D.A., Zeebe, R.E., Klaas, C., Körtzinger, A., Dickson, A.G.: Total alkalinity: the explicit conservative expression and its application to biogeochemical processes. Marine Chem. **106**(1–2), 287–300 (2017)
11. Dickson, A.G.: An exact definition of total alkalinity and a procedure for the estimation of alkalinity and total inorganic carbon from titration data. Deep Sea Res. Part A Oceanogr. Res. Papers **28**(6), 609–623 (1981)
12. Hiscock, W.T., Millero, F.J.: Alkalinity of the anoxic waters in the Western Black Sea. Deep Sea Res. Part II Topical Stud. Oceanogr. **53**(17–19), 1787–1801 (2006)

Joint Analysis of the Variability of Dissolved Organic and Inorganic Carbon Concentrations as Markers of Biogeochemical Processes in Kara Sea

Y. V. Murzakova(✉), N. A. Belyaev, A. V. Kostyleva, V. Y. Fedulov, and A. A. Polukhin

Shirshov Institute of Oceanology RAS, 117997 Moscow, Russia
julia.murz@gmail.com

Abstract. The purpose of the work is to use a joint analysis of the dependences of DOC and DIC on salinity as an indicator of hydrochemical and biological processes. The work is based on data obtained during expeditions of the IO RAS to the Kara Sea from 2007 to 2021. The study covered the entire water area of the sea, including the runoff zone of the Ob and Yenisei. The paper analyzes the dependence of dissolved organic and inorganic types of carbon in the surface layer of the sea on salinity. It is shown that the joint analysis of DOC and ROC can reflect the process of ice melting and intrusion of river waters. The limits of interseasonal variability for DOC and DIC have been determined.

Keywords: Kara Sea · Dissolved organic carbon (DOC) · Dissolved inorganic carbon (DIC) · Surface layer · Seasonal variability · River runoff · Biogeochemistry

1 Introduction

The uniqueness of the Kara Sea water area lies in the combination of such factors as the high volume of continental runoff (more than a third of the river runoff of the entire Arctic region [1]), the influence of modified Atlantic waters, exchange with the waters of the central part of the Arctic Ocean through the system of cross-slope transport, regime, heterogeneity of the spatial distribution of primary production and low water productivity against the background of high concentrations of biogenic elements. The most important factor determining the structure of the waters of the Kara Sea is the continental runoff.

Studies carried out during a lot of expeditions made it possible to identify the main trends in the distribution of dissolved organic carbon (DOC). Articles [2, 3] show the decisive influence of river runoff on the distribution of organic carbon in its dissolved and suspended form. The work of 2015 shows the conservative behavior of DOC, as well as the upward deviation from conservatism at the boundary of sea and river waters. An analysis of the distribution of DOC and POC made it possible to identify areas

T. Chaplina (Ed.): CIWO 2023, SPEES, pp. 287–294, 2023.
https://doi.org/10.1007/978-3-031-47851-2_34

of increased biological productivity. In [4], the average concentrations and limits of interannual and interseasonal variability for organic carbon in dissolved and suspended form were established. Features of the distribution and dynamics of inorganic carbon (DIC) for the open sea and zones of mixing of sea and river waters were described in [5, 6]. The dynamics of dissolved inorganic carbon in 1993, 2011, and 2016 was described in [7]. It was shown that in the zone of mixing of sea and river waters, the concentration of DIC increases towards sea waters, as well as with depth.

However, the study of these dependences separately does not allow one to reliably explain a number of observed deviations. For example, an increase in the concentration of DOC in the zones of destruction of organic matter or a decrease in DOC during the melting of one-year sea ice [8].

This paper presents the main results of a joint analysis of DOC and DIC based on field data obtained during expeditions of the IO RAS from 2007 to 2021. A joint analysis of DOC and DIC was used to test the following hypothesis: whether the deviation of DOC up and down from the average trend line is reflected in the content of DIC.

2 Material and Methods

The work is based on data obtained during expeditions of the IO RAS to the Kara Sea from 2007 to 2021. In the runoff zones of the Ob and Yenisei, measurements were made more frequently; in other areas of the sea, the stations are evenly distributed over the water area. The general map of stations is shown in Fig. 5. Data on expeditions are presented in Table 1.

An integral database of DOC and DIC values in the surface layer of the Kara Sea was formed based on measurements carried out on cruises AMK54, AMK59, PSH128, AMK66, AMK76, AMK81, AMK83 (2007–2021) (Fig. 1). The database includes data from 139 stations, 82 of which refer to the autumn period (from early August to the end of September), and 57 – to the spring-summer period (from mid-June to early August).

Table 1. Dates of expeditions and the number of stations in the Kara Sea

Expedition	Season	Number of stations
AMK54	September 2007	50
AMK59	September 2011	49
PSH125	September 2013	53
PSH128	August 2014	62
AMK66	July–August 2016	93
AMK76	July 2019	38
AMK81	September 2020	16
AMK83	June–July 21	25

The method for determining the dissolved inorganic carbon in sea water is based on its calculation based on the measured values of pH and total alkalinity [9]. Samples for determination of pH and total alkalinity were taken in 0.5 l plastic containers. Measurement of the studied parameters was carried out in the onboard laboratory by standard hydrochemical methods no later than 6 h from the time of sampling [10].

Dissolved organic carbon was determined by high-temperature catalytic oxidation to carbon dioxide with preliminary acidification of the sample to remove inorganic carbon [11]. Water samples were taken with 30 l bottles and then filtered through GF/F filters. The filtrate for the determination of DOC was taken into glass vessels with a capacity of 22 ml, acidified to pH 2, and stored in a refrigerator. The DOC was determined on a Shimadzu TOC-Vcph analyzer with an SSM-5000A attachment. The volume of the injected sample is 100 μl, the measurement range is 0.05–25000 mg/l. Instrument error 1%.

Linear regression analysis was applied for joint analysis of dissolved organic and inorganic carbon. The significance of the regression model was tested using the Fisher F-test, the calculated value of which is found as the ratio of the variance of the initial series of observations of the studied indicator and the unbiased estimate of the variance of the residual sequence for this model.

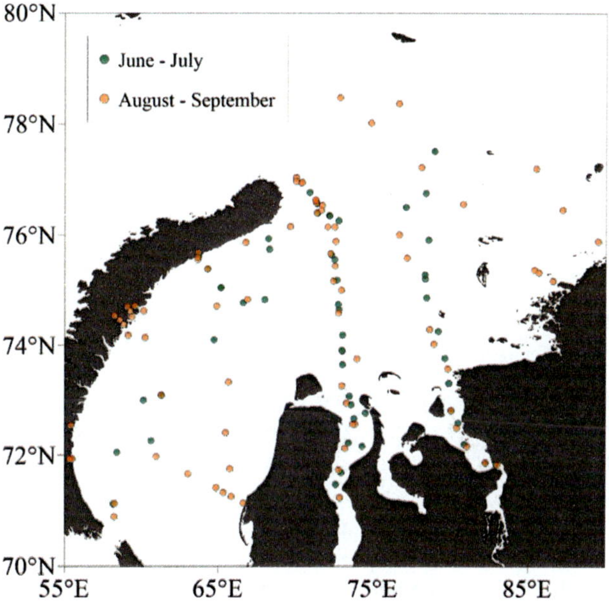

Fig. 1. Location of stations (2007–2021)

3 Results and Discussion

For joint analysis of dissolved organic and inorganic carbon, linear dependences of both forms of carbon on salinity were determined. Only the upper horizon (0–5 m) was used in the analysis to assess the impact of river runoff and ice melt on the studied parameters.

Figure 2 shows a quasi-conservative ratio of DOC and DIC to salinity in the surface layer of the mixing zone of sea and river waters, however, the content of organic carbon in the spring-summer period (from mid-June to early August) is increased compared to autumn (from early August to late September) values. In a further study, a joint analysis of two types of carbon was carried out taking into account the division into seasons - autumn and spring-summer.

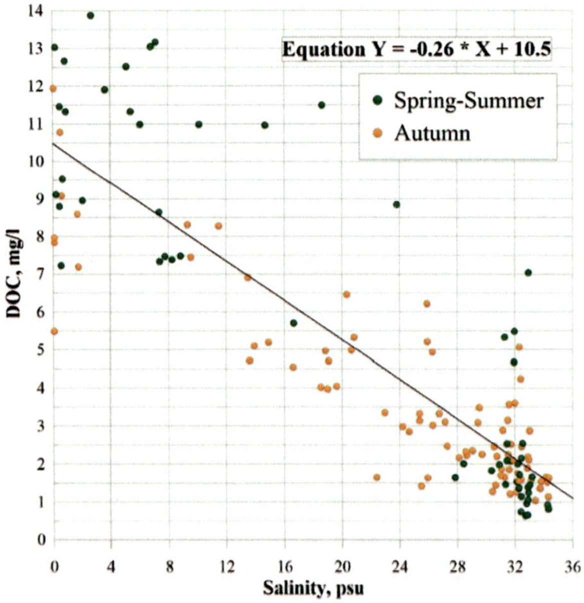

Fig. 2. Dependence of DOC on salinity in different seasons of the year for 2007–2021

Figure 3 shows the dependence of DOC on salinity in autumn. A confidence interval was determined, with respect to which three conditions were determined - DOC normal, DOC <min, DOC >max. The same was determined for DIC in each season. According to the six conditions described above, 9 groups of points were formed, each group corresponds to one condition for DOC and one for DIC.

In further work, DOC was chosen as the main parameter relative to which the changes in carbon concentrations were studied. Based on the analysis of the generated database, the dependences of DOC and DIC on salinity were studied. Linear dependencies and deviations from the average trend were determined.

In autumn, the maximum value of dissolved organic carbon - 11.94 mg/l was recorded at station 4993 (AMK54) in the Gulf of Ob, the minimum - 1.05 mg/l at station 5045 (AMK59) in the St. Anna trough. The mean value is 3.67 mg/l (number of determinations $n = 82$, standard deviation $\sigma = 2.4$). In autumn, the minimum value of dissolved inorganic carbon - 3.71 μM (st. 12818, PSH128) was recorded in the Gulf of Ob, the maximum - 26.28 μM (st. 4954, AMK54) off the coast of the Yamal Peninsula in the eastern part of the sea. The mean value is 21.57 μM ($n = 82$, $\sigma = 5.08$).

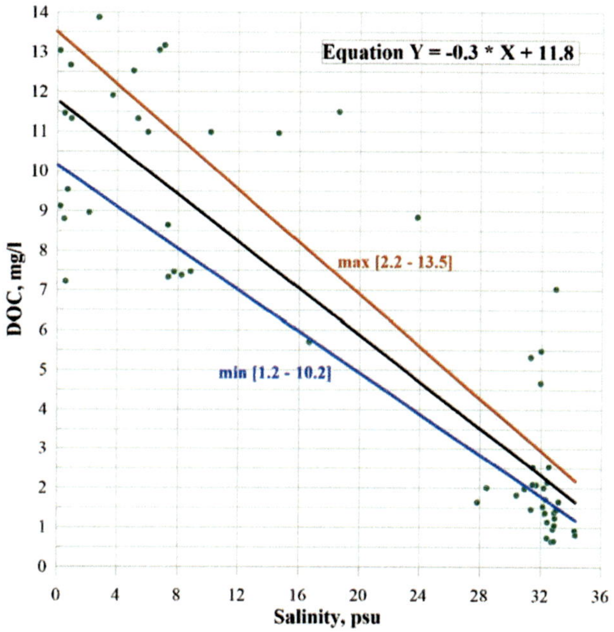

Fig. 3. Dependence of DOC on salinity in spring and boundaries of the confidence interval

DOC values measured in the spring-summer period vary from 0.64 mg/l in the seaward zone of the Kara Sea (station 7015, AMK83) to 13.87 mg/l in the Gulf of Ob (station 5356, AMK66). The mean value is 5.77 mg/l ($n = 57$, $\sigma = 4.51$). The calculated RIC values in the spring-summer period varied from 6.37 μM (station 5342, AMK66) in the Yenisei Bay to 25.81 μM (station 7016, AMK83) in the eastern part of the Kara Sea. The mean value is 20.07 μM (number of determinations $n = 57$, standard deviation $\sigma = 5.59$).

Deviations from the trend line can be caused by the following reasons:

1) Deviation from the conservative trend towards a decrease in the DOC content can be caused by the intrusion of water masses of a different origin, in particular, the presence of water masses formed as a result of ice melting, as well as the degradation of allochthonous organic matter.
2) The deviation of DOC from conservatism to a greater extent may be associated with the supply of autochthonous organic matter newly formed as a result of photosynthetic processes to the Kara Sea ecosystem, or it may be associated with a biotic factor (lysis of phytoplankton cells, an increase in primary production). It can also be associated with the intrusion of waters of another origin.
3) Downward deviation of RCL from the average trend may be due to the presence of water masses formed as a result of ice melting, or as a result of CO2 consumption by phytoplankton.
4) Deviation from the trend towards an increase in the DOC content may be associated with the destruction of DOC.

Figure 4 shows points that meet two conditions - each type of carbon is lowered relative to normal values. This may reflect the intrusion of water masses of a different origin, in this case, the melting of ice. The right picture shows summer points, the line shows the boundary of seasonal ice. On the left picture (autumn period), the points are approximately the same, which may be indirect evidence of the presence of melt water.

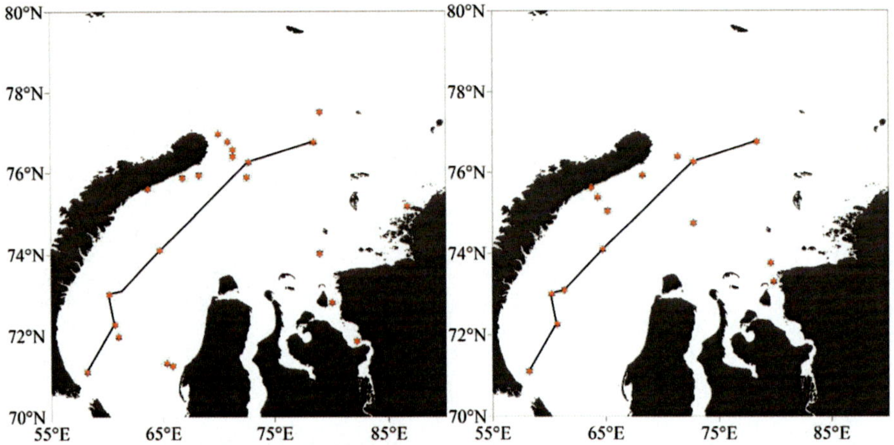

Fig. 4. DIC < min, DOC < min (autumn period on the left, summer on the right, 2021)

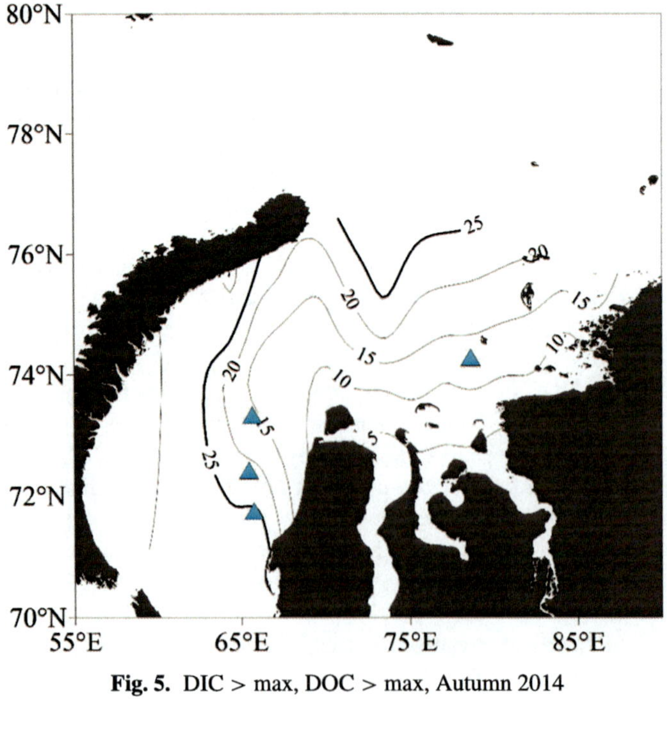

Fig. 5. DIC > max, DOC > max, Autumn 2014

The points in Fig. 5 to the west of the Yamal Peninsula in autumn differ from the average trend upwards for both parameters, these points refer to cruise PSh128, their values can be explained by the intrusion of waters of river origin, since in that year the Ob waters spread along the surface of the Kara Sea to the west [12]. To confirm this conclusion, isohalines are plotted on the map.

Figure 6 shows the most productive part of the Kara Sea in summer, the central part. Increased values of organic carbon showed active photosynthetic activity in the central part of the sea, but this result was not reflected in the concentrations of inorganic carbon.

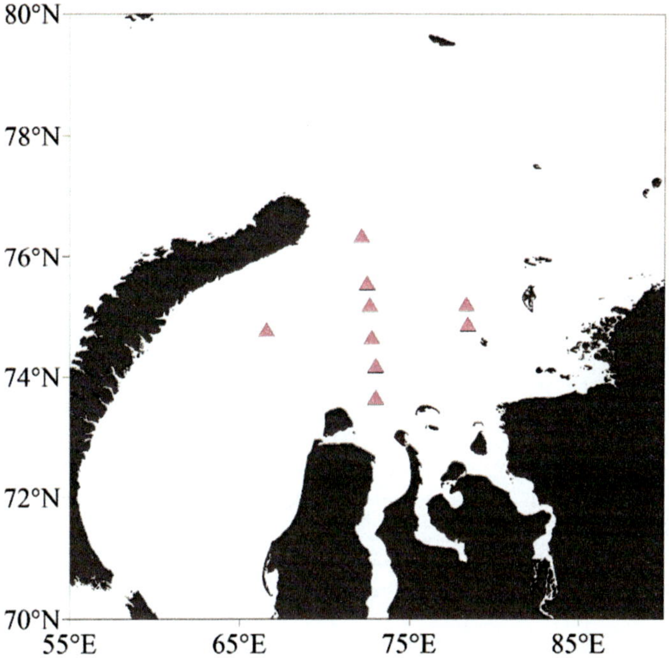

Fig. 6. DOC > max, summer

4 Conclusion

The study of biogeochemical processes on the example of individual water areas of the Kara Sea is necessary to identify common patterns in global climate change. The distribution of dissolved organic and inorganic types of carbon in the Kara Sea is closely related to river runoff. The distribution of dissolved inorganic carbon is generally conservative, with an increase in concentration directly proportional to an increase in salinity in both estuaries. The maximum DOC values were recorded in river waters, the concentration decreases with increasing influence of sea waters.

Based on the results of the analysis of the generalized data, it can be concluded that:

- The downward deviation of DOC from conservative behavior in 2021 reflected the melting of ice in the north of the Kara Sea.

- The zones of formation of autochthonous production maxima are distinguished by the content of dissolved organic carbon, but are not reflected in the concentrations of DIC.
- A joint analysis of DOC and DIC may reflect the intrusion of water masses of a different origin (river waters).

Acknowledgements. This work was carried out within the framework of the topic of State Assignment (FMWE-2023-0002) of the P.P. Shirshov Institute of Oceanology of the Russian Academy of Sciences.

References

1. Mikhailov, V.N.: Mouths of the Rivers of Russia and Neighboring Countries: Past, Present and Future, p. 413. GEOS, Moscow (1997)
2. Belyaev, N.A., Peresypkin, V.I., Ponyaev, M.S.: Organic carbon of water, suspension and upper layer of bottom sediments in the western part of the Kara Sea. Oceanology **50**(5), 748–757 (2010)
3. Belyaev, N.A., Ponyaev, M.S., Kiryutin, A.M.: Organic carbon of water, suspension and upper layer of bottom sediments in the central part of the Kara Sea. Oceanology **55**(4), 563 (2015)
4. Fedulov, V.Yu., Belyaev, N.A.: Organic carbon of the water column of the Kara Sea. Geology of Seas and Oceans, pp. 325–329 (2022)
5. Makkaveev, P.N.: Dissolved inorganic carbon in the waters of the Kara Sea and the mouths of the Ob and Yenisei rivers. Oceanology **34**(5), 668–672 (1994)
6. Makkaveev, P.N., Stunzhas, P.A., Melnikova, Z.G., Khlebopashev, P.V., Yakubov, S.: Hydrochemical characteristics of the waters of the western part of the Kara Sea. Oceanology **50**(5), 730–739 (2010)
7. Makkaveev, P.N., Nalbandov, Y., Polukhin, A.A., Schuka, S.A.: Dynamics of dissolved inorganic carbon in the Yenisei Bay during the open water period. Oceanology **59**(5), 701–713 (2019)
8. Nedashkovsky, A.P.: Study of the chemical composition of Arctic sea ice. Russia's Contribution to the International Polar Year 2007/08. Oceanography and Sea Ice, p. 407 (2011)
9. Millero, F.J.: Thermodynamics of the carbon dioxide system in the oceans. Geochim. Cosmochim. Acta. Cosmochim. Acta **59**(4), 661–677 (1995)
10. Sapozhnikov, V.V., et al.: Guidance on the chemical analysis of marine and fresh waters in the ecological monitoring of fishery reservoirs and areas of the World Ocean that are promising for fishing (2003)
11. Sugimura, Y., Suzuki, Y.: A high temperature catalytic oxidation method for the determination of non-volatile dissolved organic carbon in seawater by direct injection of a liquid sample. Mar. Chem. **24**(2), 105–131 (1988)
12. Polukhin, A.A.: Dynamics of hydrochemical characteristics of the surface freshened layer of the Kara Sea in August–September 2014. In: Ecosystem of the Kara Sea - New Data from Expeditionary Studies, pp. 58–62 (2015)

Distribution of Hydrochemical Parameters in the Surface Water Layer of the Zone of Mixing River and Sea Waters of the Amazon River Basin During the 52nd Voyage of the R/V "Academik Boris Petrov"

A. M. Seliverstova(✉), O. A. Zuev, and A. L. Chultsova

Shirshov Institute of Oceanology, Russian Academy of Sciences, 117997 Moscow, Russia
201219941315ann@gmail.com

Abstract. The article is written based on the results of 52 voyage on the R/V "Akademik Boris Petrov" from 22.11.2022 to 28.11.2022. During this period the work was carried out in the zone of mixing of river and sea waters in the estuary of the Amazon River. Hydrophysical and hydrochemical studies were carried out. According to the results of the work, it can be stated that these measurements were carried out during the "low" water season. River waters are well distinguished in terms of dissolved silicate distribution, total alkalinity and pH. Amazon River waters carry a lot of dissolved silicate, and very little phosphorus and nitrogen. The high oxygen saturation of the waters suggests an active photosynthesis process. The influence of the river plume was minimal.

Keywords: Amazon River · hydrochemistry · nutrients · dissolved silicate · dissolved oxygen · oxygen saturation

1 Introduction

The Amazon is the largest river in the world in terms of length, flow and basin area. The Amazon River System is located in the equatorial part of South America, starting from the confluence of the mountainous Marañon and Ucayali Rivers, flowing - mainly through the Amazonian lowlands - and flows into the Atlantic Ocean. Its average annual flow exceeds 209,000 m^3/s and its influence can be traced as far as the Cape Verde Islands and the Atlantic coast of Africa. The Amazon is full-flowing all year round, as seasonal fluctuations in runoff are smoothed out by different times of the rainy season on its right and left side; however, according to modeling and satellite observations, we can say that its runoff is more voluminous in July-August compared to other months and minimal in November-December [1]. Due to the high anthropogenic load on the banks of the river, as well as the increasing amount of microplastics brought to the World Ocean by the Amazon plume, it is necessary to conduct environmental monitoring of waters, including hydrochemical studies of nutrient parameters [2, 3]. These parameters reflect the intensity of production and destruction processes and allow us to judge the impact of anthropogenic pollution on the water area.

T. Chaplina (Ed.): CIWO 2023, SPEES, pp. 295–300, 2023.
https://doi.org/10.1007/978-3-031-47851-2_35

The last large-scale studies of the river were conducted by the Institute of Oceanology only in 1983 (the 9th cruise of the R/V "Professor Shtokman") and the most important tasks of the 52nd cruise of the R/V "Akademik Boris Petrov" were not only to study the features of the structure of thermohaline, biogeochemical and hydrophysical fields in the zone of interaction between the coastal ocean circulation and mainland freshwater runoff and to update the available data, but also to develop international cooperation with Brazil - a partner country of BRICS. The main purpose of this work is to characterize the abiotic component in the zone of mixing of river and sea waters of the Amazon River basin.

2 Material and Methods

Hydrochemical works in the voyage "ABP 52" in the zone of mixing of river and sea waters on the shelf of the Amazon River were carried out from 22.11.2022 to 28.11.2022. Twenty-seven stations perpendicular to the shoreline were performed in transects from 2 to 5 stations (Fig. 1). All samples were pre-filtered through nuclear filters with a pore diameter of 0.45 μm. Such parameters as dissolved oxygen (O_2), pH, total alkalinity (Alk), dissolved silicate (SiO_2), phosphate (or mineral) phosphorus ($P\text{-}PO_4$), mineral forms of nitrogen nitrate ($N\text{-}NO_3$) and nitrite ($N\text{-}NO_2$) were determined according to the methods adopted in the modern hydrochemistry [4].

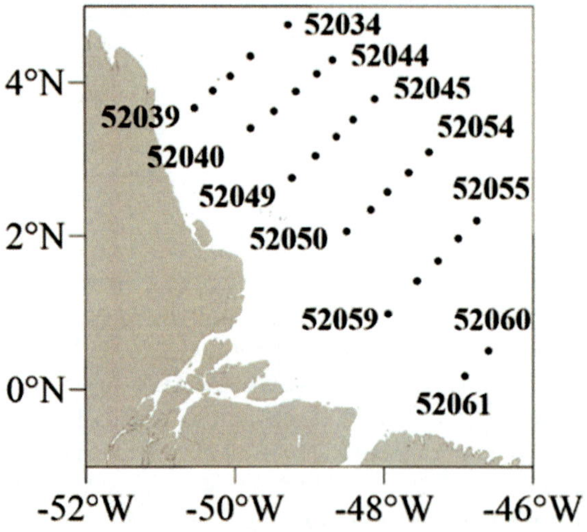

Fig. 1. Layout of stations at the estuary of the Amazon River

Dissolved oxygen in seawater was analyzed using a modified Winkler method. Oxygen saturation was calculated using the Weiss formula.

pH measurements were performed at a temperature of 25+—0.5 °C with the help of a pH electrode SI Analytics according to RD 52.24.495-2005 «Hydrogen Index and electrical conductivity of water. Methodology for Electrometric Measurements». Next, pH was recalculated to in situ values on the NBS scale. Total alkalinity was analyzed by direct open-cell titration with hydrochloric acid (0.02 M). During titration, water samples were blown with an air stream freed of carbon dioxide and ammonia. The temperature correction for pH was introduced and the elements of the carbonate system were calculated using the CO2SYS program [5].

Determination of the content of dissolved inorganic phosphorus was carried out according to the method of Morphy and Riley. Determination of dissolved inorganic silicate - according to the method of Koroleff [6].

The determination of nitrite was carried out according to the method of Bendschneider and Robinson. The determination of nitrates was based on the method of reducing them to nitrites with cadmium and measuring colorimetrically.

3 Results and Discussion

According to the thermohaline characteristics of the surface water layer, two zones can be distinguished: a southwestern, coastal zone with increased temperature and decreased salinity (up to 29.7 °C and 30.18 PSU) and a northeastern, marine zone with decreased temperature and increased salinity (up to 27.7 °C and 36.61 PSU). The observed distribution is probably related to the shallow depth of coastal stations, coastal runoff of the Amazon and its inflows, and the influence of the warm Guiana Current moving northwestward [7] (Fig. 2).

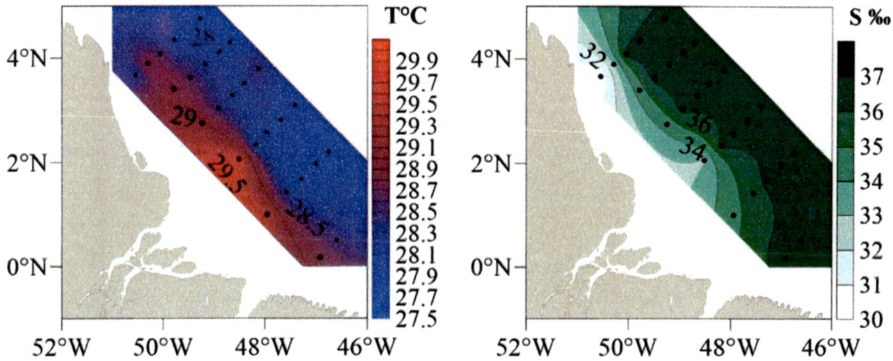

Fig. 2. Temperature (T) and salinity (S) distribution on a polygon at the estuary of the Amazon River

The distribution of total alkalinity and dissolved silicate correlate well with the distribution of temperature and salinity (Fig. 3). These parameters are markers of river discharge, as reflected in this polygon. The lowest values of total alkalinity (2038–2200 μM) and highest dissolved silicate (8.95–12.32 μM) are observed at the shore stations. As distance from the shore and from the river water discharge tongue, the values of total alkalinity increased and silicates decreased to 2433 μM and 0.45–0.55 μM, respectively. The pH values ranged from 8.18 (at shore stations) to 8.23 (in the marine part). Since Amazon River waters are characterized by low pH values of 4–5 scale units

[8], the distribution of this parameter in this case, as well as the distribution of total alkalinity and dissolved silicate, reflects the influence of river discharge.

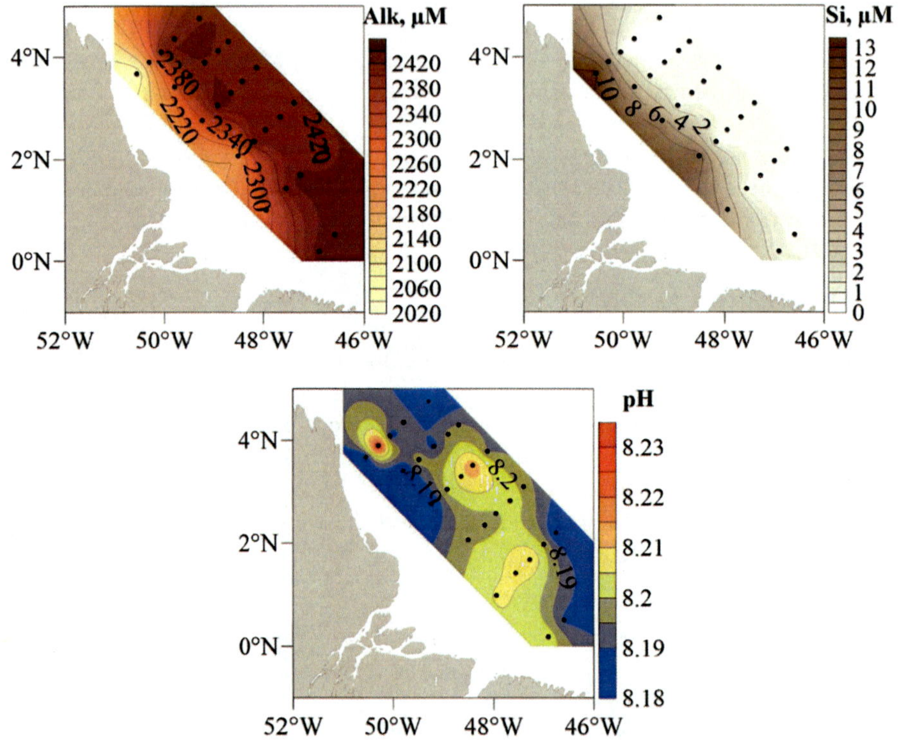

Fig. 3. Distribution of total alkalinity (Alk), dissolved silicate (Si), and potential of hydrogen (pH) on a polygon at the estuary of the Amazon River

Dissolved oxygen content at the polygon ranged from 4.35 to 4.95 ml/l (94–111% saturation) (Fig. 4). Maximums were observed at stations in the northwestern part of the polygon, where the influence of the Amazon River is most pronounced (4.94–4.95 ml/l, at 110–111% saturation), while minimums were observed at stations in the southeastern part, where the influence of rivers belonging to its basin is also observed (4.35 ml/l, at 94–98% saturation). Such low dissolved oxygen concentrations are typical for this region [9], but the high percentage of water oxygen saturation indicates active photosynthesis processes.

The contents of phosphate phosphorus, nitrate and nitrite nitrogen ions at the polygon were extremely low, averaging 0.10, 0.14 and 0 µM, respectively (Fig. 5). Local maxima of their content tended to the coastal stations, with the advancement to the seaside part the values dropped to zero values, which is associated with the decreasing influence of the Amazon River runoff.

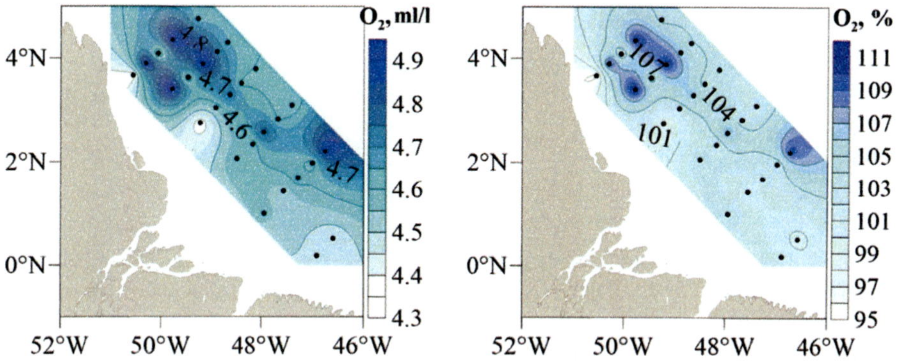

Fig. 4. Dissolved oxygen content (O2, ml/l) and oxygen saturation (O2, %) on a polygon at the estuary of the Amazon River

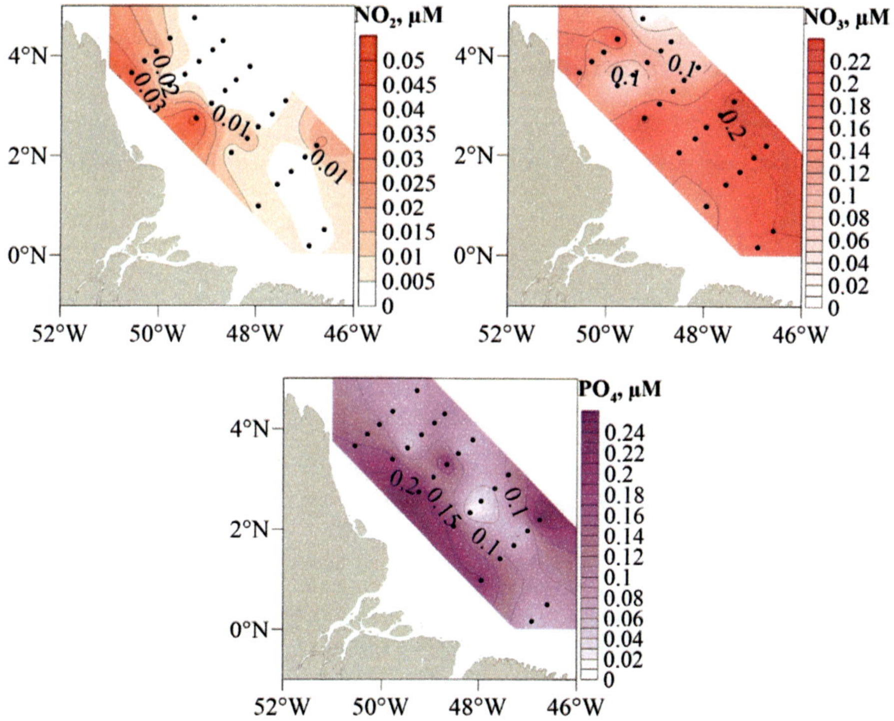

Fig. 5. Distribution of phosphate phosphorus (P-PO4, µM), nitrate nitrogen (N-NO3, µM) and nitrite (N-NO2, µM) on a polygon at the estuary of the Amazon River

It should be noted that at two neighboring stations in the northwestern part of the polygon, maxima dissolved oxygen and local minima of nitrate and nitrite (due to photosynthesis processes) were observed in one case and maxima phosphate, nitrate and nitrite with minima dissolved oxygen (due to organic matter oxidation processes) in the second case.

4 Conclusion

Despite the fullness of the Amazon River throughout the year, the periods from July to August, when higher water is observed and from November to December, with lower flows, stand out. The work in the ABP 52 cruise on November 2022 fell during the "low" water season, as evidenced by the distributions of thermohaline characteristics in the surface water layer, as well as total alkalinity and dissolved silicate.

Amazon waters are considered to be among the most chemically poor waters on Earth [7]. The low content of phosphorus and nitrogen in river runoff and elevated silicate content have been previously noted [1, 10], which is confirmed by our results.

Low concentrations of nitrogen forms in the surface water layer in both the river and marine parts of the polygon may indicate a low anthropogenic impact on the river basin. To more fully explore the impact of the Amazon plume, measurements are needed at different times of the year, as well as closer to shore, perhaps even moving further inland along the river.

Funding. The study was carried out within the state task of IO RAS, FMWE-2021-0016.

References

1. Hu, C., et al.: The dispersal of the Amazon and Orinoco River water in the tropical Atlantic and Caribbean Sea: observation from space and S-PALACE floats. Deep Sea Res. Part II Top. Stud. Oceanogr. **51**(10–11), 1151–1171 (2004)
2. Santos Queiroz, A.F., et al.: First assessment of microplastic and artificial microfiber contamination in surface waters of the Amazon Continental Shelf. Sci. Tot. Environ. **839**, 156259 (2022). https://doi.org/10.1016/j.scitotenv.2022.156259
3. Schmidt, N., et al.: The Amazon River: a major source of organic plastic additives to the tropical North Atlantic. Environ. Sci. Technol. **53**(13), 7513–7521 (2019). https://doi.org/10.1021/acs.est.9b01585
4. Bordovskii, O.K., et al.: Modern Methods of Hydrochemical Studies of the Ocean. Okeanol. Akad. Nauk SSSR, Moscow (1992). (in Russian)
5. Lewis, E.R., Wallace, D.W.R.: Program developed for CO2 system calculations. In: Environmental System Science Data Infrastructure for a Virtual Ecosystem (ESS-DIVE) (United States, 1998), CDIAC-105 (1998)
6. Grasshoff, K., Kremling, K., Ehrhardt, M. (eds.): Methods of Seawater Analysis. Wiley, Hoboken (2009)
7. Arzhanova, N.V., Burkaltseva, M.A.: Hydrochemical characterization and primary production of waters of the Brazil-Guinea region in the summer of 1969 year. In: VNIRO Proceedings, pp. 70–76 (1974). (in Russian)
8. Monin, A.S., Gordeev, V.V.: Amazonia, 217 p. Nauka, Moscow (1988). (in Russian)
9. Lyakhin, Yu.I.: Hydrochemistry of tropical areas of the World Ocean, 212 p. Gidrometeoizdat, Leningrad (1990). (in Russian)
10. Ryther, J.H., Menzel, D.W., Corwin, N.: Influence of the Amazon River outflow on the ecology of the western tropical Atlantic I. Hydrography and nutrient chemistry (1967)

Marine Geology

Features of the Behavior of Oil Hydrocarbons in the Barents Sea in 2019–2020

E. V. Koltovskaya(✉) and I. A. Nemirovskaya

Shirshov Institute of Oceanology RAS, 117218 Moscow, Russia
katyayaya15@gmail.com

Abstract. The results of studies of the distribution and composition of aliphatic hydrocarbons (AHCs) in the suspended particulate matter and bottom sediments of the Barents Sea are presented based on the data of expeditions in 2019 and 2020. R/V Akademik Mstislav Keldysh. An increase in HC concentrations in suspended matter in surface waters has been established up to 20–23 μg/l climate change. In bottom sediments, the distribution of HCs depends not only on the conditions of sedimentation and their particle size distribution, but also on the variability of redox conditions and endogenous fluxes from the sedimentary sequence.

Keywords: aliphatic hydrocarbons · suspended particulate matter · bottom sediments · Barents Sea · Arctic ecosystem

1 Introduction

The study of aliphatic hydrocarbons (AHCs) in suspended particulate matter (SPM) and bottom sediments in high-latitude water areas is largely due to the high oil and gas potential of the Arctic shelf, which, according to recent estimates, exceeds 100 billion tons in oil equivalent and is unevenly distributed over the shelf area. The peculiarity of this distribution lies in the confinement of the dominant part of the recoverable hydrocarbon resources mainly to the Barents Sea, which will play a decisive role in supporting the country's future domestic and export needs for oil and gas. This sea accounts for a total of 28% of the total hydrocarbons reserves of the Russian shelf [1].

With the development of deposits and an increase in fuel transportation, the risk of pollution of the Arctic basin with oil hydrocarbons increases. In order to develop effective measures to combat marine pollution, it is necessary to form an understanding of the sources of entry into the marine environment, since a certain contribution to the hydrocarbon content is made by their natural inputs from the sedimentary sequence [2].

At the same time, under the conditions of a changing climate in recent years, significant changes have occurred in the ecosystem of the Arctic seas, under the influence of large inflows of Atlantic waters and a significant reduction in the area of multi-year ice [3–5]. As a result, the primary production of phytoplankton increased on average by 28% [3, 4]. Changes in phytoplankton community should affect the concentrations of AHCs, which are constant components of organic matter in SPM [6].

T. Chaplina (Ed.): CIWO 2023, SPEES, pp. 303–307, 2023.
https://doi.org/10.1007/978-3-031-47851-2_36

The main goal of research is to assess the contribution of anthropogenic AHCs to the total content of the fraction in particulate matter and bottom sediments, taking into account natural background biogeochemical processes that produce oil HCs.

2 Material and Methods

To study the variability in the distribution of AHCs in the process of sedimentation and bottom sediments, we used data from expeditions of the R/V «Akademik Mstislav Keldysh» in the summer-autumn period from 2019 to 2021.

Water for the study of AHCs was sampled from the surface with a bucket when the ship approached the stations, and the water column was sampled with Niskin bathometers (Rosette complex). SPM for the analysis of the components of organic matter: C_{org}, lipids, AHCs, chlorophyll-*a* (chl-a), was isolated on glass fiber filters GF/F (pore size 0.7 μm, Whatman), and for the study of SPM, on preliminarily washed with hydrochloric acid and weighed polycarbonate nuclear filters: size pore size 0.45 μm (Dubna, Russia). The bottom sediments (upper layer) was sampled with a bottom grab «Okean-25» (Russia) (layer 0–2 cm), and intact cores were sampled with a Mini Muc K/MT 410, KUM, multicorer (Germany). The sediment column was divided into layers (0–1, 1–2, etc.), dried at a temperature of 50 °C, and a fraction of 0.25 mm was sifted out [7].

All solvents were qualified as high purity (o. s. h.). For lipid extraction methylene chloride was used in an ultrasonic bath. Fractions AHC were separated with hexane using silica gel column chromatography. Concentrations of lipid and AHC were determined by IRAffinity-1 instrument spectroscopy (Shimadzu, Japan) before and after chromatography, respectively. As a reference, benzene, hexadecane and a mixture of isooctane (25, 37.5 and 37.5, and 25 vol. %, respectively) was used with standard GSO 7248-96. The sensitivity of the method was 3 μg/mL of the extract [9].

3 Results and Discussion

Surface Water. The distribution of AHCs in SPM and the SPM itself usually coincides with the same sources [8]. Indeed, in general terms, the distribution of AHCs and SPM was of the same type. However, for the entire data set in 2019, only a weak relationship was observed between the concentrations of SPM and AHCs: $r = 0.175$, $n = 53$. A relationship was observed between SPM and chl-a: $r = 0.50$, with the highest values in the arch. Svalbard, where $r = 0.75$.

In the southern area of the Barents Sea, there were no relationships between chl-a and SPM, chl-a and AHCs: r (chl a – AHCs) = −0.22, r (chl-a – SPM) = −0.14, which shows their different sources. In 2020, on shipping lanes in the southern region of the Barents Sea, AHC concentrations exceeding 50 μg/l (limit value for oil HCs), or close to this value, were observed. The distribution is shown in Fig. 1. According to satellite data, It is possible that AHCs come with oil pollution, since the maximum amount of oil films is concentrated in this part of the Barents Sea [9]. At the same time, the maximum content in this area in 2020 (58 μg/l, 170 μg/mg of SPM) was noted at station. 6870 in the zone of cocolithophore flowering. Increase in primary production in the Barents Sea, apparently caused by climate change, also led to an increase in HC concentrations. The increased AHC concentration (up to 58 μg/l, 2020) in the southern area of the Barents Sea is associated not only with increased shipping, but also with cocolithophorid blooms. This fact allow us to conclude, that natural processes lead to the formation of higher concentration of AHC.

In the Barents Sea's surface waters in 2019 and 2020 the average concentration of AHCs for the entire dataset in SPM is estimated to be 23 μg/l (total range 6–62 μg/l) and 20 μg/l (total range 15–157 μg/l), respectively. Compared to 2016 and 2017, at an average mixture concentration of 2.8 and 8.3 μg/l [10], the concentrations of AHCs in SPM increased by an average of 8.2.

a)

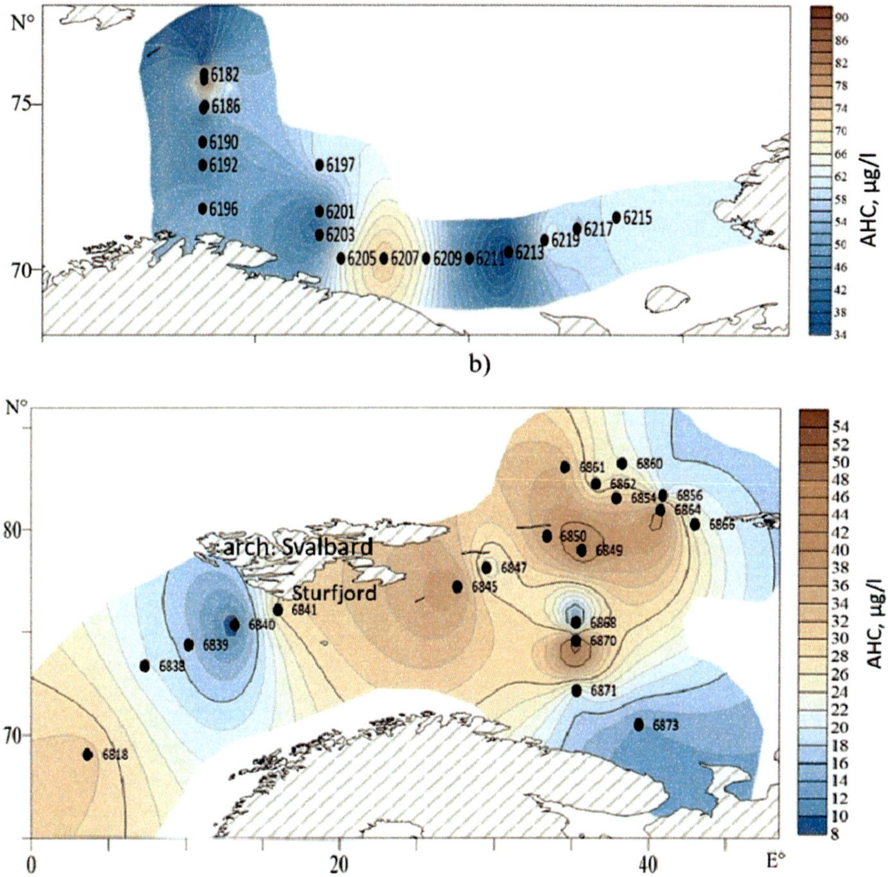

Fig. 1. Distribution of AHCs in the Barents Sea (surface waters) in 2019 (a) and 2020 (b)

Bottom Sediments. At the transition from the surface to the near-bottom horizon, the AHC concentration decreased at almost all stations. An exception was established only at three stations: 6166, 6181 and 6205 (2019). In this area the AHC concentration in the near-bottom in compare with the surface layer was 1.5–5.6 times higher. The maximum concentration at the bottom was established at a depth of 249 m in the southern part of the Barents Sea at st. 6205 – 76 μg/g.

In 2019, a high AHCs concentration was found at station 6196 on the Eastern Shelf of Svalbard - 51μg/g. The maximum was 64 μg/g in its southern part (station 6213). In 2020, for different regions of the Barents Sea, the range of concentrations was larger: 3–186 μg/g. Maximum was in Sturfjord area (st. 6842). The part of AHCs in the composition of Corg was 1.18%. At station 6841 in the surface layer the AHC content was only 36.7 μg/g (0.24% of Corg). According to hydrophysical data in this area, the most significant fluid flow was established.

Composition of AHC in SPM and Bottom Sediments. The composition of AHC in most samples was dominated by high-molecular homologues. CPI values (ratio odd to even alkanes) varied in a wide range: 1.98–5.40. Almost all samples were dominated by high molecular homologues. Usually, CPI values are close to 1 in the composition of alkanes in SPM, especially in surface waters, where their composition forms phytoplankton [10]. In many sediments in the composition of alkanes in the low-molecular region, an increase in the concentrations of n-C_{17}, the main homologue of phytoplankton, was recorded. The isoprenoid coefficient Ki is ratio of low- to high-molecular homologues: (Ki = (Pr + Ph)/(C17 + C18)). Its value increase indicated the intensity of microbial transformation of organic matter. The CPI values changed to a greater extent in the surface layer of sediments on the shelf of Svalbard and in the Medvezhinsky Trench, which was also confirmed by the low values of the ratio of n-alkanes to naphtheno-aromatic compounds, varying in the range of 0.03–0.22.

At the same time, the composition of alkanes at station 6841 in Sturfjord, in concentration and distribution of low molecular weight homologues, differed sharply from that at the neighboring station. 6840. Only at this station, among alkanes, low-molecular homologues dominate. At st. 6840 the value of CPI was average 1.8 (maximum 2.6), that higher than station 6841 (average 1.4, maximum 1.8). Its indicated a lower transformation of high molecular homologues in the area of the seep.

4 Conclusion

In the surface waters of the Barents Sea, the concentrations of AHCs in SPM increased compared to 2016–2017, due to near-marginal blooms. Climate change, which caused an increase in primary production in the Barents Sea, led to an increase in AHC concentrations. The increased content of AHCs in the southern part of the Barents Sea is associated not only with increased shipping, but also with cocolithophorid blooms, that is, natural processes. The nepheloid layers affect the distribution of AHCs in the near-bottom horizon, and the intensity of processes at the suspended-bottom sediment boundary affects their composition.

In the bottom sediments surface layer, an increase in AHC concentrations occurs in the zones of faults and hydrothermal activity (the Sturfjord, the Medvezhinsky Trench). The different genesis of AHCs confirmed by to the absence of correlation in the distributions of AHCs, SPM and C_{org}.

In the thickness of sediments, the unloading of endogenous fluid flows, the variability of the redox environment contributes not only to the utilization of individual components in the AHC composition, but also to the formation of autochthonous low molecular weight alkanes.

At present, natural processes occurring in the Barents Sea have a greater influence on the formation of the composition and levels of AHCs than anthropogenic ones.

Acknowledgements. The expeditions were carried out within the framework of the state task of the Ministry of Education and Science of Russia (№ FMWE-2021-0016), participation in the expedition of employees was carried out with the financial support of the Russian Science Foundation (project №. 19-17-00234-P), generalization of materials and geochemical studies - within the framework of the RFBR grant № 20-35–90025.

References

1. Kaminskij, V.D., Suprunenko, O.I., Smirnov, A.N., Medvedeva, T., Chernyh, A.A., Aleksandrova, A.G.: Modern resource status and prospects for the development of raw materials in the Russian shelf region of the Arctic. Explor. Protect. Mineral Resourc. **9**, 136–142 (2016)
2. AMAP (Arctic Monitoring and Assessment Programme): Sources, Inputs and Concentrations of Petroleum Hydrocarbons, Polycyclic Aromatic Hydrocarbons, and other Contaminants Related to Oil and Gas Activities in the Arctic. Chapter 4. Oslo, Norway, 87 p. (2007)
3. Dalpadado, P., et al.: Climate effects on Barents Sea ecosystem dynamics ICES. J. Mar. Sci. **69**, 1303–1316 (2012)
4. Arrigo, K.R., van Dijken, G.L.: Continued increases in Arctic Ocean primary production. Progr. Oceanog. **136**, 60–70 (2015)
5. Lapina, N.M., Torgunova, N.I., Agatova, A.I.: Organic matter in the ice of the Arctic Ocean. Probl. Commer. Oceanol. **2**, 156–172 (2011)
6. Vernadskij, V.I.: Chemical structure of the Earth's biosphere and its environment, 376 p., Nauka (2001)
7. Korchagina, Yu.I., CHetverikova, O.P.: Methods for Investigation of Dispersed Organic Matter in Sedimentary Rocks, 228 p., Nedra (1976)
8. Koltovskaya, E.V., Nemirovskaya, I.A.: Suspended matter and hydrocarbons fluxes in the Kara and Laptev Seas. Water **14**(14), 2278 (2022)
9. Kuchejko, A.A., et al.: Film pollution of the Barents Sea according to radar monitoring data 2017–2019. Ecol. Ind. Russia **24**, 48–55 (2020)
10. Nemirovskaya, I.A., Khramtsova, A.V., Khalikov, I.S., Koltovskaya, E.V., Solomatina, A.S.: Hydrocarbons in water and sediments of the Norwegian and Barents Seas. Proc. Karelian Sci. Center Russ. Acad. Sci. **4**, 94–107 (2021)

Features of the Distribution of Dispersed Sedimentary Matter Composition in the Nordic Seas

I. A. Migdisova(✉), D. P. Starodymova, A. A. Klyuvitkin, A. N. Novigatsky, and A. V. Bulokhov

Shirshov Institute of Oceanology RAS, 117997 Moscow, Russia
ira-mig@yandex.ru

Abstract. Two geodynamic sedimentary settings were studied in the sedimentary system of the Norwegian Sea – in the Lofoten Basin and in the zone of the junction of the Mohns and Knipovich ridges. Dispersed sediment samples were collected from different horizons of the water column, including the subsurface and benthic layers. Elemental and constituent compositions were studied, and the fluxes of matter were analyzed. The vertical and seasonal variability of individual components of the sedimentation system was also assessed. The distribution of the main sediment components in the two sedimentary settings differs insignificantly, several general patterns were identified. The presence of authigenic barite of different morphology in the subsurface horizon of the Lofoten Basin water column was noted.

Keywords: dispersed sedimentary matter · fluxes · elemental composition · vertical and seasonal variability · ocean geochemistry · sedimentation · authigenic barite

1 Introduction

Modern ocean sedimentary systems are the most interesting objects for research because of their close relationship with hydrological and biogeochemical processes. Current climatic changes are transforming the intensity of sedimentation processes and the chemical composition of sediment. There are several sources of sedimentary matter input to the open ocean, ranging from material transported by currents, aeolian and ice-iceberg transport products, to hydrothermal vents and the development of local biogenic systems [1].

The study of the sedimentary system of the Norwegian Sea was conducted as part of the continuation of the work carried out in this region for quite a long time. In 1995, during the 36th voyage of the R/V Akademik Mstislav Keldysh, a cycle of oceanological studies was carried out in the northern part of the Mohns Ridge, where a hydrothermal plume was detected. The collected sediment samples were enriched in Fe, Mn, Cu, Co, and Ni, but no signs of modern hydrothermal activity were found during diving with submersibles [2]. Near the study area in 2008, Norwegian oceanologists discovered the

T. Chaplina (Ed.): CIWO 2023, SPEES, pp. 308–318, 2023.
https://doi.org/10.1007/978-3-031-47851-2_37

Loki Castle hydrothermal field located in the zone of the junction of the Mohns and Knipovich ridges in the spreading axis [3].

It was decided to continue the series of studies of hydrothermal fields in this region, so the material was sampled at approximately the same point where the research took place during the 36th cruise of the R/V Akademik Mstislav Keldysh. The main goal of the work was to study the variability of elemental and component composition of sedimentary matter in two different geodynamic systems during sediment genesis processes. In order to discover the nature of suspended sediment in the water column, the following series of studies were carried out: substance fluxes with different variability (seasonal – by months during the year, vertical – at different depths of the water column) were estimated; the component composition of the material was determined; the element composition of sedimentary matter was investigated; authigenic minerals (barite) in the upper layer of the water column were studied.

2 Material and Methods

This paper presents the results of the study of sedimentary matter from the Nordic Seas in two geodynamically different areas: in the Lofoten Basin and at the junction of the Mohns and Knipovich ridges near the Loki Castle hydrothermal field. The material was sampled using automated deep-sea sedimentation observatories [4] installed in 2018 and 2019 during the voyages of the R/V Akademik Mstislav Keldysh (Fig. 1). The equipment of the substance sampling unit included sedimentation traps of two types: large conical 12-cup rotary traps «Lotos-3», used to assess seasonal variability of matter, located in the subsurface and near-bottom horizons (500 m and 2980 m in the Lofoten Basin; 500 m and 2490 m in the Mohns ridge), as well as small cylindrical integral traps "MSL-110", which allow estimating the depth variability of matter, located in intermediate layers of the water column (500, 1250, 2110, 2550 and 2980 m in the Lofoten Basin; 500, 1630, 2060, 2490, 2665, 2770 m in the Mohns ridge) [5].

The element analysis of precipitated particles was performed by inductively coupled plasma mass spectrometry on an «Agilent 7500a» analyzer (IO RAS) after decomposition of the substance with a mixture of concentrated acids (HNO_3, HF, H_2O_2, HCl). The content of total (TC) and organic carbon (POM) was measured on a «Shimadzu TOC-VCPH» analyzer (IO RAS) by direct determination of total carbon concentration. Organic carbon was determined by measurement after acidification of the sample with 1n HCl, the content of inorganic carbonate carbon was calculated by the difference of TC and POM. The contents of Si and Al were determined by photometric method. According to the method of Bychkova Ya. V. [6] to assess the completeness of sample decomposition in each series of analyses used state standard samples (SSM) with certified values of concentration of chemical elements. The following GEOs were used for this decomposition method: SDO-3 (terrigenous carbonate silt) and BCR-CRM414 (plankton). In each series of acid decomposition blank samples of detection limit were laid.

Additionally, Fe and heavy metals (Cu, Ni, Zn) were analyzed by atomic absorption spectrometry (AAS) on the "Kvant-2MT" analyzer. For analysis by AAS method (acetylene-air mixture) sample dilutions were prepared separately for each determined element depending on its content in the sample (from 10 to 50 times).

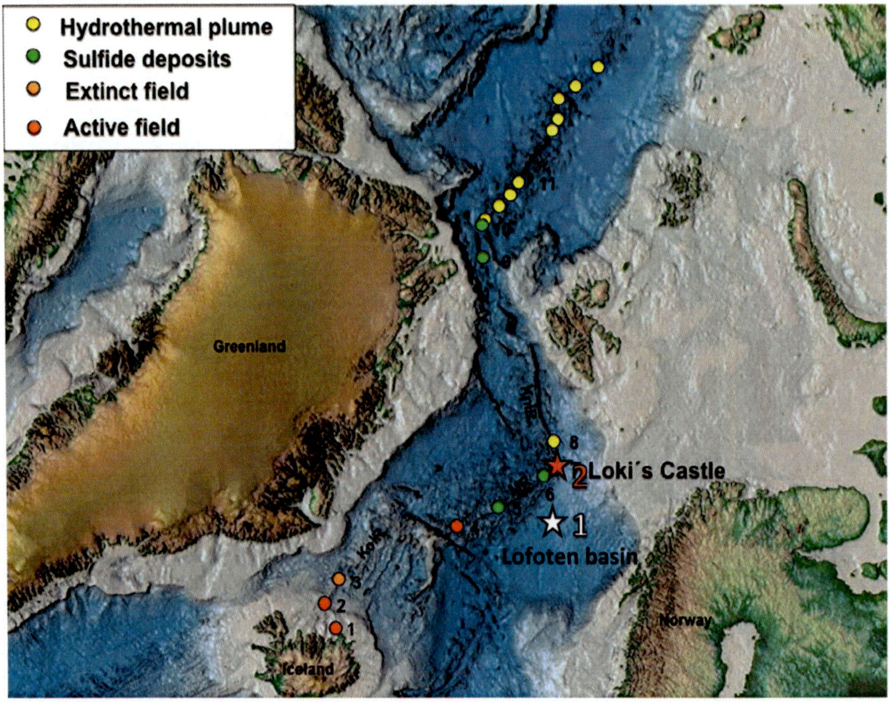

Fig. 1. Sampling stations: 1 – installed in 2018 in the Lofoten Basin of the Norwegian Sea; 2 – installed in 2019 in the Loki Castle hydrothermal field near the Mohns ridge [3]

3 Results and Discussion. Sediment Fluxes and Components

In both the Lofoten Basin and the Mohns ridge area, vertical zonality is manifested by increasing fluxes downward in depth. Figure 2 presents the vertical distribution of fluxes and components in the Lofoten Basin and Mohns ridge area. Sediment fluxes in the water column in the Lofoten Basin region gradually increase with depth, reaching a maximum in the near-bottom horizon: in the subsurface at 500 m depth the flux was 40.6 mg/m^2/day; in the intermediate layers at 1250, 2110 and 2550 m–42.4, 42.7 and 47.7 mg/m^2/day, respectively; and in the bottom layer at 2980 m–326.5 mg/m^2/day (Fig. 2). The vertical distribution of sediment fluxes in the area of the junction of the Mohns and Knipovich ridges is characterized by two extremes: in the subsurface horizon (500 m), the mean annual flux amounted to 89.4 mg/m^2/day, and in the near-bottom horizon (2770 m) – 244.0 mg/m^2/day. At intermediate horizons, flux values ranged from 41.2; 53.7; 125.5 and 221.1 mg/m^2/day at depths of 1630, 2060, 2490 and 2665 m, respectively. The flux maximum at 500 m reflects sediment input of predominantly biogenic origin from the active layer. In the near-bottom horizons (2665 and 2770 m), the fluxes significantly exceed the values obtained for the intermediate layer, which reflects the influence of the nepheloid layer [4]. The drastic flux increases in the benthic nepheloid layer both in the basin and in the ridge area can probably be explained by the influence of benthic currents puddling the sediment, resulting in a much larger amount of material enters the traps [5].

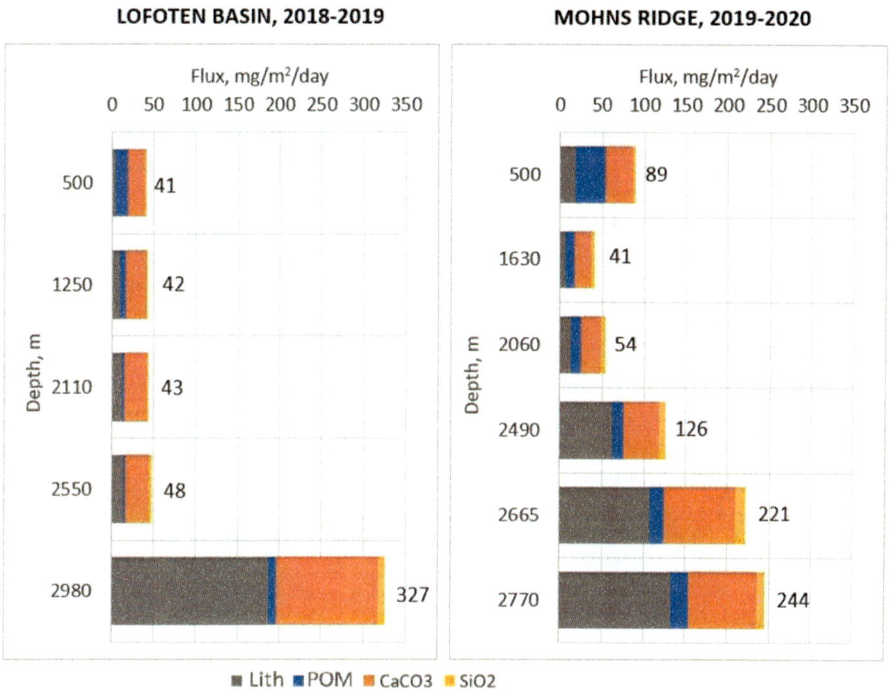

Fig. 2. Vertical distribution of fluxes and components in the Lofoten Basin and Mohns ridge area

The seasonal variability of sediment fluxes in the Lofoten Basin was assessed over a ten-month period from August to May (Fig. 3). In the layer located at a depth of 500 m under the active euphotic one, an increase in the flux was recorded in the warm season, which is reflected in a smooth peak in the fluxes in the autumn: 98.8 mg/m^2/day in August, a maximum in September (131.3 mg/m^2/day) and a smooth decrease in October (12.4 mg/m^2/day) and November (20.7 mg/m^2/day). With the onset of winter, and then spring period (6.6 mg/m^2/day in December), the flux of substance begins to decrease sharply, reaching a minimum in April (0.5 mg/m^2/day). In the bottom layer (2980 m depth) significantly higher fluxes of matter were recorded than in the subsurface horizon. The highest fluxes at this depth are observed in the spring period, reaching a maximum in April-May (554.6 and 669.7 mg/m^2/day), while the minimum occurs in the winter period, particularly in December (52.6 mg/m^2/day). The increase in fluxes is characteristic of spring, which can probably be attributed to the intensification of near-bottom currents during this period. In the area of the junction of the Mohns and Knipovich Ridges, the exposure of sediment traps was twelve months – from June to May (Fig. 3). Maximum fluxes in the subsurface horizon were recorded in the early summer months (233.1 and 271.3 mg/m^2/day), then sediment fluxes decrease very sharply by autumn and even more so by winter. A sharp decrease in fluxes to 71.9 mg/m^2/day was recorded in August, followed by a local increase in September and October to 89.1 and 85.1 mg/m^2/day respectively, and then a decrease in November to 25.1 mg/m^2/day. The lowest fluxes were recorded in spring, with values ranging from 1.8 to 3.7 mg/m^2/day. In the near-bottom horizon, the peaks of substance flux coincide with the peaks in the subsurface in summer and reach values of 255.1 and 274.3 mg/m^2/day (June and July), then gradually decrease with the onset of the cooler season. Similarly, to the Lofoten Basin, an increase of fluxes in the nepheloid layer is observed on the Mohns ridge in April–May. The subsurface sediment throughout the entire exposure contains significantly more biogenic matter as compared to the deep horizons, which is especially noticeable at the station in the vicinity of the Mohns ridge. As noted earlier [7], the similarities in the peaks of subsurface sediment fluxes both in the basin and on the ridge in September are explained by the activation of pteropod activity, in November – by foraminifera, while in the near-bottom horizon from August to October the observed flux increases are attributed to the intensification of near-bottom currents.

The distribution of water column sediment components in the Mohns ridge area is almost identical to the distribution of components at the station located in the Lofoten Basin – the proportion of lithogenic matter increases uniformly down the water column profile, the proportion of organic and carbonate matter decreases. The sedimentary matter of the subsurface horizon contains much more biogenic matter compared to the deep horizons; this is especially well seen at the station in the Mohns ridge area [7, 8].

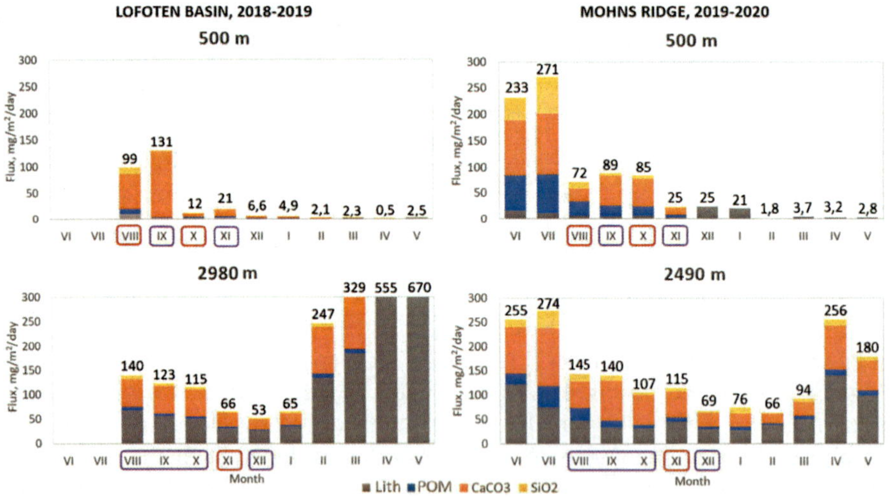

Fig. 3. Seasonal distribution of fluxes and components in the Lofoten Basin and Mohns ridge area

4 Elemental Composition of Sedimentary Matter

To assess the degree of material enrichment with elements, the obtained values were compared with the certified values of element contents in the upper continental crust (UCC) [9]. In addition to normalization by UCC, enrichment factors (EF) were calculated for the dispersed sedimentary matter (1). The use of the enrichment factor allows us to estimate the sources of elemental input into the sample. It is assumed that if the EF value is less than 3, the source of element input is lithogenic. If the EF value for an element is greater than 3, then there are other sources of its input [10].

$$EF = ([C]_{element}/Al)_{sample} / ([C]_{element}/Al)_{UCC}. \quad (1)$$

Based on the calculated DO values, it was concluded that the elemental composition of sedimentary matter in the two sedimentary environments has similar patterns. The upper horizon is enriched relative to the average composition of the UCC with a series of elements: Zn, Sr, As, Cd, Ba (Fig. 4). Substance of the bottom horizon is enriched with Mn, Sr, Zn, As, Cd. At the intermediate horizons enrichment of Mn and Pb is revealed. We also note that moderate with Fe enrichment increases with depth, and the Fe/Al ratio reaches values of 0.76, which is 1.5 times higher than the values obtained for sediment from intermediate horizons sampled in the eastern Norwegian Sea [11]. This distribution of sediment composition may be related to the distribution of particles of hydrothermal origin within hydrothermal plumes of neutral buoyancy, which are transported by currents at tens of kilometers from the source [12, 13].

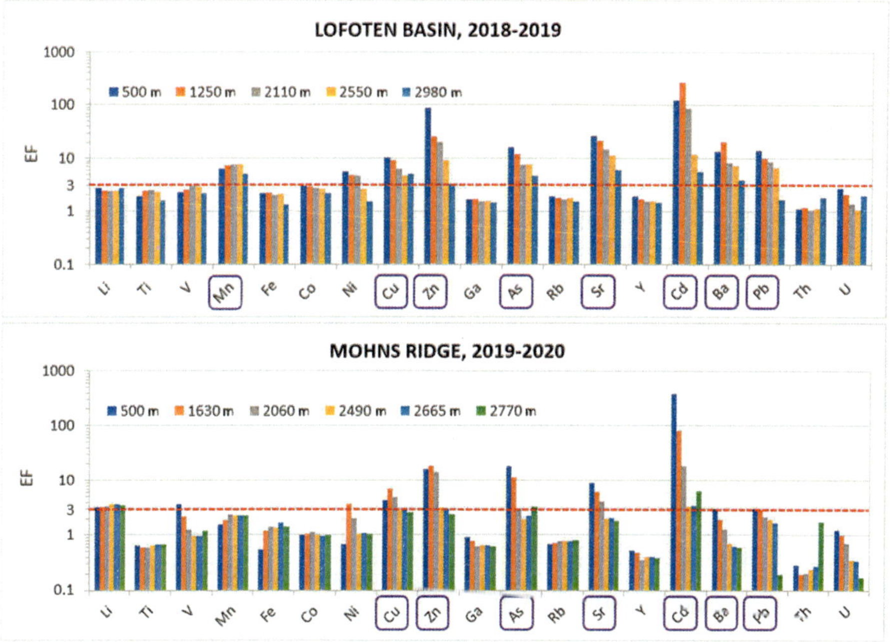

Fig. 4. Elemental enrichment factors (EF) in two sedimentation systems

5 Authigenic Barite of the Subsurface Horizon of the Water Column

A local peak of barium content in the winter period (December–February) in the subsurface horizon was registered in the Lofoten Basin; a similar pattern was observed for strontium in the studied region. According to [7], an increase in the contents of Ba and Sr during this period was also noted in the Lofoten Basin a year earlier.

Barium is an important component used as a proxy for paleoproductivity, so it is important to study its dynamics in more detail. Figure 5 shows the seasonal variability of barium over three years: studied in the Lofoten Basin by our colleagues between 2017 and 2018 [7] and between 2018 and 2019 (presented in this paper), and between 2019 and 2020 at the junction of the Mohns and Knipovich ridges. For the two years from 2017 to 2019 a peak in winter barium content was shown in the Lofoten Basin sedimentary system. This pattern is quite unusual, so it was decided to study the reason for the sharp increase in the content of this element. It is interesting to note that a similar distribution is characteristic of strontium, which, like barium, is an alkaline-earth metal and in some minerals (e.g., sulfates) is able to substitute the position of barium. It is known from literature sources that barium in suspended sedimentary matter is represented in the form of the mineral barite [14–17]. There are several theories of its origin, one of the main theories is barite crystallization in saturated microenvironments formed during dissolution of organic matter, mainly of silicate composition [14, 17]. However, according to our data, it becomes clear that the correlation of barium content with the content of organic matter and the content of amorphous silica was not found.

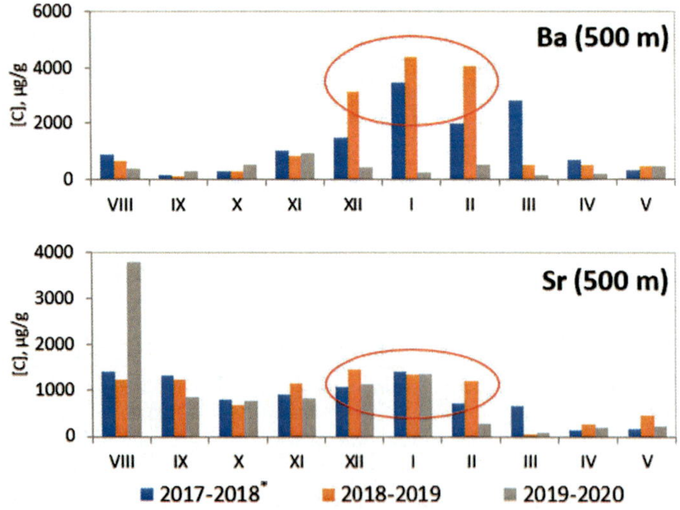

Fig. 5. Seasonal distribution of Ba and Sr in the Lofoten Basin and Mohns ridge area

To determine the source of barium input into the sediment, the fraction of excess barium was calculated using the Eq. (2):

$$Ba_{ex} = Ba_{total} - \left(Al_{total} * (Ba/Al)_{UCC}\right). \tag{2}$$

The excess barium is usually referred to barium, which is not included in the lithogenic fraction [14]. We have found that in the subsurface horizons both in the Lofoten Basin and on the Mohns Ridge, the fraction of Ba_{ex} varies from 73% (minimum in May) to 96–98% (maximum in the period from December to February), which indicates an active process of authigenic (presumably biogenic) barite formation in-situ in barium-saturated microenvironments in the upper layers of the water column. In [17], the opposite situation was noted: in the Bay of Biscay at a depth of 600 m, the fraction of Ba_{ex} reaches its maximum (97–100%) in summer (June-August). In the winter months it also remains quite high (90–93%), and in January a minimum was recorded – 67%. The reason for such discrepancies is still debatable.

Since in the months when the peak barium content was recorded, the bulk of it is represented by excess barium, it was decided to study the sediment suspension under the electron microscope. Electron microscope photographs of the biota found in the upper layer of the water column in the Lofoten Basin are shown in Fig. 6. The biota is represented by phytoplankton (radiolarians, diatom algae and coccolithophorids) and zooplankton (foraminifera). It is interesting to note that authigenic barite is often confined to calcic forms of phytoplankton - coccolithophorids (*Coccolithus pelagicus*).

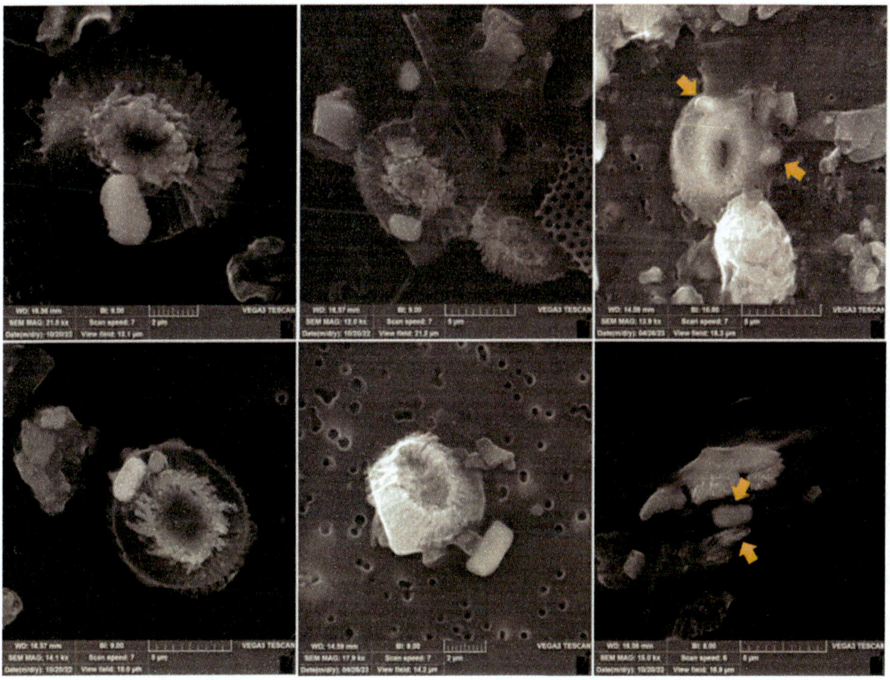

Fig. 6. Authigenic barite grains confined to calcic forms of phytoplankton (*Coccolithus pelagicus*)

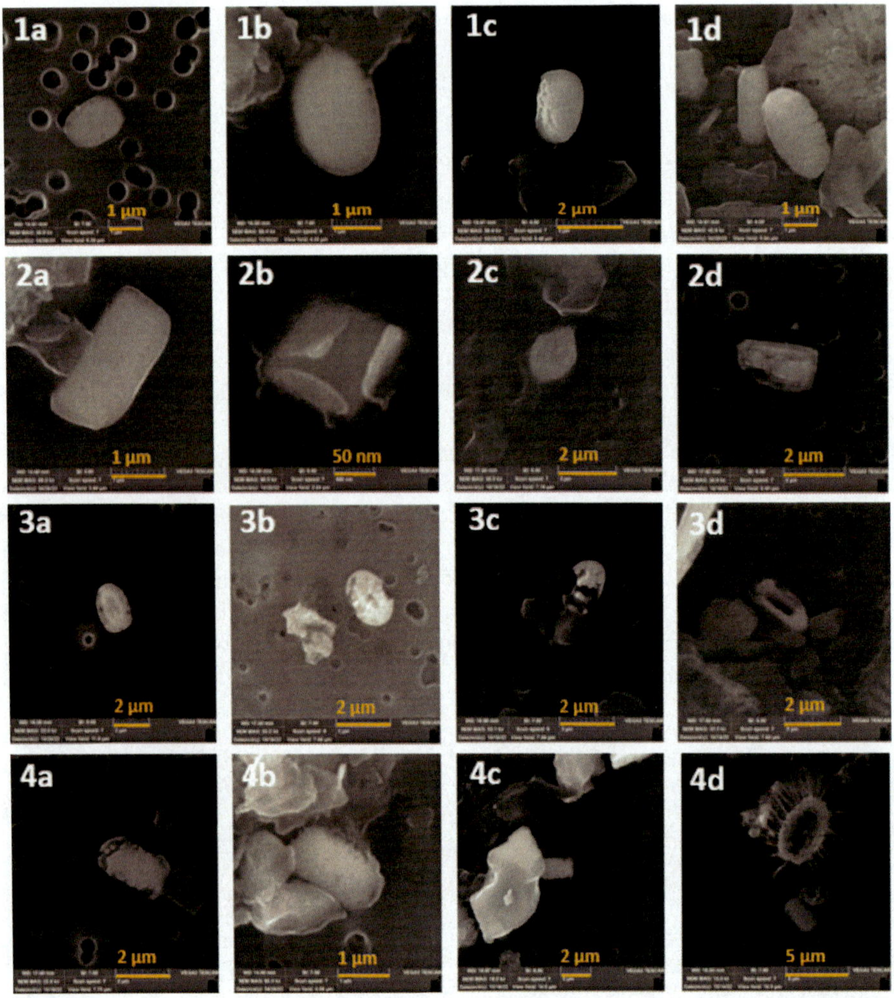

Fig. 7. Authigenic barite grains: group 1 – ellipsoidal barite; group 2 – crystalline barite; group 3 – ellipsoidal barite with traces of dissolution; group 4 – rectangular barite with «tendrils»

Microscopic analysis revealed barite grains of different morphology. General regularities characteristic for all grains of authigenic barite – grain size 1–2 microns, edges of crystals are rounded, traces of "etching" are observed in a part of grains. It was decided to divide the discovered authigenic barite into 4 groups (Fig. 7): 1 – barite of ellipsoidal shape, 2 – barite of crystalline shape, 3 – ellipsoidal barite particles with traces of reaming, 4 – barite of rectangular shape with «tendrils». In Dehairs [16], the authors also studied authigenic barite: the paper also identified 4 types of barite grains – ellipsoidal particles, crystalline, etched and amorphous.

The intermediate conclusion that we can make at this stage of studying authigenic barite is that the detected barite grains are different stages of «life» of barite crystals. Rectangular particles are closest to the crystalline form, in grains of more rounded form there are signs of dissolution first of the ribs and tops of crystals. Since the size of barite grains is very small, their settling to the bottom is only by formation of aggregates and fecal pellets, which is not so common in wintertime. At the same time, the maximum barium flux was recorded in summer during the period of intensive algal bloom, which coincides with the maximum fluxes of organic and carbonate carbon. Thus, under conditions of low productivity and low particle flux, a substantial fraction of sediment is authigenic barite, which by all indications is probably dissolved there without reaching the bottom.

6 Conclusion

The vertical distribution of mean annual sediment fluxes is characterized by a bimodal distribution. Flux maxima are observed in the subsurface horizon (summer algae bloom) and near-bottom horizon (substance resuscitation, formation of a nepheloid layer).

The component composition changes regularly with depth: the proportion of lithogenic matter increases with depth, while the proportion of organic and carbonate matter decreases. The seasonal dynamics of matter is explained by cycles of biological activity.

The elemental composition of the study area on the Mohns ridge is similar to the distribution of elements in the Lofoten Basin. The same trends of elemental behavior were revealed. The sedimentary matter of the subsurface horizon contains more nutrients and is enriched in Ba, Cd, Sr, Zn and As.

In subsurface horizons both in the Lofoten Basin and in the Mohns ridge area, the proportion of Ba_{ex} varies from 73% (minimum in May) to 96–98% (maximum in the period from December to February), which indicates an active process of in-situ formation of authigenic barite in barium-saturated microenvironments in the upper layers of the water column. Additional studies of authigenic barite are required.

Acknowledgements. The authors thank the team of the R/V Akademik Mstislav Keldysh, as well as the analysts E.O. Zolotykh and E.A. Streltsova. The study was supported by the Russian Science Foundation, project No. 20-17-00157. The interpretation of the materials was partially carried out within the framework of the State Assignment of the IO RAS, subject FMWE-2021-0006.

References

1. Lisitzin, A.P.: Oceanic Sedimentation: Lithology and Geochemistry, 400 p. American Geophysical Union Press, Washington, DC, USA (1996)
2. Bogdanov, Y.A., Sagalevitch, A.M., Galkin, S.M., Tchernyacv, E.S., Lukashin, V.N., Ivanov, G.V.: Geological and biological investigation in the Northern part of the Mohn's Ridge (Norwegian Sea). Oceanology **37**, 548–555 (1997)
3. Baumberger, T., Fruh-Green, G.L., Thorseth, I.H. Fluid composition of the sediment-influenced Loki's Castle vent field at the ultra-slow spreading Arctic Mid-Ocean Ridge. Geochimica et Cosmochimica Acta **187**, 156–178 (2016)
4. Lukashin, V.N., Bogdanov, Y.A., Shevchenko, V.P., Rusakov, V.Y., Isaeva, A.B.: A study of vertical fluxes of sedimentary material and its composition in the Norwegian Sea in the summer periods of 1991–1995. Geochem. Int. **38**, 169–183 (2000)

5. Lukashin, V.N.: Geochemistry of suspended matter and sinking material in the eastern Norwegian Sea. Geochem. Int. **46**, 711–723 (2008)
6. Bychkova, Ya.V., Bychkov, D.A., Bychkov, A.Yu.: Methods of Geochemical Research, 88 p. Publishing House Book-Memoir (2019). (in Russian)
7. Drits, A.V., Klyuvitkin, A.A., Kravchishina, M.D., Karmanov, V.A., Novigatsky, A.N.: Fluxes of sedimentary matter in the Lofoten basin of the Norwegian Sea: seasonal dynamics and the role of zooplankton. Oceanology **60**(4), 501–517 (2020)
8. Migdisova, I.A., Starodymova, D.P., Novigatsky, A.N., et al.: Variability of the composition of dispersed sedimentary matter in the water column of the Lofoten basin of the Norwegian Sea. In: Marine Research and Education: Proceedings of the X International Scientific and Practical Conference, Tver, 25–29 October 2021, pp. 163–166. LLC "PoliPRESS", Tver (2020). (in Russian)
9. Rudnick, R.L., Gao, S.: The composition of the continental crust. Treatise Geochem. **3**. The Crust. (Holland, H.D., Turekian, K.K., (eds.), pp. 1–64, Elsevier-Pergamon, Oxford (2003)
10. Rudnick, R.L.: Composition of the continental crust. (Ed. by, R.L. Rudnick, S. Gao) Treatise on Geochemistry, vol. 3, pp. 1–63. Elsevier, Amsterdam (2003)
11. Lukashin, V.N.: Sedimentary matter fluxes in the contour current sedimentation system over the continental slope of the Norwegian Sea. Oceanology **48**, 701–709 (2008)
12. Dubinin, A.V.: Geochemistry of rare earth elements in the ocean. Lithol. Mineral Resourc. **39**(4), 289–307 (2004)
13. German, C.R., Sparks, R.S.J.: Particle recycling in the TAG hydrothermal plume. Earth Planet. Sci. Lett. **116**(1–4), 129–134 (1993)
14. Lein, A.Y., Kravchishina, M.D.: Barium geochemical cycle in the ocean. Lithol. Mineral Resourc **56**(4), 293–308 (2021)
15. Francois, R., Honjo, S., Manganini, S.J., Ravizza, G.E.: Biogenic barium fluxes to the deep sea: implications for paleoproductivity reconstruction. Glob. Biogeochem. Cycles **9**, 289–303 (1995)
16. Dehairs, F., Chesselet, R., Jedwab, J.: Discrete suspended particles of barite and the barium cycle in the open Ocean. Earth Planet. Sci. Lett. **49**, 528–550 (1980)
17. Dehairs, F., Fagel, N., Anita, A.N., Peinert, R.: Export production in the Bay of Biscay as estimated from barium - barite in settling material: a comparison with new production. Deep Sea Res. Part I Oceanogr. Res. Papers **47**, 583–601 (2000)

New Data on the Age of the Khvalynian Strait

D. V. Semikolennykh[1,2](✉) and A. V. Panin[1,2]

[1] Moscow State University, 119991 Moscow, Russia
dasha.semikolennykh@gmail.com
[2] Institute of Geography RAS, 119590 Moscow, Russia

Abstract. The Early Khvalynian transgression developed in the Caspian Sea at the end of MIS 2. Waters of the transgression were discharged through the Manych Depression into the Azov-Black Sea basin. Deposits of the Early Khvalynian strait are poorly studied, and the time of the strait functioning is determined on the basis of scattered radiocarbon dates. We studied the coastal outcrop of the Chogray reservoir (riv. Vostochny Manych), where the deposits of the ancient strait are exposed, and we obtained a series of AMS dates on mollusc fauna.

Keywords: Paleogeography · Late Pleistocene · Geochronology · AMS dating · Khvalynian transgression · Manych Depression · Caspian · Malacofauna

1 Introduction

The Manych Depression, where the Khvalynian Strait functioned, is located in the South of the European part of Russia, this is a vast poorly dissected lowland plain. The bottom of the depression is occupied by systems of marshy valleys of the Zapadny and Vostochny Manych rivers, as well as partially overgrow and saline relict-estuary lakes [1].

The Early Khvalynian transgression developed in the Caspian Sea at the end of MIS 2. Waters of the transgression were discharged through the Manych Depression into the Azov-Black Sea basin. The history of the study of the Khvalynian deposits of the Manych Depression has more than a century of history. The Early Khvalynian molluscs species in the Manych deposits were described in a coastal outcrop near Lake Gruzskoe by V.V. Bogachev [2] for the first time, later studies of the deposits of the Early Khvalynian Strait were carried out by G.I. Goretsky [3], G.I. Popov [4, 5], P.V. Fedorov [6], A.A. Svitoch [7, 8], T.A. Yanina [9], E.N. Badyukova [10], A.L. Chepalyga [11] and others. Characteristic species of molluscs from the Lower Khvalynian deposits are *Didacna ebersini, D. subcatillus, D. protracta, Hipanis plicata.*

The opinion of researchers regarding the stratigraphic division of the deposits differs. Thus, P.V. Fedorov [6], G.I. Popov [5], and T.A. Yanina [9] subdivide the Lower Khvalynian deposits into two horizons, which correspond to two terrace levels: the early (Abeskunsky) horizon, overlying the Burtass deposits, and the late one, leaning against the Burtass terrace. According to A.A. Svitoch et al. [8], the Khvalynian deposits of the Manych represent a single stratum of different facies located at a height of up to 35 m.

T. Chaplina (Ed.): CIWO 2023, SPEES, pp. 319–325, 2023.
https://doi.org/10.1007/978-3-031-47851-2_38

A.L. Chepalyga et al. [11] identify three episodes of Caspian water discharge, which correspond to three generations of accumulative forms in the Manych Depression.

There are also disagreements between researchers when correlating the time of existence of the Early Khvalynian strait with the global chronostratigraphic scheme. G.I. Popov [5] and A.V. Mamedov [12] attributed this stage to MIS 3 (Briansk interstadial of the Valdai glaciation). G.I. Rychagov [13] compared the Early Khvalynian transgression of the Caspian Sea with MIS 4 (the first stage of the Valdai glaciation).

The first radiocarbon dates on fauna of the Early Khvalynian molluscs of the Manych were obtained in 2000 at the Research Laboratory of Recent Sediments and Pleistocene Paleogeography of Moscow State University [14, 15].

In 2008–2009 H.A. Arslanov (St. Petersburg State University) obtained a series of dates from a number of natural sections [16, 17] and for the first time one date was obtained by AMS radiocarbon dating at the University of Groningen (Netherlands) [7, 8].

A.A. Svitoch and R.R. Makshaev [18], based on the analysis of the available geochronological data established that the penetration of the waters of the Khvalynian transgression of the Caspian into the Azov-Black Sea began between 16 and 14 ka years ago and ended no earlier than 12.8 ka years ago.

At present, the geochronology of the Early Khvalynian stage of the Manych Depression is based on disparate dating of layers with the Khvalynian fauna of various coastal outcrops, which makes it difficult to compare the time of the opening of the Early Khvalynian strait with the global geochronological scale and to divide the Khvalynian deposits into horizons, if any.

In the course of field work in the Manych Depression in 2022 we studied in detail the section ZT-3 of the Khvalynian strait deposits, located in a natural outcrop on the southwestern shore of the Chogray reservoir (Stavropol Krai).

2 Material and Methods

The section ZT-3 (N45°36′15″ E44°12′57″; H 28.2 м) is located on the southwestern coast of the Chogray reservoir in the Stavropol Krai in the area of the watershed between the Pontus and the Caspian (Fig. 1). The section represents a coastal outcrop of the Khvalynian terrace with high 28–33 m above sea level. The width of the valley of the Vostochny Manych River reaches here 8 km.

Following deposits are exposed in the section ZT-3 from top to bottom (Fig. 2): (1) brown loam (modern soil) 0.45 m thick with plant roots and carbonate formations, gradual transition to the underlying layer; (2) pale yellow loess-like loam 0.95 m thick, dense, with a large columnar unit, the boundary with the underlying layer is indistinct; (3) subhorizontally interbedded pale-yellow loam with chocolate-like clays 0.40 m thick with inclusions of carbonate nodules and spots of ferruginization and manganese, the boundary with the underlying layer is clear; (4) horizontally and wavy layered brown (chocolate-like) clays 0.50 m thick with inclusions of carbonates, ferruginization and manganese spots, the boundary with the underlying layer is indistinct; (5) poorly sorted brown sand 0.50 thick implicit horizontal and oblique layering, clear transition to the underlying layer; (6) brown loam 0.45 m thick with the inclusion of the fauna of the

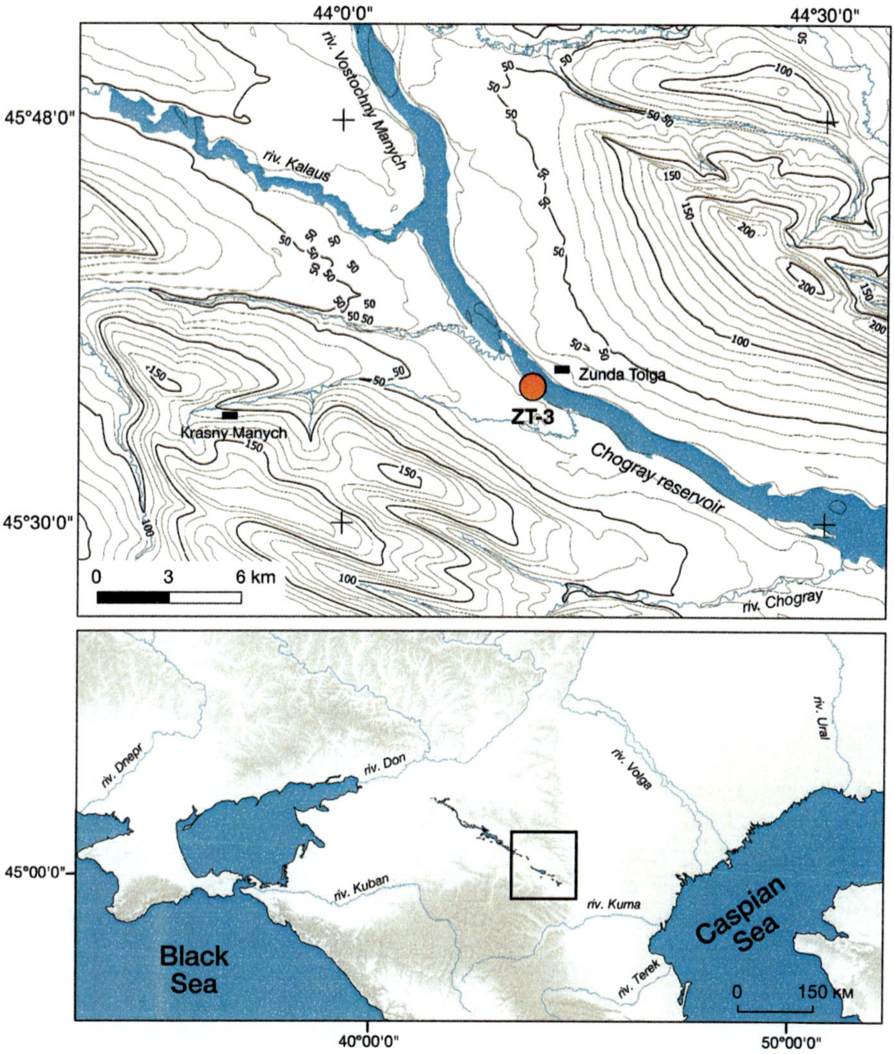

Fig. 1. Location of the section ZT-3 in the Manych Depression

Caspian molluscs, implicit horizontal and oblique layering, the boundary with the underlying layer is clear; (7) poorly sorted brown sand 0.60 m thick with the inclusion of the Caspian molluscs fauna, ferruginous and manganese spots, implicit horizontal and oblique layering, the transition to the underlying layer is clear; (8) bluish-brown heavy loam with numerous ferruginous spots, visible thickness up to 0.6 m.

Analysis of the inclusion of molluscs fauna was carried out at the Research Laboratory of Recent Sediments and Pleistocene Paleogeography of Moscow State University for the purpose of biostratigraphic substantiation of the geological age of sediments and

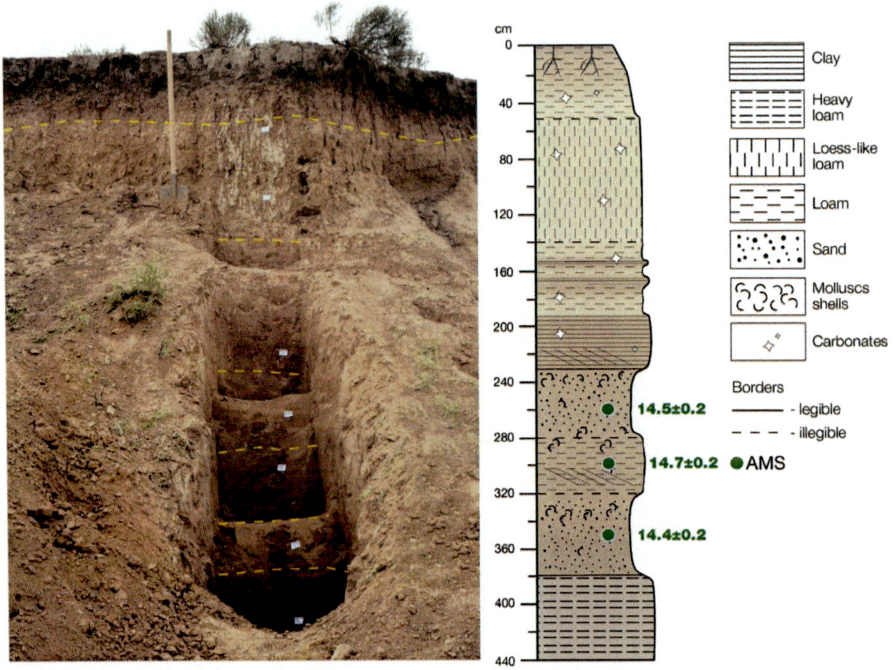

Fig. 2. Structure of section ZT-3

understanding of the paleoecological conditions of their accumulation. The malacofaunistic complex from the section ZT-3 consisted of two types of molluscs: *Monodacna caspia* and *Hypanis plicata* (Fig. 3), which are characteristic of the Khvalynian complex of the Caspian Sea. The malacofaunistic composition of the sediments indicates that the salinity of the strait was close to that of the Early Khvalynian basin in the Caspian (up to 12–14%).

Fig. 3. Species of molluscs from the section ZT-3 (1 – *Hypanis plicata*, 2 – *Monodacna caspia*)

Based on shells of mollusks taken at depths of 2.5, 3.0, 3.5 m, three dates were obtained using AMS radiocarbon dating at the University of Georgia (USA) in the range of 14.2–14.9 ka years ago.

3 Results and Discussion

The structure of the section reflects the sedimentation conditions of the final (accumulative) phase of the functioning of the Early Khvalynian strait. The malacofaunistic composition of the sediments indicates a similar or slightly lower (compared to the Early Khvalynian basin) salinity of the strait.

Based on the results of AMS dating, the final stage of the functioning of the Early Khvalynian Strait in the Manych Depression was determined in the interval of 14.2–14.9 ka years ago, which is consistent with our earlier OSL dating data for the Early Khvalynian deposits of the Ostrov Levy section [19].

Thus, the warming that began after the last glacial maximum and a significant increase in atmospheric precipitation within the East European Plain ~19–17 ka years ago and ~15–13 ka years ago led to active snowmelt and permafrost thawing, which in turn caused a significant increase in water and runoff [20–23].

Due to a significant increase in runoff from the East European Plain, the largest in the Late Pleistocene history, the Early Khvalynian transgression of the Caspian began to develop within the Caspian Sea [24] with a maximum level of about +45–50 m.

AMS dates obtained by us correlate with the final stage of the strait functioning ~15–14 ka years ago, which is in good agreement with the age of the Early Khvalynian deposits of the Lower Volga [24] and the central part of the Manych Depression [19].

The salinity of the strait in the area of the current Chogray reservoir, judging by the malacofaunistic complexes, could not exceed 12–14%. The climate of the Manych Depression, based on an analysis of the fauna of rodents that lived in the era of the existence of the strait, was determined as cool and arid [25].

4 Conclusion

- The time interval for the functioning of the strait in its final phase, according to the results of AMS dating, is set at 15–14 ka years ago. The salinity of the resulting strait did not exceed 12–14%.
- The waters of the Early Khvalynian transgression were discharged in the second half of MIS 2 into the New Euxinian transgressive basin of Pontus.
- AMS dates obtained by us is consistent with OSL dating data for the Khvalynian deposits of the Ostrov Levy section (central part of the Manych Depression) [19].

Acknowledgements. The work was supported by the RSF project 22-17-00259.

References

1. Svitoch, A.A.: Pleistocene history of the Zunda Tolga (Manych) structure. Doklady AN. **435**(1), 215–220 (2010)
2. Bogachov, V.V.: Steppes of the Manych river basin. Izvestiya Geologicheskogo komiteta. **22**(9), 73–162 (1903)
3. Goreckiy, G.I.: About paleogeography of Priazov and Western Manych regions in uzunlaro-girkanian and burtass epochs. Vopr. Geografii. **33**, 190–221 (1953)
4. Popov, G.I.: History of the Manych Strait in connection with the stratigraphy of the Black Sea and Caspian deposits. Bull. MOIP. Otd. geolog. **20**(2), 31–49 (1955)
5. Popov, G.I.: Pleistocene of the Black Sea and Caspian Sea passages, 216 p., Nauka (1983)
6. Fedorov, P.V.: Pleistocene of the Ponto-Caspian, 165 p., Nauka (1978)
7. Svitoch, A., Yanina, T., Antonova, M., van der Plicht, I.: Khvalynian fauna of Manych. Doklady Academii nauk. 689–693 (2008)
8. Svitoch, A.A., Yanina, T.A., Khomenko, A.A., Novikova, N.G.: Khvalynian sediments of the Manych Depression. Doklady Earth Sciences – Nauka. **428**(1), 70–74 (2009)
9. Yanina, T.A.: Neo-Pleistocene of the Ponto- Caspian: biostratigraphy, paleogeography, correlation, 264 p. MGU (2012)
10. Badyukova, E.N.: Some questions of the history of Manych development in the late Pleistocene-Holocene. In: Humanity and the Coastal Zone of the World Ocean in the 21st Century, pp. 326–333. Geos (2011)
11. Chepalyga, A.L., Pirogov, A.N., Sadchikova, T.A.: The fault of Caspian waters of Khvalynian basin by Manych valley during the extreme flood epoch (the Deluge). In: Problemy paleontologii I arkheologii uga Rossii I sopredel'nyh territorii. – Rostov-na-Dony: Izdatel'stvo OOO "CVVP"., 138 p. (2005)
12. Mamedov, A.V.: The Late Pleistocene-Holocene history of the Caspian Sea. Quaternary Int. **41–42**(96), 161–166 (1997)
13. Rychagov, G.I.: Khvalynian stage in the history of the Caspian Sea. Geomorphology **4**, 3–9 (2014)
14. Svitoch, A.A., Parunin, O.B.: Radiocarbon age of paleogeographic events of the late Pleistocene in the Northern Caspian Sea. Doklady Akademii Nauk. Nauka **371**(4), 535 (2000)
15. Svitoch, A.A., Yanina, T.A.: New data on malacofauna of the Marine Pleistocene of Manych. Doklady Akademii Nauk, vol. 380, no. 4, pp. 570–573. Nauka (2001)
16. Arslanov, Kh.A., Yanina, T.A.: Radiocarbon age of the Khvalynian Manych passage. In: Black Sea – Mediterranean Corridor During the Last 30 ky: Sea Level Change and Human Adaptation, pp. 10–13. EcoGeoMar., Bukharest (2008)
17. Chepalyga, A., Arslanov, Kh., Yanina, T.: Detailed age control of the Khvalynian basin history. In: Black Sea –Mediterranean Corridor During the Last 30 ky: Sea Level Change and Human Adaptation, pp. 23–26, Istanbul (2009)
18. Svitoch, A.A., Makshaev, R.R.: Interrelations of the paleogeographical events in the Pont-Manych-Caspian system during the Late Pleistocene and Holocene. Vestnik MGU Seriya 5, no. 2, pp. 24–32, Geografiya. (2017)
19. Semikolennykh, D.V., Kurbanov, R.N., Yanina, T.A.: Age of the Khvalyn strait in the late pleistocene history of the Manych depression. Vestnik MGU. Seriya 5, no 5, pp. 103–112, Geografiya (2022)
20. Tudryn, A., et al.: The Ponto-Caspian basin as a final trap for Southeastern Scandinavian Ice-Sheet meltwater. Quaternary Sci. Rev. **148**, 29–43 (2016)
21. Panin, A., Grzegorz, A., Buylaert, J.-P., Matlakhova, E., Moska, P., Novenko, E.: Two Late Pleistocene climate-driven incision/aggradation rhythms in the middle Dnieper River basin, west-central Russian Plain. Quaternary Sci. Rev. **166**, 266–288 (2017)

22. Sidorchuk, A.Yu., Panin, A.V., Borisova, O.K.: River runoff in the East European Plain over the past 20 thousand years and the problem of changes in the levels of the Southern seas. In: Voprosy geographii. Sb. 145. Gidrologicheskie izmeneniya, pp. 144–168. Izd. dom "Kodeks". (2018)
23. Borisova, O.K.: Landscape and climatic conditions in the central part of the East European Plain in the last 22 thousand years (reconstruction based on paleobotanical data). In: Vodnye resursy, vol. 48, no. 6, pp. 664–675. Nauka (2021)
24. Kurbanov, R.N., Murray, A.S., Thompson, W., Svistunov, M.I., Taratunina, N.A., Yanina, T.A.: First reliable chronology for the Early Khvalynian Caspian Sea transgression in the Lower Volga River valley, vol. 50, pp. 134–146. Boreas (2021)
25. Chepalyga, A.L., Markova, A.K.: New data on the history of the fauna and landscapes of the Manych Valley in the Late Pleistocene. In: Izvestiya Rossiiskoy akademii nauk. Seriya geographicheskaya, no. 3, pp. 57–63. Nauka (2019)

A CSM-Based Python Bot for Estimating the Stability Zone of Gas Hydrates

Yu.Yu. Smirnov(✉), T. V. Matveeva, N. A. Shchur, A. A. Shchur, and A. V. Bochkarev

Federal State Budgetary Institution "Academician I. S. GRAMBERG All-Russia Scientific Research Institute for Geology and Mineral Resources of the Ocean", Saint-Petersburg 190121, Russia

y.y.smirnov@mail.ru

Abstract. This text describes a study and its findings that utilized the "GHSZ_VNIIO" robot program to determine the thickness of the Gas Hydrate Stability Zone (GHSZ) in the Sea of Okhotsk. The program was developed using the Python language and offers automated calculation and visualization of gas hydrate distribution data. The results of the calculations revealed the presence of an extensive GHSZ in the Sea of Okhotsk, with a thickness of up to 720 m. A correlation analysis between the calculated and actual data, based on the position of the Bottom Simulating Reflector (BSR), demonstrated a high level of agreement (94%) between the predicted and observed data. This robot program enables researchers to obtain preliminary assessments of gas hydrate distribution in marine areas and possesses versatility for application in any oceanic region.

Keywords: gas hydrates · gas hydrate stability zone · Python · subprocesses · Okhotsk Sea

1 Introduction

As part of the work on quantitative assessment of natural gas resources in the gas hydrates of Russian waters, carried out in 2021 by the Federal State Budgetary Institution "VNIIOkeangeology" [1], the task was set to forecast the mapping of conditions for gas hydrate occurrence. These gas hydrates (GH) are unstable compounds of water and natural gas that form an icelike crystalline lattice through hydrogen bonding. The conditions of their existence refer to the stability zone of gas hydrates (GHSZ) – the part of the lithosphere (and ocean) within which previously formed GH remain in a stable state.

The existence of the GHSZ is primarily determined by the thermobaric conditions of the environment, referred to as "equilibrium" conditions. These equilibrium conditions are represented as a function of pressure and temperature [2]:

$$\ln(P_D) = \sum_{n=0}^{5} a_n (T + T_D)^n,$$

T. Chaplina (Ed.): CIWO 2023, SPEES, pp. 326–331, 2023.
https://doi.org/10.1007/978-3-031-47851-2_39

where P_D is the equilibrium pressure, T is the water temperature of the enviroment, T_D is the temperature deviation due to the haline properties of the sea water, and a_n are empirical constants. This equation represents a regression model – an approximation of a series of laboratory observations. By comparing the real "in situ" environmental conditions with the equilibrium conditions, it is possible to determine the boundaries of the GHSZ, namely its thickness. The accurate selection of regression coefficients, in the absence of specific data, is a nontrivial task associated with several difficulties, primarily the significant time investment required for their determination. An alternative approach is to utilize ready-made software ("black box") that avoids methodological difficulties in the field of gas geochemistry. One such software is the freely distributed CSMHYD Hydoff program [3], developed in Fortran language at the end of the 20th century [4], which is an executable file – a console application with the ".exe" extension. It incorporates multiple regression models that enable the calculation of equilibrium pressure as a function of not only the thermohaline properties of the enviroment but also account for the multicomponent gas composition. However, this program is one-dimensional, and data input is performed manually. Each equation is solved only once, providing a single equilibrium pressure value for one equilibrium temperature (with an accuracy of up to 10^{-3} K). Depending on the range of temperature variations, the required length of the calculated equilibrium curve may reach thousands of values. Consequently, the question arises about automating the process of calculating the equilibrium curve using Hydoff capabilities.

2 Material and Methods

For these purposes, a Python-based robot program was developed to handle large coordinate grids and find the equilibrium hydrate formation curve and the thickness of the GHSZ for each grid node. Python was chosen as the programming language [5] due to its general-purpose nature, easy and convenient syntax, and virtually unlimited capabilities, except for its lower computational speed compared to compiled languages due to interpretation. In this work, computational speed is not crucial as the main calculations are performed in an external program written in the compiled language Fortran, which is executed from the Python script in the background.

The robot program, which later received the name "GHSZ_VNIIO", can be divided into three blocks: "Interface block", "Robot block" and "Processing block". The calculation is performed for a grid of stations of arbitrary size. In addition to the station coordinates, hydrological and geothermal characteristics are required as input variables. In this work, the following fields were utilized (Fig. 1): the fields of bottom water temperature (T_{bot}) and bottom salinity (S_{bot}) [6], bathymetry data from the GEBCO model [7], and the values of the thermal gradient (T_g) for grid nodes [8].

The "Interface block," written using PyQt, represents a window containing four working buttons with fields for selecting input and output file paths. Input files accepted are in Excel format or text files containing the mentioned hydrophysical and geothermal characteristic fields. The "Robot block" is based on the use of the Subprocesses library, specifically the *subprocesses.Popen* class, which allows executing a child program in a new process. The main parameters of the class are *STDIN* and *STDOUT*, responsible

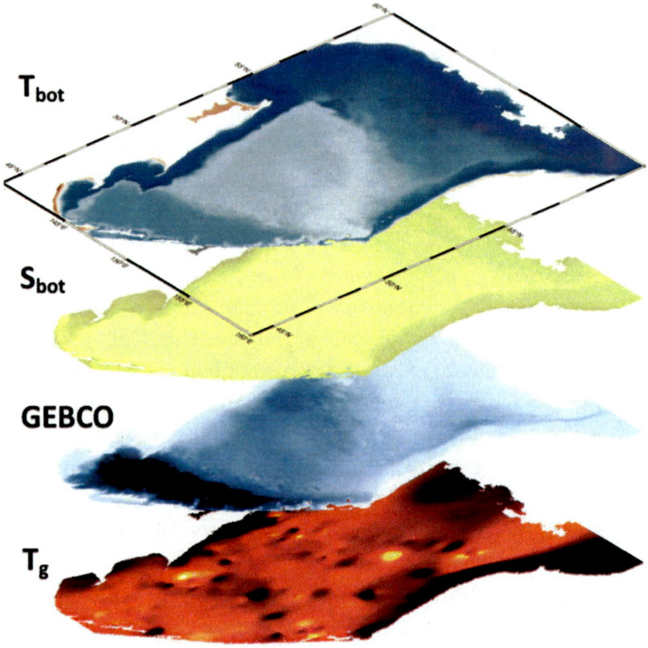

Fig. 1. The fields T_{bot} and S_{bot}, GEBCO and Tg for the Sea of Okhotsk.

for input and output data. The *STDIN* parameter takes the value of the *STDIN* = *subprocesses.PIPE* class. This assignment allows entering temperature and salinity values into the PIPE using the *stdin.write*(…) attribute, after importing the necessary data in any form, using simple while loops. The calculation is performed in the child process Hydoff.exe and the obtained results are output to a temporary text file by specifying the path. The results of the "Robot block" are represented by an equilibrium curve similar to the previously mentioned function [2]. The "Processing block" reimports the data from the temporary file and, by also incorporating arrays of environmental characteristics, constructs the "real" temperature curve, starting from the bottom water layer and ending at a certain depth in the sedimentary layer. In the final stage, a comparison is made between the two curves – the "real" and the "equilibrium" curves – to find their intersections. The distance between these intersections represents the thickness of the GHSZ.

3 Results and Discussion

The result of the robot's calculation (200 cycles with a step of 0.1 K), limited to the high-temperature region, is presented as a set of equilibrium curves for the "hydrate – water + methane + Sal% NaCl" system in Fig. 2. The calculation performed by the robot to construct the equilibrium curves has been actively applied in further studies on modeling the GHSZ, including cryogenic gas hydrates [9, 10].

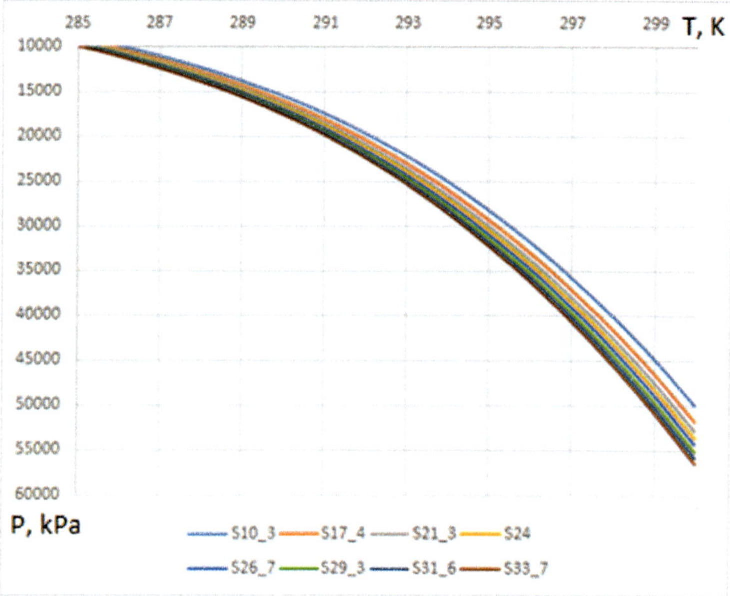

Fig. 2. Example of calculating equilibrium curves by a robot for different S, psu.

The PyQt library allows creating a graphical user interface (GUI) application for the robot itself (Fig. 3), making it user-friendly even for inexperienced users who may struggle with raw program code. It enables the creation of a convenient GUI for inputting the initial data. Essentially, all that remains for the researcher is to run the script in the directory where Hydoff.exe is located and, upon completion of the calculation, obtain the output data in the form of an Excel file in the ".xlsx" format. The table in the Excel file is fully formatted using the capabilities of the NumPy library. For each grid node coordinate, the corresponding predicted thickness of the GHSZ is provided. In addition to the robot-related parts, the script also calculates the boundaries of the GHSZ and their thickness, which is the difference between these boundaries. Essentially, the hydrological characteristic array obtained as input is transformed into a thickness array, ready for further mapping using GIS-tools.

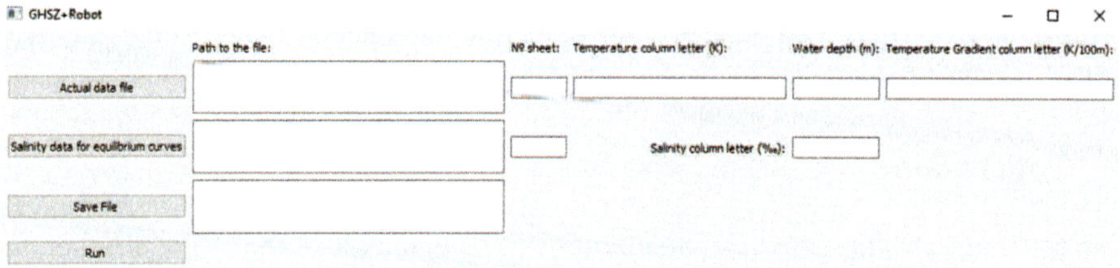

Fig. 3. Example of a GUI of the "GHSZ_VNIIO"

The visualization of the results can be done either using external GIS tools or directly within the script by incorporating mapping libraries such as Cartopy. As an example, we can provide a three-dimensional visualization of the calculated thickness of the GHSZ using the mentioned fields, performed by adding a PyGMT block to the program code (Fig. 4):

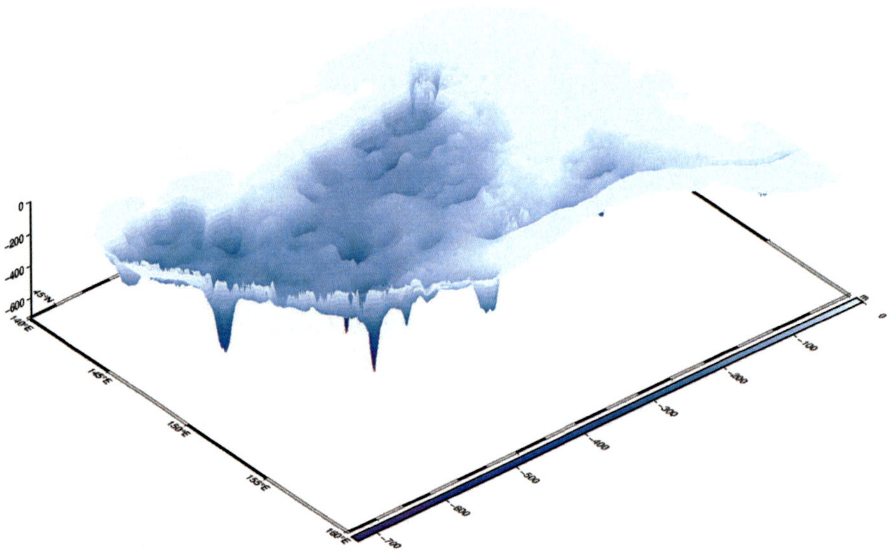

Fig. 4. Calculated GHSZ thickness (m) for the Sea of Okhotsk using fields according to the "GHSZ_VNIIO" program.

For a large part of the Sea of Okhotsk, an extensive GHSZ is predicted, starting from depths above 450 m. The highest values of GHSZ thickness (up to 720 m) are associated with the Deryugin Basin region, where low thermogradient values are present. The correlation coefficient between the base of the calculated GHSZ and the Bottom Simulating Reflector (BSR) determined at 300,000 grid nodes is slightly over 94%, indicating a reasonably high quality of calculations despite the existing simplifications. However, only 10% of all stations have a BSR discrepancy exceeding 25%. It is important to note that the robot can only predict the thermobaric GHSZ, while in reality the presence of hydrates is influenced by various factors, with the main one being the presence of a natural gas source. Therefore, such estimates of GHSZ should be considered approximate.

4 Conclusion

The paper presents the principle of determining the thickness of the GHSZ for thermogenic gas hydrates using the "GHSZ_VNIIO" program, which is built on top of CSMHyd Hydoff. The paper provides a description of the structure and operation of the created robot program to automate the process of one-dimensional calculations. It highlights the capabilities of the Python language in solving non-standard engineering problems.

Among some disadvantages of the programming language encountered during the work, the inability to create an executable file due to the interpreted nature of the language is mentioned.

According to the calculation results for the Sea of Okhotsk, an extensive GHSZ with a thickness of up to 720 m is predicted. The assessment of the prediction quality based on the position of the Bottom Simulating Reflector (BSR) revealed a correlation of 94% between the calculated and actual data.

Despite its approximate nature, the robot program allows for preliminary assessments of the distribution of GHSZ in marine areas. Its main advantage is its universality, as researchers only need to select the necessary hydrophysical and geothermal characteristic fields to calculate the GHSZ for any oceanic area.

In the future, the algorithm can be further developed to handle gas compositions of arbitrary nature. The resulting file, in addition to the GHSZ thickness for each station with coordinate referencing, also contains the hydrological and geothermal parameters used in the calculation. This allows for the construction of oceanographic fields for all the listed characteristics and the analysis of the obtained data.

References

1. Matveeva, T.V.: Quantitative assessment of gas hydrate resources in the water areas of the seas of the Russian Federation (stage 2). In: Report on the results of works performed in 2020 within the framework of the State Task of the Federal Agency for Subsoil Use dated October 13, 2020 No. 049-00018-20-03 for 2020 and for the planned period of 2021 and 2022. SPb. VNIIOkeangeologiya (2020). (in Russian)
2. Malakhova, V.V.: The response of the Arctic Ocean gas hydrate associated with subsea permafrost to natural and anthropogenic climate changes. IOP Conf. Ser.: Earth Environ. Sci. **606**(1), 012035 (2020)
3. Sloan, E.D.: Gas hydrates: review of physical/chemical properties. Energy Fuels **12**(2), 191–196 (1998)
4. Babakhani, S.M., Bahmani, M., Shariati, J., Badr, K., Balouchi, Y.: Comparing the capability of artificial neural network (ANN) and CSMHYD program for predicting of hydrate formation pressure in binary mixtures. J. Pet. Sci. Eng. **136**, 78–87 (2015)
5. Plas, J.V.: Python for Complex tasks: data science and machine learning. Series "Bestsellers O'Reilly", p. 576 (2018). (In Russian)
6. Lellouche, J.-M., et al.: The Copernicus global 1/12° oceanic and sea ice GLORYS12 reanalysis. Front. Earth Sci. **9**, 698876 (2021)
7. GEBCO URL: http://www.gebco.net/data_and_products/gridded_Bathymetry_Data. Accessed 18 Jan 2021
8. Bochkarev A.V., Matveeva T.V.: Estimation of heat flow on the Eurasian margin of the Arctic Ocean for calculating the stability conditions of gas hydrates. Geotectonics (2023) (in print)
9. Matveeva, T.V., Kaminsky, V.D., Semenova, A.A., Shchur, N.A.: Factors affecting the formation and evolution of permafrost and stability zone of gas hydrates: case study of the Laptev Sea. Geosciences **10**(12), 504 (2020)
10. Smirnov, Y.Y., Shchur, A.A., Matveeva, T.V., Shchur, N.A., Bochkarev, A.V.: Modeling of the cryogenic gas hydrates stability zone of the Russian arctic shelf: challenges and responses. In: Proceedings of X International Scientific and Practical Conference "Marine Research and Education (MARESEDU-2021)", PolyPRESS, pp. 91–95 (2021). (in Russian)

Study of Background and Focused Hydrocarbon Discharge, NE Kara Sea

A. K. Yumasheva[1] (✉), E. D. Basova[2], E. N. Poludetkina[2], M. Y. Tokarev[2], A. Y. Rybalko[3], and A. O. Kil[3]

[1] Oil and Gas Research Institute RAS, Moscow 119333, Russia
anastasiayumasheva@yandex.ru
[2] M. V. Lomonosov Moscow State University, Moscow 119991, Russia
[3] All-Russian Scientific and Research Institute for Geology and Mineral Resources of the World Ocean, Saint Petersburg 190121, Russia

Abstract. The work in the TTR-21 expedition was aimed at searching for areas of focused fluid discharge and the identifying of background and abnormal concentrations of fluids and organic matter in near-surface sediments. Seismoacoustic research methods have made it possible to identify places that are characterized by a high probability of the presence of fluid-saturated deposits. Their testing showed both increased concentrations of methane and the presence of homologues, and an increased content of organic matter, including light bitumen, indicating a possible migration flow of hydrocarbons from deep intervals of the sedimentary section.

Keywords: Kara Sea · fluid discharge · hydrocarbons · seismoacoustic studies

1 Introduction

July 10 – August 23, 2022, Lomonosov Moscow State University with the support of the Ministry of Science and Higher Education of the Russian Federation, organized a multidisciplinary scientific expedition TTR-21 (program "Training-through-Research (Floating University)"). The research was carried out in the north-eastern part of the Kara Sea from the vessel of the research vessel "Academician Boris Petrov".

One of the most promising potential oil and gas regions of Russia is the territory of the Kara Sea with two huge separate sedimentary basins distinguished. They are characterized by different geological structure, age range of the sedimentary cover, volumes of hydrocarbon resources [1]. The South Kara part has been studied well enough, a number of large and unique deposits have been discovered – in general, this is a petroleum-bearing region that is part of the West Siberian oil and gas province. The level of the northern part study, which is a separate North Kara promising oil and gas province, is extremely low [2]. Therefore, the geological and geophysical surveys carried out are extremely important and relevant for further study of the area, refinement of the geological model and assessment of oil and gas potential. [3, 4].

Currently, there is a growing interest to the problem of gas release in the bottom sediments and seawater, therefore, one of the objectives of the expedition research was

T. Chaplina (Ed.): CIWO 2023, SPEES, pp. 332–338, 2023.
https://doi.org/10.1007/978-3-031-47851-2_40

a set of studies aimed at obtaining data on the features of gas-saturated sediments, gas migration in the ground and water column. Thus, experimental surveys were conducted that aimed at the fundamental scientific substantiation of methodological approaches to geological, geophysical and geochemical mapping of zones promising for hydrocarbon deposits, as well as the interpretation of geological-geochemical testing of Arctic bottom sediments in order to characterize the zones of focused and spatial fluid discharge.

2 Material and Methods

During the TTR-21 cruise, the following set of marine research methods was used on board the RV «Academic Boris Petrov»: 2D seismic exploration of high resolution and ultra-high resolution, multibeam echosounder, acoustic profiling with an on-board profiler, bottom sampling using a gravity tube.

Within the framework of regional studies, more than 3,000 km of high-resolution seismic exploration, 500 km of ultra-high-resolution seismic exploration, more than 5,000 km of multipath sonar and acoustic profiling were performed. 102 sampling stations were selected within 5 key sites (Fig. 1) to characterize the composition, structure and properties of bottom sediments and the fluids saturating them.

Extensive areas of fluid discharge were identified according to the data of ultra-high resolution seismic survey and acoustic profiling, the top of gas-saturated deposits characterized by increased amplitudes, a sharp change in the wave pattern to a chaotic type of recording and screening of the underlying section. Also, anomalies of the "bright spot" type were detected formed due to changes in amplitudes at the boundary of gas-saturated and overlying sediments. Seismoacoustic methods have made it possible to determine the places of potential fluid conductivity from the depths of the sedimentary cover to the seabed surface.

Bottom sampling was carried out on the structures previously isolated by geophysical data, to clarify the structure of the sedimentary cover and to define background and abnormal concentrations of migratory gaseous and liquid fluids.

The main geochemical studies carried out in this work were: gas chromatography of hydrocarbon and non-hydrocarbon gases dissolved in sediment, isotopic gas studies; fluorescent analysis of sediments.

Gas chromatography (GC) was carried out on gas chromatograph "Chromatek-Crystal 5000" equipped with flame ionization and thermal conductivity detector. Content of alkane hydrocarbons C1 to C6, including saturated and unsaturated, CO2, N2, O2, H2 and He was defined.

The performed fluorescent analysis is based on the dependence existing between the content and group composition of bitumen. As a result bitumen contents were determined in the sediment. It was stated that sediments have low concentrations of syngenetic bitumen, and the increase in its concentrations especially in the low-molecular part (light hydrocarbons, waxes) is directly related to the admixture of migratory genesis. Fluorescence and GC method showed the presence of elevated concentrations of methane and homologues, as well as high organic matter contents.

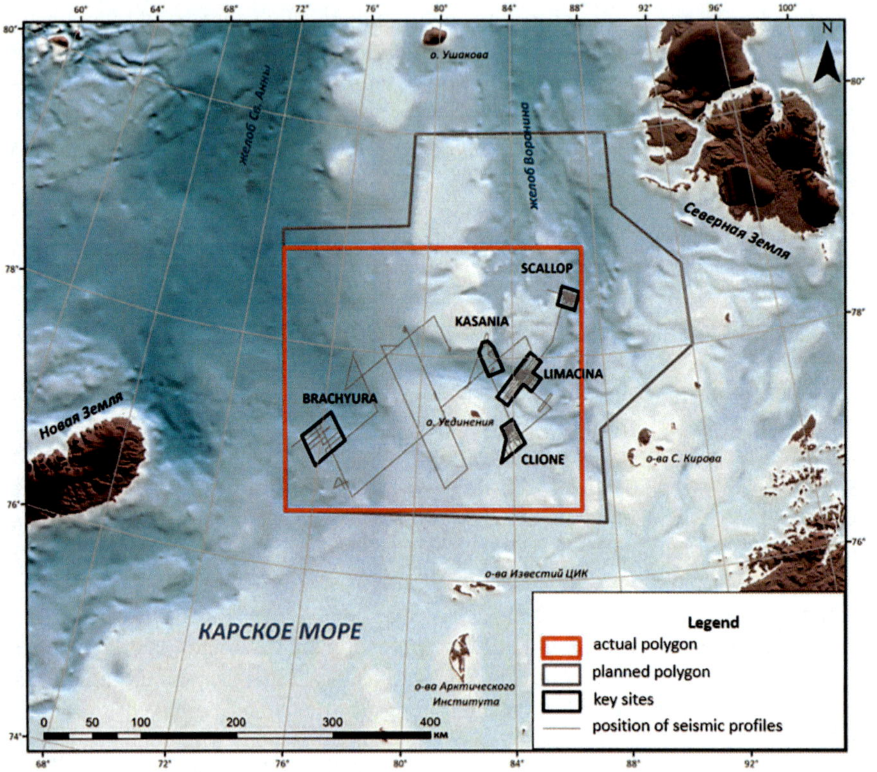

Fig. 1. Map of the actual material of the research area

3 Results and Discussion

5 key sites with signs of fluid discharge were studied: "Brachyura", "Clione", "Limacina", "Scallop", "Kasania". The key areas "Brachyura" and "Kasania" are characterized by the presence of fields of pockmarks that stand out in the bottom relief, detected using a multipath echo sounder. Probably, the formation of these structures occurred as a result of focused fluid discharge from the depths of the sedimentary cover.

Key site Brachyura characterizes areas of gas saturation, which were distinguished by a sharp change in the wave pattern to a chaotic type of recording and screening of the underlying section. In addition, local isometric negative landforms, similar to the pockmark, are observed on the multipath sonar data inside the distribution contour of gas-saturated sediments. The width of these objects varies from 25 to 70 m, depth – up to 2 m. Component and isotopic composition of the gas indicate a powerful influx of thermogenic fluids, probably along the zones of active fracturing (Fig. 2) [5].

Field of 49 pockmarks was detected within the Kasania area using multipath sonar. They have depths of about several meters. The largest discovered pockmark has a width of about 27 m, depth of 5 m or more (Fig. 3). The formation of structures occurs in fairly soft marine sediments. According to gas-geochemical studies, the bacterial origin of the gas phase has been determined [5, 6].

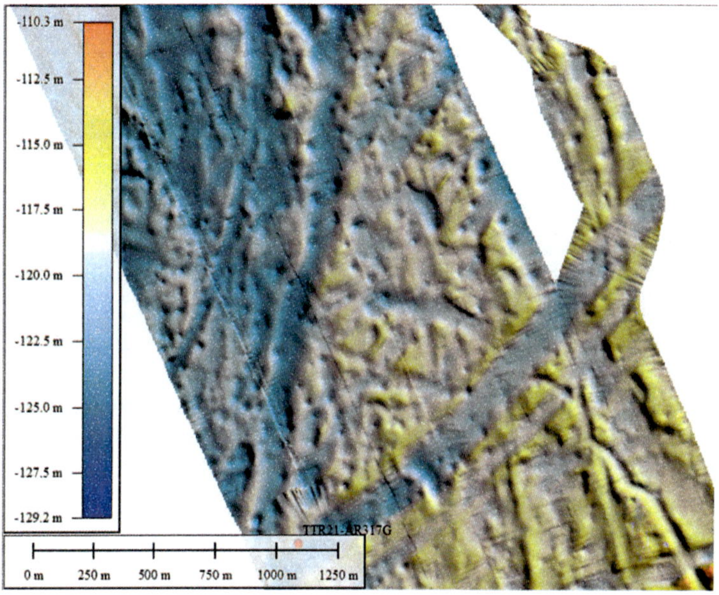

Fig. 2. Fields of pokmarks, Brachyura polygon

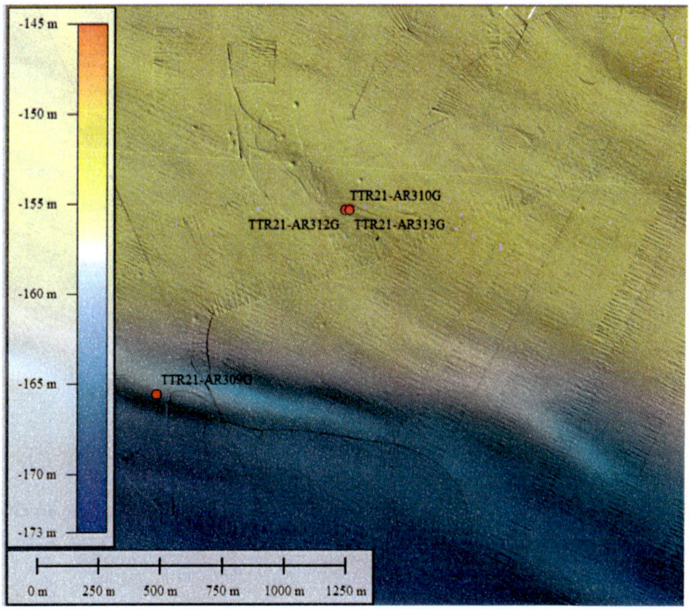

Fig. 3. Fields of pokmarks, Kasania polygon

Clione area is characterized by two suspected discontinuous faults, along which the channels of erosion valleys were laid, fluid unloading in these areas can be assumed, and in the upper part of the section, according to acoustic data, intensive fluid migration zones were detected, accompanied by an extensive gas front in the first tens of meters of the section, which are presumably associated with the degradation of permafrost strata, which, perhaps, could be a fluid barrier at depth.

At the Limacina site, according to high-frequency seismic data, presumably gas fronts were found that fit to the surface of the seabed. Acoustically transparent strata stand out well during high-frequency seismic exploration. At the same time, signs of fluid discharge were observed in the water column. 2 faults were also identified, which are manifested in the modern bottom relief, such faults can be fluid-conducting channels. No anomalies were identified in gas geochemistry.

In the "Scallop" section, a structure similar to a diapir has been identified, presumably limited by discontinuous faults (Fig. 4) along which fluid discharge can occur, and incisions characterized by the presence of an acoustically transparent thickness have also been identified, which may indicate the gas saturation of sediments. During the survey of the bottom relief, local depressions of a rounded shape were noted. For example, the depth of one recess is about 20 m, the diameter is about 375 m [8].

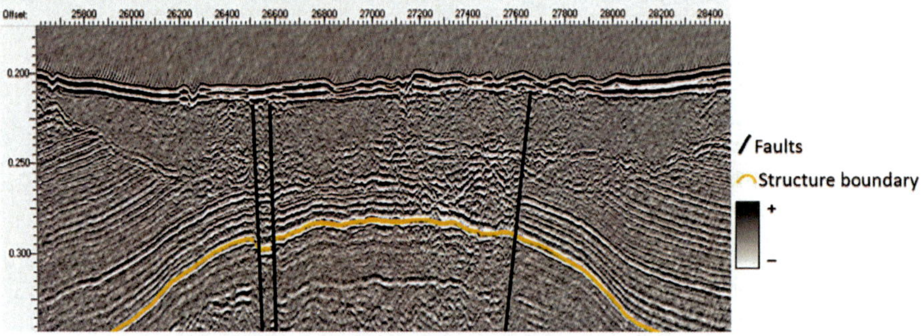

Fig. 4. Time cross-section, Scallop polygon

Testing of such structures showed the presence of thermogenic gaseous hydrocarbons C1-C5. The concentration of methane varies from 3 to 54 ppm, the concentration of homologues reaches 9 ppm. Fluorescence analysis of the organic matter showed the presence of light, probably migratory liquid hydrocarbons [9].

Detailed sampling stations were selected directly above the expected areas of the gas front discharge. As a result of geochemical research methods, zones with the presence of elevated and abnormal concentrations of the hydrocarbon gas phase were identified, the component and isotopic composition of gases was determined, and the composition of bitumen in bottom sediments was characterized.

The performed studies of the gas phase revealed elevated methane contents (up to 730 ppm) and homologues up to and including pentane (1–10 ppm).[6] Such values indicate a deep, thermogenic source of HCs, which is confirmed by the data of isotope studies. Fluorescence studies confirm the presence of elevated concentrations of light oily hydrocarbons in sediments, which may be epigenetic to sediment, migratory from the depths of the sedimentary cover.

4 Conclusion

An important task of the expedition work was to study the features of background and focused unloading of hydrocarbons on the seabed surface. Geochemical survey of bottom sediments is one of the main methods for assessing the effectiveness of hydrocarbon systems in the basin, but its methodological approaches for the Arctic waters, characterized by the presence of numerous litho-geochemical barriers on the routes of fluids, require careful development, testing and periodic adjustments.

Thus, it has been established that the zones of focused unloading detected during gas-geochemical studies of bottom silts are associated in most cases with various anomalies on the data of seismoacoustic profiling and with characteristic structures in the bottom relief. Presumably, the source of the fluid is deep Paleozoic sedimentary complexes. These violations of the fluid barrier lead to the rise and further redistribution of the fluid in the upper part of the section and it's unloading in the form of gas flares in the water column.

It is necessary to map such areas in order to identify processes that can be classified as dangerous, as well as for further industrial operation.

Acknowledgements. The expedition work was carried out with the support of the Ministry of Science and Higher Education of the Russian Federation within the framework of the plan-program of expeditionary research of Lomonosov Moscow State University on the topic "Features of Quaternary sedimentogenesis, relief formation and natural fluid discharge on the seabed in the northeastern part of the Kara Sea" and "Training-through-research on the Arctic shelf".

References:

1. Dolgunov, K.A., Martirosyan, V.N., Vasilyeva, E.A., Sapozhnikov, B.G.: Structural and tectonic features of the structure and prospects for the oil and gas potential of the northern part of the Barents-Kara region. Geol. Oil Gas **6**, 70–83 (2011)
2. Pirozhkova, M.A., Yurkina, K.V.: On the issue of the prospects of oil and gas accumulation within the passive continental margin on the example of the sedimentary basin of the North-Eastern part of the Kara Sea. In: Collection of articles based on the materials of the All-Russian Youth scientific and Technical Conference of the oil and gas industry "Young Oil", 351 p. (2015)
3. Kontorovich, V.A.: Oil and gas content of the Kara Sea. Neftegaz. Ru Business J. (11), 34–43 (2018)
4. Kontorovich, V.A., Kontorovich, A.E.: Geological structure and petroleum potential of the Kara Sea shelf. Rep. Acad. Sci. **489**(3), 272–276 (2019)
5. Hovland, M., Judd, A.G.: Seabed Pockmarks and Seepages: Impact on Geology, Biology and Marine Environment, 293 p. (1988)
6. Yumasheva, A.K., Poludetkina, E.N., Tokarev, M.Y., Rybalko, A.E., Kil, A.O.: Zones of focused hydrocarbon unloading as an indicator of the oil and gas potential of the subsoil of the North Kara Basin. In: Relief and Quaternary Formations of the Arctic, Subarctic and North-West of Russia, pp. 308–313 (2022)
7. Malyshev, N.A., et al.: A new model of the geological structure and evolution of the North Kara Sedimentary Basin. Dokl. Earth Sci. **445**(1), 791–795 (2012)

8. Nikishin, V.A.: Evaporate deposits and salt diapirs of the Urvantsev trough on the North Kara Sea. Moscow Univ. Geol. Bull. **67**(4), 265–268 (2012)
9. Poludetkina, E.N. Tokarev, M.Y., Yumasheva, A.K., Kil, A.O., Basova, E.D., Rybalko, A.E.: Features of fluid discharge to the seabed and in the upper part of the section in the northeastern part of the Kara shelf based on the results of the TTR-21 expedition. In: Marine Research and Education (MARESEDU), pp. 264–268 (2022)
10. Sokolov, S.Y., Moroz, E.A., Sukhikh, E.A., Razumovskiy, A.A., Levchenko, O.V.: Manifestations of deep degasing into the water column and upper part of the Pechora sea sedimentary section. Georesursy **21**(4), 68–76 (2019). https://doi.org/10.18599/grs.2019.4.68-76

Marine Geophysics

Features of Multiple Waves Suppression in the Seismoacoustic Data Processing Obtained During Near Surface Geophysical Surveys in the Northern Part of the Kara Sea

I. O. Paltsev(✉), V. V. Polovkov, A. A. Kudinov, and A. E. Rybalko

St. Petersburg State University, St. Petersburg 199034, Russia
st096144@student.spbu.ru

Abstract. At present, high resolution seismic survey methods are widely used for engineering surveys in the offshore area. At the same time, there are specific features such as multiple waves and ghosts that reduce the data quality. Therefore, the purpose of this work was to select the optimal procedures and parameters for the primary processing of data obtained in expeditions with a certain observation system. For this purpose, modeling of seismic data with the same observation system was performed. After that, the processing procedures were tested on these data. The selection of accurate values was carried out after fine-tuning on real data. While analyzing the results, it can be concluded that the methods of adaptive subtraction of interference waves from the free surface can effectively improve the quality of primary processing.

Keywords: high-resolution marine seismic survey · multiple waves · processing flow · ghost waves · resolution · geohazards

1 Introduction

In the summer of 2022, the TTR-21 ("Training Through Research") research expedition took place in the Kara Sea. One of the expedition's tasks was geological and geophysical mapping of geohazards. The complex of research included geological sampling, high-resolution seismic survey (HRS), acoustic profiling and multibeam echo sounding. The HRS data were quickly processed according to a typical processing flow. After that, geologic objects were identified on seismic sections in combination with materials from other methods and points for bottom sampling were selected. The initial HRS data were complicated by the presence of multiples and ghost waves, which were not suppressed in the typical processing flow. Modern processing procedures can successfully eliminate this type of noise, resulting in increased resolution, improved seismic stratigraphic analysis capabilities, and more unambiguous identification of geologic features [1].

The specifics of engineering seismic surveys (near-surface towing, short streamers, no depth control, etc.) impose limitations on the range of classical processing procedures that can be borrowed from petroleum seismic surveys [2]. Multiple wave suppression

T. Chaplina (Ed.): CIWO 2023, SPEES, pp. 341–349, 2023.
https://doi.org/10.1007/978-3-031-47851-2_41

procedures based on the difference in the shape of the single and multiple reflection hyperbolas are ineffective for surveys with short observing systems [3] and near-surface towing complicates the signature of the receiving and emitting system and leads to a decrease in the resolution of the method. The elimination of multiple reflections at short distances can be quite effectively realized by adaptive subtraction of the multiple wave field model obtained by static shift or calculation of the autocorrelation function [4]. On the other hand, accounting for the signature of the receiving and emitting system and dealing with ghost waves can be done by adaptive subtraction of the ghost wave model or by predictive deconvolution [5].

The processing of HRS data must consider the features of the acquired data and, in some cases, is an iterative process consisting of testing different algorithms and selecting the most suitable ones [6]. Therefore, the work involved the selection of procedures and the selection of optimal parameters for effective suppression of interfering waves.

2 Material and Methods

The expedition included a high-resolution seismic survey using the common depth point method (CDP). Most of the survey was carried out using an observation system consisting of a single source (sparker) and 16 channels with a spacing of 2 m. This system ensured minimal time spent on launching operations and minimal vessel turning radius (reduced circulation time). The tight coupling of source and receiver simplifies the processing, especially in bad weather). The quality of the data collected supports the task of studying near-surface sediments. In the process of processing a constant velocity law was applied, according to which the stack is summarized.

Since only pressure waves can be excited when working in the water area, the sources of such waves should be placed at some depth below the water surface. Such an arrangement leads to the formation of ghost reflections. Waves of this type interfere with the useful signal weakening or amplifying it and increase the total length of the signal [7].

On the expedition, the depth of the receiving and emitting equipment was chosen to create the highest amplitude and least long signal possible. The optimum depth, which ensures constructive interference of the signal at the minimum length, is equal to a quarter of a wavelength [8].

To calculate the source and receiver depths, the center frequency of the sparker was determined to be 220 Hz. Thus, the depth was 1.6 m. In accordance with these parameters the rafts for the receiving and radiating equipment were assembled.

The source energy was selected during the testing process. Most of the survey was conducted with an energy of 4 kJ at 58 electrodes. As the electrodes burned out, the energy was reduced to 3.5 kJ to preserve the central frequency.

The survey was conducted at a vessel speed of 6 knots, which provided a shot interval of about 3 m.

Signal processing of seismic materials, input of kinematic corrections and summation were performed in the program “RadExPro” in one flow.

When analyzing the seismogram and its spectrum, the parameters of the bandpass filter were selected to suppress low-frequency noise and other random noises. The “Bandpass Filter” module was used for filtering. This module applies frequency filtering to every input trace. The filtering algorithm operates in frequency domain and is realized by means of multiplying trace Fourier transform on digital filter frequency characteristic [9]. Afterwards, a correction for spherical divergence was introduced using the "Amplitude Correction" module. The coefficient 1 was used.

The "NMO/NMI" module was used to input normal moveout corrections (1.46 km/s or other values of wave velocity in water). The "Ensemble Stack" module summarized the obtained CDP gathers.

Execution of the migration procedure and unloading of the final seismic section was also performed in one flow.

The migration procedure was performed using the "Stolt F-K Migration" module. The migration velocity was set equal to the speed of sound in water (1.46 km per second). The distance between the traces was set equal to the size of the CDP site (5 m). The maximum frequency involved in image formation was equal to 5000 Hz. Other parameters were set according to standard values.

High-resolution seismic data after operational processing were complicated by multiple waves and ghosts, which were not suppressed in the typical processing flow.

Processing to create the modernized processing flow was performed in the "RadExPro" program. Before testing, the optimal methods of interference wave suppression for this observation system were determined. In this program they are presented in the modules: "DeGhosting", "Zero-Offset DeMultiple", "Custom Impulse Trace Transforms".

At the first stage the modules were tested on model data.

For this purpose, a simulation with precisely known position of the boundaries and equipment burial was performed (Fig. 1a). Three-layer models with different water column thicknesses were used. The horizontal extent of each model was 1000 m, and the vertical extent was 500 m.

The modeling was carried out with a sparker signal. The dominant frequency was 220 Hz. The observation system corresponded to the most frequently used in the expedition (1 source, 16 channels with an interval of 2 m, step between the excitation points – 3 m, minimum distance between the source and receiver – 10 m, source, and receiver burial – 1.6 m).

For comparison, an additional set of model data was obtained with no waves reflected from the free surface. The criteria for the optimality of parameter selection during further processing will be the highest similarity to this section (Fig. 1f).

The stack section obtained from the model data is shown in Fig. 1b. Primary processing was applied to the data, including amplitude correction for spherical divergence, stacking with a constant velocity law, and migration.

Multiple wave suppression was performed using the "Zero-Offset DeMultiple" module. The algorithm is based on adaptive subtraction of a model of multiples from the original wave field. The model is obtained from the data itself, either by static shift of the original traces or by autoconvolution [9]. Adaptive subtraction of the multiple wave model allows to get rid of unwanted energy without affecting single reflections [5].

Some parameters of the module were chosen based on the knowledge of the impulse length. Other parameters were selected by testing different values. The result of the module operation is presented in Fig. 1c.

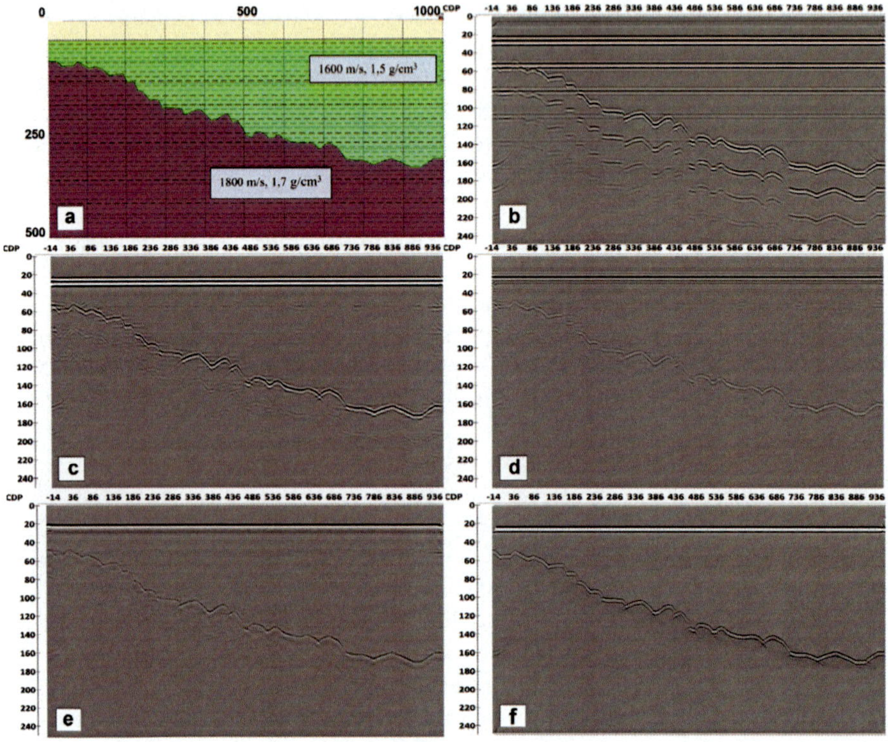

Fig. 1. Testing modules on model data

The next step was to test the ghost suppression procedures on model data. (Fig. 1d). First, the "DeGhosting" module was applied. The algorithm implemented in the module is based on adaptive subtraction of the ghost wave model from the source wave field. The ghost wave model is created from the data itself by means of static shifting of the original traces by the specified time. A shaping filter that minimizes the results of the ghost wave model subtraction from the original field is constructed for each trace based on the original and model data. The filter searches for similar reflections present both in the original data and in the ghost wave model and minimized their root mean square amplitudes in the resulting field. Additional nonstationarity that allows the filter to adapt to events that are similar but not identical is introduced in the process of the filter calculation [9].

This allows the subtraction algorithm to work efficiently to a certain degree even with approximate ghost wave models wherein the arrival times and reflection amplitude differ somewhat from the observed ones. However, the closer the ghost wave model is to the observed ghost waves, the better it will be subtracted. Therefore, the best results can be achieved with minimum source-receiver distances, since zero offset allows obtaining the most accurate model. For the same reason the module works less efficiently when data are substantially affected by sea disturbance [9].

On the model data, the ghost delay was calculated using the following equation:

$$\tau = 2dcos\theta/V, \tag{1}$$

where V – speed of sound in water, d – depth, θ – angle of incidence.

The other parameters were selected based on the testing of different values.

Further, the "Custom Impulse Trace Transforms" module was tested to perform deterministic deconvolution. A reflection from the bottom with a time-limited impulse was chosen as the waveform. After that, in the module settings, the division operation was selected for the amplitude spectrum and the subtraction operation for the phase spectrum. The result of the module operation is presented in Fig. 1e.

In the second step, testing was performed on a part of the expedition data for finer tuning.

3 Results and Discussion

To demonstrate the result of the "Zero-Offset DeMultiple" module, a profile obtained in shallow sea conditions was selected (Fig. 2).

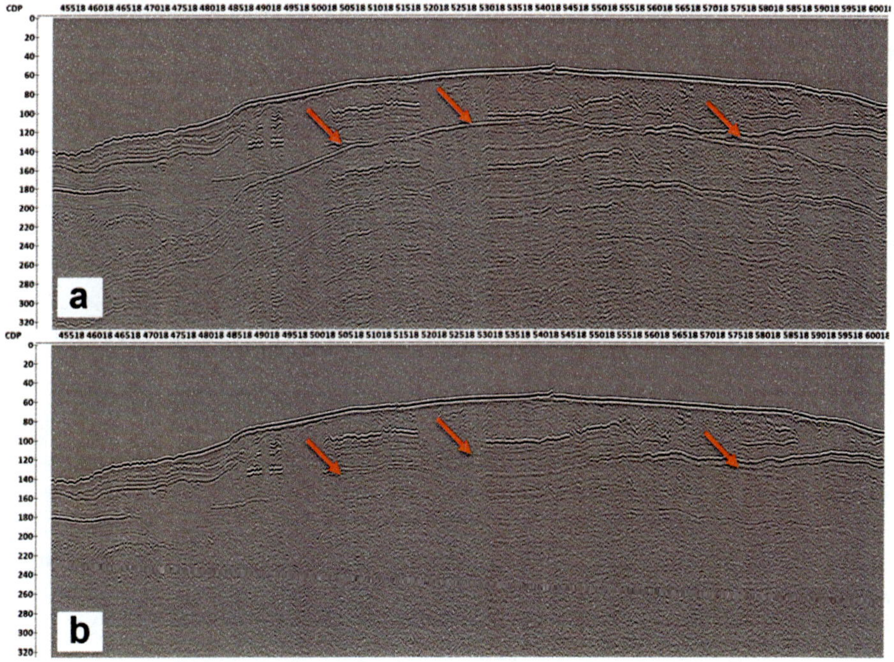

Fig. 2. Seismic section before processing (a) and after (b)

The module worked quite effectively, it was possible to suppress multiple waves of the first and higher orders, including partial multiple waves (Fig. 2b).

Then, the "DeGhosting" module was tested on real data. The values obtained during testing on model data were used as starting parameters of the module. During the fine tuning of parameters, the amplitude-frequency spectrum of the signal was analyzed. The goal was to obtain the spectrum with the least pronounced notches. Figure 3 show after application of the "Zero-Offset DeMultiple" and "DeGhosting" modules, as well as their amplitude-frequency spectra.

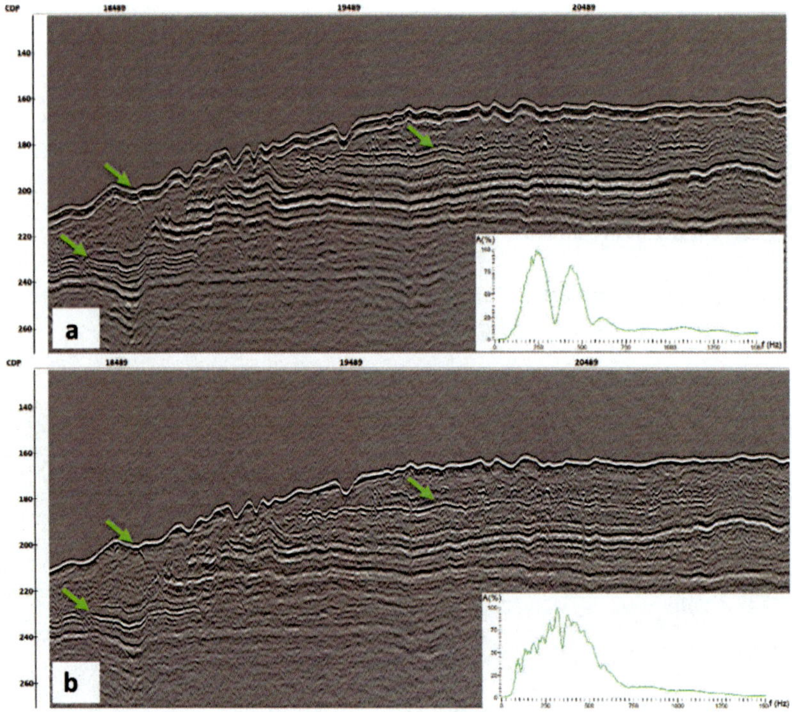

Fig. 3. Seismic section before the module was applied (a) and after (b)

Testing of the "Custom Impulse Trace Transforms" module for performing deterministic deconvolution was performed on the same profile section after the multiple wave suppression procedure. A straightened and summed reflection from the bottom with a time-limited impulse was chosen as the signal waveform.

After that, in the module settings, the division operation was selected for the amplitude spectrum and the subtraction operation for the phase spectrum. The result of the module operation on real data is presented in Fig. 4.

The "Predictive Deconvolution" module was also tested (Fig. 5). This module can be used to perform spiking deconvolution to increase the resolution of the record [10]. The length of the prediction interval was set equal to the sampling step (0.0625 ms), and the length of the deconvolution operator was set equal to the length of the probing pulse (10 ms).

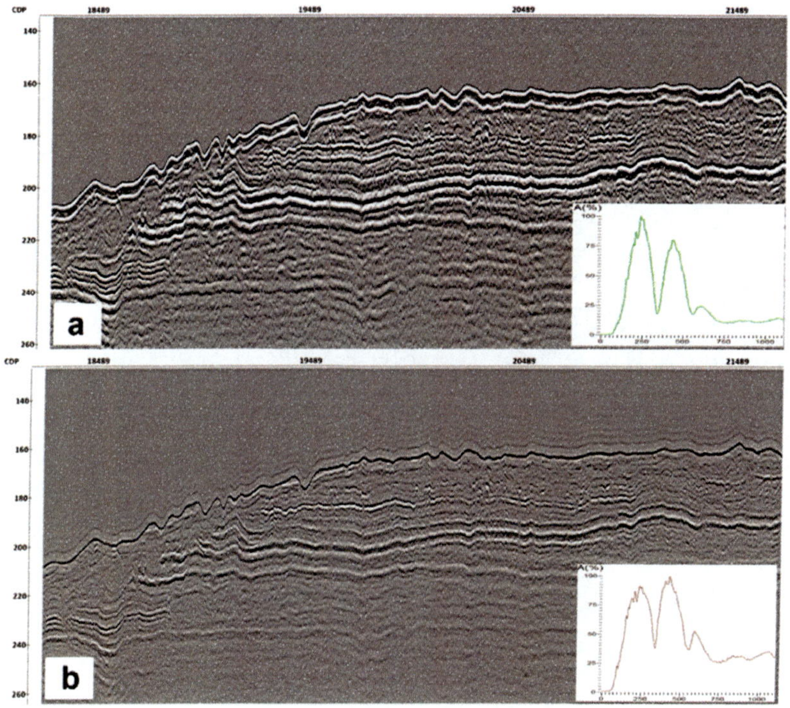

Fig. 4. Seismic section before the module was applied (a) and after (b)

Since a large amount of data in the expedition was acquired with a short streamer (30 m) and at relatively large sea depths, the "Zero-Offset DeMultiple" module can be effectively used to suppress multiples with a long repetition period. At the same time, the model of multiple waves can be obtained by autosurveys of seismic traces.

Of the three tested modules for ghosts suppression ("DeGhosting", "Custom Impulse Trace Transforms", "Predictive Deconvolution"), the first two procedures were the most effective. The data after predictive deconvolution, applied as impulse deconvolution, had the lowest ratio of useful signal amplitude to noise amplitude.

With similar results, the "DeGhosting" module can be recommended as a ghost suppression procedure for the operational processing flow since it does not require additional actions to obtain the waveform.

Thus, to eliminate multiple waves and ghosts, the "Zero-Offset DeMultiple" and "DeGhosting" modules can be added to the typical processing flow.

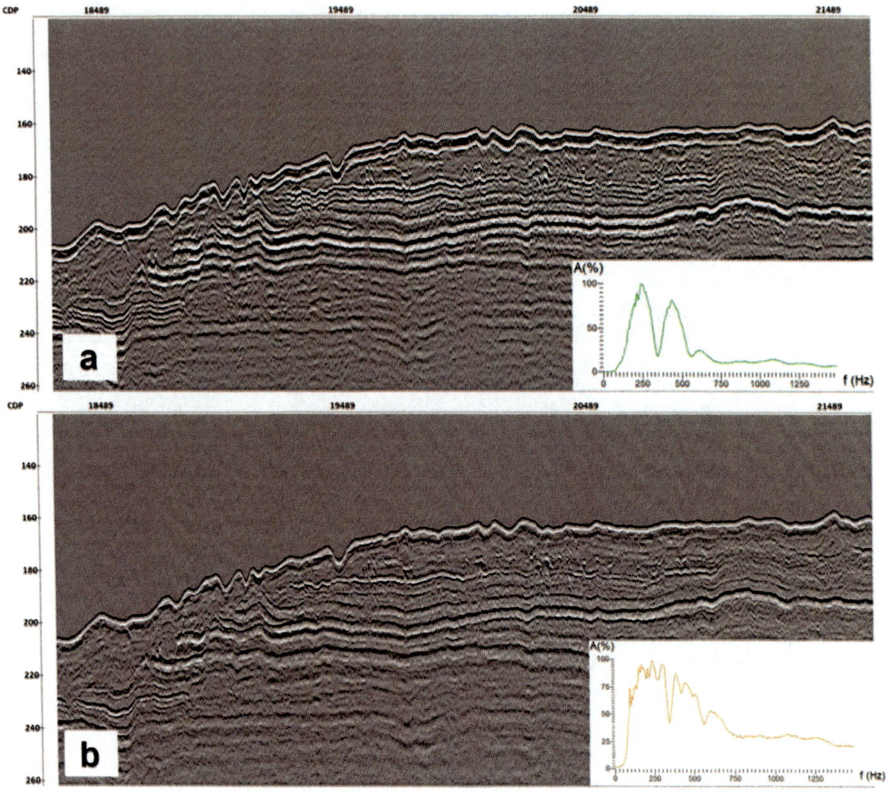

Fig. 5. Seismic section before the module was applied (a) and after (b)

4 Conclusion

During the work, an optimal processing flow was developed for the data obtained with the following observation system: 1 source (sparker with a dominant frequency of 220 Hz), 16 channels with an interval of 2 m, step between excitation points – 3 m, minimum distance between source and receiver – 10 m, source, and receiver burial – 1.6 m.

"Zero-Offset DeMultiple" module was selected as a procedure of multiples suppression with a long repetition period. Autoconvolution of traces was chosen as the method for specifying the multiple wave field model, the length of the shaping filter was set equal to 500 samples, the level of white noise was equal to 0.001%, the step of calculation of filters was equal to 10 traces.

The "DeGhosting" module was selected for suppression of ghosts. The length of the shaping filter was 15 ms, the white noise level was 0.01%, the constant shift was 3 ms, the number of neighboring traces considered in the filter construction was 3, the number of traces over which the calculated filters would be averaged was 20, the filters were calculated for each trace.

Processing of the whole data set obtained during the expedition using the modernized processing flow allowed us to improve the data quality by suppressing ghosts and eliminating multiple reflections. The seismic records better highlighted the already known objects, and new features of the wave field became recognizable.

This processing flow can be used in future surveys with a similar observing system to improve the quality primary processing.

References

1. Dondurur, D.: Acquisition and Processing of Marine Seismic Data, 598 p. Elsevier (2018)
2. Verschuur, D.J.: Seismic Multiple Removal Techniques: Past, Present and Future, 174 p. EAGE Publications, Houten, The Netherlands (2006)
3. Boganik, G.N., Gurvich, I.I.: Seismic Survey: Textbook for Universities, 744 p. AIS Publishing House, Tver (2006)
4. Bondarev, V.I.: Seismic survey: textbook for universities. In: 3 books, 620 p. Publishing house UGGU, Ekaterinburg (2006)
5. Vakulenko, S.A.: Modern methods of marine engineering seismic processing. In: Proceedings of the International Geological and Geophysical Conference "GeoEurasia 2018. Modern Methods of Study and Development of Eurasian subsoil", pp. 892–897. LLC "PolyPRESS", Moscow (2018)
6. Vakulenko, S.A., Buryak, S.V., Shuvalov, A.A., Alekhin, A.A.: Processing of high-resolution seismic data – suppression of the receiving and emitting system signature. In: Proceedings of the 17th 7 Scientific and Practical Conference and Exhibition "Engineering and Ore Geophysics 2021", pp. 1–7. OOO "EAGE GEOMODEL", Moscow (2021)
7. Claerbout, J.F.: Fundamentals of Geophysical Data Processing, 274 p. McGraw-Hill, New York (1976)
8. Shalaeva, N.V., Starovoitov, A.V.: Fundamentals of Seismoacoustics in Shallow Water Areas, 256 p. MGU Publishing House, Moscow (2010)
9. RadExPro 2016.1 User Manual [Electronic resource]. http://radexpro.com/wpcontent/uploads/2016/09/RU_RadExPro2016.1%20manual.pdf
10. Yilmaz, Ö.: Seismic Data Analysis: Processing, Inversion, and Interpretation of Seismic Data. Society of Exploration Geophysicists (2001)

Joint Processing Technology of MCS Data (600 m Streamer) and Wide-Angle Reflection/refraction Seismic Sonobuoy Soundings Acquired in Ice Conditions During Expedition "ARCTIKA-2014"

S. N. Tabyrtsa[1,2](✉), A. S. Zholondz[2], and V. V. Polovkov[1]

[1] Saint Petersburg State University, Saint Petersburg, Russia
sn_tabyrtsa@list.ru
[2] The All-Russia Scientific Research Institute Geology and Mineral Resources of the World Ocean (VNIIOkeangeologia), Saint Petersburg, Russia

Abstract. The regional seismic data – 2D MCS (multichannel seismic) with a short streamer of 600 m and wide-angle reflection/refraction seismic sonobuoy soundings were obtained in the Arctic Ocean in 2014. The subject of study is the eastern part of the profile AR1407 (expedition "ARCTIKA-2014"). This study is related to the joint reprocessing of two types of data. The processing of this data was made in 2014–2015. The reprocessing started in 2023 and it will have been completed by the end of 2023 year. Now it is in progress.

Keywords: Arctic Ocean · seismic · sonobuoy soundings · MCS data · WARRP · velocity model

1 Introduction

Regional seismic data, including 2D MCS (multichannel seismic survey) and wide-angle reflection/refraction seismic sonobuoy soundings, was obtained in the Arctic Ocean in 2011, 2012 and 2014 (Fig. 1).

Depending on the ice conditions, various lengths of seismic streamers were used during the MCS: 4500 m—in relatively clear water, and 600 m—in difficult ice conditions. It is the main reason to combine these two methods: MCS with short streamer (600 m) and wide angle reflected/ refracted profiling (WARRP). Different studies on this technology were also carried out by foreign researchers [1–8], in addition, the authors of this paper considered various options for joint processing of refracted and reflected waves obtained in studies with ocean bottom seismic stations [9–12].

Profile AR1407 analyzed in this paper (see Fig. 1) is a unique profile with a length of more than 1500 km, crossing all the main structures: it starts in the Barents-Kara shelf, goes across the Nansen Basin, the Gakkel Ridge, the Amundsen Basin, the Lomonosov

T. Chaplina (Ed.): CIWO 2023, SPEES, pp. 350–359, 2023.
https://doi.org/10.1007/978-3-031-47851-2_42

Ridge, and the Makarov Basin. Soundings were accomplished as an inverse flank observation system with common reception point. The interval between sonobuoys was about 50 km. The step between the shot points (SP) was about 50 m. 28 soundings of the sonobuoy data were received on profile AR1407.

The subject of study is the eastern part of the profile from the Amundsen Basin to the Makarov Basin.

Processing of the MCS and sonobuoy data was carried out after receiving field materials, in 2014–2015. In 2023, reprocessing of the data on profile AR1407 began in order to update the previous results. What is it for? As a rule, a customer wants to get a result very quickly after field observations. And this could affect the quality of the result. The first study was done by several different organizations. At the moment, the processing is carried out from the beginning to the end by one company.

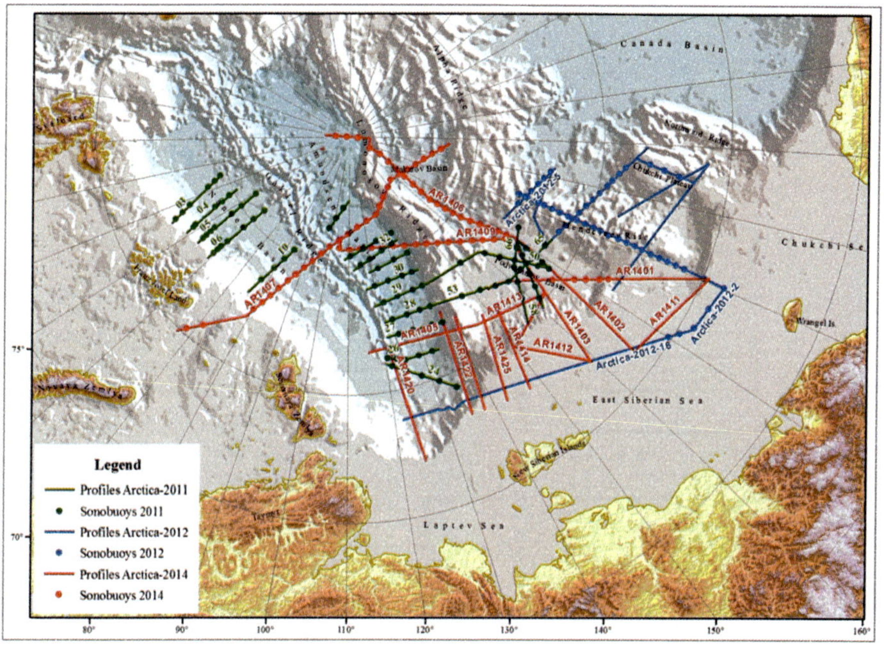

Fig. 1. Map of exploration area with MCS survey and sonobuoy soundings

A typical sonobuoy seismogram and a MCS seismogram (common shot point) are shown in Fig. 2. This picture presents the difference between seismograms – the right seismogram is too short to determine reliable velocities. But technology of sonobuoy soundings is inverted flank system that makes it difficult to define the slope of the boundary. In this case, MCS data can help.

Therefore, combining MCS data and sonobuoys data, i.e. reflection and refraction seismic data allowed to construct a reliable velocity model [13].

The processing sequence is shown in Fig. 3. This figure illustrates the main stages of processing of different types of waves with colors (reflected waves of MCS data are

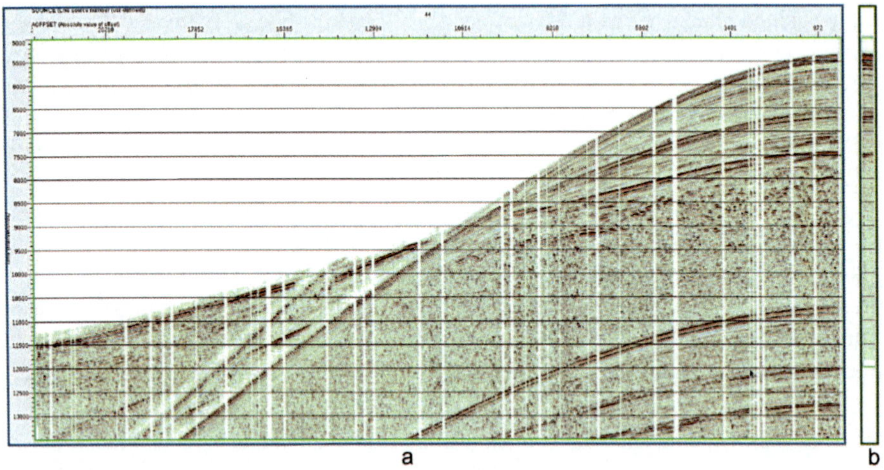

Fig. 2. A typical sonobuoy seismogram (a) and MCS seismogram (b) at one location

in green, refracted waves of WARRP data are in yellow, joint processing of both types waves and data is in purple).

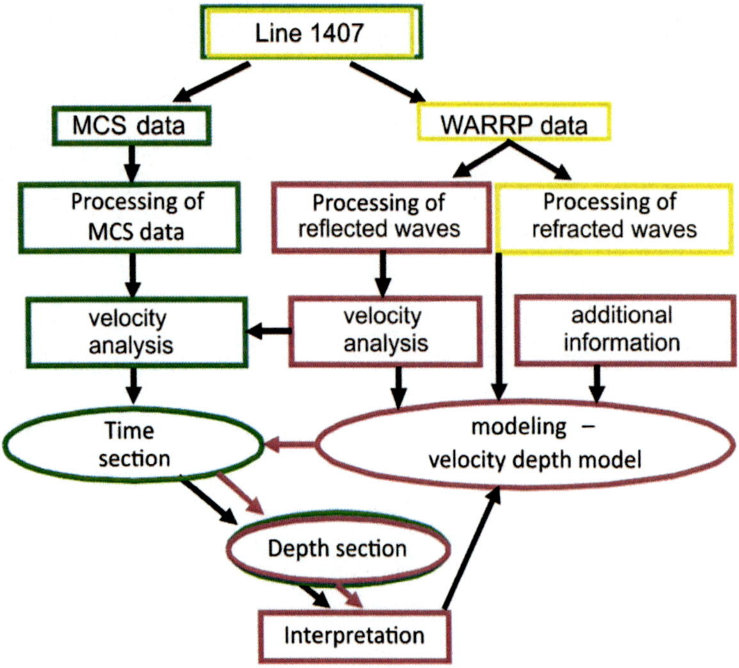

Fig. 3. Block diagram of the processing sequence

2 Material and Methods (Data and Methods, Experiments and Data Representation)

The main sequence of joint processing of the MCS and sonobuoy data is [14]:

1. Processing of short streamer MCS data;
2. Processing of sonobuoy data. Spectral analysis of the reflected waves according to the MCS data with comparison of stratigraphic complexes in the time section. Creating a file with velocities;
3. Time-Depth conversion using velocities from the previous stage. Loading of major interpreted boundaries in the Seiswide software. Formation of an initial depth velocity Seiswide model;
4. Ray-tracing modelling of wide-angle refraction/reflection sonobuoy data. Interpretation of reflection and refraction P-waves on sonobuoy records according to a interpreted MCS section.
5. Matching of the interval velocities obtained from the vertical velocity spectrum with the interval velocities of the depth velocity model. Estimation of a interval velocity dispersion. Correction of the seiswide velocity model.
6. Conversion of the final depth velocity model to SEG-Y files of interval velocities in the time domain. Recalculation of the initial time section into an updated depth section.

   This reprocessing is still in progress. Therefore, here we will compare only a part of the results of the old and new processing.

3 Results and Discussion

Let's consider some of the stages of joint processing in more detail.

Processing of MCS Data with Short Streamer of 600 m
Figure 4 shows a comparison of time sections of the profile in the area from the Amundsen Basin to the Makarov Basin.

After the reprocessing of 2023, the spectrum is represented by a wide frequency range, which indicates the preservation of amplitudes and a less loss of the useful signal as a result of processing. In the lower section, the structures are also more clearly visible. This is also observed in Fig. 5.

Processing of Sonobuoy Data
In 2014, the sounding data were processed. In 2023, reprocessing was performed in order to improve the quality of waves on the sonobuoy soundings and increase the signal-to-noise ratio.

Figure 6 illustrates the initial sonobuoy sounding. The sounding shows that as the source-receiver distance increases, signal attenuates and single amplitude bursts appear—the spectrum in this area is presented in the blue rectangle. It is necessary to increase a signal-to-noise ratio in this region and suppress the bursts so that the refracted waves can be traced and at the same time the reflected waves can be improved. The latter is one of the difficulties in processing MCS data.

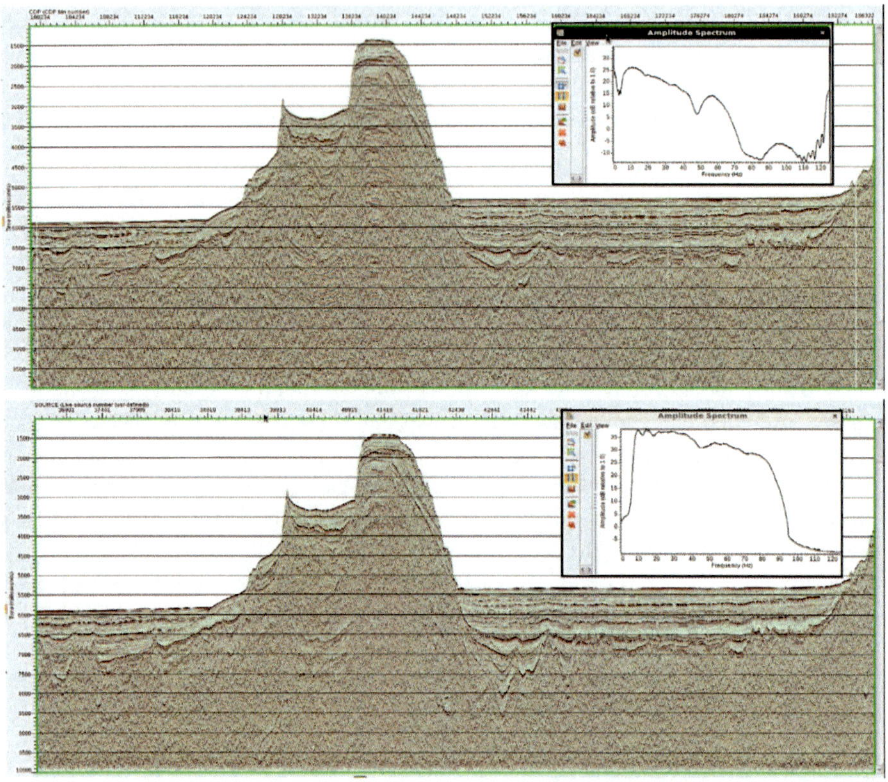

Fig. 4. Time sections with spectrums (top – 2014, bottom – 2023).

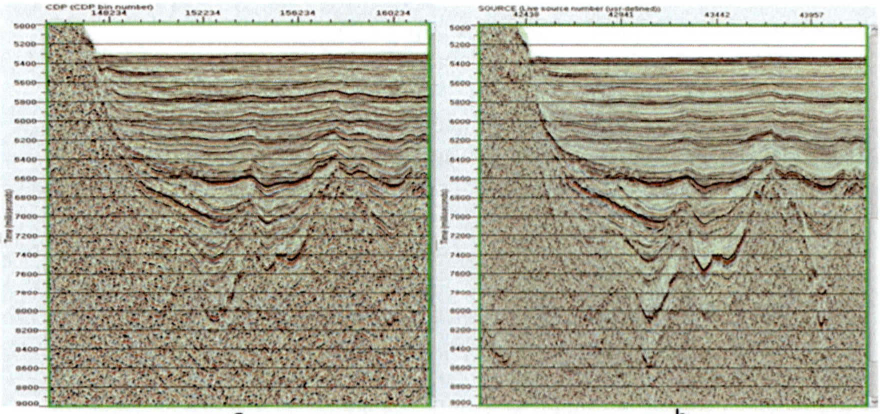

a b

Fig. 5. Parts of time section (left – 2014, right – 2023)

Sonobuoy data is often analyzed in two different processing graphs considering different types of waves – the reflected and refracted. As a result, there are two different soundings. In this case, it is necessary to improve both types of waves at the same time, because all types of waves are used in a velocity model. It is necessary to be able to trace refracted and reflected waves from the same boundary on the same soundings.

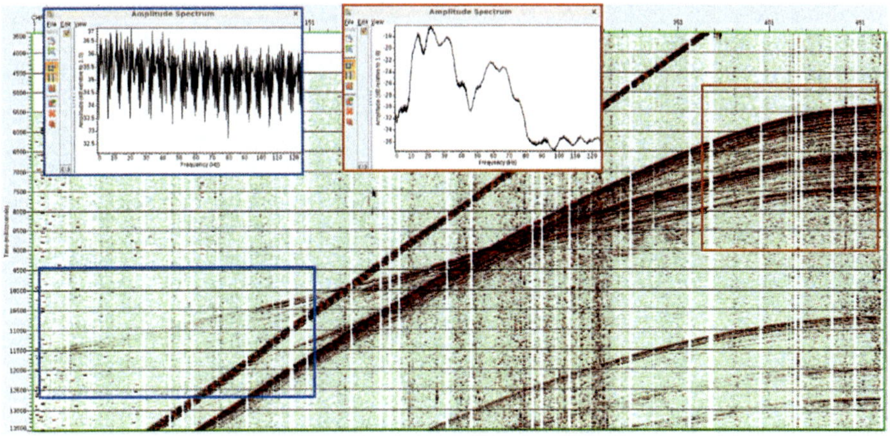

Fig. 6. Initial sonobuoy sounding with amplitude spectrums

The processing of 2014 was performed in the FOCUS system with some minimum procedures, which helped to amplify the original signal without raising noise. The processing of 2023 was performed in the ProMAX / SeisSpace system and, for some sonobuoys, the useful signal is slightly better. Figure 7 shows a comparison of two seismograms after processing in 2014 and 2023 using the procedure of the automatic gain control (AGC with a window of 500 ms). In this figure, the processing of 2023 looks better, but it also depends on the program used.

Thus, the processing of sonobuoy soundings of 2014 and 2023 is comparable and the best one can be used during the modeling.

The RMS velocity section (see Fig. 2) according to the MCS data is incorrect due to small offsets. The length of the traveltimes is not enough to obtain clear maximum coherence on velocity spectrum.

The velocities of WARRP data are used for further recalculation of the time section into the depth section,

Figure 8 shows a comparison of the velocity analysis at the same sounding. The figure shows that the maximum coherence and waves on the seismogram look more clearly as a result of processing in 2023.

The Fig. 9 shows RMS velocity sections based on the MCS data and sonobuoys data. There are a strong decrease in velocities in this section, which corresponds to the velocity of sonobuoys data at the sonobuoy location. As a result, only smoothed velocities from sonobuoys were used to convert the time section into the first version of the depth section.

Figure 10 illustrates the result of recalculation of the time section into a depth one with the MCS velocities and WARRP velocities.

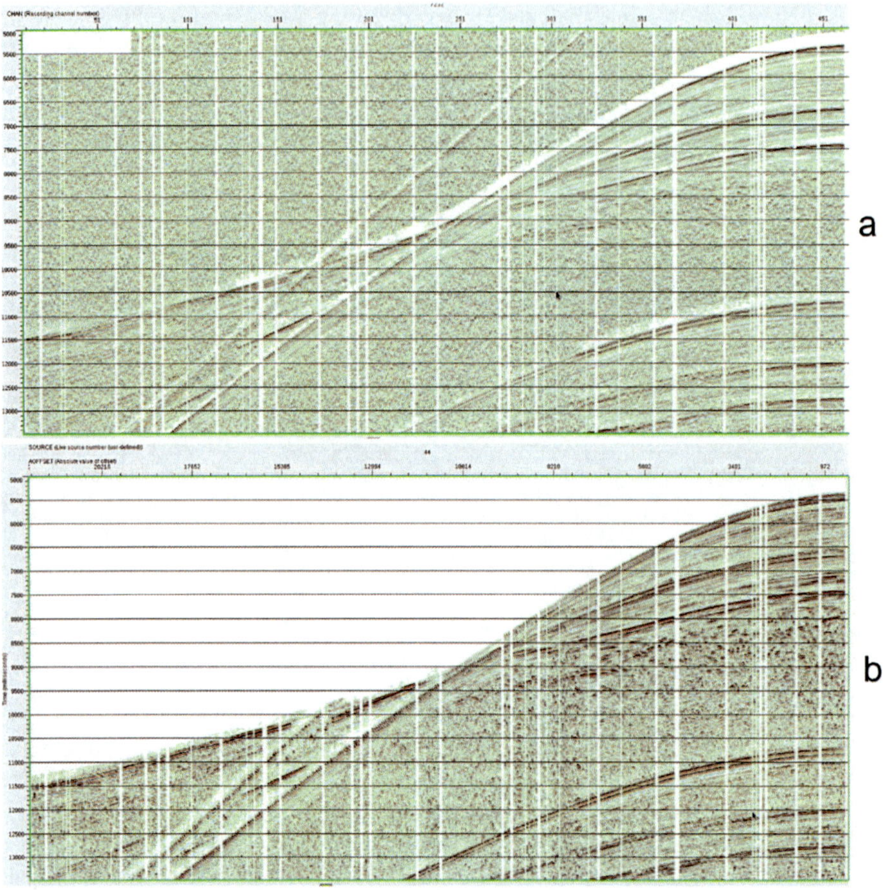

Fig. 7. The sounding after processing (a: 2014; b: 2023)

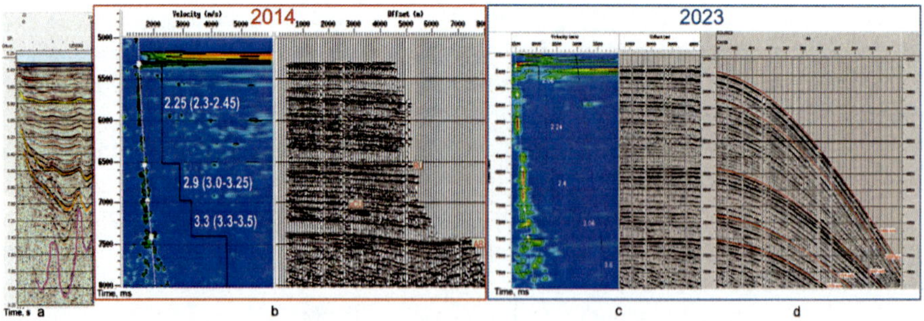

Fig. 8. Comparison of the velocity analysis at one sonobuoy sounding of 2014 (b) and 2023 (c, d)

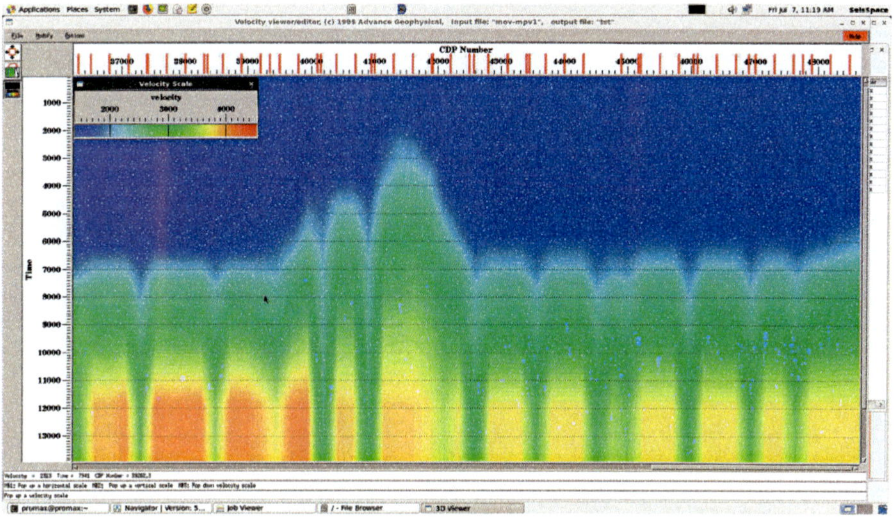

Fig. 9. RMS velocity sections based on MCS data and sonobuoys data

There is a difference between the same horizons about 1 km in the area of the dark green rectangle, which means that higher MSC velocities stretch horizons and contribute to a clearer display of intermediate boundaries (or multiples). For example, at a depth of approximately 5700 m (see Fig. 10, a), there is a horizon at the top of the rectangle, which, when using WARRP velocities (see Fig. 10, b), is not distinguished as a separate horizon. All this can lead to significant errors when creating a velocity depth model.

Creating a Depth Velocity Model in SeisWide

The next stage of processing is a construction (updating) of a velocity depth model.

The recalculated depth seismic section imports to the SeisWide software as an initial velocity model. It is a forward solution with model fitting by correcting velocities and boundaries. The goal is to match calculated and observed traveltime curve by changing of velocities and boundaries. During this stage, the interval velocities on vertical spectrums are compared with interval velocities from Seiswide model.

The time of reflection waves on seismogram must be matched with time of boundaries of the main seismic complexes on the seismic time section.

The old velocity model (2014) is presented in the publication [15]. The remodeling process takes sufficient time and now is in progress.

Final Depth Velocity Model

At the last stage, the depth SeisWide model is converted into a file of vertical velocities in time by a special conversion program. The time section is recalculated into a new depth section using new interval velocities. Then interpretation of the depth section is made again. As a result of the interpretation, the boundaries depth may change.

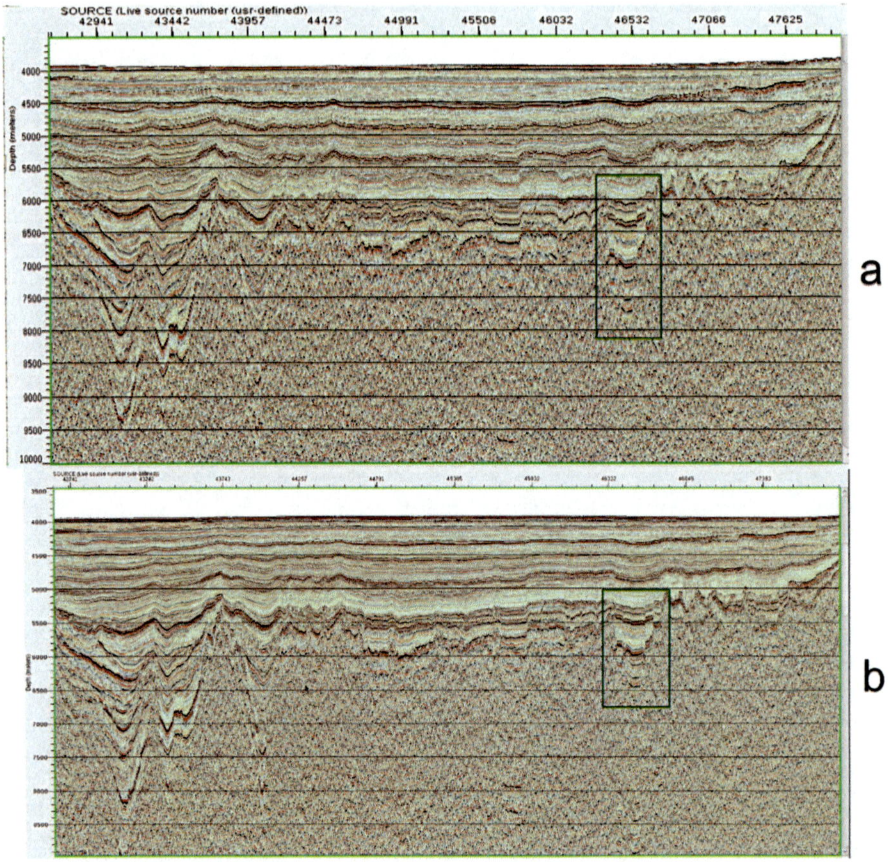

Fig. 10. Part of depth sections as the result of recalculation of the time section with the MCS velocities (top) and WARRP velocities (bottom)

4 Conclusion

A joint processing technology of MCS data (600 m streamer) and wide-angle reflection/refraction seismic sonobuoy soundings consists of several stages, where all types of waves of both methods are used. Data processing is a sequential cycle, which also involves additional information about velocities.

The MCS data (600 m streamer) and wide-angle reflection/refraction seismic sonobuoy soundings acquired in ice conditions during the expedition "ARCTIKA-2014" were processed together in 2014. The 2023 reprocessing has already improved seismic sections and, moreover, may change the velocity model and, accordingly, the position of the reflectors.

Therefore, combination of two methods are a good approach for getting reliable information in difficult ice conditions.

Acknowledgements. We thank the Saint Petersburg University for the opportunity to make this publication and FSBI "VNIIOkeangeologia" for the data. And also we appreciate the help of colleagues: head of marine seismic department "VNIIOkeangeologia" Oleg Smirnov and lead engineer Natalia Leonova.

References

1. Chian, D., Lebedeva-Ivanova, N.: Atlas of Sonobuoy Velocity Analyses in Canada Basin, 55 p. Geological Survey of Canada Open File 7661 (2015). https://doi.org/10.4095/295857
2. Coakley B., Kristoffersen Y., Hopper J.: Cruise Report for Underway Geophysics Program HLY 05-03, 84 p. University of New Hampshire (2005)
3. Coakley, B., Brumley, K., et al.: Exploring the geology of the central Arctic Ocean; understanding the basin features in place and time. J. Geol. Soc. **173**, 967–987 (2016)
4. Mayer, L., Armstrong, A.: U.S. Law of the Sea cruise to map the foot of the slope and 2500-m isobath of the US Arctic Ocean margin, 179 p. University of New Hampshire (2008)
5. Evangelatos, J., Mosher, D.: Seismic stratigraphy, structure and morphology of Makarov Basin and surrounding regions: tectonic implications. Marine Geol. **374**, 1–13 (2016)
6. Evangelatos, J., Funck, T., et al.: The sedimentary and crustal velocity structure of Makarov Basin and adjacent Alpha Ridge. Tectonophysics **696–697**, 99–114 (2017)
7. Jokat, W., Schmidt-Aursch, M.C.: Geophysical characteristics of the ultraslow spreading Gakkel ridge. Arctic Ocean Geophys. J. Int. **168**, 983–998 (2007)
8. Jokat, W., Ickrath, M.: Structure of ridges and basins off East Siberia along 81°N, Arctic Ocean. Marine Petrol. Geol. **64**, 222–232 (2015)
9. Roslov, Y.V., Polovkov, V.V.: Prestack depth migration and velocity model derived from first break tomography at long offsets. In: 4th EAGE St. Petersburg International Conference and Exhibition on Geosciences – New Discoveries through Integration of Geosciences, cp-156-00075 (2010). https://doi.org/10.3997/2214-4609.20145484
10. Polovkov, V.V.: Complexing of the reflected and refracted waves in the processing and interpretation of multichannel marine seismic data. In: 5th EAGE St. Petersburg International Conference and Exhibition on Geosciences – Making the Most of the Earths Resources, cp-283-00042 (2012). https://doi.org/10.3997/2214-4609.20143632
11. Roslov, Y.V., Merezhko, A.A., Polovkov, V.V., Popov, D.A., Zhemchuzhnikov, E.G.: Multicomponent seismic survey in transition zone of Pechora bay with node system turtle-500. In: 6th EAGE Saint Petersburg International Conference and Exhibition, vol. 2014, pp. 1–5 (2014). https://doi.org/10.3997/2214-4609.20140213
12. Ryzhkov, V.I., Roslov, Y.V., Sergeev, K.S., Polovkov, V.V., Elistratov, A.V.: Engineering surveys by the method of the cableless ocean bottom seismic. In: Geomodel 2015 – 17th science and Applied Research Conference on Oil and Gas Geological Exploration and Development, vol. 2015, cp-448–00003 (2015). https://doi.org/10.3997/2214-4609.201412223
13. Butsenko, V.V., Bezumov, D., Tabyrtsa, S., Leonova, N.: Matched 2D velocity models of the sedimentary cover and the acoustic basement of the Arctic Ocean based on wide-angle reflection/refraction seismic sonobuoy soundings and MCS data. In: EGU General Assembly 2018. Geophysical Research Abstracts, 20, EGU2018–2221 (2018)
14. Bezumov, D.V., Butsenko, V.V., Leonova, N.E., Tabyrtsa, S.N.: Construction of matched velocity models based on seismic data of MCS and sonobouy soundings in the Eurasian basin of the Arctic Ocean. In: New in Geology and geophysics of the arctic, antarctic and oceans. VNIIOkeangeologiya, pp. 8–9. VSEGEI, St. Petersburg State University, JSC "Polyargeo","AMIGE" (2018)
15. Piskarev, A.L., Poselov, V.A., Kaminsky, V.D., et al.: Geologic Structures of the Arctic Basin, 386 p. VNIIOkeangeologia, St-Petersburg (2018)

Marine Ecology and Environmental Management

Persistent Organic Pollutants in Rhinoceros Auklet (*Cerorhinca Monocerata*) from the Sea of Japan

M. A. Belanov[1(✉)], M. Yu. Shchelkanov[2,3,4,5], D. V. Pankratov[2], M. M. Donets[1], and V. Yu. Tsygankov[1]

[1] School of Advanced Engineering Studies "Institute of Biotechnology, Bioengineering and Food Systems", Far Eastern Federal University, Vladivostok 690922, Russia
belanov.ma@gmail.com

[2] Research Institute of Epidemiology and Microbiology Named After G.P. Somov Rospotrebnadzor, Vladivostok 690087, Russia

[3] School of Medicine, Far Eastern Federal University, Vladivostok 690922, Russia

[4] Federal Scientific Center of the East Asia Terrestrial Biodiversity Far Eastern Branch of the Russian Academy of Sciences, Vladivostok 690022, Russia

[5] National Scientific Center of Marine Biology Far Eastern Branch, Russian Academy of Sciences, Vladivostok 690041, Russia

Abstract. The study presents the levels of organochlorine pesticides and polychlorinated biphenyls in the organs of the rhinoceros auklet (*Cerorhinca monocerata*) from the coast of Boysman Bay (Sea of Japan). A priority group of toxic chemical compounds for this species and in the region has been identified.

Keywords: OCP · PCB · POP · the Sea of Japan · seabirds

1 Introduction

The global decline in the populations of many bird species has various causes, including habitat degradation, climate change and environmental pollution. Among the most dangerous factors affecting the organism of the birds, toxic compounds of anthropogenic origin stand out. They can have a long-term impact on ecosystems, and accumulate in living organisms, including various bird species [1].

Among all man-made chemicals, persistent organic pollutants (POPs) are the focus of researchers' attention. The best-known representatives of this group of compounds are polychlorinated biphenyls (PCBs) and organochlorine pesticides (OCPs), characterized by high persistence, toxicity, bioaccumulation, biomagnification and negative effects on humans and wildlife [2, 3]. POPs were widely used throughout the world in agricultural and industrial products until they were banned under the Stockholm Convention on POPs [4]. Despite this, they are still used in some developing countries due to their low cost and effectiveness [5]. PCBs have been banned worldwide for several decades, but even today they are a serious problem due to leakage from storage sites, illegal disposal

T. Chaplina (Ed.): CIWO 2023, SPEES, pp. 363–369, 2023.
https://doi.org/10.1007/978-3-031-47851-2_43

and use, and "historical" sediment residues [6]. In addition, they are produced by fuel combustion and because of heavy industry (including steelmaking, cement production, etc.) [7].

A good example of the impact of POPs on ecosystems is the effect of DDT on birds. Studies have shown that the most stable metabolite of this compound (DDE) caused the thinning of the eggshell of birds of prey, which highly affected the stability of their population and reduced the success of survival and hatching of juveniles [8]. The high toxicity of POPs to birds marked the beginning of their ban following the publication of Rachel Carson's book Silent Spring [9]. Thus, birds are sensitive objects for monitoring of POPs and may reflect their presence in the environment.

Seabirds are effective indicators of marine pollution because they aggregate the effects of pollutants across time and space. Due to their wide distribution, trophic status and ecological characteristics, different species of birds allow a wide range of ecological niches to be explored and habitats to be explored, from continental forests to seacoasts and islands. In this case, samples of eggs, blood, feathers, and other tissues can be used for research, the selection of which is possible in nesting colonies [8].

Seabirds are effective indicators of marine pollution because they combine the effects of pollutants across time and space. Due to their wide distribution, trophic status and ecological characteristics, different species of birds allow a wide range of ecological niches and habitats to be explored, from continental forests to sea coasts and islands. In this case, samples of eggs, blood, feathers, and other tissues can be used for research, the selection of which is possible in nesting colonies [10].

The Far East is one of the least researched regions of Russia in term of POPs. The largest number of studies here are devoted to the mass commercial species of marine fish, and birds have been studied fragmentarily [11].

Thus, the aim of this work is to present data on the content of organochlorine pesticides (isomers of HCH, DDT and its metabolites) and polychlorinated biphenyls in the organs of the rhinoceros auklet (*Cerorhinca monocerata* Pallas, 1811) of Boisman Bay (Sea of Japan) and to determine the priority group of pollutants affecting these birds.

2 Material and Methods

A mass death of rhinoceros auklet was recorded on July 13, 2021, on the coast of the Sea of Japan in the south of the Khasansky district of Primorsky Krai. Later it will be established that the cause of death of birds is infectious peritonitis with endogenous intoxication syndrome and multiple organ failure caused by a complex of pathogenic microorganisms: enteropathogenic variants of *E. coli*, *P. vulgaris*, *P. mirabilis*, *P. pennery*, *En. Faecalis* and *W. anomalus* (*Candia pelliculosa*) [12]. Prior to this, one of the versions of the death of birds was toxic poisoning with POPs, which was investigated by the authors.

Samples (n = 5) were collected from the coast of Boysman Bay, Sea of Japan (Primorsky Territory). The individuals were dissected, the muscles, liver, and feathers were taken for analysis, frozen at –20 °C, and delivered to the laboratory. Prior to chemical analysis, the muscles and liver were homogenized. Feathers were washed with distilled water, dried, and milled to 1 mm particles. POPs were extracted from homogenates and feather material with a mixture of hexane: acetone, followed by the destruction of co-extractive components (lipids) with concentrated sulfuric acid and separated with polar (OCPs) and non-polar (PCBs) solvents on chromatographic columns using an activated sorbent Florisil© 100–200 mesh [13]. Determination of the content of organic toxicants in the biomaterial was carried out on a Shimadzu GC MS-QP 2010 Ultra gas chromatography-mass spectrometer. Among the POPs in the samples, the following were determined: p,p′-DDT, o,p′-DDT, p,p′-DDD, o,p′-DDD, p,p′-DDE, o,p′-DDE, α-, β-, γ-, δ-HCH and 28, 52, 101, 118, 153, 138, 180 PCB congeners.

Statistical analysis of the results was performed using IBM SPSS Statistics software. The reliability of the data was assessed using the two-tailed Kruskal–Wallis test with a significance level of $p \leq 0.05$.

3 Results and Discussion

Persistent organic pollutants were found in all samples of rhinoceros auklet. The content of ΣPOPs (ΣOCP + ΣPCB) in various organs varied from 446 to 19 929 ng/g lipid weight (lw). In feathers, the range was from 1326 to 4 971, in the muscles – from 449 to 15 823, in the liver – from 492 to 19 929 ng/g lw.

Organochlorine pesticides were found in all samples of the rhinoceros auklet. The levels of ΣOCP (ΣHCH + ΣDDT) ranged from 30 to 1 827 ng/g lw. In feathers, the range was from 214 to 760, in the muscles – from 32 to 779, in the liver – from 125 to 1 826 ng/g lw.

β-HCH was found in all samples of feather material. The detection rate of γ- and δ-isomers was 80%. α-HCH was below detection limits. Among DDT metabolites, p,p′-DDE was found in all samples, o,p′-DDE was determined in 80% of samples.

β- and γ-forms of HCH were found in the liver with the detection rate 100% and 80%, respectively. DDE was found in all samples. The detection rate of o,p′-DDD, o,p′-DDT and p,p′-DDT was 40%.

In samples of the muscles, β- and γ-HCH were found in 60% of the samples. The α- and δ-isomers were below the detection limits. p,p′-DDE was determined in all studied samples, o,p′-DDE – in 80% of the samples. o,p′-DDD, p,p′-DDD and o,p′-DDT were found in one sample.

To determine the time of pollutants circulation in the ecosystem, the ratios of the initial (maternal) compaunds to their degradation products [DDT/(DDD + DDE) and γ-HCH/(α- + β-HCH)] are used [11]. If the value is greater than 1, this indicates a recent entry of toxicant into the ecosystem. Our calculated coefficients were below 1 in all samples studied, which indicates the absence of fresh contamination.

Polychlorinated biphenyls were found in all samples. ΣPCB levels ranged from 419 to 18 103 ng/g lw. The ranges of total PCB concentrations were: in feathers 898–14 784, in muscles 419–21 202, in liver 444–18 102 ng/g lw, respectively. All studied PCB congeners were identified.

Bioaccumulative processes in birds are directly related to migratory behavior, nutritional characteristics, and life strategy of a particular species. The rhinoceros auklet is a seabird. In the summer season, it mainly nests in the coastal-island territories of the South Primorye and Sakhalin Island. The food preferences of the species are focused on fish: anchovies, herring, sand lances, etc. A secondary role is given to marine invertebrates: squids and crustaceans. The species winters off the coast of the Japanese Islands and in the southern part of the Korean Peninsula [14]. As part of feeding, the rhinoceros auklet dives for prey, so it directly contacts with water. To preserve the condition of their feathers, waterfowl birds preen the feathers with a preen oil produced by the coccygeal gland. It can protect the bird from getting wet and other adverse effects of moisture, and facilitates movement through the water. Due to the presence of such a lipid coating on bird feathers, it is able to accumulate organic contaminants, acting as a "glue" for external contaminants [15]. While considering the concentrations of PCBs in bird feathers, there is a dominance of the levels of highly chlorinated congeners (101, 118, 153, 138 and 180) (Fig. 1), which correlates with our past data on high levels of this group of pollutants in fish from the Sea of Japan [16]. This indicates the priority of PCBs as a toxicant polluting the Sea of Japan, as well as the existence of other local sources of PCBs entering the waters and atmosphere of the Sea of Japan.

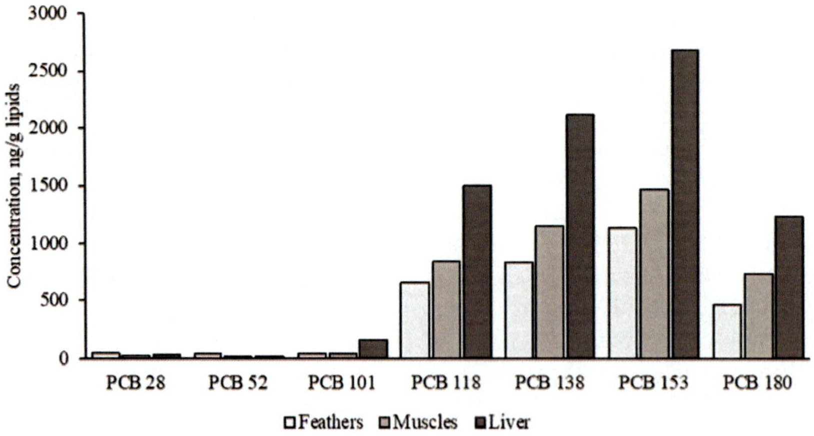

Fig. 1. Concentrations of PCB congeners in the organs of rhinoceros auklet

Among the organs of the rhinoceros auklet, the maximum total concentrations of DDT and PCBs were observed in the liver (Fig. 2), which confirms the detoxication properties of this organ. At the same time, the highest levels of HCH are found in bird feathers. This may be due to both – to the higher concentrations of this pesticide in water (and its accumulation from water), and to high volatility of HCH (and its accumulation from the atmosphere) [17]. In general, feather material can indicate both – internal and external contamination. The first probably occurs during the growth of the feather while it is connected to the bloodstream. Thus, an increase in POPs concentrations in the newly grown feathers of adult birds and nestlings is associated with the levels of pollutants in the blood [18]. There is also a direct correlation between concentrations in muscle tissue and fat. Some studies also observed a correlation between POPs in feathers and liver [19]. Once a feather fully regrows, it is disconnected from the bloodstream, and hence internal contamination has less of an effect on POPs concentrations in feathers [20]. Since adults were used in our study, it should be assumed that internal pollution makes the smallest contribution to the total content of pollutants.

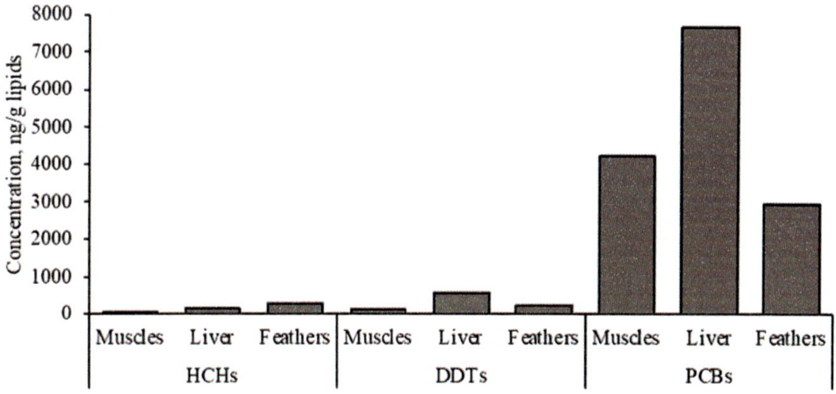

Fig. 2. Total levels of POPs in individual organs

Comparison of average concentrations of POPs with birds of similar ecological groups in the Asia-Pacific Region (Fig. 3) showed that the levels of HCH in the rhinoceros auklet correspond the levels in Vietnam and the Korean Peninsula, but lower than in India, the Lake Baikal, and the Sea of Okhotsk. DDT concentrations in birds of the Sea of Japan are lower than in Korea, Vietnam, India, and continental Russia. In the case of PCBs, the levels in rhinoceros auklet are higher than in Korea and Vietnam, but significantly lower than in India and Lake Baikal [11, 21]. Thus, birds may show regional patterns of POPs distribution.

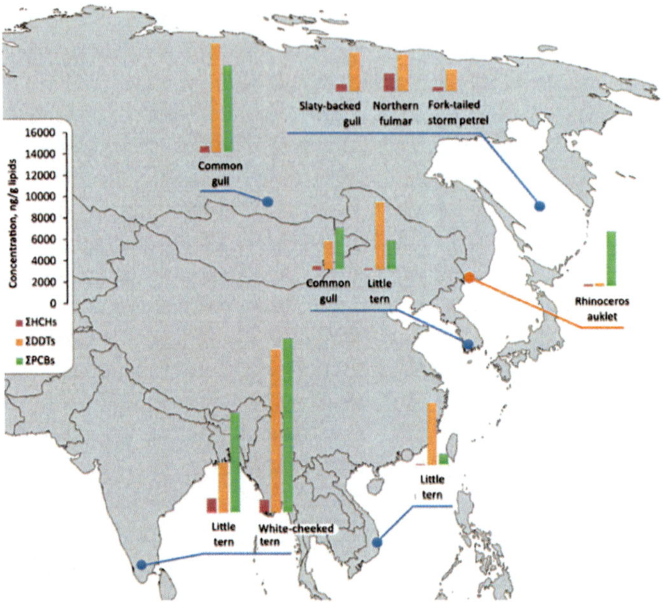

Fig. 3. Regional concentrations of POPs in seabirds

4 Conclusion

Thus, birds are good indicators of the state of the environment and reflect the specifics of the pollution of their habitat. The concentrations of HCH in the rhinoceros auklet from the Sea of Japan probably reflect the background concentrations of this compound in the Asia-Pacific region. For DDT and PCB there are regional specific concentrations that are typical for each study area. In general, the detected concentrations of pollutants reflect the regional specificity of the Sea of Japan in relation to POPs pollution and confirm the priority of PCB monitoring in the ecosystems of the Sea of Japan.

Acknowledgements. The work was supported by the Russian Ministry of Education and Science (project No. FZNS-2023-0011).

References:

1. Hao, Y., et al.: Ecotoxicology of persistent organic pollutants in birds. Environ. Sci.: Processes Impacts **23**(3), 400–416 (2021)
2. Grung, M., et al.: Pesticide levels and environmental risk in aquatic environments in China – A review. Env. Int. **81**, 87–97 (2015)
3. Ding, Y., et al.: Organochlorines in fish from the coastal coral reefs of Weizhou Island, South China sea: Levels, sources, and bioaccumulation. Chemosphere **232**, 1–8 (2019)
4. UNEP: Ridding the World of POPs: a Guide to the Stockholm Convention on Persistent Organic Pollutants, 18 p. UNEP (United Nations Environmental Program), Switzerland, Geneva (2002)

5. Xia, C.: Persistent organic pollutants and trace elements in marine fish from Chinese coastal waters: Doctoral Thesis, 243 p. City University of Hong Kong, Hong Kong (2011)
6. Lakshmanan, D., et al.: Spatial and temporal variation of polychlorinated biphenyls in the Houston Ship Channel. Chemosphere **80**(2), 100–112 (2010)
7. Cui, S., et al.: Levels, congener profile and inventory of polychlorinated biphenyls in sediment from the Songhua River in the vicinity of cement plant, China: a case study. Environ. Sci. Pollut. Res. **23**(16), 15952–15962 (2016)
8. Tanabe, S., Subramanian, A.: Bioindicators of POPs: Monitoring in Developing Countries, 190 p. Kyoto University Press; Trans Pacific Press, Kyoto, Japan, Melbourne (2006)
9. Carson, R.: Silent Spring, 216 p. Houghton Mifflin Harcourt (1962)
10. Furness, R.: Seabirds as monitors of the marine environment. ICES J. Marine Sci. **54**(4), 726–737 (1997)
11. Tsygankov, V.Y., Lukyanova, O.N., Boyarova, M.D.: Organochlorine pesticide accumulation in seabirds and marine mammals from the Northwest Pacific. Mar. Pollut. Bull. **128**, 208–213 (2018)
12. Dunaeva, M.N., et al.: Reconstruction of epizootic outbreak provoked the largescale death of Rhinoceros auklet on the coast of the Japan Sea in the Southern part of Primorsky Krai (July, 2021). Acta Biomedica Scientifica **7**(3), 90–97 (2022)
13. Donets, M.M., et al.: Organochlorine pesticides (OCPs) and polychlorinated biphenyls (PCBs) in Pacific salmon from the Kamchatka Peninsula and Sakhalin Island, Northwest Pacific. Marine Pollut. Bull. **169**, 112498 (2021)
14. Glushchenko, Y.N., Nechaev, V.A., Redkin, Y.A.: Birds of Primorsky Krai: a brief faunistic review, 523 p. Association of Scientific Publications KMK, Moscow (2016)
15. Jaspers, V.L.B., et al.: Evaluation of the usefulness of bird feathers as a non-destructive biomonitoring tool for organic pollutants: a comparative and meta-analytical approach. Env. Int. **33**(3), 328–337 (2007)
16. Donets, M.M., et al.: Flounders as indicators of environmental contamination by persistent organic pollutants and health risk. Marine Pollut. Bull. **164**, 111977 (2021)
17. ATSDR: Toxicological Profile for Hexachlorocyclohexane (HCH), 457 p. U.S. Department of Health and Human Services, ATSDR. Atlanta, GA (2023)
18. Van den Steen, E., et al.: Experimental evaluation of the usefulness of feathers as a non-destructive biomonitor for polychlorinated biphenyls (PCBs) using silastic implants as a novel method of exposure. Env. Int. **33**(2), 257–264 (2007)
19. Eulaers, I., et al.: Legacy and current-use brominated flame retardants in the Barn Owl. Sci. Total Env. **472**, 454–462 (2014)
20. Dauwe, T., et al.: Feathers as a nondestructive biomonitor for persistent organic pollutants. Env. Toxicol. Chem. **24**(2), 442–449 (2005)
21. Kunisue, T., et al.: Accumulation features of persistent organochlorines in resident and migratory birds from Asia. Env. Pollut. **125**(2), 157–172 (2003)

Ichthyofauna of the Karkinitsky Gulf (Black Sea) and Its Changes Under the Anthropogenic Impact

R. E. Belogurova(✉) and E. P. Karpova

A. O. Kovalevsky Institute of Biology of the Southern Seas of RAS, Sevastopol 299011, Russia
prishchepa.raisa@yandex.ru

Abstract. The ichthyofauna of the Karkinitsky Gulf includes 99 species of fish belonging to 42 families, 16 orders and 2 classes. The basis of taxonomic diversity is formed by representatives of the families Gobiidae (16 species), Cyprinidae (8 species), Syngnatidae (7 species), Labridae and Blenniidae (6 species each), Clupeidae (5 species), Mugilidae and (4 species each). In the eastern part of the Karkinitsky Gulf, changes in the composition of the ichthyofauna are determined by the restructuring of the hydrochemical characteristics of the water area associated with the functioning of the North-Crimean Channal, as a result of which the fluctuations in water salinity ranged from 0.55 to 26.80 ‰. Changes in the taxonomic composition, the nature of dominance and localization of fish from different ecological groups affected mainly fish of the freshwater complex and Ponto-Caspian endemics. At present, a fundamental restructuring of historically established ecosystems is taking place in the Karkinitsky Gulf and the process of formation of specific ichthyocenes is underway; species composition and its ecological structure are unstable and mainly depend on human activities. The starting of the Channel in 2022, obviously, will cause a new round of transformations in the ecosystem of the eastern part of the Karkinitsky Gulf.

Keywords: Black Sea · Karkinitsky Gulf · salinity · Ichthyofauna · anthropogenic impact · transformation

1 Introduction

The significance of coastal marine ecosystems for humanity is colossal: they are characterized by high productivity and biodiversity; spawning and feeding of juveniles of many fish species occurs in the thicket biocenoses of coastal areas, most of the world's biological resource extraction comes from coastal zones [1]. At the same time, coastal marine ecosystems are becoming less stable due to increasing anthropogenic impact: this area is one of the most densely populated and exploited worldwide. Economic activity involves a constant influx of pollutants from land into coastal marine ecosystems, eutrophication, construction of hydraulic structures, dumping, mining, overexploitation of bioresources, biodiversity loss, invasion of alien species, and many other negative impacts [2, 3].

T. Chaplina (Ed.): CIWO 2023, SPEES, pp. 370–375, 2023.
https://doi.org/10.1007/978-3-031-47851-2_44

The physico-geographical and biocenotic conditions of the Black Sea coastal zone are quite heterogeneous; they are characterized by a variety of qualitative and quantitative distribution of life; this made it possible to identify five natural regions within the Crimean Peninsula [4]. We focus our survey on one of such areas, the Karkinitsky Gulf, located in the northwestern part of the Black Sea; the gulf has unique physical and geographical, hydrological and hydrochemical, biological, and commercial characteristics. Bakalskaya Spit and Bakalskaya Bank divide the gulf into two subregions, each is specific in its ecological conditions: the western deep-water area has a free water exchange with the open sea and the relatively closed eastern shallow-water part is cut by numerous bays with depths that do not exceed 1 m [2, 5].

Anthropogenic impacts in the Karkinitsky Gulf of the Black Sea are unique in their scale. Until 2014, the ecosystem of the eastern part of the bay was under the desalination effect of the North-Crimean Channal (NCC). The operation and then the termination of it in the Crimea led to radical restructuring of the ichthyofauna of this water area [6]. The developed poaching in the Karkinitsky Bay was one of the reasons for the almost complete disappearance of the sturgeon Acipenseridae. The large-scale fishery for the grass shrimp Palaemon adspersus Rathke, 1837, which is carried out in the seagrass thickets of the bay, annually damages juveniles of commercial fish and protected species [7].

Taking into account the insufficient knowledge of the ichthyofauna of the Karkinitsky Bay, as well as the cardinal changes in its hydrochemical conditions, which led to changes in the composition of the ichthyofauna, the refinement of its modern composition becomes relevant.

2 Material and Methods

The material for the study was collected during expeditionary surveys in the Karkinitsky Gulf from Cape Tarkhankut to Dzharylgach Bay. Samples were collected at 34 stations, as shown at Fig. 1, during the warm seasons of 2008–2014 and 2016–2018, as well as in November 2015.

The fishing gear used for collecting ichthyological material were hand nets with a mesh diameter of 2–5 mm; a towed shrimp landing net with a semicircular inlet (1 m^2) with a hamsaros net (a mesh size of 6.5 mm); gill nets with a mesh size of 12–14 and 18–20 mm, and a shrimp wing net ("venter" or "versha") (a mesh size of 6.5–8.0 mm). The obtained ichthyological material was identified to the species level using identification guides [8]. The systematical data were presented in accordance with the review of [9].

3 Results and Discussion

The modern ichthyofauna of the Karkinitsky Gulf is represented by 99 species of fish, of which 80 species are found in catches. The basis of taxonomic diversity is the Gobiidae family (16 species); the Cyprinidae family is represented by 8 fish species, and the Syngnatidae family by 7 species. The families of Labridae and Blenniidae each comprise 6 species, the Clupeidae family is represented by 5 species. The families of Mugilidae and Gobiesocidae are represented by 4 species, Acipenseridae and Scombridae – 3. Two species of fish were recorded for families Atherinidae, Gasterosteidae, Sparidae and Sciaenidae. The remaining 29 families include one species each.

The dynamics of the composition of the ichthyofauna of the Karkinitsky Gulf is most fully traced by the example of the processes taking place in its eastern part, as shown at Fig. 2. The species composition of fish communities of two time periods was analyzed: the first period (2008–2014) was the active discharge of fresh water from the NCC into the innermost part of the gulf, and the second (2015–2018) was the ending of the NCC in Crimea.

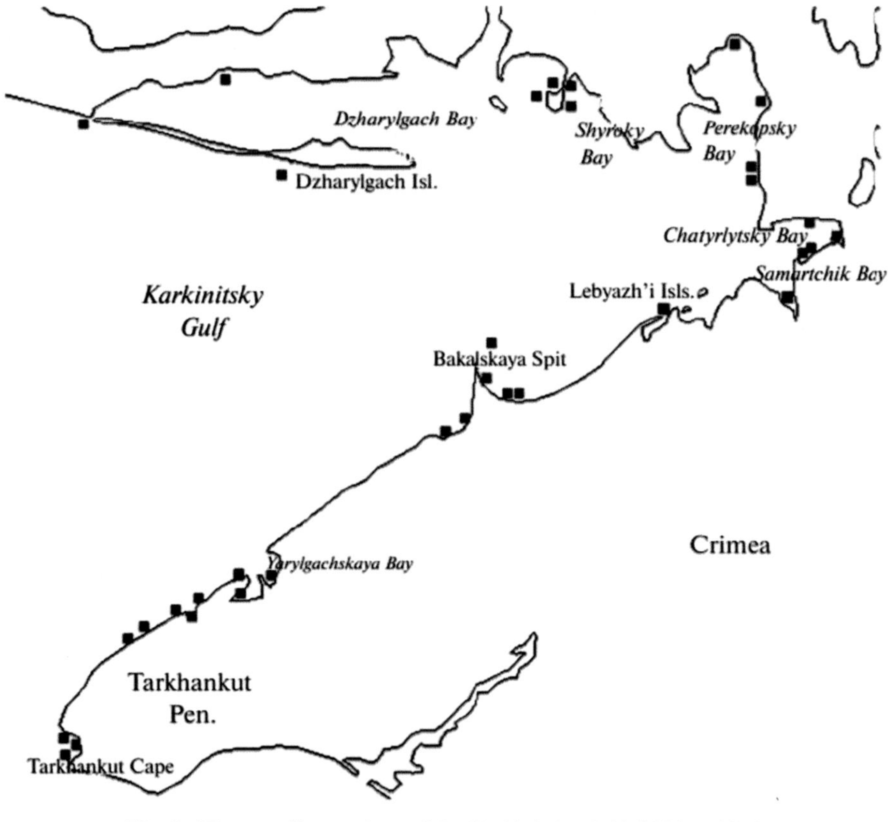

Fig. 1. The sampling stations of the Karkinitsky Gulf (2008 – 2018)

Changes in the ichthyofauna are clearly seen when comparing the proportion of species in catches. Until 2014, in Chartyrlykskу Bay, gobies dominated in abundance and biomass (round goby *Neogobius melanostomus* (Pallas, 1814) and monkey goby *Neogobius fluviatilis* (Pallas, 1814)). Cyprinidae fish (silver Prussian carp and common bleak) were also recorded here. Diverse in species composition, but few in number are other representatives of this family (roach, Amur bitterling, common rudd), as well as pumpkinseed from the Centrarchidae family. In the Samarchik Bay, silver Prussian carp predominated in terms of biomass (more than 50%). After the ending of the NCC in the

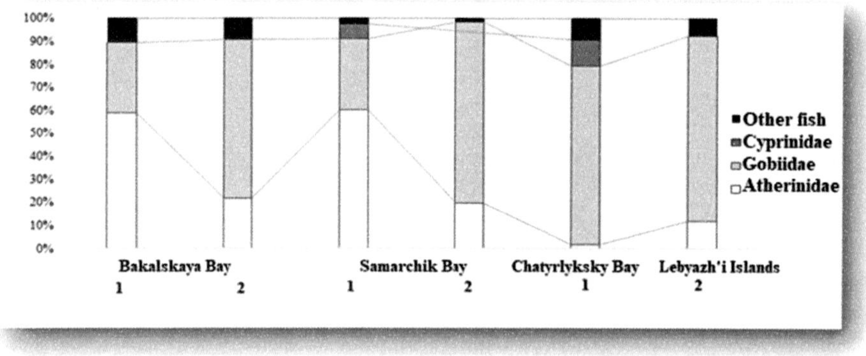

Fig. 2. Relative abundance of the main fish groups in catches in the eastern part of the Karkinitsky Gulf in the period 2008–2014 (1) and 2015–2018 (2)

Crimea, representatives of the freshwater ichthyofauna disappeared in these bays, but the proportion of brackish-water gobies (monkey goby and round goby) is still high. Thus, the composition of the ichthyofauna of the eastern part of the Karkinitsky Gulf in the period 2008–2014 generally remained fairly stable and was characterized by only some spatial changes, as well as the relative abundance of different species in the catches. After 2015, the structure of fish communities in the eastern part of Karkinitsky Bay changed markedly due to the transformation of the hydrochemical regime.

In the Karkinitsky Gulf, vast areas of the bottom in the coastal zone are occupied by thickets of sea grasses, where active spawning, growth and feeding of various fish species take place. Figure 3 presents the most complete dynamics of the diversity of fish communities in the seagrass biocenosis of the eastern part of the Karkinitsky Bay for Bakalskaya Bay.

In 2015–2018 the proportion of grass goby significantly ($U = 53.0$ at a significance level of $p < 0.05$, $N_1 = 20$, $N_2 = 17$) increased from 9.57% to 33.01% in number and from 20.54% to 50% by weight compared to 2012–2014 catches. However, in 2015–2018 the proportion of the Big-scale sand smelt *Atherina boyeri* (Eichwald, 1831) decreased. There was also a significant increase at $p < 0.05$ in the proportion of the black goby *Gobius niger* Linnaeus, 1758 from 5.16% to 7.71% by number ($U = 28.5$, $N\ 1 = 13$, $N\ 2 = 12$). There is a statistically significant decrease in the proportion of Ponto-Caspian gobies: round goby ($U = 29.5$ at a significance level of $p < 0.05$, $N\ 1 = 19$, $N\ 2 = 17$) and tubenose goby *Proterorhinus marmoratus* (Pallas, 1814). Among other fish species, an increase in catches of representatives of the Mediterranean faunistic complex, the Black Sea horse mackerel and mullet *Mullus barbatus ponticus* Essipov, 1927, was noted.

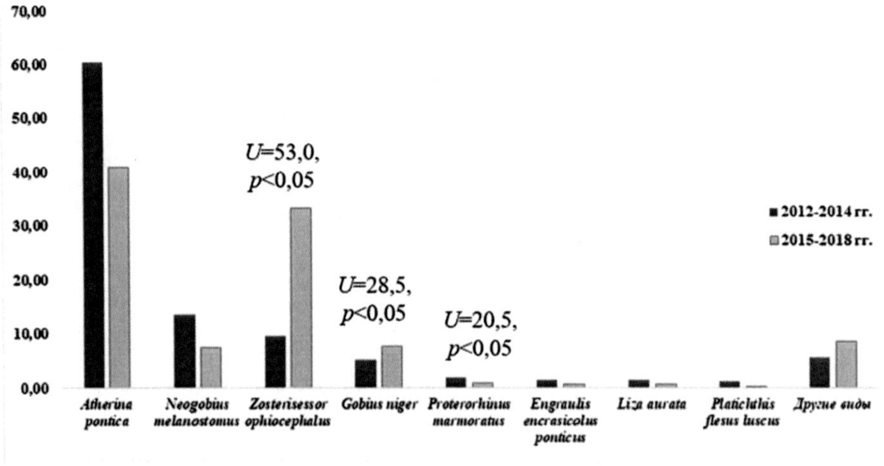

Fig. 3. Dynamics of the composition of fish communities in the sea grass biocenosis of the Bakalskaya Bay by abundance

4 Conclusion

Thus, among other regions of the Azov–Black Sea basin, the fish fauna of the Karkinitsky Gulf is remarkable because of its high species richness; it is only slightly inferior in the species number in comparison to the coastal zone of the southwestern Crimea [10]. This richness of the fish fauna depends on the wide variety of biocenotic conditions, including those formed under the influence of an anthropogenic factor, that is, redistribution of the Dnieper runoff through the North Crimean Canal.

In the western part of the Karkinitsky Gulf, long-term changes in the composition of the ichthyofauna are associated with the general processes for the Black Sea caused by overfishing, the difficulty of fish migration through the Bosporus and the general decline in abundance within the range. Currently, 5 previously encountered species (swordfish, John Dory, bluefin tuna, bastard sturgeon and Atlantic sturgeon) are not found in Karkinitsky Bay.

Changes were revealed in the structural characteristics of fish communities in the biocenosis of sea grasses in the eastern part of the Karkinitsky Gulf: in the catches, the share of marine fish species by origin increased – grass goby (almost 3 times in number) and black goby (from 5.16% to 7.71% in terms of abundance), the proportion of brackish-water round goby decreased by 2 times.

Favorable conditions in the eastern part of the Karkinitsky Gulf have developed for the habitat of the Red Book representatives of the Syngnatidae family, however, large-scale fishing for grass shrimp in thickets of sea grasses can damage the populations of these species. Spontaneous recreational activity, developed mainly in the region of the Bakalskaya Spit, poses a threat to its natural complexes, leads to the destruction of biotopes and the oppression of the populations of protected and valuable fish species.

At present, a fundamental restructuring of historically established ecosystems is taking place in the Karkinitsky Gulf and the process of formation of specific ichthyocenes is underway; species composition and its ecological structure are unstable and mainly depend on human activities. The starting of the Channel in 2022, obviously, will cause a new round of transformations in the ecosystem of the eastern part of the Karkinitsky Gulf.

Acknowledgements. The study was carried out in accordance with the Patterns of formation and anthropogenic transformation of biodiversity and biological resources of the Azov–Black Sea basin and other regions of the World Ocean State Research Project, State Registration Number 121030100028-0; as well as on the Fundamental research of population biology of marine animals, their morphological and genetic diversity Research Topic, State Registration Number 121040500247-0.

References

1. Pauly, D., Alder, J., Booth, S., et al.: Fisheries in large marine ecosystems: descriptions and diagnoses. In: Sherman, K., Hempel, G. (eds.) The UNEP Large Marine Ecosystems Report: A Perspective on Changing Conditions in LMEs of the World's Regional Seas, pp. 23–40. United Nations Environment Programme, Nairobi (2008)
2. Eremeev, V.N., Boltachev, A.R., Aleksandrov, B.G., et al.: Biological diversity of the coastal zone of the Crimean Peninsula: problems, preservation and restoration pathways, 92 p. NAS Ukraine, A. O. Kovalevsky Institute of Biology of the Southern Seas. Sevastopol (2012)
3. Karpova, E.P., Boltachev, A.R., Ablyazov, E.R., Prishchepa, R.E.: Fish communities of the Karkinitsky Gulf. In: Proceedings of the All-Russ. Conf. "Marine Biological Research: Achievements and Prospects", Sevastopol, 19–24 Sep 2016, vol. 2, pp. 86–89. EKOSI-Gidrofizika, Sevastopol (2016)
4. Vodyanitskii, V.A.: On the natural historical zoning of the Black Sea and, in particular, off the Crimea coast. In: Proceedings of the Sevastopol Biological Station, vol. VII, pp. 249–255. Akad. Nauk SSSR, Moscow (1949)
5. Boltachev, A.R., Karpova, E.P.: Marine Fishes of the Crimean Peninsula, 2nd edn., 376 p. Biznes-Inform, Simferopol (2017)
6. Belogurova, R.E., Karpova, E.P., Ablyazov, E.R.: Long-term changes in the fish fauna of the Karkinitsky gulf of the Black Sea. Russian J. Mar. Biol. **46**(6), 452–460 (2020). https://doi.org/10.1134/S1063074020060036
7. Boltachev, A.R., Statkevich, S.V., Karpova, E.P., Khutorenko, I.V.: Black Sea grass prawn *Palaemon adspersus* (Decapoda, Palaemonidae): biology, fisheries and problems. Vopr. Rybolov. **18**(3), 313–327 (2017)
8. Vasil'eva, E.D.: Fishes of the Black Sea: A Key to Marine, Brackish-Water, Euryhaline, and Anadromous Species, with Color Illustrations, collected by S.V. Bogorodsky. VNIRO, Moscow (2007)
9. Parin, N.V., Evseenko, S.A., Vasil'eva, E.D.: Fishes of Russian Seas: Annotated Catalogue. KMK, Moscow (2014)
10. Boltachev, A.R., Karpova, E.P.: The ichthyofauna of the Sevastopol coastal zone (the Black Sea). Morsk. Ekol. Zh. **11**(2), 10–27 (2012)

Comparative Analysis of Contamination with Microplastic Particles of Ice in the Curonian Lagoon (Baltic Sea) and Novik Bay (Sea of Japan)

I. Yu. Bocherikova[1,2](✉), O. I. Lobchuk[1], and I. P. Chubarenko[1]

[1] Shirshov Institute of Oceanology, Russian Academy of Sciences, Moscow 117997, Russia
ibocherikova@yandex.ru

[2] Immanuel Kant Baltic Federal University, Kaliningrad 236016, Russia

Abstract. In this work, ice cores from the Novik Bay (Sea of Japan) and the Curonian Lagoon (Baltic Sea, Kaliningrad region) are examined for contamination with microplastics (0.3–5 mm). The method of sampling and processing samples in a similar. The average concentration of MP particles (0.3–5 mm) in the ice of Novik Bay was 204.3 items/L of meltwater, and in the Curonian Lagoon – 60.3 items/L of meltwater. The number of fibers, fragments, and films somewhat increased with decreasing particle size.

Keywords: microplastic · ice · microplastic distribution

1 Introduction

In recent years, plastic production has exceeded 390 million tons [1], while, according to available data, about 3% of the plastic produced ends up in the ocean [2]. Based on the already available data, it can be argued that the spread of microplastics (MP) in the oceans occurs everywhere, including the Arctic [3]. There is more and more information indicating that sea ice contains several orders of magnitude more plastic particles than seawater [4]. Thus, ice can be a temporary storage medium and carrier of plastic particles [5].

2 Materials and Methods

Two field studies were carried out on sea ice contamination with microplastic particles: in the Novik Bay, the Sea of Japan, in Primorsky Krai [6] and in the Curonian Lagoon, the Baltic Sea, in the Kaliningrad Region [7]. To study the concentration of MP in the Novik Bay, 2 ice cores were taken, and 5 ice cores were taken in the Curonian Lagoon. Working methods for the both aquatories were similar. In the Curonian Lagoon, the cores were cut with a PI-8 manual ice corer (manufactured by Meteopribor LLC) with an internal and external diameter of 17 cm and 22 cm, respectively. The cut-out cores

T. Chaplina (Ed.): CIWO 2023, SPEES, pp. 376–380, 2023.
https://doi.org/10.1007/978-3-031-47851-2_45

had a diameter of 15–16 cm; the cores were not always of the same length. The ice cores for MP analysis were taken at a distance of about 100 m from one another in the center of the bay. When the core was taken out of the ice cover (see Fig. 1 for the core example), its length (ice thickness) was measured [7]. Then the cores were wrapped with aluminum foil, packed into a new labeled plastic bag, put into a plastic container, and transported to the laboratory for storage in a refrigerator (−18 °C) [6]. A clean Petri dish with a new filter was exposed nearby on the surface of the ice cover to control possible external contamination. Further work with MPs ice cores was carried out in a clean chemical laboratory and included their layer-by-layer fragmentation, followed by the extraction of inclusions (sediments, MPs particles, etc.) [7].

Fig. 1. Ice core from the Curonian Lagoon

The cores were segmented on a special stand installed in a fume hood, and starting from the lower ice layer in laboratory conditions, the cores were cut into fragments 4–5 cm thick. The pieces of ice were melted at room temperature during the day. Melt water was poured through filters using a vacuum filtration device (PVF) with three funnels. Ice slices were melted at room temperature for a day. Portions of melt water were passed through a filter unit [7]. Then, each filter (diameter 47 mm, pores 0.45 μm or 100 μm) with a precipitate was separated with tweezers from the elementary cell stone of the filter and placed in a Petri dish (diameter 60 mm). Filters were dried for 2–3 days. Petri dishes with dried filters were covered with a lid, sealed and labeled [6]. After filtration, the filters were packed in sterile Petri dishes and labelled. The volume of melted water was measured with a chemical measuring cup with an accuracy of 10 ml. The filters were viewed on a stereoscopic digital microscope, Micromed MS-2-ZOM Digital, with tenfold magnification. MP (particles in the size range of 0.3–5 mm) were detected in all layers of ice. No clear patterns of MP distribution were found.

There was a lot of organic matter on the filters from the Curonian Lagoon; to remove the remnants of organic matter and biota, the filters were oxidized. The filter was placed in a beaker, hydrogen peroxide H2O2 (30%), distilled water, and catalyst Fe (II) were added, and all this was kept for 24 h on a water bath at a temperature of 75 °C. After oxidation, HCl solution was added to the sample to remove calcite fractions, and the sample with a filter was left to settle for about a day [8]. On the next day, the cooled solution from the glass was poured through the PVF onto the same sample filters with which it was oxidized. After filtration, the filters were removed and placed in dry, labelled Petri dishes.

3 Results and Discussion

Identification of the polymeric composition (chemical features, polymer type) of the microparticles (n = 80, or 4% of putative MPs particles) was performed using μ-Raman spectroscopy. Raman spectra were obtained using a Raman Centaur U (LTD «NanoScanTechnology», Russia) spectrometer with spectral resolution 2.5 cm^{-1}. The details are provided [6]. In total, 23 out of 72 particles (32%) from the Novik bay appeared to be non-plastics, while, for samples from the Curonian Lagoon – 25 out of 80 particles appeared to be plastics particles.

4 Novik Bay

Identification of large MPs particles (0.3–5 mm) in the Ice Core 1 from the Novik Bay was carried out on 7 out of 8 filters [6]. The upper filter was not included in the analysis due to heavy load with (presumably) coal particles – widely reported contaminations in the region [9]. For μ-Raman verification, 80 particles (about 4%) were selected. All types of dyes, non-removable chemical compounds of an anthropogenic nature, and cellulose particles were excluded from further analysis. As a result, 49 items (68%) were confirmed to be MPs. Of these, PS, PET and PA/PAA made about 19% each, other plastic types (PDMS, PVAC, PF, PTFE, NC,) made together about 22%. This way, 68% (out of 1932 particles identified as putative MPs) were confirmed by μ-Raman spectrometry as MPs particles, resulting in 1325 MPs items. MPs (0.3–5 mm) were found in all the ice layers [6].

Six filters of the Ice Core were analyzed for the content of MPs particles (25–300 μm) on ice layers [6]. The upper filters 1 and 2 were excluded in the analysis due to heavy load with a black. The most abundant polymers were presented by polyolefins (51.8%). And polyesters (35.7%). Fragments were composed of polyolefins (62%), whereas fibers were composed of polyesters (73%) and polyamides (14%). Considering particles size distribution along the ice core, we observed that most of the smallest size particles accumulated in the external layers, whereas the biggest particles accumulated in the central layers [6] (Fig. 2).

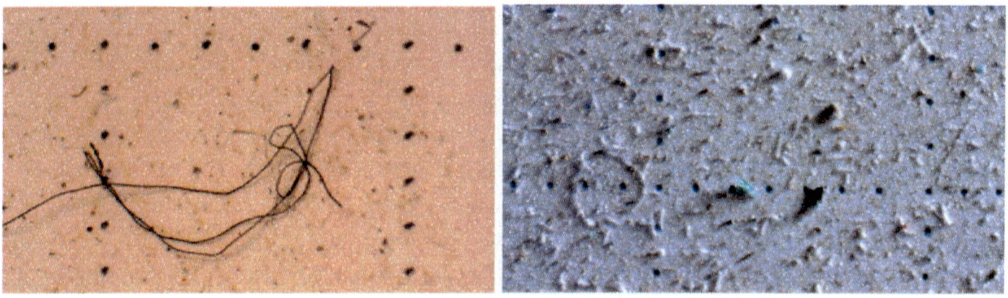

Fig. 2. Examples of particles found on filters from the Novik Bay

5 Curonian Lagoon

A total of 27 filters were obtained from 5 ice cores, in which 1322 MPs particles were found [7]. The average concentration of MPs (0.3–5 mm) in the core was 67.9 items/L. More than half of the detected fibers were transparent (53%); slightly less were blue (43%), followed by green, yellow, and red (about 1% each color). The average concentration of MP in the upper parts of the cores (2–5 cm) is 188.6 items/L; in the middle and lower parts, it is 64.7 and 74.2 items/L, respectively. In all cores, except for one taken at one station, the highest concentrations of MPs were noted in the upper part, the lowest in separate layers in the middle of the core, and a slight increase in concentration in the lower part (Fig. 3).

Fig. 3. Example of particles found on filters from the Curonian Lagoon

6 Conclusion

The number of filaments, fragments, and films somewhat increased with decreasing particle size. The average concentration of MP particles (0.3–5 mm) in the ice of Novik Bay was 204.3 items/L of meltwater [6], and in the Curonian Lagoon, it was 60.3 items/L of meltwater [7]. This is an order of magnitude larger than that obtained in the Chukchi Sea ice (20 μm–5 mm [5]), and several orders of magnitude larger than that in the water of open coastal zones in the region [10] The measurement of the salinity of melted and subglacial waters showed that the ice and waters of the Curonian Lagoon are practically fresh, while in the Novik Bay there is a sea ice (the subglacial waters had a salinity of 34 psu, the core meltwater 3–9 psu). Sea ice is formed by intense vertical (thermal and haline) convection, unlike freshwater ice, which forms on the surface of a stably stratified water column. Although theoretically, on this basis, it can be assumed that sea ice should be more contaminated than fresh water (as in our case), the available data are not yet sufficient to confirm this. Despite the fact that both bays are semi-closed and receive a comparable volume of wastewater from surrounding settlements, the overall level of contamination in their waters may vary significantly.

Acknowledgements. Investigations are financially supported by RSF via grant No 19-17-0041. Sincere thanks are expressed to Alexander Lazaryuk for ice core sampling in the Novik Bay, to Sergey Fetisov and Alexey Grave for field sampling in the Curonian Lagoon, and to Andrey Zubin for μ-Raman spectroscopy.

References

1. Plastics – the Facts. https://plasticseurope.org/knowledge-hub/plastics-the-facts-2022/ (2022)
2. Jambeck, J.R., et al.: Plastic waste inputs from land into the ocean. Science **347**, 768–771 (2015). https://doi.org/10.1126/science.1260352
3. Kanhai, L.D.K., Gardfeldt, K., Lyashevska, O., Hassellov, M., Thompson, R.C., O'Connor, I.: Microplastics in sub-surface waters of the Arctic Central Basin. Marine Pollut. Bull. **130**, 8–18 (2018). https://doi.org/10.1016/j.marpolbul.2018.03.011
4. Peeken, I., et al.: Arctic Sea ice is an important temporal sink and means of transport for microplastic. Nat. Commun. **9**(1), 1505 (2018). https://doi.org/10.1038/s41467-018-03825-5
5. Kim, S.K., et al.: Importance of seasonal sea ice in the western Arctic Ocean to the Arctic and global microplastic budgets. J. Hazard. Mater. **418**, 125971 (2021). https://doi.org/10.1016/j.jhazmat.2021.125971
6. Chubarenko, I., et al.: Microplastics in the first-year sea ice of the Novik Bay, Sea of Japan. Marine Pollut. Bull. **185**, 114236 (2022). https://doi.org/10.1016/j.marpolbul.2022.114236
7. Bocherikova, I.Y., Chubarenko, I.P.: Microplastics content in ice, snow and under-ice water of the Curonian Lagoon in winter 2021. J. Oceanol. Res. **50**(3), 102–117 (2022). https://doi.org/10.29006/1564-2291.JOR-2022.50(3).6
8. Zobkov, M.B., Esiukova, E.E.: Microplastics in the Marine Environment: Review of Methods of Selection, Preparation and Analysis of Samples of Water, Sediments and Coastal Sediments. Oceanology **58**(1), 149–157 (2018). https://doi.org/10.7868/S0030157418010148
9. Lebedev, A.A., et al. Coal terminal impact on marine suspention composition: Nakhodka Gulf (Japan Sea). In: Proceedings of the Russian State Hydrometeorological University, vol. 48, pp. 195–201 (2017)
10. Kojima, F., et al.: Microplastic abundance in river runoff in the NOWPAP region. In: Regional Overview, 80 p. PGI FEB RAS, Vladivostok (2022)

Chlorinated Organic Compounds in Mytilidae Mollusks from Nakhodka Bay (the Sea of Japan)

A. D. Borovkova(✉) and V. Yu. Tsygankov

Far Eastern Federal University, Vladivostok 690091, Russia
borovkova.adm@students.dvfu.ru

Abstract. Persistent organic pollutants (POPs) are considered to be one of the most hazardous and underinvestigated organic compound class in the context of environmental and human health effects. POPs characteristics incorporate high toxicity, ability to bioaccumulate and convey through food chains. Due to transferring within air and water masses, POPs are able to relocate and deposit accumulating in aquatic and terrestrial ecosystems outermost the source of the pollution. Despite the Stockholm Convention signing in 2001 by a number of countries, some states still retain the right to use DDT for the purposes of sanitary and epidemiological control. This study presents the current levels of POPs content in the soft tissues of bivalves from Nakhodka Bay (the Sea of Japan).

Keywords: Bivalves · Mytilidae · Persistent organic pollutants · POPs · Organochlorine pesticides · OCPs · Dichlorodiphenyltrichloroethane · DDT and its metabolites · Hexachlorocyclohexanes · HCH isomers · Polychlorinated biphenyls · PCBs

1 Introduction

At the time being, each natural ecosystem is abandoned to human intervention. Increasingly growing the World Ocean's contamination is one of the essential and urgent issues. Continental shelf areas are subjected to the greatest load associated with a vast number of terrigenous runoff, discharge of bilge water and intensive seafaring [1].

Persistent organic pollutants (POPs) constitute one of the most understudied and harmful chemical substances. In addition, POPs are highly toxic and feature the ability to store in living organisms' adipose tissues. Likewise, the compounds extremely slowly disintegrate in natural environment. In response to transboundary transport with the aid of water and air masses, toxicants are able to roam and precipitate accumulating in aquatic and terrestrial ecosystems outermost the source of the pollution. Therefore, the majority of these substances negatively affect the pond life and human health [2].

Principal difference between POPs and pollutants that are constantly presented in the natural environment and manifest their toxicity only when ambient concentrations are exceeded, lies in their origin – these substances are artificially created compounds introduced into the biosphere by humans and alien to it. As a result, in such a short period of time (since the 60s of the last century), biological beings of all levels of life

T. Chaplina (Ed.): CIWO 2023, SPEES, pp. 381–386, 2023.
https://doi.org/10.1007/978-3-031-47851-2_46

did not have the opportunity to develop adaptations to previously unknown pollutants [3]. POPs exhibit carcinogenic, cytotoxic, teratogenic and genotoxic properties, and, in addition, have a negative effect on the nervous and cardiovascular systems [4, 5].

In general POPs can be divided into two types: intentionally and unintentionally POPs. First group contains compounds producing as wanted products by different chemical reactions that include chlorine (organochlorine pesticides (OCPs)), industrial chemicals). Unintentionally POPs are produced as unwanted by-products of combustion or chemical process that take places in the presence of chlorine compounds (polycyclic aromatic hydrocarbons (PAHs), dioxin and furan compounds) [6].

It is reported that the most commonly used POPs are dichlorodiphenyltrichloroethane (DDT) and hexachlorocyclohexane (HCH). At the present time, the majority of countries have forbidden production and utilization of these pesticides. However, some states, such as China and India, still retain the right to produce and use DDT, which leads to contamination of the territories not only of the users themselves, but also of their neighbors.

As of this date, there are a quantity of studies devoted to determining the levels of chlorinated hydrocarbons in different areas of the World Ocean. To the contrary, Russian Far-Eastern Seas' contamination issue is not elucidated enough and therefor is currently central. This investigation aims to scrutinize the present content of POPs in bivalves' soft tissues from Nakhodka Bay (the Sea of Japan).

2 Material and Methods

In our investigation we used three species of bivalves from Mytilidae family (*Crenomytilus grayanus*, *Mytilus trossulus* and *Modiolus modiolus*) as bioindicators. Mollusks samples were collected at five stations from different sites of Nakhodka Bay in the summer of 2018 (Fig. 1). Five individuals of *M. trossulus* were deprived from the Nakhodka Gulf, nine – from the Cape Kozmina. Six *C. grayanus* mollusks were sampled from Kozmina Bay, and twelve *M. modiolus* bivalves were collected from Musatova Bay and the Cape Krasny.

Mollusks' soft tissues were analyzed for levels of OCPs and PCBs. Lipids were extracted from homogenate (10–20 g) using a n-hexane: acetone mixture, with subsequent disintegration of fat components with concentrated sulphuric acid [7]. The obtained POPs extract was separated with non-polar (for PCB) and polar (for OCP) solvents on a chromatographic column with Florisil®. Concentrations of organochlorine pesticides (OCPs) (α-HCH, β-HCH, γ-HCH, δ-HCH, o,p′-DDT, p,p′-DDT, o,p′-DDD, p,p′-DDD, o,p′-DDE, and p,p′-DDE) and polychlorinated biphenyls (PCBs) (congeners 28, 52, 155, 101, 118, 143, 153, 138, 180, 207) in samples were measured by gas chromatography–mass spectrometry (GC–MS), and gas-chromatography with electron capture detector (GC–ECD) for validation of results.

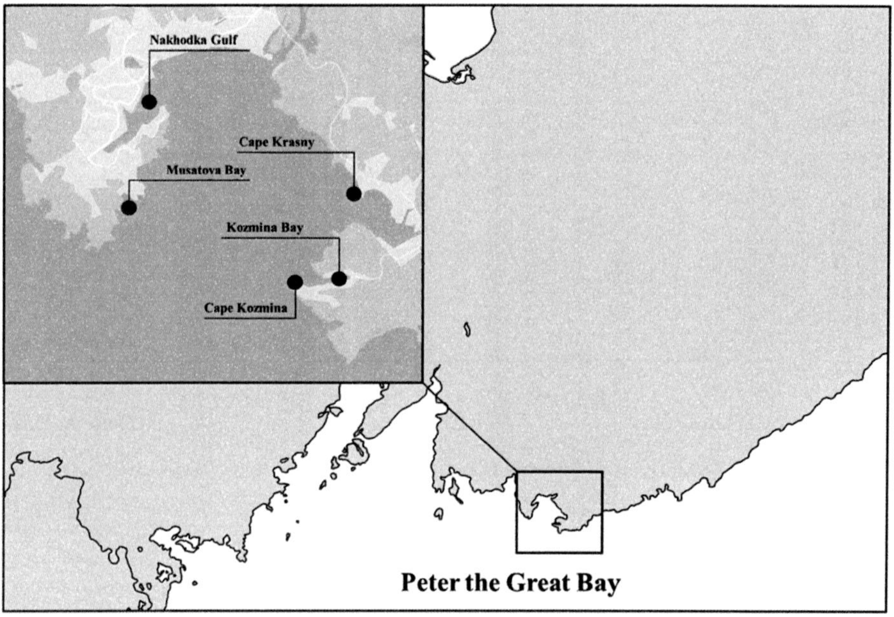

Fig. 1. Map sampling location

3 Results and Discussion

The result showed the presence of persistent organic pollutants ($\sum$OCPs + $\sum$PCBs) in all soft tissue samples with the range from 6.4 to 2769.7 ng/g lipid weight. Organochlorine pesticides prevailed over PCBs. The highest POPs levels were elicited in *M. modiolus* from the Cape Krasny in an amount of 7.4 to 2769.7 ng/g lipid weight. The lowest concentrations were in *M. trossulus* from the Nakhodka Gulf (mean 32.53 $\pm$ 11.99 ng/g lipid weight). The distribution of toxicants between studied areas is showed in Fig. 2.

Among OCPs DDT group dominated in all samples. Preponderate metabolite – DDE, showing the disintegration of the original compound and, in addition, chronic toxicant residence in the environment. The highest OCP content was recorded in mollusks' soft tissues sampled from the Cape Krasny, among which DDT group prevailed. The predominant metabolite was p,p′-DDE, found in the concentration range from 703.94 to 2769.7 ng/g lipid weight. Not far from the Cape Krasny, the Partizanskaya River flows into Nakhodka Bay, carrying runoff from fields where pesticides were actively used for agricultural crops treatment in the latter half of the 20th century. This circumstances might have led to basin contamination with forms that characterize the long-standing influx of pollutants into the water area.

Between HCH isomers α-, β- and δ-forms were detected with a predominance of β-HCH, which is considered to be the most stable isomer and indicate a long-term intake of pollutants. The presence of forms, demonstrating recent contamination, might be implicated by transboundary transport within air and water masses from territories where pesticides still are in the use.

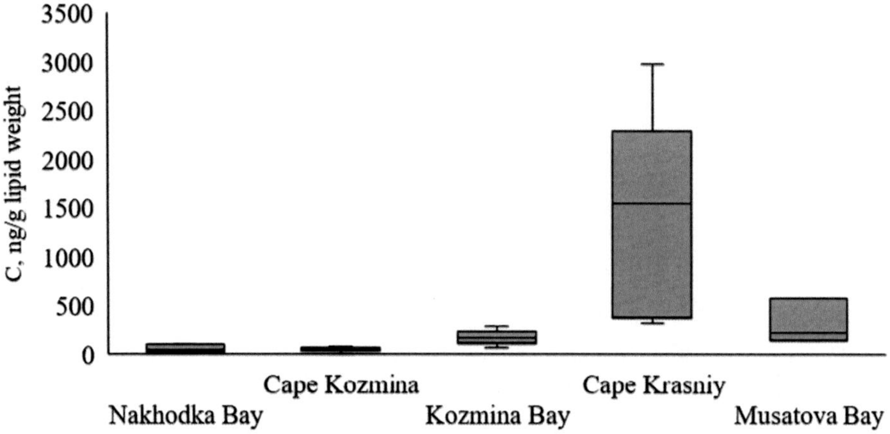

Fig. 2. Mean levels of ∑POPs in mollusks' soft tissues from studied areas

In regard to PCBs in all samples 28 congener was recorded with the range from 7.7 to 314.1 ng/g lipid weight. Highly chlorinated 153 PCB was found on one occasion only in mollusk's soft tissue from Musatova Bay in an amount of 11.5 ng/g lipid weight. Samples from Kozmina Bay showed the upmost PCB concentrations among studied areas within the limits from 68.0 to 151.2 ng/g lipid weight. In the 1990s, the bay was applicable as a marine vessels dismantling center and an airfield for hydroplanes. Furthermore, an oil loading terminal has been placed in commission since 2009. Presently the water area is the end point of the East Siberia – Pacific Ocean oil pipeline. Likewise, the bay is located in the zone of active navigation, given that within the precincts of Wrangel Bay and Kozmina Bay one of the largest ports of Primorsky Krai, Vostochny Port, is located. These factors might have resulted in basil contaminating with PCBs.

Despite of relevance and globality of persistent organic pollutants contamination issue, there are not so many studies devoted to estimating chlorinated organic compounds levels in aquatic ecosystems in Russia. Since POPs are characterized by high toxicity and resistance to natural factors, it is important to regularly monitor the content of toxicants in the water area and commercial species of hydrobionts.

In Peter the Great Bay, analysis of OCPs content in mollusks soft tissue was carried out until the year of 2004. The data from various researches dedicated to pesticides content determination in the soft tissues of mollusks of the family Mytilidae are presented in Table 1. Due to the fact that the results of previous studies are presented as ng/g ww, our results were also converted to ng/g ww.

Data comparing the content of pesticides from 2002 to 2018 demonstrate a decrease in toxicants levels in the waters of Peter the Great Bay. In all studies in the soft tissues of mollusks, the predominance of the concentrations of the decomposition products of the initial compounds (DDE and DDD, β-HCH) was recorded. The results characterize a gradual reduction in the release of pollutants into the environment, both with river runoff from the fields of Primorsky Krai, where pesticides were previously used, and through transboundary transport by water and air masses from areas where OCPs are still in use. Before the signing of the Stockholm Convention in 2001 by a number of countries, the

Table 1. Regional studies of OCPs in bivalves (ng/g ww)

Specie	Area	Data	∑OCPs	∑HCH	∑DDT	References
Mezuhopecten yessoensis	Posyet Bay	2002	395	395	–	[8]
			616	170	446	
Crenomytilus grayanus	Reinecke Island	2004	25.8	–	–	[9]
	Amurskiy Bay		48	30	22.9	
	Ussuriyskiy Bay		22	–	–	
	Nakhodka Bay	2018	0.15	0.06	0.19	Current study
Modiolus modiolus			0.91	0.12	1.56	
Mytilus trossulus			0.12	0.10	0.14	

production and usage of organochlorine pesticides was not limited, which might have caused the detection of higher levels of DDT in 2002 compared to 2017.

Analysis of the content of polychlorinated biphenyls in the soft tissues of bivalve mollusks from Peter the Great Bay was held in the year of 1999 [10]. The results of the study showed the presence of PCBs in all *C. grayanus* individuals in the range from 2700 to 3700 ng/g lipid weight. The data obtained during the work in 2017 indicate a decrease in PCBs concentrations. Nevertheless, despite the absence of a developed industrial complex on the territory of Primorsky Krai, mollusks' soft tissues are still liable to contamination with low-chlorinated congeners. Such results could be associated with the impossibility of full control of PCBs release into the environment and disposal, since, according to the Stockholm Convention, polychlorinated biphenyls belong to the category of unintentionally produced compounds.

4 Conclusion

Thus, received data demonstrate the continuing contamination of Nakhodka Bay water area with persistent organic pollutants. In Russia production and utilization of OCPs are prohibited, however, before Stockholm Convention signing in 2002, croppers from Primorsky Krai exploited pesticides in agriculture, which explains the presence of toxicants' products of degradation. The entry of forms in small concentrations, indicating recent basil contamination, might be associated with the phenomenon of transboundary transport of pollutants within water and air masses from areas where pesticides continue to be used for agricultural and medical purposes. Moreover, a temporal trends analyses of POPs content in the waters of Peter the Great Bay showed a decrease in pollutants' concentrations since 1999. The phasing out of the use of OCPs in a growing number of countries and measures to prevent unintentional production of PCBs have resulted in significant reductions in POPs levels in mollusks soft tissues. Despite this, it is necessary to continue monitoring studies of the input and transformation of chlorinated hydrocarbons in the coastal regions of the Russian Far Eastern seas.

References

1. Boyarova, M.D.: Current levels of organochlorine pesticides in aquatic organisms from peter the great bay (Japanese Sea) and lake khanka: abstract, dissertation of candidate of biological science, p. 24. Far East Federal University, Vladivostok (2008)
2. Tsygankov, V.Yu. et al.: Persistent organic pollutants (POPs) in the Far East region: seas, organisms, people: monograph, p. 344. Far East Federal University, Vladivostok (2020)
3. Walker, C.H.: Organic pollutants. An ecotoxicological perspective. p. 258. Talor & Frensis Inc., C. H. Walker (2001)
4. Kaiser, J., Enserink, M.: Treaty takes a POP at the dirty zone. Science **290**, 2053 (2000)
5. Carson, R.: Silent spring. Houghton Mifflin Harcourt, 216 (1962)
6. El-Shahawi, M.S., Hamzaa, A., Bashammakh, A.S., Al-Saggaf, W.T.: An overview on the accumulation, distribution, transformations, toxicity and analytical methods for the monitoring of persistent organic pollutants. Talanta, 1587–1597 (2009)
7. Tsygankov, V.Yu.: Sample preparation method for the determination of organochlorine pesticides in aquatic organisms by gas chromatography. V.Yu. Tsygankov, M.D. Boyarova. – Achievements in the Life Sciences, 4 (2015)
8. Boyarova, M.D., Lukyanova, O.N.: Chlorinated hydrocarbons in aquatic organisms from Posyet Bay the Sea of Japan. Izvestiya TINRO **145**, 271–278 (2006)
9. Lukyanova, O.N., Boyarova, M.D., Tsygankov, V.Yu.: Persistent organic pollutants in mollusks and fish from the Sea of Japan and the Sea of Okhotsk. Sevastopol: ECOSY-Hydrophysics **3**, 136–139 (2016)
10. Tkalin, A.V., Lyshavskaya, T.S., Hills, J.W.: Organochlorine pesticides in mussels and bottom sediments from Peter the Great Bay near Vladivostok. Ocean Reserch **19**(2), 115–119 (1997)

Individual Components of Microbiological Community in Water Northeast Part of the Kara Sea

V. Yu. Budarova[1](✉), E. A. Tikhonova[2], N. V. Burdiyan[2], and Yu. V. Doroshenko[2]

[1] Astrakhan State University named after V.N. Tatishchev, Astrakhan, Russia
prosto-valerochka@mail.ru

[2] A.O. Kovalevsky Institute of Biology of the Southern Seas, Russian Academy of Sciences, Sevastopol, Russia

Abstract. The paper presents the results of microbiological studies of the coastal waters of the northeast Kara Sea in summer under low temperature conditions. In the surface water layer, the abundance of bacteria using protein substrate as a single source of carbon and energy varied from 1 to 10^6 cell/mL (from 10^2 to 10^5 cell/mL in the bottom layer), that of bacteria using carbohydrates (glucose) varied from 10 to 10^4 cell/mL (from 10^3 to 10^5 cell/mL in the bottom layer), and that of bacteria using lipids ranged from 1 to 100 cell/mL in both layers. Hydrocarbon-oxidizing bacteria were found in 53% of samples collected in the surface layer and in 36% of samples collected in the bottom layer. Their counts were from 1 to 10 cell/mL in both the surface and bottom layers. An uneven distribution of the studied bacterial groups was observed. No significant correlations were found between the abundance of the observed groups of microorganisms and the mentioned physical and chemical parameters of the marine environment.

Keywords: Surface water · Bottom water · Indicator bacteria · Pollution · Kara sea

1 Introduction

The Kara Sea is the most important area of the Siberian epicontinental Arctic. Its basin is a key area for crucial processes related to the response of Arctic ecosystems to ongoing climatic changes, the transformation of huge-volume continental runoff, and interactions in the mainland–shelf–deepwater basin system. The Kara ecosystem plays an essential role as an intermediate "home" during the introduction of alien species from the West into the Siberian Arctic seas. It also bears the largest accumulated environmental risks associated with radioactive waste repositories in the northwest part of the sea and bays of the Novaya Zemlya [1]. The Kara Sea is one of the most promising habitats to study biota [2, 3]. Microorganisms play a leading role in the multistep process of degradation of autochthonous and allochthonous organic matter in the water column and bottom sediments of seas and oceans, including all Arctic seas and the Arctic Ocean [4]. The bacteria abundance is a reliable indicator of organic pollution of aquatic ecosystems, and

T. Chaplina (Ed.): CIWO 2023, SPEES, pp. 387–395, 2023.
https://doi.org/10.1007/978-3-031-47851-2_47

microbial indication methods are a convenient tool for monitoring, which allow quick characterization not only of the quality of natural waters, but also of the biodegradation potential of the water area [5]. Experiments with heterotrophic bacteria inhabiting various environments showed that the bacteria are capable of using carbon as an energy source. It was also confirmed that this group of bacteria are capable of active degrading various hydrocarbons in shallow and deep (2000–3500 m) waters and even at low temperatures [6]. Hydrocarbon-oxidizing bacteria (HOB) are widely used as indicators of environmental pollution [7]. The start of microbiological research in the Kara Sea is associated with the name of B. L. Isachenko, who was the first to establish the ubiquity of microorganisms in the water and bottom sediments of the Arctic seas [8]. The works of A. E. Kriss provide data on counts of saprotrophic microorganisms in the water column of the Kara Sea cultured on nutrient media [9].

The paper aims to study the distribution patterns of bacteria transforming the major classes of organic compounds, including diesel fuel, in the waters of the northeast Kara Sea in summer.

2 Materials and Methods

The material for this study was sea water samples collected with water samplers from the surface and bottom layers of the water column according to the map of stations Fig. 1 during the 50th cruise of the R/V *Akademik Boris Petrov* in July–August 2022. Original numbers used directly during the expedition and the coordinates of sampling stations are shown in Table 1.

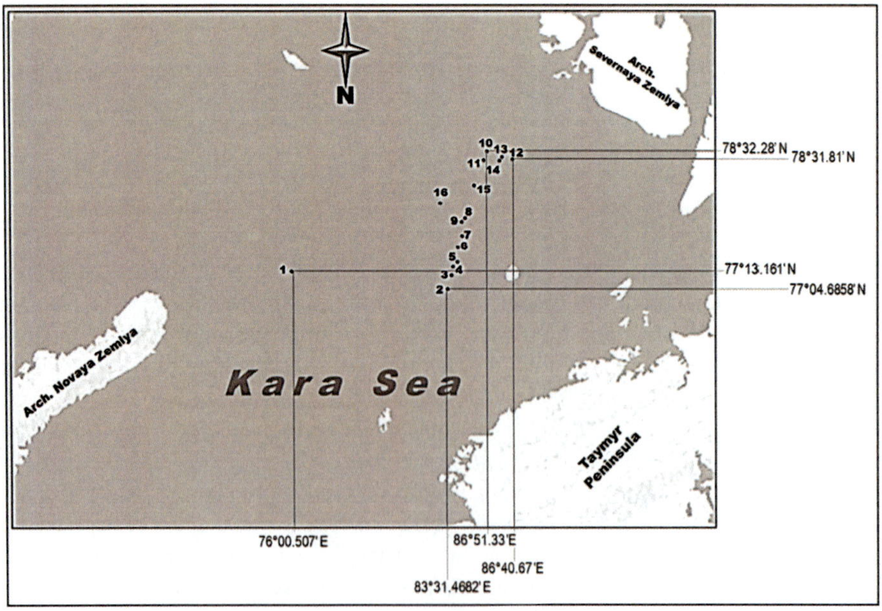

Fig. 1. Sampling stations in the Kara Sea

Sea water samples for microbiological analysis were taken from water samplers and put into sterile glass tubes with a volume of 21 cm^3. The abundance of heterotrophic bacteria (HB) capable of transforming major classes of organic compounds, including petroleum hydrocarbons, was determined in the studied samples. Lipids, carbohydrates (glucose), diesel fuel were used as an organic substrate and peptone was used as a protein substrate. Microbiological work was performed in the ship's laboratory immediately after water sampling. A total of 15 surface water samples and 9 bottom water samples were collected for microbiological analysis. The determination of bacterial abundance was carried out by the 10-fold serial dilution method [10] using liquid nutrient media. For HB, a peptone medium was used [11]: peptone – 10.0 g; yeast extract – 1.0 g; sea water – 1 L. The saccharolytic group of bacteria (SB) was grown according to [12]: glucose – 1 g; peptone – 0.5 g; K_2HPO_4 – 0.1 g; distilled water – 1.0 L; pH – 6.8. For HOB and lipolytic bacteria (LB), the Dianova–Voroshilova medium [13] of the following composition was used: NaCl – 18.0 g; MgSO4 × $7H_20$ – 0.2 g; KCl – 0.7 g; KH_2PO_4 –1.0 g; K_2HPO_4 – 1.0 g; $CaCl_2$ – 0.02 g; $FeCl_3$ – 2 drops; distilled water – 1.0 L; pH – 7.2–7.4. As the only source of carbon and energy, depending on the objectives, sterile diesel fuel or vegetable oil were added to each test tube (1% of the medium volume). The salinity of

Table 1. Coordinates, original and index numbers of sampling stations in the 50th cruise of the R/V *Akademik Boris Petrov*

Station number	Station number according to the cruise programme (original)	Geographical latitude (degrees, minutes)	Geographical longitude (degrees, minutes)
1	TTR21-AR206G	077° 13.161′ N	076° 00.507′ E
2	TTR21-AR215G	077° 04.6858′ N	083° 31.4682′ E
3	TTR21-AR226G	077° 15.57225′ N	083° 44.39815′ E
4	TTR21-AR232G	077° 22.3257′ N	083° 53.0205′ E
5	TTR21-AR233G	077° 22.8252′ N	083° 42.2896′ E
6	TTR21-AR236G	077° 23.5376′ N	083° 44.7686′ E
7	TTR21-AR245G	077° 21.5366′ N	083° 58.7959′ E
8	TTR21-AR250G	077°51.54′ N	084°27.71′ E
9	TTR21-AR254G	077°51.14′ N	084°21.68′ E
10	TTR21-AR270G	078°32.28′ N	086°44.22′ E
11	TTR21 AR275G	078°35.75′ N	085°56.50′ E
12	TTR21-AR276G	078°31.81′ N	086°51.33′ E
13	TTR21-AR280G	078°32.19′ N	086°45.70′ E
14	TTR21-AR286G	078°30.48′ N	086°40.67′ E
15	TTR21-AR289(1)G	078°30.40′ N	086°42.48′ E
16	TTR21-AR305G	078°03.19′ N	082°59.09′ E

sea water was taken into account when preparing the media. The cultures were incubated at room temperature.

The overboard water temperature at the sampling stations was measured with a thermometer lowered on a rope from the stern into the surface water column. The temperature of the bottom water layer was measured immediately after the bottom corer was raised on the deck.

Similarly, the hydrogen ion concentration (pH) was recorded with a pH-meter by pouring 50 mL of sampled sea water into a sterile container.

3 Results and Discussion

The surface layer of the water column. The surface water layer temperature during the study period varied from 1.5 to 5 °C. Hydrogen ion concentration (pH) ranged from 7.67 to 8.20 Table 2.

Table 2. Depth, temperature, pH in samples of surface and bottom water at sampling stations

Station number	Station number according to the cruise programme (original)	Depth (m)	pH of surface water	pH of bottom water	Temperature (°C)	
					of surface water	of bottom water
1	TTR21-AR206G	123	7.67	7.66	5	4
2	TTR21-AR215G	55.9	7.67	7.43	4.5	4
3	TTR21-AR226G	56.5	7.98	7.88	4	2.5
4	TTR21-AR232G	59.4	7.91	7.71	3.5	3
5	TTR21-AR233G	63.15	7.94	-*	4	-
6	TTR21-AR236G	99.41	8.2	-	4	-
7	TTR21-AR245G	54.34	7.97	7.8	4	3.5
8	TTR21-AR250G	193.39	8.08	7.9	3	2
9	TTR21-AR254G	152.53	8.07	7.88	3	2
10	TTR21-AR270G	153.31	8.03	-	2	-
11	TTR21-AR275G	96.83	7.97	-	1.5	-
12	TTR21-AR276G	139.5	8.09	-	2	-
13	TTR21-AR280G	150.1	8.13	-	2	-
14	TTR21-AR286G	174	8.08	-	2	-
15	TTR21-AR289(1)G	177.79	7.97	7.8	3	3
16	TTR21-AR305G	151.3	8.05	7.78	5	3

Note: "-" – determination was not performed

The abundance of HB using peptone as a protein substrate varied widely from 1 to 10^6 bacterial cells per milliliter of water Fig. 2. This bacterial group was isolated in

100% of the samples. The maximum HB abundance was recorded at stations 5, 10, 15, and 16. At stations 1 and 2, the HB abundance was scarce. In the remaining samples, the HB abundance ranged from 10^2 to 10^5 cell/mL.

Lipolytic bacteria were isolated in 82% of the samples. The highest LB abundance values (100 cell/mL) were found at stations 7 and 8. The lowest LB abundance was recorded at station 1. In the remaining samples, the LB abundance did not exceed 10 cell/mL. As a rule, the maximum abundance of lipolytic microorganisms is observed in places of domestic wastewater discharge. LB are observed in areas exposed to chronic petroleum pollution and their maximum abundance coincides with periods of mass development of zoo- and phytoplankton in conditionally clean water areas [14].

Bacteria using carbohydrates (glucose) as an organic substrate were isolated in 100% of the samples and ranged of 10 to 10^4 cell/mL. The highest SB abundance was obtained in the samples of stations 11 and 15, the lowest one – at stations 1, 2, and 4.

Diesel-degrading bacteria were detected in 53% of the samples with only one sample (station 7) showing the HOB abundance of 10 cell/mL, whereas at the remaining stations, the HOB abundance did not exceed one cell per milliliter of water.

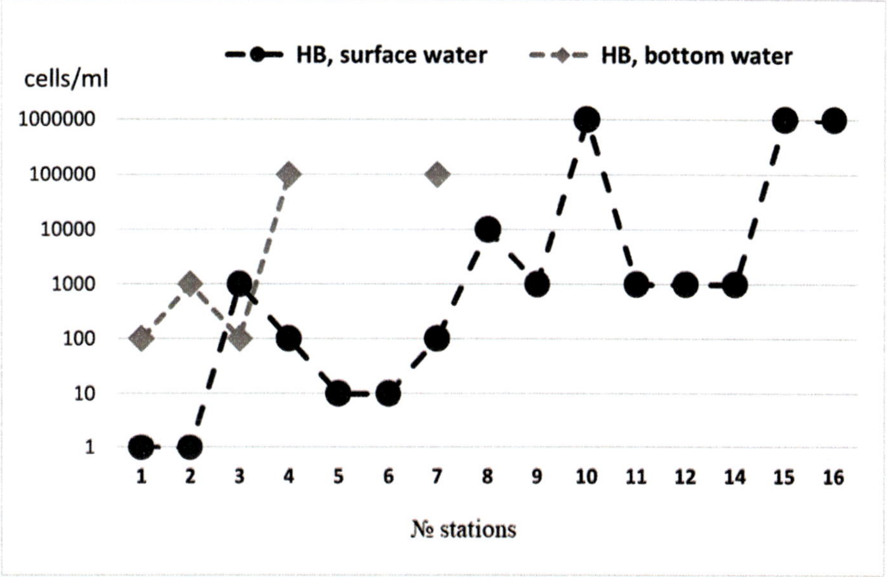

Fig. 2. Abundance (cell/mL) of heterotrophic bacteria in the surface and bottom layers of water column at the study stations

Bottom water layer. The depth at the bottom water sampling stations varied from 54 to 193 m and the bottom water temperature varied from 2 to 4 °C. Hydrogen ion concentration (pH) ranged from 7.43 to 7.90 Table 2.

The HB abundance ranged from 10 to 10^5 cell/mL. This group of bacteria was isolated in only 45% of samples from the bottom water layer Fig. 2. The highest count of HB was obtained in the samples of stations 3, 4, and 7, with the lowest count being at station 1.

The abundance of glucose-using bacteria varied from 10^3 to 10^6 cell/mL, SB were detected in 50% of the samples. The maximum was recorded in the area of station 4, in the remaining samples the SB abundance ranged from 10^3 to 10^5 cell/mL.

The abundance of bacteria capable of lipid oxidation varied widely from 1 to 100 cell/mL. The highest values were observed in the waters of station 4.

In the bottom water layer, the HOB abundance ranged from 1 to 10 cell/mL, with HOB found in only 36% of the bottom water samples. The highest HOB values were noted in the samples taken at stations 1, 4, and 7.

The factors determining the abundance of microorganisms in sea water are temperature, salinity, oxygen availability depending on water motion or stagnation [15, 16]. Based on the obtained temperature data and hydrogen ion concentration values, psychrophilic and acidotolerant microorganisms prevail in the study water area. According to the results of multifactor analysis [4], the bacterial abundance was influenced by benthic water temperature ($r = 0.8$). The results of our studies showed that there are no significant correlations between the abundance of observed microorganisms and the values of such indicators as surface and bottom water temperature, hydrogen ion concentration, and depth. Quantitative indicators of the heterotrophic bacteria in the surface and bottom waters in summer were similar to the data previously obtained in different parts of the Kara Sea. Thus, according to N. G. Teplinskaya [17], the total abundance of bacteria in the southwest Kara Sea is 18–150 thousand cell/mL. A high concentration of bacterioplankton (over 500 thousand cell/mL) was found in the surface waters of Baidaratskaya Bay [18]. According to work [4], in September 2007, at 15 stations located in three different areas of the Kara Sea water area, the bacteria abundance in the bottom water layer of the sea was relatively low and did not exceed 420 thousand cell/mL. In the mouth area of the Ob River, this parameter reached its maximum values and ranged within 344–417 thousand cell/mL, while in the sea area, it was 66–198 thousand cell/mL. At the transect along the St. Anna Trough, the abundance of bacteria in the bottom water layer ranged from 15 to 315 thousand cell/mL. The authors [4] also noted the heterogeneity of quantitative indicators of bacterial abundance in the zone of tidal currents near the Yamal coast and increase in its values in the area of influence of the Vostochno-Novozemelsk current. The studies performed by A. S. Savvichev [19] have shown that microbial abundance in Arctic waters averages 10^5–10^6 cell/mL. In particular, the total microbial abundance (cell/mL) varies in the following ranges: 10^6–10^7 in the Barents Sea, 10^5–10^6 in the White Sea, 10^5–10^6 in the Kara Sea, 10^5–10^6 in the Laptev Sea, 10^4–10^5 in the East Siberian Sea and Chukchi Sea. It is noted [20] that under Arctic water temperatures, the biochemical activity of heterotrophic microorganisms is relatively high. Based on our results, in most samples, the abundance of bacteria transforming the major classes of organic compounds in the water bottom layer exceeded that in the surface water layer. This agrees with the results given in the work [4, 21]. According to the authors [4], the bacterial abundance depended on the presence and availability of organic matter in the bottom water layer.

4 Conclusion

The results of microbiological works obtained during the 50th cruise of the R/V *Academic Boris Petrov* allowed studying the spatial distribution of bacteria capable of oxidizing proteins, fats and carbohydrates, including diesel fuel, and obtaining their quantitative characteristics in the surface and bottom water layers of the northeast Kara Sea in summer. The analysis of the obtained microbiological data showed an uneven spatial distribution and ratio of the studied bacterial groups in both surface and bottom water layers, which is apparently due to zonal distribution of organic matter and hydrometeorological factors. Hydrocarbon-oxidizing bacteria (degraders of diesel fuel) were detected in 53% of the samples collected in the surface water layer and in 36% of the samples collected in the bottom water layer. The low concentration of hydrocarbon-oxidizing bacteria indicates absence of fresh pollution of the water area with diesel fuel. No significant correlations were found between the abundance of the observed groups of microorganisms and values of such parameters as surface and bottom water temperature, hydrogen ion concentration, hydrocarbon concentration, and depth.

Expeditionary work was supported by the Ministry of Science and Higher Education of the Russian Federation in framework the expeditionary research plan of Lomonosov Moscow State University on the topic «Features quaternary sedimentogenesis, relief and natural fluidization on the seabed in the north-eastern part of the Kara Sea» and «Training through research on the Arctic shelf»; state assignment of Moscow Institute of Physics and Technology «Additional provision of the education system in the field of marine sciences - preparation of the young personnel reserve for the scientific and educational program «Floating University» on the basis of complex studies of the seas of Russia and the world ocean», state assignment of IO RAS «Support research as well as pilot development».

Acknowledgments. The expedition work was carried out with the support of the Ministry of Science and Higher Education of the Russian Federation within the framework of: the plan-program of expeditionary research of Lomonosov Moscow State University on the topic "Features of Quaternary sedimentogenesis, relief formation and natural fluid discharge on the seabed in the northeastern part of the Kara Sea" and "Training through research on the Arctic shelf"; state. tasks of the Moscow Institute of Physics and Technology (National Research University) "Additional support of the system in the field of marine sciences is the training of a young personnel reserve under the scientific and educational program "Floating University" on the basis of comprehensive studies of the seas of Russia and the World Ocean, the state assignment of the IO RAS "Ensuring scientific research, as well as experimental developments". Present work on determination of bacteria number and hydrocarbon concentration in water was carried out within framework of the State assignment of the IBSS on the theme "Molismological and biogeochemical foundations of the marine ecosystems homeostasis" (No. 121031500515-8).

References

1. Flint, M.V., et al.: Ecosystems of Siberian Arctic seas-2019: spring processes in the Kara Sea (cruise 76 of the R/V Akademik Mstislav Keldysh). Oceanology **60**, 134–137 (2020)

2. Belevich, T.A., Milyutina, I.A., Troitsky, A.V.: Seasonal variability of photosynthetic microbial eukaryotes (< 3 μm) in the Kara Sea revealed by 18S rDNA metabarcoding of sediment trap fluxes. Plants **10**, 2394 (2021). https://doi.org/10.3390/plants10112394
3. Vedenin, A.A., Galkin, S.V., Gebruk, A.V.: List of macrobenthic species: Data from the siberian seas and the adjacent area of the deep-sea central arctic. Data in Brief. **36**, 107115 (2021). https://doi.org/10.1016/j.dib.2021.107115
4. Romanova, N.D., et al.: Heterotrophic microbial communities on the water-sediment boundary in the Kara Sea. Oceanology **53**, 334–344 (2013)
5. Pashkova, T.S.: Mikrobiologicheskaya indikatsiya zagryazneniya vod prirodnykh istochnikov goroda Stavropolya. Bioraznoobrazie, bioresursy, voprosy khimii, biotekhnologii i zdorovye naseleniya Severo-Kavkazskogo regiona: mater. V (62-y) ezhegodnoy nauch.-prakt. konf. (Stavropol, 03–21 apr. 2017 g.), pp. 342—344. Stavropol (2017). (In Russ.)
6. Wang, W., Li, Zh., Zeng, L., Dong, Ch., Shao, Z.: The oxidation of hydrocarbons by diverse heterotrophic and mixotrophic bacteria that inhabit deep-sea hydrothermal ecosystems. The ISME Journal **14**, 1994–2006 (2020). https://doi.org/10.1038/s41396-020-0662-y
7. Polyaka, Yu.M., Demchuk, A.S., Sharova, A.N., Gubelit, Yu.I., Berezina, N.A.: Hydrocarbon-oxidizing bacteria in the digestive system of fish as an indicator of coastal pollution. Doklady Biological Sciences **491**, 71–74 (2020)
8. Isachenko, B.L.: Mikrobiologicheskaya kharakteristika gruntov i vody Karskogo moray. Izbr. Trudy. M.L.: Izd. AN SSSR, 334–363 (1951). (In Russ.)
9. Kriss, A.E.: Mikrobiologicheskaya okeanografiya. M.: Nauka, 269 (1976). (In Russ.)
10. Netrusov, A.I., Egorova, M.A., Zakharchuk, L.M., et al.: Practical microbiology/Edited by. A.I. Netrusova. M.. Akademiya, 608 (2005)
11. Mironov, O.G., Milovidova, N.Iu., Shchekaturina, T.L.: Biologicheskie aspekty neftianogo zagriazneniia morskoi sredy [Biological aspects of marine environment pollution by oil], p. 248. Naukova dumka Publ. Kiev (1988)
12. Practical microbiology. / Ed. N. S. Egorov. Moscow University, Moscow (1982)
13. Voroshilova, A.A., Dianova, E.V.: Okislyayushchie neft' bakterii – pokazateli intensivnosti biologicheskogo okisleniya nefti v prirodnykh usloviyakh. Mikrobiologiya **21**(4), 408–415 (1952). (In Russ.)
14. Ermakova, E.P., Gavrilova, E.I.: Mnogoletnyaya dinamika nekotorykh mikrobiologicheskikh pokazateley morskikh vod m. Sheskharis Novorossiyskoy bukhty. Izvestiya vysshikh uchebnykh zavedeniy. Severo-Kavkazskiy region. Estestvennye nauki. Seriya **2**(215), 85—88 (2004). (In Russ.)
15. Belousova, N.I., Shkidchenko, A.N.: Destruktsiya nefteproduktov razlichnoy stepeni kondensatsii mikroorganizmami pri ponizhennykh temperaturakh. Prikladnaya biokhimiya i mikrobiologiya **40**(3), 312–316 (2004). (In Russ.)
16. Junge, K., Eicken, H., Deming, J.W.: Bacterial activity at –2 to –20°C in Arctic wintertime sea ice. Appl. Environ. Microbiol. **70**(1), 550–557 (2004)
17. Teplinskaya, N.G.: Bakterioplankton i bakteriobentos Karskogo morya. Ekologiya i bioresursy Karskogo morya. – Apatity: KNTs AN SSSR, 29–37 (1989). (In Russ.)
18. Kuznetsov, L.L., Baytaz, O.N., Makarevich, P.R.: Strukturno-funktsionalnye pokazateli planktonnogo soobshchestva Baydaratskoy guby po materialam osennikh ekspeditsiy 1991–1992 gg. Biologiya i okeanografiya Karskogo i Barentseva morey. – Apatity: KNTs RAN, 88–95 (1998). (In Russ.)
19. Savvichev, A.S.: Mikrobnye protsessy tsiklov ugleroda i sery v moryakh Rossiyskoy Arktiki: Avtoref. dis. dokt. biol. nauk. – M., 48 (2011)
20. Morita, R.Y.: Starvation-survival of heterotrophs in the marine environment. Adv. Microb. Ecol. **6**, 171–198 (1982). (In Russ.)

21. Shchuka, T.A., Volodkovich, Yu.L.: Issledovanie protsessov mikrobnogo razrusheniya neftyanogo zagryazneniya i opyt monitoringa rasprostraneniya nefteokislyayushchikh mikroorganizmov v yugo-vostochnykh chastyakh Baltiyskogo i Karskogo morey. Problemy ekologicheskogo monitoringa i modelirovaniya ekosistem. **XXVI**(1), 180–204 (2015). (In Russ.)

Influence of External Factors on the Seaweeds Productivity in the Coastal Zone of Crimea

T. A. Filippova(✉) and E. F. Vasechkina

Marine Hydrophysical Institute of RAS, Sevastopol 299011, Russia
filippovata@mhi-ras.ru

Abstract. Bottom seaweeds is the main source of oxygen and carbon in coastal zone of the Crimea. The productivity of macrophytes mainly depends on a number of environmental factors, such as water temperature, solar irradiance and nutrients (nonorganic nitrogen and phosphorus). We carried out a series of simulations with extreme environmental changes to track the response of six algae that are widespread in the Black Sea on these changes. It was found that the main factor, which affect alga productivity, is light. Furthermore, macroalgae growth rate in the Crimean coastal zone is limited by phosphorus concentration in water. Increasing of nitrogen compounds content in water mainly affects the growth of green alga. Simultaneous long-time increasing of both nutrients in water leads to multiple increasing of seaweeds biomass.

Keywords: Simulation model · Black Sea · Macroalgae · Environment · Extreme changes

1 Introduction

Bottom alga is the main source of oxygen and carbon in a coastal zone. Biological productivity of marine phytocenoses depends on many environmental factors; the most significant are water temperature, irradiance at the depth of seaweeds growth, concentration of nutrients in water. It is difficult to rank the importance of each environmental factor for algae growth. We suggest to use some numerical estimates that can be obtained from the simulations with various scenarios of environmental changes to overcome this problem. The purpose of this work is to determine the degree of influence of various external factors on biological productivity of bottom seaweeds using numerical modeling.

2 Materials and Methods

The objects of this study are six species of seaweeds widespread in the coastal zone of Crimea. *Cystoseira crinita (Ericaria crinita)* is a brown perennial macroalgae. It is a basifite of phytocenosis growing on rocky ground; one of the largest alga in Black sea, its average height is 50 sm [1]. *Ulva rigida* is a green annual alga, can be epilithic or free-floating plant. The size of settled form can arise 2 m, but floating form rare grows

T. Chaplina (Ed.): CIWO 2023, SPEES, pp. 396–403, 2023.
https://doi.org/10.1007/978-3-031-47851-2_48

up to 20 sm. *Cladophora sericea* is a green annual alga, epilithic or epiphytic usually on *Cystoseira spp*. The size of thallus 5–50 sm. *Enteromorpha linza* is also a green annual alga, epilithic and epiphytic. It is small, thalli is 5–25 sm. *Enteromorpha* can form a carpets near seashore. *Vetrebrata subulifera* is a red alga, usually epiphytic. Average size of branches is 5–15 sm, a height is 20–25 sm [2]. The last seaweed is *Zostera noltii* – a higher flowering plant. This seagrass is a basifite of phytocenosis growing on a sandy bottom. *Zostera* roots form sea landscapes, which keep sand, *Zostera* leaves are a shelter for many marine organisms [3–5]. The average height of *Zostera noltii* is 25 sm [2].

We used the macroalgae biomass dynamics model [6, 7] to study the effect of environmental changes on algae growth. The model based on three differential equations for alga biomass dynamics (1), tissue content of nitrogen (2) and tissue content of phosphorus (3):

$$\frac{dB}{dt} = (P_r - E - m)B, \tag{1}$$

$$\frac{dQ_{\mathrm{N}}}{dt} = V_{nit}\frac{[\mathrm{NO_3}]}{K_{nit} + [\mathrm{NO_3}]} + V_{amm}\frac{[\mathrm{NH_4}]}{K_{\mathrm{amm}} + [\mathrm{NH_4}]} + V_{amm(b)}\frac{\left[NH_{4(b)}\right]}{K_{amm(b)} + \left[NH_{4(b)}\right]} - P_r Q_{\mathrm{N}}, \tag{2}$$

$$\frac{dQ_{\mathrm{P}}}{dt} = V_P\frac{[\mathrm{P}]}{K_P + [\mathrm{P}]} - P_r Q_P. \tag{3}$$

where B is the biomass (g/m^2); P_r is the specific growth rate (1/h); E is the part of excreted organic matter; m is the mortality coefficient (1/h); Q_{N}, Q_{P} are nitrogen and phosphorus tissue contents (mkmol/g DW); V_{nit}, V_{amm}, $V_{amm(b)}$, V_P are uptake rates of nitrates, ammonium (index '*b*' means pore water) and phosphorus (mkmol N (P)/(g DW h)); K_{nit}, K_{amm}, $K_{amm(b)}$, K_P are half-saturation constants (mkmol N (P)/l); $[NO_3]$, $[NH_4]$, $\left[NH_{4(b)}\right]$, $[P]$ are concentrations of nitrates, ammonium and phosphorus in water (mkmol N (P)/l).

In Eqs. (1–3) the specific growth rate P_r depends on the rate of photosynthesis and quotas of nitrogen and phosphorus in alga tissues:

$$P_r = \left\{(P_{\max} + R_d)\tanh\left(\alpha\frac{I_0 e^{-\beta h}}{(P_{\max} + R_d)}\right) - R_d\right\}\min(f(Q_{\mathrm{N}}), f(Q_{\mathrm{P}})),$$

$$f(Q_{\mathrm{N}}) = \frac{Q_{\mathrm{N}} - Q_{\mathrm{N}}^{\min}}{Q_{\mathrm{N}}^{\max} - Q_{\mathrm{N}}^{\min}}, \tag{4}$$

$$f(Q_{\mathrm{P}}) = \frac{Q_{\mathrm{P}} - Q_{\mathrm{P}}^{\min}}{Q_{\mathrm{P}}^{\max} - Q_{\mathrm{P}}^{\min}},$$

where P_{max} is the maximum rate of photosynthetic activity (mg O_2/g DW h^{-1}); R_d is the respiration rate (mg O_2/g DW h^{-1}); I_0 is the irradiance on sea surface (mkmol photon/m^2 s^{-1}); α is the initial slope of the photosynthetic curve; h is a depth (m); β is a light absorption coefficient; $Q_{\mathrm{N}}^{\min}$, $Q_{\mathrm{N}}^{\max}$, $Q_{\mathrm{P}}^{\min}$, $Q_{\mathrm{P}}^{\max}$ (mkmol/g DW) – are minimum and maximum quotas of nitrogen and phosphorus in alga tissues.

Real rate of nutrient uptake (in Eqs. (2) and (3)) were determined due to Michaelis–Menten kinetics and limiting functions:

$$V_{nit} = V_{nit}^{\max}(1 - f(Q_{\mathrm{N}})),$$

$$V_{amm} = V_{amm}^{\max}(1 - f(Q_{\mathrm{N}})), \tag{5}$$

$$V_{amm(b)} = V_{amm(b)}^{\max}(1 - f(Q_{\mathrm{N}})),$$

$$V_P = V_P^{\max}(1 - f(Q_{\mathrm{P}}),$$

where $V_{nit}^{\max}$, $V_{amm}^{\max}$, $V_{amm(b)}^{\max}$, $V_P^{\max}$ are maximum rates of nutrient uptake.

The control parameters of the model are water temperature, the irradiance on sea surface, concentrations of ammonium, nitrates and phosphates in water [6]. Modeling of each plant species was carried out using eighteen parameters that determine the dynamics of photosynthesis activity, nutrient uptake, mortality, excretion of dissolved organic matter, nitrogen and phosphorus content in plant tissues. Parameters, which described kinetics of nutrient uptake, were obtained from published data, the most complete table is presented in [6, 7]. Photosynthetic parameters were determined in our lab, methods were described in [8]. Table 1 shows modelling parameters for seaweeds.

To study the influence of environmental factors on the growth of algae, a series of calculations was carried out in the period July-August, the most productive growing season. To compare the growth activity of algae in the stationary state and under changing environmental conditions, the changes in algae biomass was calculated by the formula:

$$\Delta B = (B_1 - B_0) / B_0 \tag{6}$$

where B_0 is the plant biomass at the end of two months simulation of seaweeds growth in the steady state; B_1 – the biomass at the end of the simulation under changed environmental conditions. The steady state was determined by the seasonal dynamic of temperature, irradiance and nutrient concentration, typical for the South Bay, Sevastopol (Table 2). Data on the seasonal dynamics of the hydrological and hydrochemical characteristics of water were provided by the Oceanographic Data Bank of the Marine Hydrophysical Institute of RAS [9]. The seasonal variation of solar radiation on the sea surface was taken from [10], light intensity on a depth of alga growth was determined by water turbidity (total suspended matter in water column, mg/L). The simulations were constructed in such a way that only one of the environmental factors changed, the rest remained unchanged. The changes in the growth rate of plants with a multiple decrease and increase in nutrients, as well as water turbidity, was studied.

Table 1. Seaweeds parameters

Photosynthetic and general parameters

Species	P_{max}	R_d	α	E_c	E_{DOM}	Q_N^{min}	Q_N^{max}	Q_P^{min}	Q_P^{max}	m
Cystoseira barbata	3.58	0.8	0.12	18	0.39	457	3711	15	113	0.005
Ulva rigida	28	2.97	0.45	7	0.23	552	3789	15	96	0.074
Cladophora glomerata	32	2	0.1	10.4	0.23	552	3789	15	96	0.11
Vertebrata fucoides	6.5	1	0.03	23	0.38	853	3453	14	80	0.016
Enteromorpha linza	33	3	0.4	16.7	0.23	552	3789	15	96	0.123
Zostera marina	6.85	1.7	0.07	40	0.1	750	4500	30	118	0.006

Kinetic parameters

Species	V_{nit}^{max}	V_{amm}^{max}	$V_{amm(b)}^{max}$	V_P^{max}	K_{nit}	K_{amm}	$K_{amm(b)}$	K_P
Cystoseira barbata	10	50	-	1.8	7	20	-	5
Ulva rigida	116	211	-	9.31	34	20	-	4.41
Cladophora glomerata	115.72	327.83	-	3.06	5.27	32.68	-	0.33
Vertebrata fucoides	49.88	138.18	-	7.39	14.01	23.9	-	4.47
Enteromorpha linza	169	188	-	7.39	13.3	13.4	-	1.51
Zostera marina	4.9	67.9	2.1	4.59	31.47	93.1	58.1	9

Table 2. Environmental parameters of the steady state simulation

Environmental parameter	Units	min	max
T	°C	23	26.3
I at noon	mkmol photon/m^2 s^{-1}	225	1177
$[NO_3]$	mkmol N /l	3.9	5
$[NH_4]$	mkmol N /l	0.5	1.6
$[P]$	mkmol P /l	0.1	0.2
$[NH_{4(b)}]$	mkmol N /l	100	100

3 Results and Discussion

The first simulation was done to study the reaction of seaweeds to a halving of the nutrients concentration in water (Fig. 1). With a simultaneous decrease in the concentrations of nitrogen and phosphorus in water, the maximum decrease in biomass was demonstrated by green algae (*Ulva* – 45% and *Cladophora* – 43%, to a lesser extent *Enteromorpha* – 24%). With a separate reduction of nitrogen or phosphorus concentrations, *Ulva* showed its dependence on the nitrogen content in water, and *Cladophora* - on phosphorus. Perennials reacted more weakly (of *Cystoseira* – 16% and *Zostera* – 8%), phosphorus was the limiting element for *Cystoseira*, and nitrogen for *Zostera*. A twofold decrease in water temperature led to a decrease in plant biomass, which was more pronounced for *Cladophora* (–26%). An increase in water turbidity and respectively decrease of light inhibits plant growth up to 23%.

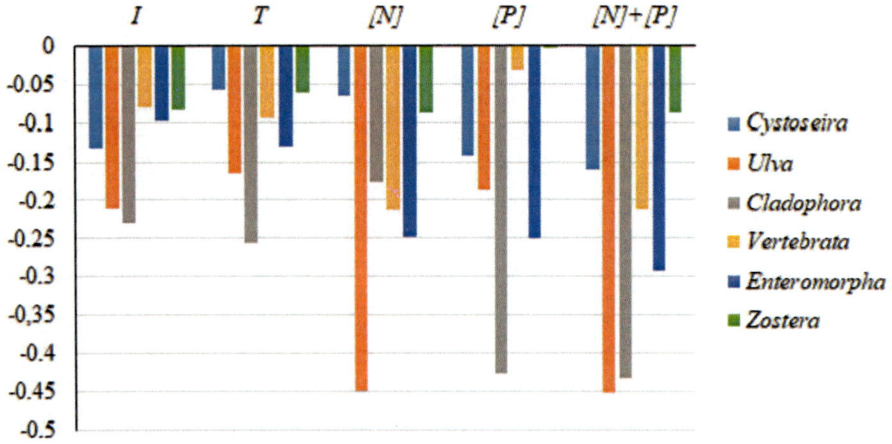

Fig. 1. Biomass dynamics in case of decrease in the concentration of biogenic elements, water temperature and insolation by half

In the second experiment, doubling nitrogen and phosphorus concentrations (simultaneously and separately), and doubling of the total suspended matter were simulated. With a simultaneous increase of nutrient concentrations in water, the growth rate of all plants increased, green algae were in the lead. The biomass changes were significant: *Ulva* – 80%, *Cladophora* – 43%, *Enteromorpha* – 33%, other plants from 12 to 20%. With a separate increase, the *Ulva* stands out sharply with 57% growth with an increase in the content of inorganic nitrogen in the water. *Cystoseira* and *Cladophora* respond more than other plants to an increase in phosphate content (19% increase in biomass). *Vertebrata* and *Zostera* did not react to the increase of phosphates concentration in water. An increase of total suspended matter in water by half, which significantly reduces the flow of light to the crown of plants, has the strongest affect on the growth rate of all plants; biomass changes were for *Cystoseira* – 40%, *Ulva* – 80%, *Cladophora* – 78%, *Enteromorpha* – 52%, *Vertebrata* – 47%, *Zostera* – 54%.

In the third experiment, a case of a multiple increase from 3 to 10 times of water nutrient concentration was simulated (Fig. 2). With an increase of nitrogen concentration in water, the maximum growth rate increments were recorded at a 3–4-fold increase. A further raise did not affect the growth rate of plants, with the exception of *Zostera*, which continued to increase its biomass. As noted earlier [11], phosphorus is not a limiting element for *Zostera*, the plant's need for this element is quite low. The same effects were observed in the simulations with an increase of phosphates without changes in the nitrogen content; the maximum response was recorded for *Cystoseira* (– 29%) and

Cladophora (– 20%). Such reaction is a consequence of the fact that with an increase in the concentration of one of the nutrients (nitrogen or phosphorus), the balance between them in plant tissues is disturbed, one of the elements becomes limiting, and respectively plant cannot increase the growth rate.

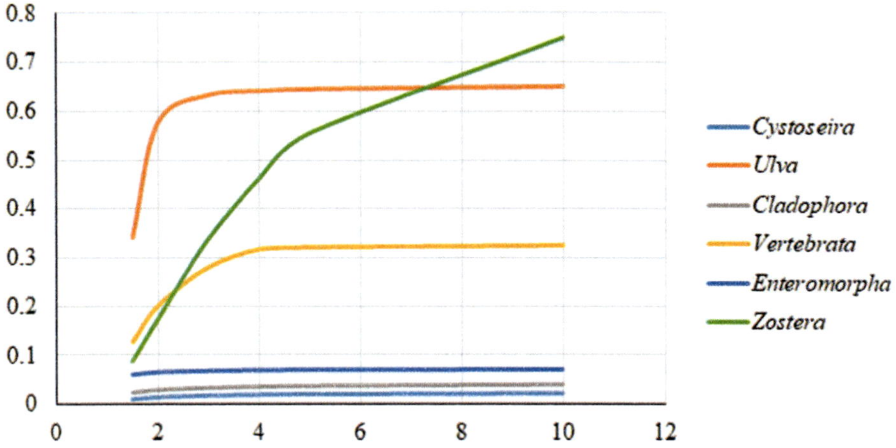

Fig. 2. Dynamics of biomass increment in case of multiple increase concentration of nitrogen in water

With a simultaneous increase of nutrients concentrations in water, the intensity of seaweeds growth increased significantly (Fig. 3). With a tenfold increase nitrogen and phosphorus concentration in water, *Ulva* showed a 7-fold increase in biomass, the rest of the plants, with the exception of the red algae *Vertebrata*, increased their biomass by 1.4 –1.7 times (*Vertebrata* – 56%).

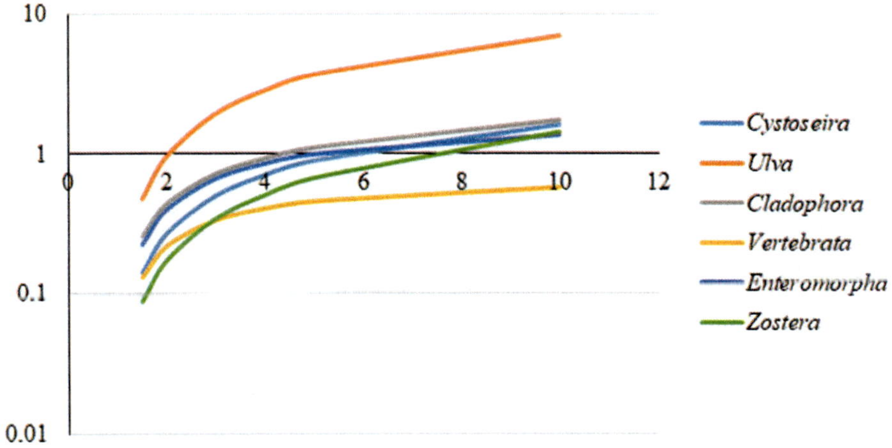

Fig. 3. Dynamics of biomass increment in case of simultaneous multiple increase of nutrients concentrations in water

4 Conclusion

We carried out a series of simulations, which allowed to study the reaction of seaweeds to changing environmental conditions. We investigated the response of macroalgae and sea grass biomass to multiple increase and decrease of the nutrients concentrations in water in comparison with the steady state of the bottom phytocenosis. It was found that the seaweeds react differently to extreme changes in environmental conditions, and green algae react the most. Obviously, such response leads to structural changes in the phytocenosis. With significant emissions of nitrogen and phosphorus with sewage into the shallow bays of Sevastopol a situation of flowering of epiphytes and free-floating green algae may occur with a high probability. The inhibition of irradiance in the crowns that accompanies this situation will have a depressing effect on perennials, which form bottom phytocenoses. In future this situation will lead to further disruption of the structure of phytocenoses and even their disappearance.

Acknowledgements. The investigation was carried out within the framework of the state assignment of the MHI RAS on theme: FNNN-2021-0005 “Complex interdisciplinary studies of oceanologic processes which determine functioning and evolution of ecosystems in the coastal zones of the Black Sea and the Sea of Azov” (“Coastal Research” code).

References:

1. Donchenko, L.V., Bitutskaya, O., Vlaschik, L., Limareva, N.: Biologically active complex with high antioxidant properties based on macrophytes of the azov-black sea basin. KnE Life Sci. **5**(1), 592–603 (2020). https://doi.org/10.18502/kls.v5i1.6133
2. Marine plants of the Black Sea. An illustrated field guide. M.: Sevastopol: Digit Print Press, N. Milchakova, 144 (2011)
3. Olsen, J.L., et al.: The genome of the seagrass *Zostera marina* reveals angiosperm adaptation to the sea. Nature **530**, 331–335 (2016). https://doi.org/10.1038/nature16548
4. York, P.H., Carter, A.B., Chartrand, K.M., Sankey, T., Wells, L., Rasheed, M.A.: Dynamics of a deep-water seagrass population on the Great Barrier Reef: annual occurrence and response to a major dredging program. Scientific Reports **5**(13167) (2015). https://doi.org/10.1038/srep13167
5. Rasmussen, J.R., Olesen, B., Krause-Jensen, D.: Effects of filamentous macroalgae mats on growth and survival of eelgrass, *Zostera Marina*, Seedlings. Aquatic Botany **99**, 41–48 (2012). https://doi.org/10.1016/j.aquabot.2012.01.005
6. Vasechkina, E.F., Filippova, T.A.: Modeling of the biochemical processes in the benthic phytocenosis of the coastal zone. Physical Oceanography **26**(1), 47–62 (2019). https://doi.org/10.22449/1573-160X-2019-1-47-62
7. Filippova, T.A., Vasechkina, E.F.: Simulation of chemical and biological processes of seagrass growth. Physical Oceanography **29**(6), 674–687 (2022). https://doi.org/10.22449/1573-160X2022-6-674-687
8. Vasechkina, E.F., Rudneva, I.I., Filippova, T.A., Naumenko, I.P., Parkhomenko, A.V., Shaida, V.G.: Photosynthetic parameters of the seaweeds widely spread near the crimean coast. Regional Studies in Marine Science (in press.). https://doi.org/10.1016/j.rsma.2023.103170
9. Bank of oceanographic data of the Marine Hydrophysical Institute of the Russian Academy of Sciences: http://bod-mhi.ru/ru/index.shtml *(*Internet resource*)*

10. Finenko, Z.Z., Churilova, T.Ya., Sosik, Kh.M., Basturk, O.: Variability of photosynthetic parameters of phytoplankton in the surface layer of the Black Sea. Oceanology **42**(1), 60–75 (2002)
11. Hansen, A.B., Pedersen, A.S., Kühl, M., Brodersen, K.E.: Temperature effects on leaf and epiphyte photosynthesis, bicarbonate use and diel O_2 budgets of the seagrass *Zostera Marina* L. Frontiers in Marine Science **9**(822485) (2022). https://doi.org/10.3389/fmars.2022.822485

Microlitter Distribution on Sandy Beaches of the Neva Bay of the Gulf of Finland (Baltic Sea) in 2018–2022

E. A. Golubeva(✉) and A. A. Ershova

PlasticLab, Russian State Hydrometeorological University, Saint-Petersburg 192007, Russia
katygolubev@gmail.com

Abstract. The city of St. Petersburg plays a significant role in the environmental state of the Neva Bay, including its contamination with wastes and marine litter. The goal of this work is to assess the distribution of microlitter on the coasts of the open part of the Gulf of Finland and the Neva Bay (Baltic Sea) based on the data of field observations of PlasticLab RSHU for the period of 2018–2022. Structure of microlitter, variability of its concentrations on the coasts, and the main factors of its accumulation were studied. The largest amount of microlitter was found on the beaches confined to the central part of the city (Dekabristov Island (2019), Kanonersky Island (2022)), and the smallest amount was found in the open part of the Gulf of Finland behind the FPB (Kotlin (South) (2022), Zelenogorsk (2021), Solnechnoye (2019)). In general, there is a variety of types of materials found on the beaches, but microplastics is the dominant one. During the study period (2018–2022), on the one hand, there is a downward trend in the average amount of microlitters on the beaches. On the other hand, a significant scatter of concentrations indicates the insufficiency of single surveys during the year and the need for repeated observations during the entire ice-free period, taking into account storm events, i.e. conducting seasonal monitoring surveys.

Keywords: Marine litter · Microlitter · Gulf of Finland · Neva Bay · Beaches · Pollution · Microplastics

1 Introduction

UNEP (United Nations Environment Program) defines marine litter as anthropogenic waste that enters the marine environment either accidentally or intentionally [11]. It not only accumulates in the water column and settles to the bottom, but is also carried out to the sea coasts as a result of tidal and surge phenomena. Marine litter accumulates in various parts of the river mouths, on the sea coasts, and its fragmentation into smaller particles called microlitter (particles less than 5 mm) occurs due to many environmental factors.

The Baltic Sea is one of the most studied seas in the world in terms of the marine litter and microplastics pollution: studies have been carried out here over the past ten years [2, 4].

T. Chaplina (Ed.): CIWO 2023, SPEES, pp. 404–411, 2023.
https://doi.org/10.1007/978-3-031-47851-2_49

The eastern part of the Gulf of Finland is subject to strong anthropogenic impact: the most vulnerable area of the Gulf of Finland is the Neva Bay, where the city of St. Petersburg is located - a metropolis, which is the largest transport hub of the North-Western Federal District of Russia. Every year, 2,3 million tons of solid communal wastes (SCW) are generated in St. Petersburg [9]. There is a system of separate collection of various wastes, which can be further processed. However, due to the imperfection of the legislative system in the field of waste management and the low workload of processing plants, recycling is not profitable and still remains at a low level here. It was estimated that annual load of plastic wastes from St. Petersburg to the adjacent water area is about 1 500 tons [3]. These can range from household goods (PET bottles, hygiene products, etc.) to industrial spills (e.g. plastic pellets).

2 Materials and Methods

Marine litter monitoring is necessary for understanding the distribution and structure of marine litter both in the marine environment and on the sea coasts. Currently, no standard document exists on how to collect and analyze beach litter. In the Baltic region, studies of marine litter and microplastics have been carried out since 2016 in the Kaliningrad region in the South-Eastern Baltic Sea (SEB) by the Atlantic Department of the P.P. Shirshov Institute of Oceanology, RAS [2]. In the city of St. Petersburg studies began in 2018 by the Russian State Hydrometeorological University (PlasticLab laboratory). Monitoring is carried out in the summer on 16 beaches of St. Petersburg and the Leningrad Region, located in the eastern part of the Gulf of Finland and in the Neva Bay. The monitored beaches sometimes alternate due to the constantly changing anthropogenic landscape of the urban coastline (dredging and land reclamation works, construction, etc.), but most of the observation stations remain the same.

In the eastern part of the Gulf of Finland studies are carried out using two international methods for beach survey ("Sand Rake" and "Frame" method [7, 8]), developed for the Baltic coasts based on the monitoring experience of the OSPAR project [13]. The OSPAR method is intended only for macrolitter study on the 100-m beach strip, but for a more detailed study of marine litter (fractions from micro- to meso-) the above methods were developed for the Baltic Sea coasts by a group of scientists from the Leibniz Institute for the Baltic Sea Research, Warnemünde, Germany [6]. They are designed to assess the accumulation of large particles of microlitter (2–5 mm), as well as mesolitter (5–25 mm) in the upper layer of sand (30–50 mm) [5]. Both methods focus on a visually distinguishable fraction of microlitter (2–5 mm), but at the same time two functionally different zones of the beach are studied [10] – the wave wrack zone and the "dry" zone of the beach. In this work, the results of the "Frame" method sampling are analyzed, i.e. the wave wrack zone - the zone of active impact of sea waves on the coast. Sampling of microliter is carried out in two squares of 1 m^2 inside a 40 m^2 polygon, where the top layer of sand (2–3 cm) is sieved through a 2 mm sieve. Collected microlitter particles are counted and classified according to the size, shape and type of material that is typical for the Baltic Sea: plastic, glass, paper, metal and other materials to determine their source [6].

3 Results and Discussion

In this paper we analyzed for the first time the material of 130 field studies on 16 beaches of the open part of the Gulf of Finland and the Neva Bay for the period 2018–2022. In 2018, 32 studies were carried out on 12 beaches (June-July), in 2019, 22 studies were carried out on 11 beaches (June-July), in 2020, 24 studies were carried out on 12 beaches (July-August), in 2021 (June-July) and in 2022 (June), 26 studies were conducted on 13 beaches of PlasticLab regular monitoring.

Average statistical characteristics were obtained for data series of microlitter concentrations for a five-year period: average ($\overline{x}$), median (Me), standard deviation (σ), maximum and minimum values. Average concentrations in 2018 (9.91 $\pm$ 6.28 particles/m^2), in 2020 (4.93 $\pm$ 3.50 particles/m^2) and in 2021 (5.12 $\pm$ 4.05 particles/m^2) exceed the standard deviation, and in 2019 (13.50 $\pm$ 14.90 particles/m^2) and in 2022 (7.35 $\pm$ 7.87 particles/m^2), on the contrary, the standard deviation exceeds the average, which indicates a strong variability in microlitter concentrations during these years (Fig. 1). Maximum values are observed in 2019. The next 3 years - 2020, 2021 and 2022 are characterized by the minimum scatter of data.

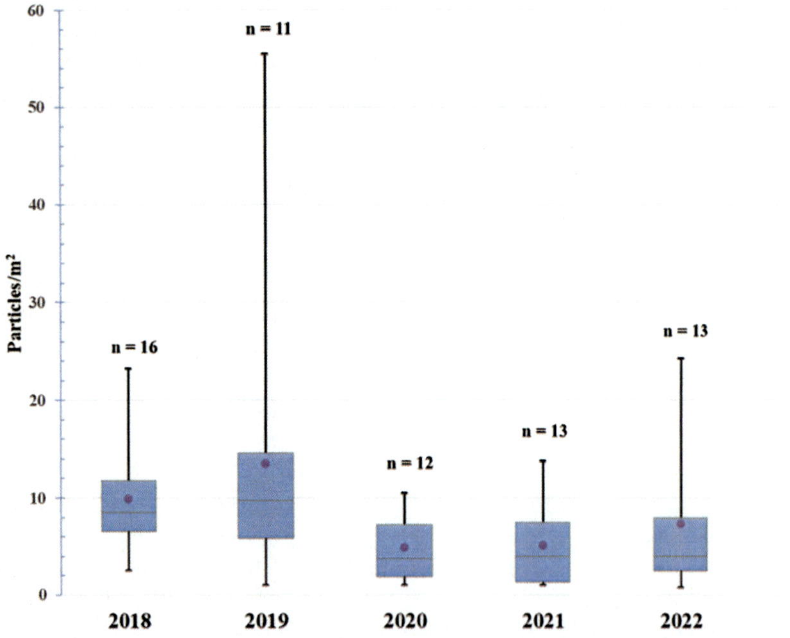

Fig. 1. Scatter plot for microlitter concentrations in 2018–2022 on the sandy coasts of the Neva Bay and the open part of the Gulf of Finland, particles/m^2 (n - is the number of observations)

In 2018 and 2019 in general, the amount of microlitter is greater than in the next 3 years. However, it is difficult to identify any significant trend towards an increase or decrease in the microliter content on the beaches in the wave wrack zone both in the Neva Bay and in the open part of the Gulf of Finland during the study period.

The spatial distribution of the microliter content on 16 beaches of the open part of the Gulf of Finland and the Neva Bay is shown on Fig. 2.

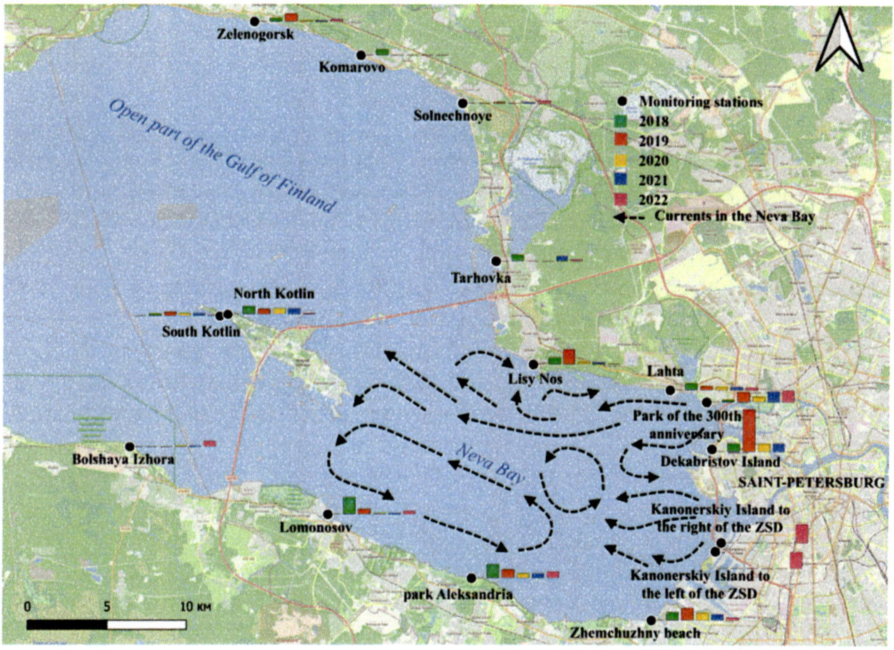

Fig. 2. The content of microlitter (particles/m^2) on the beaches of the Neva Bay and the open part of the Gulf of Finland in the period 2018–2022 (system of currents in the Neva Bay according to [12])

The largest amount of microlitter particles per square meter in the wave wrack zone was registered on the beaches in the very center of St. Petersburg - on Dekabristov Island (55,5 particles/m^2) in 2019 and on Kanonersky Island (24,3 particles/m^2- the beach to the right of the Western High-Speed Diameter (WHSD)) in 2022. The smallest amount of microlitter particles was found on the beaches in the "Western Kotlin" Nature Reserve (1 particles/m^2) in 2022, in Zelenogorsk (1 particles/m^2) in 2021 and in Lomonosov (1 particles/m^2) in 2020. In general, a significant variability of microliter concentrations primarily reflects the complex and constantly changing hydrodynamic conditions in the studied area. The accumulation of litter on the beaches depends on many factors: wind and wave action and frequent surge events wash back and away litter particles on the flat sandy beaches of the Gulf of Finland, constantly changing the pollution picture. Also, the key factors of the beach pollution are the high urban load and the runoff of the river Neva: during the entire observation period the most polluted areas remain to be the beaches in the city center near the main branches of the river Neva (Neva delta). Since the observations were carried out once a year, the relationship of microlitter content with the river flow has not been studied in detail.

Despite the fact that the maximum values of microlitter are observed every year in different zones of the reservoir, in general, the Neva Bay is characterized by significantly higher concentrations of microlitter than the open part of the gulf: this is especially true for microplastics. This can be explained by strong anthropogenic load, flow of household waste from a large city, as well as by slow water exchange within the reservoir due to a Flood Protection Barrier (FPB), which is a man-made barrier on the way of pollutants to the Gulf of Finland.

Water circulation in the Neva Bay is characterized by a system of constant multidirectional currents that affect the nature of the distribution of marine litter on the coasts, in combination with temporary wave, surge, drift and runoff phenomena. In general, there is a prevailing circulation of a cyclonic nature in the Neva Bay: a constant current is directed from East to West along the northern coast [12], and there is also a local circulation along the southern coast due to the influence of the Marine Canal, which interferes with water exchange in the water area of the Neva Bay.

Indeed, more microplastic particles are found on the northern coast during the entire observation period. At the same time, more glass and building plaster from the "other" category are observed annually on the southern coast. In general, in 2018–2022 microlitter on the beaches is mainly represented by plastic and glass, with the exception of the beach in the city of Lomonosov in 2018, where microlitter from the "other" category prevails (pieces of building plaster and glass).

Possible sources of microlitter on the sandy coasts of the eastern part of the Gulf of Finland and the Neva Bay differ from beach to beach, but are mainly associated with industrial and other economic activities carried out nearby. Thus, in Alexandria Park, glass is the predominant type of microlitter every year, a possible source of which may be the nearby Petrovsky Glass Manufactory, and the microlitter structure of the "Western Kotlin" PA, especially on the northern beach, is mainly characterized by particles of rusty metal, which are the result of active navigation in the area, as well as the possible destruction of the forts of Kronstadt. A large number of brick pieces on Kanonersky Island may be due to the fact that the beaches are in close proximity to industrial sites or places where construction waste is disposed of. The sources of microplastic particles are more difficult to identify. These can be both waste processing enterprises and treatment facilities, which, as a result of treating large volumes of wastewater, do not capture microplastic particles, even with a high degree of purification.

In general, the variety of types of materials that make up the microlitter of the Gulf of Finland and the Neva Bay is high, but microplastics are the main ones (Fig. 3). Most microplastic particles were found on the beach on Dekabristov Island in 2019 (55.25 particles/m^2), and the least in 2019 in Alexandria Park (0 particles/m^2) and in Solnechnoe (0 particles/m^2) as well as in 2020 in Bolshaya Izhora (0 particles/m^2).

In the Russian part of the Baltic Sea - in the South-East Baltic in the Curonian and Vistula lagoons (the territory of the Kaliningrad region of the Russian Federation), studies using the same methods showed that the main objects among the observed microlitter in the wave wrack zone were: pieces of polystyrene and plastic [2], in contrast to the eastern part of the Gulf of Finland and the Neva Bay, where more microplastic and glass particles were found. Such differences in the composition of microlitter can be associated both with different sources and different hydrological conditions. Studies conducted in 2018

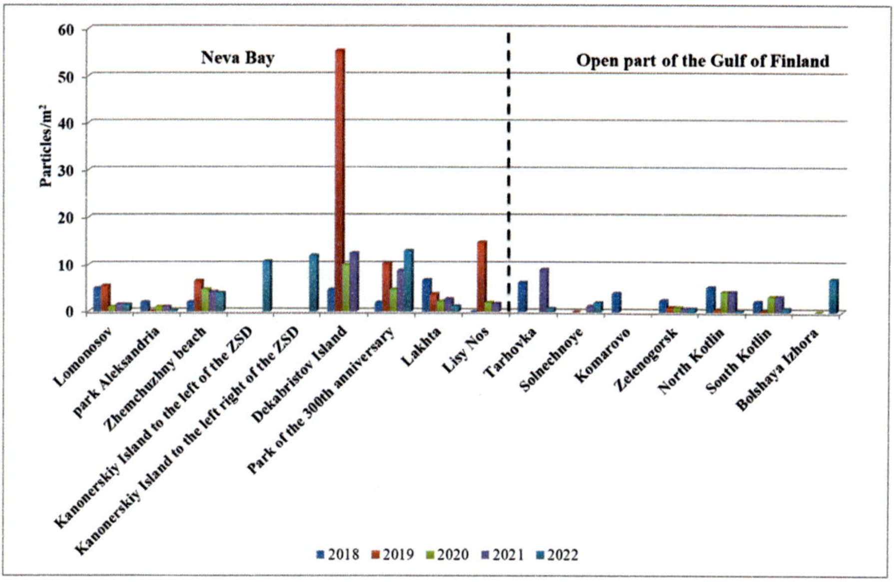

Fig. 3. Microplastic concentrations in 2018–2022 on the beaches of the Neva Bay and the open part of the Gulf of Finland, particles/m^2

on the beaches in the Curonian Spit National Park using the same method (collection of particles of 2–5 mm) showed an average content of plastic fragments of 3,35 particles/m^2 [1], while on average, 6.6 particles/m^2 of microplastics were found on the beaches of the Neva Bay and 2,5 particles/m^2 in the open part of the Gulf of Finland.

Comparison of the concentrations of microlitter and microplastics found on the Baltic coasts of Germany and Lithuania in the period from June to September 2014 [6] using the method used in this study showed that, in contrast to the beaches of Germany and Lithuania, in the amount of microliter (average of 8 particles/m^2) in the eastern part of the Gulf of Finland is 80 times higher than on the coasts of Germany (0.1 particles/m^2) and 2 times higher than on the coasts of Lithuania (3.9 particles/m^2). The main type of microlitter in the eastern part of the Gulf of Finland is plastic and glass, and on the Baltic coasts of Lithuania and Germany - paraffin (from the "other" category) and objects made of artificial polymers, respectively.

Thus, compared with studies in 2014, the Russian part of the Gulf of Finland is more polluted with microlitter than the Baltic coast of Lithuania and Germany. Compared to the results of studies of the Curonian Spit National Park in 2018, the sandy coasts of the eastern part of the Gulf of Finland, especially the Neva Bay, are more contaminated with microplastics than the sandy coasts of the South-East Baltic.

4 Conclusion

1. During the study period of 2018–2022 more microlitter particles, especially microplastics, were found on the beaches of the Neva Bay than in the open part of the Gulf of Finland. This can be explained by stronger anthropogenic load due to the Neva river

flow and household wastes from a large city, as well as due to the isolation of this water area.

2. Over the past five years of observations, the most polluted beaches are the beaches in close proximity to the center of the city of St. Petersburg (Dekabristov Island) and in the area of the industrial zone (Kanonersky Island).

3. In the period 2018–2022 it is impossible to identify any significant trend towards an increase or decrease in the content of microlitter on the beaches in the wave wrack zone both in the Neva Bay and in the open part of the Gulf of Finland. Significant variability in microlitter content on the beaches primarily reflects the complex and constantly changing hydrodynamic conditions in the studied reservoir. The accumulation of litter on the beaches depends on many factors: wind and wave action and frequent surge events wash back and away particles on the beaches, constantly changing the pollution picture. Thus, observations carried only once a year are not enough to reveal any trend of microlitter accumulation in the Neva Bay of the Gulf of Finland. Seasonal surveys are needed to study the relationship with changes in river flow and microlitter content on the beaches in the city center.

4. The eastern part of the Gulf of Finland is more polluted than the Baltic coasts of other countries (Lithuania, Germany) and the coasts of the Kaliningrad region (South-East Baltic).

References

1. Chubarenko, I., et al.: From macro to micro, from patchy to uniform: analyzing plastic contamination along and across a sandy tide-less coast. Mar. Pollut. Bull. **156**, 111198 (2020). https://doi.org/10.1016/j.marpolbul.2020.111198
2. Chubarenko, I.P.: Microplastics in the marine environment: monograph/I.P. Chubarenko, and others, p. 520. Scientific World, Moscow (2021). 20 s col. ill. ISBN 978-5-91522-513-7
3. Ecology of the Gulf of Finland and the Neva Bay: plastic pollution of the marine environment. EuroScientist: Online at: https://www.euroscientist.com/ecology-of-the-gulf-of-finland-and-the-neva-bay-plastic-pollution-of-the-marine-environment/
4. Ershova, A.A.: Marine litter and microplastics in the Gulf of Finland. Study of the global environmental problem of the XXI century at the regional level. Environment **2**(24), 28–31 (2022)
5. Ershova, A.A.: Monitoring of marine litter on the coasts of lagoons / closed bays / river estuaries (Neva Bay): Methodological guide for industrial practice, p. 5. St. Petersburg (2018)
6. Ershova, A.A.: Plastic pollution of the World Ocean, p. 170. Textbook - St. Petersburg: RSHU (2022)
7. Haseler, M., et al.: Marine litter pollution in baltic sea beaches – application of the sand rake method. Front. Environ. Sci. **8**, 599978 (2020). https://doi.org/10.3389/fenvs.2020.599978
8. Haseler, M., Schernewski, G., Balciunas, A., et al.: Monitoring methods for large micro- and meso-litter and applications at Baltic beaches. J Coast Conserv **22**, 27 (2018)
9. Kazakov, I.D.: Reforming the waste management industry. Environment **2**(24), 7–9 (2022)
10. Kuzmina, A.S., Ershova, A.A.: Contamination of sandy beaches with marine litter microparticles (the eastern part of the gulf of finland of the baltic sea). Ecological Safety of Coastal and Shelf Zones of Sea **2**, 86–100 (2022). https://doi.org/10.22449/2413-5577-2022-2-86-100
11. Marine litter. UNEP - UN Environment Programme. Online at: https://www.unep.org/ru/issleduyte-temy/okeany-i-morya/nasha-deyatelnost/rabota-po-regionalnym-moryam/morskoy-musor

12. Nezhikhovsky, R.A.: The Neva River and the Neva Bay / R. A. Nezhikhovsky, p. 112. Gidrometeoizdat, Leningrad (1981)
13. OSPAR Commission: Guideline for Monitoring Marine Litter on the Beaches in the OSPAR Maritime Area (2010). Online at: https://www.ospar.org/ospar-data/1002e_beachlitterguideline_englishonly.pdf

Data on the Microplastics Content in the Bottom Sediments of the Gulf of Finland and Southeastern Part of the Baltic Sea According to the Data of 39th Cruise of the R/v "Akademik Nikolai Strakhov"

P. N. Krivoshlyk[1,2](✉) and I. P. Chubarenko[1]

[1] Shirshov Institute of Oceanology, Russian Academy of Sciences, Moscow 117997, Russia
Poka20106@yandex.ru

[2] Immanuel Kant Baltic Federal University, Kaliningrad 236041, Russia

Abstract. The problem of microplastic (MPs) contamination is nowadays of concern to scientists over the world. MPs accumulation pattern in bottom sediments is shown to be rather chaotic and diverse. MPs content (0.3–5 mm) was analyzed in 19 samples of bottom sediments collected using the Van Veen grab in the Gulf of Finland and southeastern part of the Baltic Sea. A modified NOAA method was used for the extraction of MPs particles; zinc chloride was used for separation by density. As many as 10 421 suspected MPs particles were detected, more than 90% were fibers. The relationship between the content of MPs, the type of bottom sediments and sampling depth was not found. The results correlate well with previous research of MPs contamination in the Baltic Sea.

Keywords: Microplastic contamination · Type of bottom sediments · Southeastern part of the Baltic Sea · Gulf of Finland

1 Introduction

Currently, microplastics particles (MPs) (conventionally, less than 5 mm in size 5) are detected in all environments of the World Ocean, e.g., surf zone of the beach, water surface, water column, bottom or beach sediments [1, 2, 9, 10], and the Baltic Sea is no exception [3]. It is not fully understood which factors have the greatest influence on the accumulation of MPs in bottom sediments, where the pattern of contamination is extremely heterogeneous [3]. This makes monitoring of marine environment and evaluation of the mass of plastic very difficult. Many researches have shown already that it is the bottom sediments of the open sea where the maximum concentrations of MP particles are observed. For example, the surf zone of the beaches of the southeastern part of the Baltic Sea contains from 9 to 258 pieces of MPs particles per kg of dry weight (pcs/kgDW) of sand [4, 5]. In the layer of algae covering boulders on the underwater slope there are 376 ± 404 pcs/kgDW [6]. At the same time, in the sea sediments of the Gotland Deep, the mean contamination level for 53 samples was 863 ± 1371 pcs/kgDW, with maximum value of 10,179 pcs/kgDW [3].

T. Chaplina (Ed.): CIWO 2023, SPEES, pp. 412–417, 2023.
https://doi.org/10.1007/978-3-031-47851-2_50

The content of MPs particles (0.3–5 mm) was revealed based on the analysis of 19 samples of bottom sediments (sampling depth from 37 to 106 m) obtained during the 39 cruise of the research vessel "Akademik Nikolay Strakhov", which passed through the Baltic Sea (southeastern part and the Gulf of Finland) in May 2019 (Fig. 1). Samples were taken by the Van Veen grab. The sediments were packed in clean plastic bags immediately after sampling on the ship board. To control possible external contamination, open Petri dishes were exposed during every sampling.

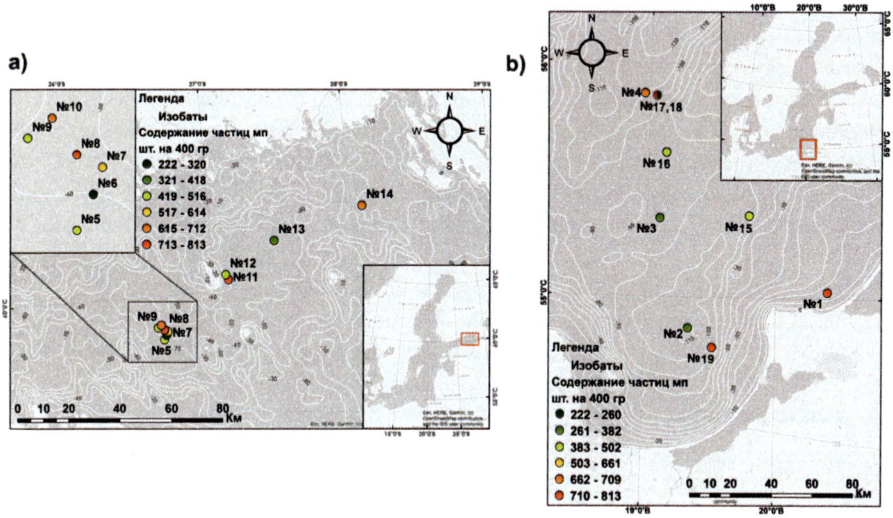

Fig. 1. Map of the location of bottom sediment sampling stations in the 39th cruise of the r/v "Ak. N. Strakhov" and the content of MPs particles (0.3–5 mm): a) the Gulf of Finland, b) the southeastern part of the Baltic Sea (see also Table 1, and online version for colours).

2 Material and Methods

Before processing, each sample was carefully mixed, and 400 g of sediment was taken from it. Sediments were treated using a modified laboratory method for the analysis of MPs in the marine environment: recommendations for quantifying synthetic particles in waters and sediments - NOAA Technical Memorandum NOS-OR&R-48 [7, 8]. Methods of sample preparation and analysis were kept the same as in [3–6, 11], which makes the results fully comparable. The method includes: 1) washing the sediment through sieves with mesh 333, 175, and 100 microns with the addition of bright, easily distinguishable control plastic fluorescent artificial reference particles (ARPs) to each sample as an additional measure to control the extraction efficiency. Washing is carried out in order to reduce the volume of the sample and get rid of fine silty fraction. The second stage is heating in a water bath with a temperature no higher than 75 °C with the addition of hydrogen peroxide (H_2O_2, 30%) and a catalyst (Fe(II)). This way, organic particles are removed from the sample. At the next stage, a hydrochloric acid solution (HCl, 4.5%) is added to the sample for the final removal of organic matter and dissolution of the calcite fraction (shells, insect parts). At the fourth stage, plastic particles are separated from the bottom sediment residues using the zinc chloride solution ($ZnCl_2$) (solution density 1.5 g/cm^3). When there is not much sediment in the sample, the funnel method is used to extract MPs particles, the sample is settled in solution for about an hour, after which the settled sediment is washed off throughthe 100 microns filter, and the solution with MPs particles is drained through 175 and 100 microns sieves. If there is a lot of sediment left on the sieves after washing, separation takes place in a glass. The precipitate remaining in the glass is thoroughly washed from the acid residues, then 200 ml of zinc chloride is poured and mixed. The resulting solution with MPs is drained through filters 175 and 100 microns; this procedure is repeated 3–4 times. The fifth stage, before the filters are put to dry, they are washed in a soda solution and distilled water. The filters are dried at room temperature in a closed Petri dishes (completely protected from external contamination). After drying, the samples are viewed on a microscope with magnification from 10x to 40x (Fig. 2). Verification of the synthetic origin of the particle' material is currently being carried out using the Raman spectroscopy method (LTD "NanoScanTechnology", Russia).

To determine the grain size composition of sediment samples, about 100 g of the material was taken from each sample. The analysis was carried out by sieving in accordance with the logarithmic phi (φ) scale of Wentworth [12].

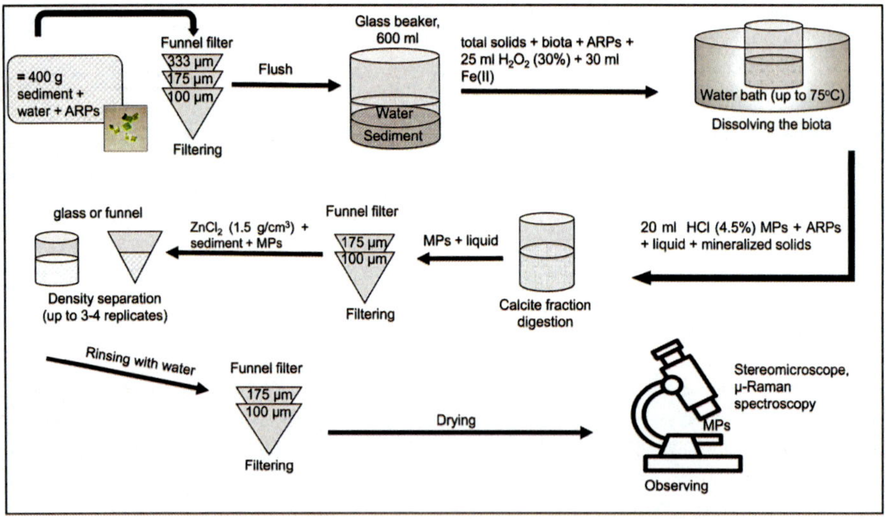

Fig. 2. Stages of processing of samples of bottom sediments

3 Results and Discussion

A total of 10,421 suspected MPs particles have been detected. The content of MPs at the stations varied from 222 (station 6) to 813 (Station 19) particles in the sample, with an average value of 548 ± 190 pieces for sample. The mean contamination of the sample in the laboratory was 21 ± 9 items/sample (fibers only), the mean contamination on the ship was 3 ± 2 items/sample (fibers only). The quantity of fibers in the control samples was not statistically significant in comparison with the concentrations of MPs found in the samples of bottom sediments. Subsequently, no corrections for external contamination were applied.

No correlation was found between the content of MPs particles and the type of bottom sediment. Sandy and silty sediments showed the same variations in the concentrations of MPs particles. There was also no dependence of the concentration of MPs particles on the depth. At a depth of more than 100 m, both maximum and minimum concentrations of MPs particles were detected.

In shape, 96.46% of the MPs particles were fibers, 3.5% of fragments were detected, the smallest fraction was films, only 0.04% (Table 1). The size distribution showed that most particles have a length of 0.5–2 mm (69.4%), particles with a size of 0.3–0.5 mm were detected by 16.2%, particles with a length of 2–5 mm were detected by 12.9%, particles larger than 5 mm were detected by only 1.6%. The majority of MPs particles turned out to be transparent (78.1%); blue, green, black, red, white, yellow particles turned out to be accordingly 7,8%, 5%, 3,3%, 2,5%, 1,5%, 1,4%; the smallest amount of particles had brown color, 0.4%.

Table 1. Data on the content of MPs in the collected samples. Gray shading denotes stations in the Gulf of Finland

№ sampling stations	Vessel-station	Sediment type	Depth, m	Fragments, pcs	Films, pcs	Fibres, pcs	Total, pcs
1	*ANS39002_b.s.-1*	Fine Sand	105	12	0	791	803
2	*ANS39003_b.s.-2*	Fine Silt	0	8	0	349	357
3	*ANS39004_b.s.-3*	Very Coarse Silt	80	20	0	362	382
4	*ANS39005_b.s.-4*	Medium Silt	80	19	0	690	709
5	*ANS39006_b.s.-5*	Very Fine Silt	106	9	0	428	437
6	*ANS39007_b.s.-6*	Very Fine Silt	61	9	0	213	222
7	*ANS39008_b.s.-7*	Fine Sand	47	45	0	560	605
8	*ANS39009_b.s.-8*	Fine Sand	31	26	0	746	772
9	*ANS39011_b.s.-9*	Very Fine Silt	63	20	0	466	486

(*continued*)

Table 1. (*continued*)

№ sampling stations	Vessel-station	Sediment type	Depth, m	Fragments, pcs	Films, pcs	Fibres, pcs	Total, pcs
10	*ANS39012_b.s.-10*	Coarse Sand	51	20	0	641	661
11	*ANS39013_b.s.-11*	Fine Sand	59	16	0	775	791
12	*ANS39014_b.s.-12*	Coarse Silt	59	12	0	438	450
13	*ANS39015_b.s.-13*	Medium Sand	57	9	0	343	352
14	*ANS39017_b.s.-14*	Medium Sand	37	33	1	603	637
15	*ANS39020_b.s.-16*	Very Fine Silt	97	8	1	449	458
16	*ANS39021_b.s.-17*	Coarse Silt	87	18	0	484	502
17	*ANS39023_b.s.-18*	Coarse Sand	105	7	0	253	260
18	*ANS39023_b.s.-18*	Very Fine Silt	105	14	0	710	724
19	*ANS39033_b.s.-19*	Very Fine Silt	95	60	2	751	813
Total:				365	4	10052	10421
Percent:				3,5	0,04	96,46	100

Contamination of sediments in the Gulf of Finland and southeastern part of the Baltic Sea is of the same level of about 200–800 items/kgDW. At the same time, the Gulf of Finland receives waters from the Neva River, flowing through large city of St. Peterburg, while the region in the southeastern Baltic is an open-sea area, quite far from potential sources of MPs - the mouth of the Vistula River and inlets of the Vistula and the Curonian lagoons. This may indicate that the majority of MPs income from the Neva River is retained in the Neva Bay, which is currently bordered by a dam.

4 Conclusion

Thus, even if we assume that the synthetic nature will be confirmed for 90% of the particles (as it was revealed by earlier studies for this operator), the resulting contamination will be about 400 MP particles (0.3–5 mm), which corresponds well to the values previously obtained [3].

Acknowledgements. The study was supported by the state assignment of AB IO RAS (theme № FMWE-2021-0012). We express our gratitude to Alexander Krek, the head of the expedition on r/v "ANS39", Ruslan Zakirov, the engineer of the Laboratory of Coastal Systems of AO IO RAS, responsible for collecting samples, as well as Anastasia Kupriyanova, a junior researcher at the Laboratory of Marine Physics of AO IO RAS, for the cameral processing of samples.

References:

1. Turan, N.B., Erkan, H.S., Engin, G.O.: Current status of studies on microplastics in the world's marine environments. J. Clean. Produ. **327**, 129394 (2021)
2. Peng, X., et al.: Microplastics contaminate the deepest part of the world's ocean. Geochemical Perspectives Letters **9**(1), 1–5 (2018)
3. Chubarenko, I., Esiukova, E., Zobkov, M., Isachenko, I.: Microplastics distribution in bottom sediments of the Baltic Sea Proper. Marine Pollution Bulletin **179**, 113743 (2022)
4. Krivoshlyk, P., Chubarenco, I.: Seasonal variations in the content of microplastic particles in the sands of the beach swash zone. J. Oceanolog. Res. **49**(2), 57–66 (2021)
5. Chubarenko, I., et al.: From macro to micro, from patchy to uniform: Analyzing plastic contamination along and across a sandy tide-less coast. Marine Pollution Bulletin **156**, 111198 (2020)
6. Esiukova, E., Lobchuk, O., Haseler, M., Chubarenko, I.: Marine macrophytes retain microplastics. Marine Pollution Bulletin **171**, 112738 (2021)
7. Masura, J., Baker, J., Foster, G., Arthur, C.: Laboratory. Methods for the Analysis of Microplastics in the Marine Environment: Recommendations for quantifying synthetic particles in waters and sediments. NOAA Technical Memorandum NOS-OR&R-48, p. 31 (2015)
8. Zobkov, M., Esiukova, E.: Microplastics in the marine environment: a review of methods for sampling, preparation and analysis of water samples, bottom sediments and coastal sediments. Journal of Oceanology **58**(1), 149–157 (2018)
9. Andrady, A.L.: Microplastics in the marine environment. Marine Pollution Bulletin **62**(8), 1596–1605 (2011)
10. Strafella, P., López Correa, M., Pyko, I., Teichert, S., Gomiero, A.: Distribution of microplastics in the marine environment. Handbook of microplastics in the environment, 1–35 (2020)
11. Esiukova, E.: Plastic pollution on the Baltic beaches of Kaliningrad region, Russia. Marine Pollution Bulletin **114**(2), 1072–1080 (2017)
12. Wentworth, C.K.: A scale of grade and class terms for clastic sediments. The Journal of Geology **30**(5), 377–392 (1922)

^{137}Cs Latitudinal Distribution in the Surface Waters of the Atlantic Ocean: Results of Expeditionary Research in 2020

O. N. Miroshnichenko(✉), I. G. Sidorov, and A. A. Paraskiv

A.O. Kovalevsky Institute of Biology of the Southern Seas of RAS, 299011 Sevastopol, Russia
oksaniya_89@mail.ru

Abstract. The latitudinal distribution of the anthropogenic radionuclide ^{137}Cs in the Atlantic surface waters was studied according to the results of expeditionary research carried out during 79 cruise of the R/V «Akademik Mstislav Keldysh» (March-May 2020). Radiospectrometric analysis of the samples showed that the ^{137}Cs activity concentration in the Atlantic Ocean water was in the range of 0.65 ± 0.10–1.44 ± 0.21 Bq/m^3. It was found that the ^{137}Cs activity peaks corresponded to the global fallout peaks in the middle latitudes of the Earth. Comparison of the data obtained in 2020 with the results of the 1988/1989, 1992/1993, 2002 expeditions revealed a tendency to smooth out the ^{137}Cs activity peaks in the middle latitudes due to hydrodynamic processes in the World Ocean.

Keywords: ^{137}Cs · global fallout · Atlantic Ocean · sea water

1 Introduction

The main sources of anthropogenic radionuclides in the Atlantic Ocean were global atmospheric fallout after nuclear weapons testing, the maximum of which occurred in the 60s of the 20th century [1, 2]. Additional sources of anthropogenic radionuclides were liquid radioactive discharges from nuclear reprocessing plants in western Europe and the accident at the Chernobyl nuclear power plant, pollution from which was largely transported to the marginal seas of the North Atlantic – the Irish Sea, Northern Sea, Baltic Sea [3, 4]. The long-lived radionuclide ^{137}Cs (half-life = 30.17 years) is one of the main fission products formed as a result of nuclear weapons testing and operation of nuclear reactors. In addition, due to its chemical properties, ^{137}Cs is highly soluble in seawater, thereby being one of the most common anthropogenic radionuclide in the marine environment. Thus, ^{137}Cs inventory in the World Ocean from all major sources in 2000 was estimated at 348.2 PBq, while other anthropogenic radionuclides ^{90}Sr – 199.7 PBq, ^{239,240}Pu – 8.6 PBq [2]. In this regard, ^{137}Cs is one of the main artificial radioactivity sources in the World Ocean [2, 5].

T. Chaplina (Ed.): CIWO 2023, SPEES, pp. 418–424, 2023.
https://doi.org/10.1007/978-3-031-47851-2_51

In contrast to the Pacific Ocean, nuclear weapons tests were not conducted in the Atlantic Ocean, therefore, there was no local radioactive fallout in the Atlantic Ocean [6]. The input of ^{137}Cs to the Atlantic Ocean as a result of nuclear tests occurred due to global fallout through the radioactive aerosols circulation in the atmosphere. The global radioactive precipitation prevailed in the middle latitudes of the Northern hemisphere and, to a lesser extent, in the middle latitudes of the Southern hemisphere, due to a relatively small number of nuclear test explosions, as well as limited atmospheric exchange between the northern and southern stratospheres [6]. About 201.1 PBq ^{137}Cs was input to the Atlantic Ocean during the period of nuclear tests –1/3 part of the total ^{137}Cs inventory in the World Ocean from global fallout [1, 2].

The aim of this research is to study the latitudinal transatlantic distribution of anthropogenic radionuclide ^{137}Cs in the modern period.

2 Material and Methods

Surface seawater were taken at the 79th cruise (4th stage) of the R/V «Akademik Mstislav Keldysh» (March–May 2020). Expedition was organized and managed by Shirshov Institute of Oceanology, Russian Academy of Sciences.

Samples were taken in the open part of the Atlantic Ocean in the range from 50° south latitude to 46° north latitude. A schematic map of sampling stations is shown in Fig. 1.

Water samples with a volume of 100 L were taken in expedition conditions using the flow system of a research vessel. Previously, the selected volume of water was separated from the suspended matter by filtration through a filter-cartridge with a nominal pore diameter of 0.5 μm.

To determine the ^{137}Cs activity concentration in seawater, a sorption method based on passing water samples through two sequentially connected adsorbers impregnated with nickel-potassium ferrocyanide was used [7, 8]. The ^{137}Cs activity in the samples was measured on a scintillation NaI(Tl) gamma-ray spectrometer by gamma-emitting daughter radionuclide ^{137m}Ba, taking into account the external radioactive background and the efficiency of radiation registration. The relative error of the obtained results did not exceed 20%. The sorption efficiency was determined by the difference in activity in the first and second adsorbers, after which the ^{137}Cs activity concentration in the water sample was calculated [7, 8].

3 Results and Discussion

The results of ^{137}Cs determining in the surface water of the Atlantic Ocean are presented in Table 1. The ^{137}Cs activity concentration ranged from 0.65 ± 0.10 to 1.44 ± 0.21 Bq/m^3 and, on average, amounted to 1.1 ± 0.2 Bq/m^3.

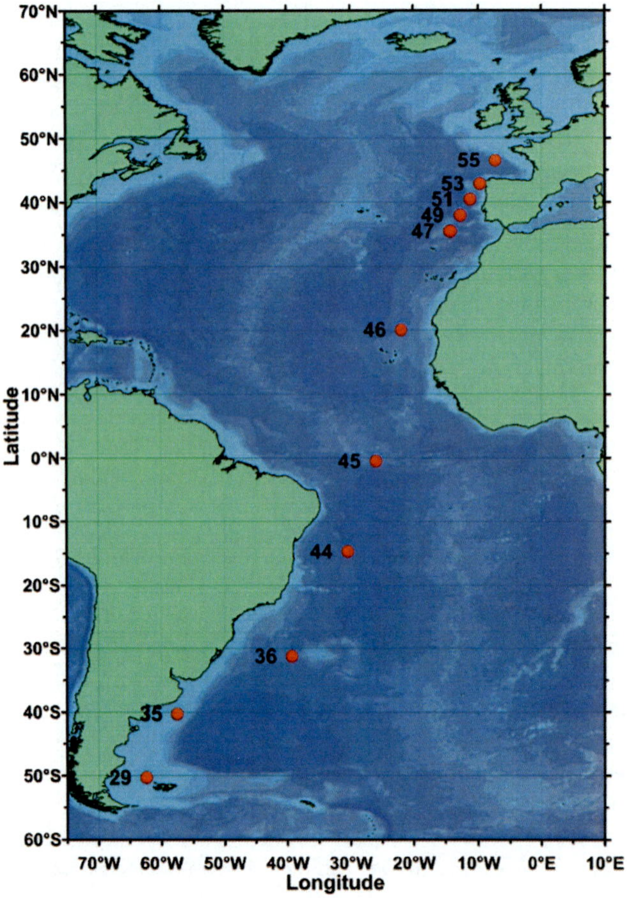

Fig. 1. Station map at the 79th cruise of the R/V «Akademik Mstislav Keldysh»

An analysis of the data obtained in 2020 showed that the southern part of the Atlantic Ocean remains the region with the least pollution by anthropogenic radionuclide ^{137}Cs in surface seawater, compared with other regions of the World Ocean [7, 9].

The transatlantic distribution of ^{137}Cs in the ocean surface layer was latitudinal. The ^{137}Cs maximum concentrations were related to the middle latitudes of the Northern and Southern hemispheres, respectively (Table 1), which is consistent with the patterns of global radioactive fallout distribution in the Atlantic Ocean.

A comparative analysis of the obtained results with previously performed studies on the latitude ^{137}Cs distribution in the Atlantic Ocean – the Swedish Antarctic expedition SWEDARP in 1988/1989 [10], the French expedition of the R/V «Jeanne D'Arc» in 1992/1993 [11], the 7th Ukrainian Antarctic expedition aboard the R/V «Gorizont» in 2002 [12].

Table 1. ^{137}Cs activity concentration in the surface water of the Atlantic Ocean according to the results of expeditionary research in 2020

Sample station	Coordinates	Bottom depth, m	Sampling depth, m	$^{137}Cs \pm \sigma$, Bq/m^3
29	50°14.915 S 62°21.801 W	142	6	0.68 ± 0.10
35	40°14.769 S 57°28.530 W	90	6	0.65 ± 0.10
36	31°12.014 S 39°15.717 W	4208	6	1.32 ± 0.20
44	14°40.610 S 30°25.066 W	4989	6	0.80 ± 0.13
45	00°30.171 S 25°57.783 W	3171	6	1.29 ± 0.19
46	20°04.718 N 21°57.875 W	3800	6	1.05 ± 0.19
47	35°30.966 N 14°18.846 W	4206	6	1.02 ± 0.17
49	37°59.022 N 12°49.453 W	4489	6	1.38 ± 0.20
51	40°27.891 N 11°16.491 W	4580	6	1.30 ± 0.20
53	42°52.970 N 09°45.951 W	640	6	1.08 ± 0.18
55	46°32.905 N 07°18.913 W	4330	6	1.44 ± 0.21

As shown in Fig. 2, during the period after the end of the nuclear weapon tests, the activity peaks of ^{137}Cs that entered the Atlantic waters with global precipitation were smoothed out. The observed trend is due to a number of factors. On the one hand, it is a decrease of ^{137}Cs concentration levels in surface water due to radioactive decay, as well as vertical water exchange processes in the marine environment. On the other hand, these are hydrodynamic processes of spatial radionuclide transport over the ocean area, which cause the alignment of ^{137}Cs concentration gradients. Thus, the current results demonstrate the regularity of the global ^{137}Cs spatial distribution in the Atlantic Ocean on a half-century time scale.

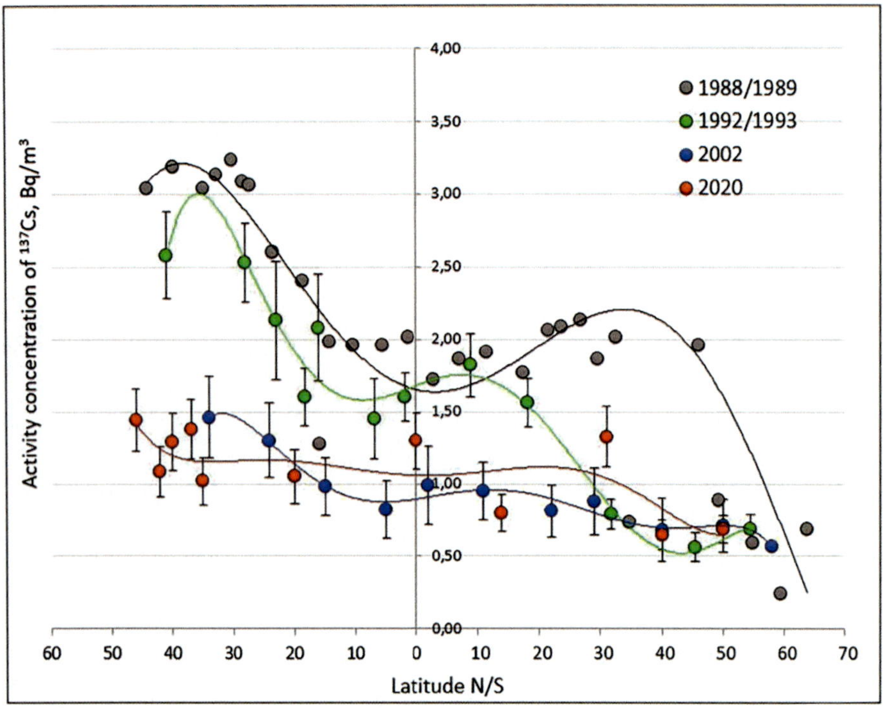

Fig. 2. Latitudinal distribution of ^{137}Cs activity concentrations in the surface Atlantic waters in 1988/1989 [10], 1992/1993 [11], 2002 [12] and 2020

The average ^{137}Cs activity concentration in the studied latitude range of the Atlantic Ocean decreased by 2 times compared to the data of the 1988/1989 expedition [10], and by 1.5 times compared to the data of the 1992/1993 expedition [11]. At the same time, in comparison with the results obtained in 2002 [12], the average ^{137}Cs activity concentration in the surface water of the Atlantic Ocean for 18 years remained almost at the same level, and at some sampling stations in the Southern hemisphere, there was an excess of ^{137}Cs activity levels. Relative change in the ^{137}Cs activity concentration according to the data of the expedition studies obtained in 2020 and 2002 presented on Fig. 3.

The ratio presented in Fig. 3 indicates the arrival of additional amounts of ^{137}Cs in the southern Atlantic Ocean. It is shown in [13–15] that bomb ^{137}Cs is transported by oceanic currents from the North Pacific Ocean, where its content is more significant, through the Indian Ocean to the southern regions of the Atlantic. Migration of ^{137}Cs from the southwestern Indian Ocean to the southeastern Atlantic occurs by transfer along the western branch of the Agulhas current [13–15].

Thus, the observed ^{137}Cs activity distribution in the South Atlantic surface waters north of 40°S is associated with the World Ocean global circulation and the transfer of this radionuclide from the North Pacific Ocean.

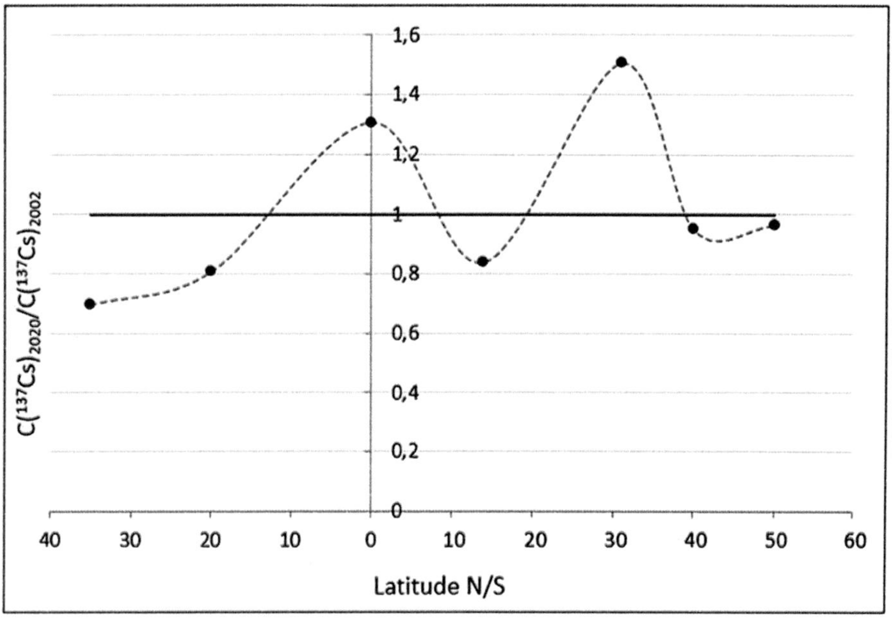

Fig. 3. The relative change of ^{137}Cs activity concentration C(^{137}Cs) in the surface waters of the Atlantic Ocean according to the 79th R/V «Akademik Mstislav Keldysh» (2020) and the Antarctic expedition on the R/V «Gorizont» (2002)

4 Conclusion

According to the results of expeditionary research aboard the R/V «Akademik Mstislav Keldysh» in 2020, it was determined that the average ^{137}Cs activity concentration in the Atlantic Ocean was 1.1 ± 0.2 Bq/m^3. In general, the transatlantic distribution of ^{137}Cs had a latitudinal character known from previous studies. Analysis of data on the ^{137}Cs distribution in the Atlantic Ocean on a scale of three decades has shown a tendency to smooth out the peaks of ^{137}Cs due to hydrodynamic processes and radioactive decay.

In the South Atlantic Ocean, north of 40°S, the ^{137}Cs content was higher than would be expected based on the rate of its decrease obtained in previous studies. Probably, this is a result of ^{137}Cs additional amounts transfer through the World Ocean global circulation system from the North Pacific Ocean.

Acknowledgements. The work was carried out within the framework of the State Assignment of the A.O. Kovalevsky Institute of Biology of the Southern Seas of RAS – «Molismological and biogeochemical fundamentals of marine ecosystems homeostasis», state registration number 121031500515-8.

Research in the Atlantic Ocean was carried out within the framework of the State Assignment of the A.O. Kovalevsky Institute of Biology of the Southern Seas of RAS – «Comprehensive studies of the current state of the ecosystem of the Atlantic sector of Antarctica», state registration number 121090800137-6.

The authors are sincerely grateful to the Shirshov Institute of Oceanology, Russian Academy of Sciences, the crew of the R/V «Akademik Mstislav Keldysh» for the unique opportunity to conduct investigations reported in this paper and for assistance in sampling operations.

References

1. UNSCEAR: Sources and Effects of Ionizing Radiation – United Nations: New York (2000)
2. Aarkrog, A.: Input of anthropogenic radionuclides into the World Ocean. Deep Sea Res. Part II Top. Stud. Oceanogr. **50**, 2597–2606 (2003)
3. Dahlgaard, H., et al.: On the background level of ^{99}Tc, ^{90}Sr and ^{137}Cs in the North Atlantic. J. Marine Syst. **6**, 571–578 (1995)
4. Povinec, P.P., et al.: Temporal and spatial trends in the distribution of 137Cs in surface waters of Northern European Seas – a record of 40 years of investigations. Deep Sea Res. Part II Top. Stud. Oceanogr. **50**, 2785–2801 (2003)
5. Buesseler, K.O.: Fukushima and ocean radioactivity. Oceanography **27**, 92–105 (2014)
6. Worldwide Marine Radioactivity Studies (WOMARS): Radionuclide Levels in Oceans and Seas, 1287 p. IAEA, Vienna (2005)
7. Miroshnichenko, O.N., et al.: Cesium-137 concentration in the surface waters of Eurasian seas: evidence from the expedition research of 2017. Geochem. Int. **57**, 1349–1354 (2019)
8. Miroshnichenko, O.N., Paraskiv, A.A.: ^{137}Cs concentration in surface waters of Far Eastern seas: results of expeditionary research in 2018. Marine Biol. J. **5**, 55–63 (2020)
9. Aoyama, M., et al.: Cross equator transport of ^{137}Cs from North Pacific ocean to South Pacific ocean (BEAGLE2003 cruises). Progr. Oceanogr. **89**, 7–16 (2011)
10. Holm, E., et al.: Radioactivity from the North Pole to the Antarctic. In: Guéguéniat, P., Germain P., Métivier, H. (eds.) Radionuclides in the Oceans: Inputs and Inventories. IPSN, Editions de Physique, pp. 59–74 (1996)
11. Bourlat, Y., et al.: ^{90}Sr, ^{137}Cs and $^{239+240}$Pu in world ocean water samples collected from 1992 to 1994. In: Guéguéniat, P., Germain P., Métivier, H. (eds.) Radionuclides in the Oceans: Inputs and Inventories. IPSN, Editions de Physique, pp. 75–93 (1996)
12. Gulin, S.B., Stokozov, N.A.: ^{137}Cs concentrations in Atlantic and Western Antarctic surface waters: results of the 7th Ukrainian antarctic expedition, 2002. J. Environ. Radioactivity **83**, 1–7 (2005)
13. Sanchez-Cabeza, J.A., et al.: Transport of North Pacific ^{137}Cs labeled waters to the South-Eastern Atlantic Ocean. Progr. Oceanogr. **89**, 31–37 (2011)
14. Tsumune, D., et al.: Transport of ^{137}Cs to the Southern hemisphere in an ocean general circulation model. Progr. Oceanogr. **89**, 38–48 (2011)
15. Inomata, Y., Aoyama, M.: Evaluating the transport of surface seawater from 1956 to 2021 using ^{137}Cs deposited in the global ocean as a chemical tracer. Earth Syst. Sci. Data Discuss. **2022**, 1–108 (2022)

Qualitative and Quantitative Assessment of Hydrocarbons Composition in the Water of the Kara Sea Northeastern Part

Y. S. Tkachenko(✉) and E. A. Tikhonova

A.O. Kovalevsky Institute of Biology of the Southern Seas of RAS, Sevastopol 299011, Russian Federation
yulechkatkachenko.90@mail.ru

Abstract. As part of the voyage of the research vessel «Akademik Boris Petrov», studies were carried out on organic pollution of surface and bottom waters in the Kara Sea northeastern part. Determination of the qualitative and quantitative composition of aliphatic hydrocarbons (AHC) was carried out by gas chromatography on chromatograph «Kristall 5000.2» with a flame ionization detector. The total content and individual composition of AHC in the surface and bottom waters of the Kara Sea northeastern part was studied. The content of AHC in the surface waters of the Kara Sea averages 0.06 ± 0.04 mg/l, which slightly exceeds the maximum permissible concentration (MPC) for oil hydrocarbons –0.05 mg/l. And in the bottom waters, the average AHC concentration was 0.1 ± 0.12 mg/l, which exceeds the MPC by 2 times. In general, it can be concluded that the concentrations of hydrocarbons in the water are quite high, exceeding the maximum allowable concentration in 44% of samples in the surface water and in 40% of samples in the bottom water, however, in accordance with the calculated markers (CPI_2, ACL, LWH/HWH, TAR and C_{31}/C_{29}) found that most of the hydrocarbons are of autochthonous nature and terrigenous origin.

Keywords: Aliphatic hydrocarbons · Diagnostic indices · Oil pollution · Sea water · Kara Sea

1 Introduction

The Kara Sea is one of the Arctic seas of Russia and belongs to the Arctic Ocean. According to the degree of resistance to anthropogenic pollution, it is classified as unstable, it is strongly influenced by river runoff, atmospheric and diffuse pollution with low indicators of the assimilation capacity of the marine environment [1]. The natural processes of decomposition of anthropogenic hydrocarbons (HC) are slowed down in high-latitude water areas due to a wide range of natural and climatic conditions. At low temperatures, the rate of evaporation of light fractions of oil slows down, fluidity decreases, due to absorption on the ice surface, oil hydrocarbons accumulate in porous layers, channels and voids of the ice cover, which can lead to irreversible consequences [2].

T. Chaplina (Ed.): CIWO 2023, SPEES, pp. 425–435, 2023.
https://doi.org/10.1007/978-3-031-47851-2_52

In connection with the activation of oil and gas production on the Arctic shelf, the study of hydrocarbons in the ecosystem of the Arctic seas is of particular relevance [3]. Oil and oil products are among the most common marine pollutants. Considering that hydrocarbons form the basis of oils, it is advisable to conduct studies to determine the qualitative and quantitative composition of aliphatic hydrocarbons (AHC) in sea water. It is also an important task to identify the probable sources of their entry into the environment. This analysis can be carried out by calculating and interpreting various diagnostic indices that indicate one or another nature of the origin of AHC.

The purpose of this work was to determine the qualitative and quantitative composition of AHC in sea water in the Kara Sea northeastern part, and to identify the probable sources of their origin using diagnostic indices.

In this work, in order to differentiate the allochthonous and autochthonous origin of AHC, such markers as the ratio of terrigenous matter to autochthonous compounds (*terrigenous/aquatic ratio, TAR*), the ratio of low molecular weight to high molecular weight alkanes (*Low-molecular Weight to High-molecular Weight Homologies ratio, LWH/HWH*), average chain length (*ACL*) [4–8]. To clarify the biogenic component of the compounds, the ratio of n-alkanes characteristic of herbaceous to woody vegetation (C_{31}/C_{29}) is used, and to identify the oil and biogenic origin of AHC, the odd parity index *CPI* is used, in particular, CPI_2 calculated for the high-molecular part of the spectrum [4, 9–14].

2 Materials and Methods

The material for the study was water samples taken with a bathometer from the bottom and surface water as part of the cruise of the R/V «Akademik Boris Petrov» in the summer of 2022 in various parts of the water area of the Kara Sea northeastern part (Fig. 1). The 16 samples were taken from the surface water layer and 5 samples from the bottom layer. The depth of sampling from the bottom horizon varied from 54 m to 151 m. In surface waters, the water temperature ranged from 1.5 °C to 5 °C, the temperature range in the bottom horizon was from 2.5 °C to 4 °C. The pH value of both the surface and bottom water layers varied within 7–8.

Sample preparation was carried out in accordance with the procedure [15]. A sample of water (250 ml) acidified with sulfuric acid (1:1) (1.5 ml) is extracted twice with n-hexane (25 ml each). The extraction was carried out in a glass separating funnel, 1 L in volume, for 5 min. The contents of the funnel were vigorously shaken for 10 min. After separating the mixture, the n-hexane layer containing non-polar and low-polar organic compounds was separated from the aqueous layer. Then, the resulting layer was passed through a glass column with a drawn lower end filled with alumina. The resulting sample volume was concentrated to a volume of 1 ml at room temperature in a fume hood.

Determination of the qualitative and quantitative composition of AHC was carried out on the basis of the collective use center "Spectrometry and Chromatography" of the Federal Research Center of the Institute of Biology of the Southern Seas by gas chromatography on a "Crystal 5000.2" chromatograph with a flame ionization detector (FID).

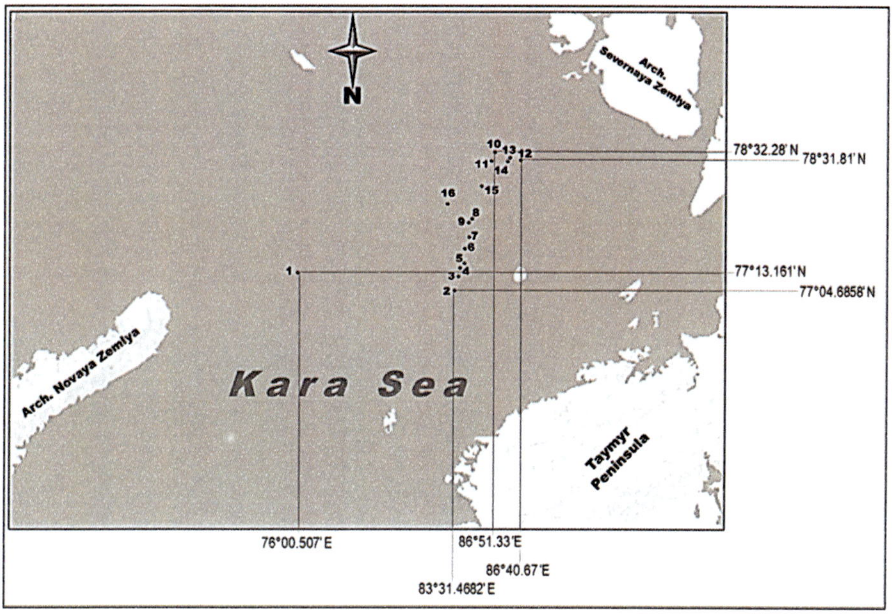

Fig. 1. Map of water sampling stations for the analysis of aliphatic hydrocarbons in the Kara Sea northeastern part

An aliquot part of the concentrated extract (40 ml) was injected into a gas chromatograph evaporator heated to 250 °C using a micro-syringe. The separation of hydrocarbons was carried out on a capillary column HT8 25 m × 0.32 mm with a fixed phase thickness of 0.25 microns (SGE Analytical Science). The separation of the AHC was ensured by the following conditions of the chromatographic system: the temperature of the column was programmed from 40 to 330 °C (the rate of temperature rise is 10 °C/min); the flow of carrier gas (helium) in the column was 2.5 ml/min without dividing the flow; the temperature of the detector was 320 °C.

The quantitative determination of the total AHC content was determined by the method of absolute calibration of the FID using a mixture of AHC prepared by the gravimetric method. A working standard solution with a concentration of 1 mg/ml was prepared from the initial standard solution by diluting it with n-hexane. The standard sample ASTM D2887 was used to separate the hydrocarbon mixture. Percentage normalization was used to process the results.

To confirm the reliability of the difference in values, two samples were compared based on the Student's t-test ($p < 0.05$). The data was previously normalized by base 10 logarithm. Statistical analysis was performed using the Statistica 12 program.

3 Results and Discussions

The total concentration of AHC in different parts of the water area in the bottom horizon ranged from 0.029 mg/l to 0.32 mg/l, in the surface – from 0.008 mg/l to 0.14 mg/l (Fig. 2).

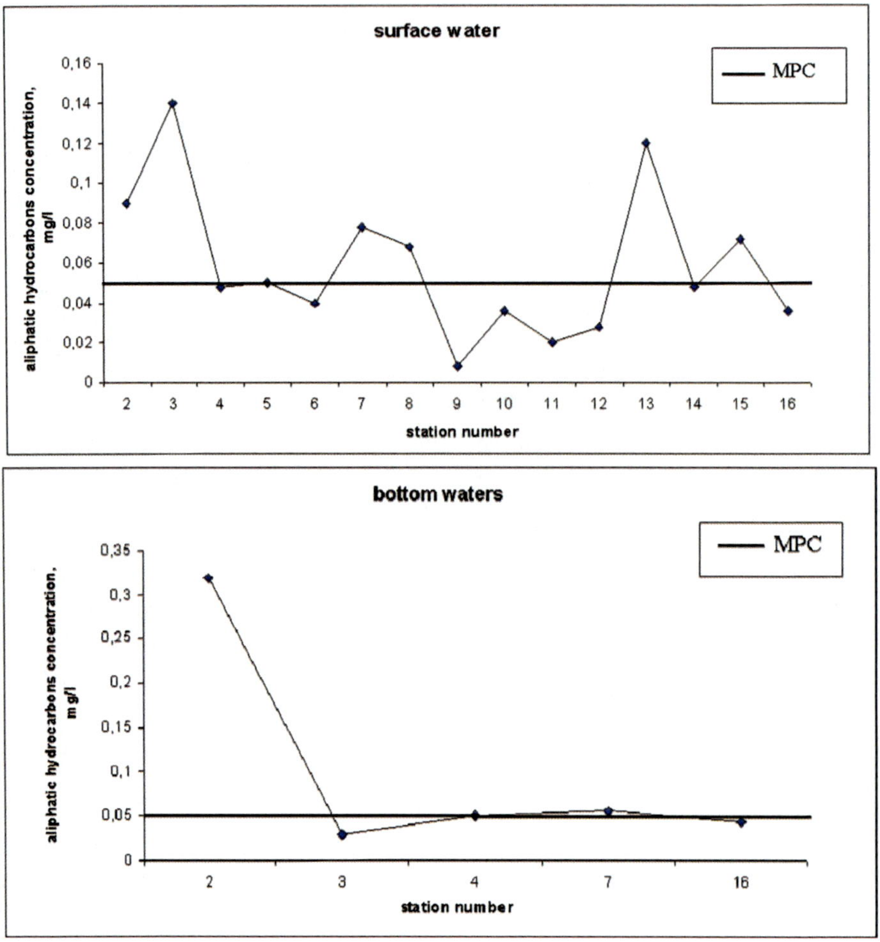

Fig. 2. Hydrocarbon concentrations in the surface and bottom waters of the Kara Sea northeastern part

The AHC content in the open surface waters of the Kara Sea averages 0.06 ± 0.01 mg/l, which slightly exceeds the MPC for petroleum hydrocarbons – 0.05 mg/l. And in the bottom horizon, the average concentration of AHC was 0.1 ± 0.05 mg/l, which exceeds the MPC by 2 times. In the surface layer of water, the maximum concentration of AHC exceeds the MPC by 3 times, and in the bottom horizon – by 6 times. However, statistical analysis showed no significant differences between the content of AHC in the surface and bottom horizons (tV = – 0.624, pv = 0.66). Therefore, we cannot compare the values of the bottom and surface horizons and judge in which layer the concentration is higher and in which it is lower.

A comparative characteristic of the results obtained with the results of other authors was carried out [16–23]. The data are summarized in a table (Table 1).

Table 1. AHC content (mg/l) in surface waters of different areas of the Kara, Barents, White and other northern seas

Water area	aliphatic hydrocarbons, mg/l	
	surface water	
	interval	average
Northeastern part of the Kara Sea (2022)	0,08–0,14	0,1
Eastern part of the Barents Sea (2016, 2017)	0,0033–0,011 0,0036–0,0087	0,0053 0,0065
Southern part of the Barents Sea (2017)	0,0058–0,0096	0,0083
South of the Barents Sea (2019)	0,011–0,062	0,032
Barents Sea (2020)	0,014–0,058	0,024
White Sea (2015)	0–0,08	–
White Sea (2017)	–	0,041
Laptev Sea (2016)	0,004–0,19	0,039
Norwegian Sea (2019, 2020)	0,006–0,062 0,009–0,038	0,023 0,02
r. Indigirka -East Siberian Sea	0,001–0,061	0,016
Mezen Bay	–	0,01

According to the table, it can be seen that the concentrations of AHC in the surface waters of the northeastern part of the Kara Sea are insignificant, but higher than in other northern seas. The average concentration of AHC in the surface horizon of the Kara Sea exceeds the MPC by 0.01 mg/l, in turn, the average concentrations of other northern seas do not exceed the MPC.

From 2016 to 2022, there is an increase in the concentrations of AHC, both in the surface and bottom waters of the northern seas of Russia [18, 22, 23], but in general, their excess is insignificant and there is no chronic oil pollution [18].

The origin of AHC reflects the composition of alkanes [2]. N-alkanes in the range C_{17}–C_{33} have been identified in both surface and bottom waters of the northeastern part of the Kara Sea. High-molecular homologues dominated in all samples of surface and bottom waters. Almost all chromatograms in the low molecular weight region showed an increase in concentrations of C_{17}, the main homologue of phytoplankton. Such data indicate the content of biodegradation products of phytoplankton organisms in both surface and bottom waters, as well as microbial transformation of organic substances.

The maximum concentrations occurred on homologues C_{25}–C_{29} in the high molecular weight part of the spectrum (Fig. 3). Odd compounds in this area are predominantly of allochthonous origin.

The analysis of chromatograms of the surface and bottom horizons of the northeastern part of the Kara Sea indicates the entry of allochthonous compounds into the water area, possibly under the influence of rivers, as well as the presence in the waters of autochthonous substances formed as a result of biosynthesis and biodestruction of phytoplankton organisms.

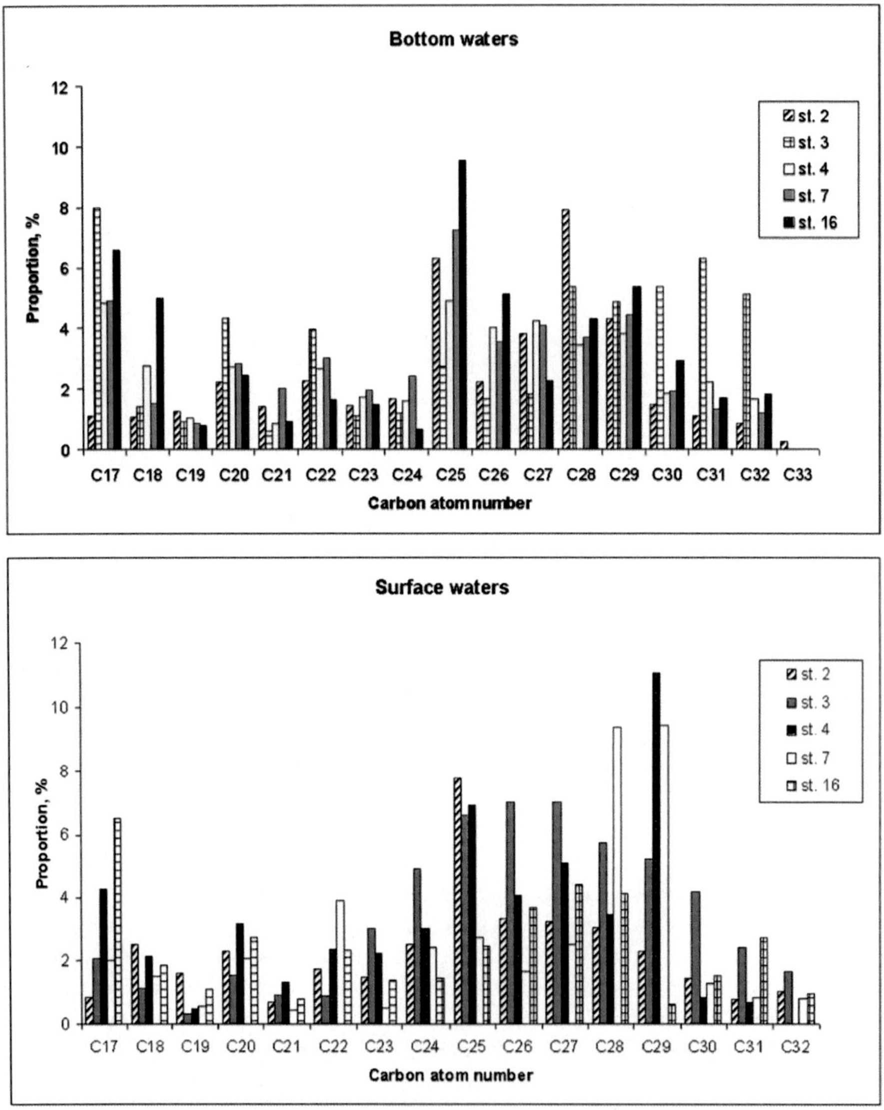

Fig. 3. Distribution of n-alkanes in the waters of the Kara Sea northeastern part

In order to differentiate the origin of AHC, diagnostic indices are used. To determine the allochthonous and autochthonous origin of AHC, the TAR index (the ratio of individual compounds with an odd number of atoms) is used. This indicator varied in the surface horizon from 0.125 to 5.71, and in the bottom layer – from 1.12 to 3.13 (Tables 2 and 3). The average values of the index were 2.24 ± 1.6 and 1.8 ± 0.8, respectively. This marker characterizes the dominance of allochthonous matter coming from the land. This is most likely due to river runoff, which has a great impact on the water area [24–26].

Another marker for identifying the allochthonous and autochthonous origin of AHC is *LWH/HWH* (the ratio of low-molecular alkanes to high-molecular ones). This indicator was in the range of 0.19–0.54 in the surface water layer and 0.22–0.46 (Tables 2 and 3) in the bottom layer. On average, this indicator was 0.36 ± 0.14 and 0.35 ± 0.09, respectively. The obtained value in all samples was less than 1, which characterizes the allochthonous flow of AHC into the water area.

An important indicator is the C_{31}/C_{29} ratio, which allows us to assess the contribution of the biogenic component of compounds. The main peak associated with woody vegetation falls on C_{29}, and with grassy vegetation – C_{31} [10, 11]. In the surface horizon, the value of this indicator at all stations, except station 16, is below 1. And in the bottom horizon, the ratio C_{31}/C_{29} was also below 1 at all stations, except Station 3 (Tables 2 and 3). Based on the data obtained, it can be concluded that both in the bottom and surface horizons, woody vegetation prevails over herbaceous vegetation in the composition of allochthonous components.

The average length of the *ACL* hydrocarbon chain is also used to identify the genesis of alkanes. High *ACL* values indicate the dominance of herbaceous vegetation in the formation of AUV [9]. In the presence of oil pollution, the *ACL* value is reduced [27]. In surface waters, the value of the calculated marker varied between 26.36–28.45, which averaged 27.45 ± 0.51 (Tables 2 and 3). In the bottom layer, the value of the *ACL* index fluctuated in the range of 26.84–28.95, averaging 27.44 ± 0.89 (Tables 2 and 3). The calculated marker shows approximately the same role of herbaceous and woody plants in the formation of AHC of surface and bottom waters of the water area. It is also worth noting that the *ACL* index in all the studied areas of the water area was above 0.5, which indicates the autochthonous nature of the AHC.

It is known that during the transformation of oil, light n-alkanes disappear first, and the content of more stable high-molecular compounds increases. Therefore, the ratio of n-alkanes in the high-molecular region is more informative for determining oil pollution [2]. To identify the oil and biogenic origin of the AHC, the odd *CPI* index is used. In the presence of oil pollution, the ratio of odd to even compounds in the high molecular weight region is close to one [2, 28].

The values of the CPI_2 index calculated for the high molecular weight region in surface waters ranged from 0.9 to 3.17, and in the bottom horizon the interval was 0.92–1.42 (Tables 2 and 3). On average, the CPI_2 index value in the surface horizon was 1.57 ± 0.64, and for the bottom horizon – 1.17 ± 0.2. In surface waters, the value of this marker at all stations, except for stations 3 and 8, was greater than one, which indicates the biogenic origin of the AHC. Biogenic sources of AHC intake into seawater are the processes of vital activity and metabolism of marine organisms [2]. In the bottom horizon, the CPI_2 index value at stations 4, 7 and 16 is greater than one, which also indicates the biogenic origin of the AHC. At the remaining stations (stations 3, 8 of the surface horizon and stations 2, 3 of the bottom horizon), the index approached one, which is an indicator of the presence of an AHC of oil origin.

In general, it can be concluded that the concentrations of AHC are quite large: the maximum permissible concentration in the bottom water is 40% of the samples, in the surface water – 44%, but in accordance with the calculated diagnostic indices, most AHC are of autochthonous and allochthonous origin.

Table 2. Concentration and diagnostic indices of n-alkanes in bottom waters of the Kara Sea northeastern part

Diagnostic indices	st. 2	st. 3	st. 4	st. 7	st. 16
D	17–33	17–32	17–32	17–32	17–32
K	0.32	0.029	0.05	0.056	0.044
LWH/HWH	0.22	0.37	0.32	0.37	0.46
Paq	0.64	0.27	0.49	0.65	0.63
TAR	3.13	1.52	2	1.34	1.12
ACL	27	28.95	27.57	26.86	26.84
TMD	2.68	1.53	1.84	1.58	1.79
C_{31}/C_{19}	0.88	6.66	2	1.4	2.33
CPI_1	0.95	1.41	1.2	1.5	1.29
CPI_2	1.05	0.92	1.15	1.42	1.29
C_{31}/C_{29}	0.35	2	0.5	0.35	0.35

Note: D – range of identified n-alkanes; K – total concentration of identified n-alkanes, mg/l

Table 3. Concentration and diagnostic indices of n-alkanes in the surface waters of the Kara Sea northeastern part

Diagnostic indices	st. 1	st. 2	st. 3	st. 4	st. 5	st. 6	st. 7	st. 8	st. 9	st. 10	st. 11	st. 12	st. 13	st. 14	st. 15	st. 16
D	17–32	17–33	17–32	17–31	17–32	17–32	17–32	17–32	17–32	17–32	17–32	17–32	17–32	17–33	17–32	17–32
K	0.1	0.09	0.14	0.048	0.05	0.04	0.078	0.068	0.008	0.036	0.02	0.028	0.12	0.048	0.072	0.036
LWH/HWH	0.28	0.13	0.13	0.3	0.5	0.29	0.19	0.4	0.47	0.53	0.45	0.54	0.4	0.27	0.39	0.53
Paq	0.76	0.53	0.56	0.43	0.48	0.58	0.24	0.28	0.45	0.64	0.6	0.4	0.45	0.26	0.52	0.54
TAR	2.42	4.7	5.71	3.32	1	1.86	5.1	1.25	1.43	0.9	1.5	0.93	1.79	2.04	1.73	0.125
ACL	26.36	27.5	27.27	27.5	27.24	27.5	28.1	28.11	27.4	26.9	27	27.82	27.48	28.45	27.2	27.7
TMD	5.31	4.47	3.41	2.71	1.36	1.83	4.5	1.14	1.76	1.19	2	1.06	1.67	2.35	2.1	1.125
C_{31}/C_{19}	0.53	10	6	1.5	0.05	3.33	1.75	2.5	1	0.67	1.25	2	0.57	7.5	1.4	2.5
CPI_1	0.64	1.24	1.01	0.96	4.15	1.5	0.55	3.33	1.46	1.52	1.5	3.18	1.84	3.04	0.58	1.33
CPI_2	1.52	1.35	0.99	2.12	2.9	1.26	1.11	0.9	1.5	1.67	3.17	1.18	1.59	1.19	1.51	1.17
C_{31}/C_{29}	0.4	0.61	0.43	0.06	0.05	0.25	0.1	0.25	0.17	0.4	0.5	0.4	0.2	0.75	0.28	5

Note: D – range of identified n-alkanes; K – total concentration of identified n-alkanes, mg/l

4 Conclusion

1. The total concentration of AHC in different parts of the water area in the bottom horizon ranged from 0.029 mg/l to 0.32 mg/l, in the surface – from 0.008 mg/l to 0.14 mg/l. The content of AHC in the open surface waters of the Kara Sea is on average

0.06 ± 0.01 mg/l, which slightly exceeds the maximum permissible concentration for oil AHC – 0.05 mg/l. And in the bottom horizon, the average concentration of AHC was 0.1 ± 0.05 mg/l, which exceeds the MPC by 2 times.
2. The distribution of n-alkanes on chromatograms indicates the mixed nature of AHC. The increased levels of AHC are caused by the presence in the waters of the products of biodegradation of phytoplankton organisms and microbial transformation of organic substances. Also, the high content of AHC may be associated with the melting of ice and natural oil seepage. The calculated diagnostic indices show that most hydrocarbons have an autochthonous nature and allochthonous origin.
3. Variability in the distribution of AHC in the surface and bottom waters of the northeastern part of the Kara Sea is mainly due to natural processes.

Acknowledgements. The expedition work was carried out with the support of the Ministry of Science and Higher Education of the Russian Federation within the framework of: the plan-program of expeditionary research of Lomonosov Moscow State University on the topic "Features of Quaternary sediment genesis, relief formation and natural fluid discharge on the seabed in the northeastern part of the Kara Sea" and "Training through research on the Arctic shelf"; state tasks of the Moscow Institute of Physics and Technology (National Research University) "Additional support of the system in the field of marine sciences is the training of a young personnel reserve under the scientific and educational program "Floating University" on the basis of comprehensive studies of the seas of Russia and the World Ocean, the state assignment of the IO RAS "Ensuring scientific research, as well as experimental developments". Laboratory work was carried out within framework of the State assignment of the IBSS on the theme "Molismological and biogeochemical foundations of the marine ecosystems homeostasis" (No. 121031500515-8).

References

1. Morgunov, B.A. (ed.): Diagnostic Analysis of the State of the Environment of the Arctic Zone of the Russian Federation (Extended Summary), 200p. Scientific World (2011)
2. Nemirovskaya, I.A.: Oil in the ocean (pollution and natural flows). In: Nemirovskaya, I.A. (ed.) A. P. Lisitsyn, 432p. Scientific World (2013)
3. Nemirovskaya, I.A.: Variability of concentrations and composition of hydrocarbons in the frontal zones of the Kara Sea. Oceanology **55**(4), 552–562 (2015)
4. Silliman, J.E., Schelske, C.L.: Saturated hydrocarbons in the sediments of Lake Apopka, Florida. Organ. Geochem. **34**(2), 253–260 (2003). https://doi.org/10.1016/S0146-6380(02)00169-9
5. Zhang, S., et al.: An analysis of organic matter sources for surface sediments in the central South Yellow Sea, China: evidence based on macroelements and n-alkanes. Marine Pollut. Bullet. **88**(1–2), 389–397 (2014). https://doi.org/10.1016/j.marpolbul.2014.07.064
6. Duan, Y.: Organic geochemistry of recent marine sediments from the Nansha Sea, China. Organ. Geochem. **31**(2–3), 159–167 (2000). https://doi.org/10.1016/S0146-6380(99)00135-7
7. Blumer, M., Guillard, R.R.L., Chase, T.: Hydrocarbons of marine phytoplankton. Mar. Biol. **8**(3), 183–189 (1971). https://doi.org/10.1007/BF00355214
8. Eglinton, G., Hamilton, R.J.: Leaf epicuticular waxes. Science **156**(3780), 1322–1335 (1967). https://doi.org/10.1126/science.156.3780.1322

9. Huang, X., et al.: Significance of long chain iso and anteiso monomethyl alkanes in the Lamiaceae (mint family). Organ. Geochem. **42**(2), 156–165 (2011). https://doi.org/10.1016/j.orggeochem.2010.11.008
10. Ficken, K.J., et al.: An n-alkane proxy for the sedimentary input of submerged/floating fresh water aquatic macrophytes. Organ. Geochem. **31**(7–8), 745–749 (2000). https://doi.org/10.1016/S0146-6380(00)00081-4
11. Mead, R., et al.: Sediment and soil organic matter source assessment as revealed by the molecular distribution and carbon isotopic composition of n -alkanes. Organ. Geochem. **36**(3), 363–370 (2005). https://doi.org/10.1016/j.orggeochem.2004.10.003
12. Lu, X., Zhai, S.: The distribution and environmental significance of n-alkanes in the ChangjiangRiverestuary sediments. Acta Sci. Circumstan. **28**(6), 1221–1226 (2008)
13. Simoneit, B.R.T.: Characterization of organic constituents in aerosols in relation to their rigin and transport: a review. Int. J. Environ. Anal. Chem. **23**(3), 207–237 (1986). https://doi.org/10.1080/03067318608076446
14. Mazurek, M.A., Simoneit, B.R.T.: Characterization of biogenic and petroleum derived organic matter in aerosols over remote, rural and urban areas. In: Keith, L.H. (ed.) Identification and Analysis of Organic Pollutants in Air, pp. 353–370. Butterworth Publishers, Woburn (1984)
15. Drugov, Y.S., Rodin, A.A.: Environmental analyses in oil and petroleum product spills: a practical guide. In: Laboratory of Knowledge, 273p (2020)
16. Titova, A.M., Nemirovskaya, I.A.: The content of hydrocarbons in water and sediments of the Barents Sea. In: Marine Research and Education (MARESEDU 2017), Moscow, pp. 185–189 (2017)
17. Lemenkova, P.A.: Pollution of the Barents Sea with petroleum hydrocarbons. In: Oil and Gas of Western Siberia, Tyumen, pp. 113–118 (2015)
18. Nemirovskaya, I.A., Titova, A.M., Khramtsova, A.V.: Hydrocarbons in the water column and bottom sediments of the Barents Sea. In: Barents Sea System, Moscow, pp. 531–547 (2021). https://doi.org/10.29006/978-5-6045110-0-8/(35)
19. Klimovsky, N.V., Mokhova, O.N., Chernova, V.G., Kargin, M.V.: Distribution of pollutants in the waters of the White Sea. In: Natural Resources, Their Current State, Protection, Commercial and Technical Use. Petropavlovsk-Kamchatsky, pp. 93–96 (2017)
20. Klimovsky, N.V., Chernova, V.G., Petrakova, I.V.: Assessment of the content of pollutants in the ecosystem of the Kutovaya part of the Onega Bay of the White Sea. In: Marine Pollution: Ecological Monitoring, Bioindication, Rationing, pp. 111–115. "Kolorit", Sevastopol (2018)
21. Mokhova, O.N., Klimovsky, N.V., Chernova, V.G., Melnik, R.A.: On the state of pollution of the waters of the White Sea. Marine Biological Research: Achievements and Prospects, vol. 3, pp. 163–167. EKOSI-Hydrophysics, Sevastopol (2016)
22. Nemirovskaya, I.A., Rejepova, Z.Yu., Lisitsyn, A.P.: Hydrocarbons of surface waters in the transarctic section. Rep. Acad. Sci. **483**(3), 345–349 (2019). https://doi.org/10.31857/S0869-56524863345-34
23. Nemirovskaya, I.A., Artemyev, V.A., Khramtsova, A.V., Solomatina, A.S.: Hydrocarbons in the waters and bottom sediments of the seas of the Western Arctic during the period of variability of the ice cover. In: Monitoring of the State and Pollution of the Environment, Ecosystems and Climate of the Arctic Zone, Moscow, pp. 192–194 (2020)
24. Evseev, A.V., et al.: Environmental hot spots and impact zones of the Russian Arctic. In: ACOPS, GEF PDFB Project No. GF/1100-99-13, 51p. UNEP, Moscow (2000)
25. Lisitsyn, A.P., Kuptsov, V.M.: Flows of sedimentary matter and carbon in the marginal filter of the Yenisei and in the KaraSea. Oceanology **43**(4), 593–604 (2003)
26. Burenkov, V.I., Goldin, Y., Kravchishina, M.D.: Distribution of suspension concentration in the Kara Seain September 2007 according to ship and satellite data. Oceanology **50**(5), 842–849 (2010)

27. Jeng, W.L.: Higher plant n-alkane average chain length as an indicator of petrogenic hydrocarbons contamination in marine sediments. Mar. Chem. **102**(3–4), 242–251 (2006). https://doi.org/10.1016/j.marchem.2006.05.001
28. Sojinu, S.O., et al.: Assessing anthropogenic contamination in surface sediments of NigerDelta, Nigeriawith fecal sterols and n-alkanes as indicators. Sci. Total Environ. **441**, 89–96 (2012). https://doi.org/10.1016/j.scitotenv.2012.09.015

Physical and Biological Interactions

Results of Hydrooptical Field Studies in the Barents and Kara Seas in September 2022

D. I. Glukhovets[1,2(✉)], E. A. Aglova[1,2], V. A. Artemiev[1,2], O. V. Glitko[1,2], V. A. Glukhov[1,2], D. N. Deryagin[1,2], S. K. Klimenko[1,2], M. A. Pavlova[1,2], and I. V. Sahling[1,2]

[1] Shirshov Institute of Oceanology of RAS, 117997 Moscow, Russia
glukhovets@ocean.ru
[2] Moscow Institute of Physics and Technology, 141701 Dolgoprudny, Russia

Abstract. The presented results were obtained during the 1st stage of the 89th cruise of the R/V «Akademik Mstislav Keldysh» in the Barents and Kara seas in September 2022. The spatiotemporal variability of the seawater bio-optical characteristics distributions was studied in order to identify climatic and interannual changes. An array of hydrooptical measurement data was obtained, including the results of measurements performed using a flow-through complex and a polarization lidar, as well as data on light fields and seawater bio-optical characteristics. In the Kara Sea two under-aircraft measurement sessions were performed together with the 'Optik' Tu-134 flying laboratory. This array will be used to validate regional satellite algorithms.

Keywords: ocean optics · field studies · seawater bio-optical characteristics · flow-through measuring complex · lidar sounding · satellite ocean color · coccolithophore bloom · Barents Sea · Kara Sea

1 Introduction

The hydrooptical studies carried out during the 1st stage of the 89th cruise of the R/V "Akademik Mstislav Keldysh" (September 5 to 18, 2022) continues the long-term observations of the Barents Sea and the Kara Sea waters, conducted in recent years in the Shirshov Institute of Oceanology of the Russian Academy of Sciences (IO RAS) [1, 2]. The central focus of the hydrooptical studies is the long-term observation of the spatial and temporal changes in the bio-optical characteristics of seawater, aiming to spot climatic shifts and interannual fluctuations [3–7]. To accomplish this task, profiling measurements of beam attenuation coefficient, irradiance, chlorophyll-a fluorescence intensity, and temperature, as well as water sampling for determinations of fluorescence and absorption spectra were performed at the stations. Along the ship route, the characteristics of the seawater surface layer were continuously recorded using a flow-through measurement complex (chlorophyll-a and colored dissolved organic matter (CDOM) fluorescence intensities, beam attenuation coefficient, temperature and salinity). Lidar sounding of the upper layer of seawater was carried out both at the stations and along

T. Chaplina (Ed.): CIWO 2023, SPEES, pp. 439–445, 2023.
https://doi.org/10.1007/978-3-031-47851-2_53

the ship's route. Simultaneously, irradiance values at the sea surface level were continuously recorded. Express methods of fluorescence intensity and absorption coefficient determination made it possible to estimate the concentration of chlorophyll-a prior to standard laboratory measurements performed by the respective research team.

Massive coccolithophore blooms have been investigated in the southern part of the Barents Sea. We define a "bloom" as an event where the coccolithophore concentration surpasses one million cells per liter. These blooms carry significant climatic implications as they influence the ocean surface layer heat budget. This impact is related to a growth of the water column albedo, leading to a decreased absorption of solar energy into the water [8]. Additionally, coccolithophore blooms alter the carbon dioxide balance of the atmosphere-ocean system [9]. The fluctuations in bloom events, observed in recent years, can potentially serve as a critical indicator of ongoing climatic shifts.

These studies included two aircraft measurement sessions by the unique scientific facility (USF) the 'Optik' Tu-134 flying laboratory, involving comprehensive hydrooptical measurements at the stations. The aircraft made passes over the ship above the sea surface on September 9 and 10, 2022, to validate airborne algorithms for retrieving the characteristics of the underlying surface (chlorophyll, etc.). This was complemented by underway measurements along the ship route, using a flow-through measurement complex and a polarized lidar.

Satellite ocean color scanners data available in cloudless weather conditions were regularly acquired during the expedition. The comprehensive approach to study the bio-optical characteristics of the seawater surface layer combines the accuracy of shipboard direct measurements and wide coverage of the water area by satellite images. Collection and processing of satellite data for sending to the ship were performed by the staff of the Ocean Optics Laboratory of IO RAS.

2 Materials and Methods

The flow-through measurement complex [10] involved a number of instruments, including four-channel spectral and two-channel flow-through fluorimeters (developed at IO RAS). Also, the temperature and salinity meters Expert-002 and Idronaut OS314 were used. For measuring the seawater beam attenuation coefficient, we used the PUM-A transmissometer (developed at IO RAS). Surface and underwater photosynthetically active radiation were measured with the help of LI-COR devices, while BIC and RAMSES meters were used for measuring spectral values of surface and underwater irradiance.

The shipboard polarized lidar (developed at IO RAS) consisting of optical unit, power supply unit for pulsed diode-pumped solid-state laser, control and synchropulse formation unit, LeCroy HDO 4034 digital oscilloscope also was used, as well as SBE 25 SeaBird CTD probe with installed LISST-Tau transmissometer; PUM-200 transmissometer (developed at IO RAS); PRO-1 floating spectroradiometer (developed at IO RAS); deck spectroradiometer; spectral integrating cavity absorption meter (ICAM) and Secchi disk.

As remote sensing data, Level-2 remote sensing reflectance spectra was taken, downloaded from the NASA Ocean Color Web (https://oceancolor.gsfc.nasa.govn). This data was collected using MODIS and VIIRS ocean color scanners, providing spatial resolutions of 1 km and 750 m respectively. To ensure accuracy, the interval between field and remote measurements was kept under 24 h. Computations of bio-optical characteristics were carried out with the help of the regional algorithms developed by the Ocean Optics Laboratory, IO RAS [11]. Satellite data was processed using the SMCS software [12], also developed at the IO RAS. This software allows users to receive the Level-3 files and export it to various graphic formats.

The K17 empirical algorithm is based on a logarithmic regression linking the concentration of chlorophyll-a and the *Rrs* ratio at the wavelengths 531 nm and 547 nm [13]. The regional semi-analytical algorithm was developed to estimate the yellow substance absorption coefficient a_g, particle backscattering coefficient b_{bp}, and diffuse attenuation coefficient $K_d(\lambda)$ in four spectral channels 443, 490, 555, and 625 nm in the Kara and White seas using MODIS satellite data [14].

3 Results and Discussion

Aircraft measurements of the seawater surface layer energy budget components were carried out together with the Tu-134 "Optik" Flying Laboratory at two sites in the Kara Sea. Obtained results made it possible to develop a method for calculating the spectral remote sensing reflectance based on the total upwelling radiance measurements [15].

The spatial distributions of seawater hydro- and bio-optical characteristics recorded by the flow-through measurement complex are shown in Fig. 1. These data make it possible to separate the contributions of various sources to the CDOM fluorescence signal by the regression slope between salinity and the CDOM fluorescence intensity. The surface layer temperature values may be used as an additional factor for separating. The avalanche sedimentation presence at the exit from the Yugorsky Shar Strait was shown. Wherein the colored dissolved organic matter spreads into the Barents Sea for more than 20 miles.

Lidar sounding of the seawater upper layer was carried out both at the stations and along the ship route. The polarization lidar sounding signals verification of the Kara and Barents seas waters at all stations was carried out with data on vertical distribution profiles of temperature, density, salinity, chlorophyll-a fluorescence intensity and seawater beam attenuation coefficient. Lidar sounding data made it possible to register [16] in the Kara Gates Strait an internal wave with a period of 4–8 min at a 11–13 m depth.

This dataset also included records from nine stations where data was acquired using the SBE 25 CTD probe.

At one of the stations, the SBE 25 probe was equipped with two identical LISST-Tau transmissometers, one of which was located vertically and the other horizontally. At the same time, the SBE 25 probe was deployed synchronously with the PUM-200 transmissometer. During the comparison of the data obtained from two LISST-Tau sensors with the data of the PUM-200 transmissometer, orientation-induced discrepancies were revealed. The best agreement with the PUM-200 data was achieved with the horizontal arrangement of the LISST-Tau.

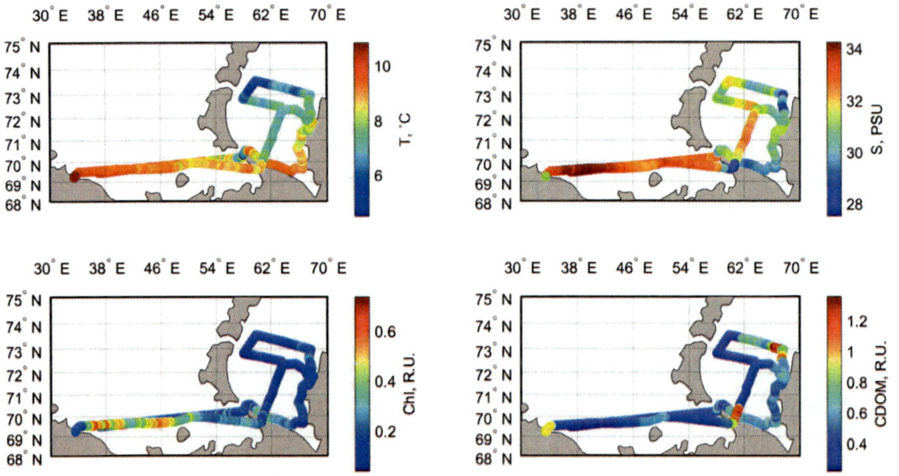

Fig. 1. Spatial distributions of temperature, salinity, as well as Chl and CDOM fluorescence intensities along the ship track during the first stage of AMK 89 cruise

A modern deck spectroradiometer prototype was made and tested; it was shown a good agreement with the floating spectroradiometer data, which has absolute radiometric calibrations of the radiance and irradiance channels.

Flow-through transects were performed in the fading coccolithophore bloom areas, determined from the data obtained on the cruise from satellite ocean color scanners in the Barents Sea southern part. According to cell concentration estimates based on continuous measurements of the seawater beam attenuation coefficient in the surface layer, the concentration changed from ~ 3 million cells/L (September 6, 2022) to ~ 1.5 million cells/L (September 16, 2022), which agrees well with the satellite estimates data.

Figure 2 visually illustrates the spatial distribution of the CDOM absorption coefficient and other bio-optical characteristics of the Kara Sea with the Yugorsky Shar Strait seawater surface layer, derived from the MODIS data, using the regional algorithms of the Ocean Optics Laboratory of the IO RAS. Generally, these distributions are consistent with the direct measurements of seawater bio-optical characteristics, which were carried out in cruise. These indicator values are influenced by the land proximity and a huge river runoff.

The K17 and RSA regional satellite algorithms were also developed at the Ocean Optics Laboratory. Their results demonstrated the most accurate correspondence with *in situ* measurements, with relative errors of 39% and 35% respectively. While the standard satellite algorithms generally yielded unsatisfactory results, with the largest discrepancies observed at the stations located within the river runoff influence zones. This could be attributed to high CDOM content, as well as non-standard ratios of CDOM and chlorophyll-a in the seawater.

The comparison results of satellite algorithm operation for obtaining $K_d(490)$ and calculations using RAMSES profiles shows (Fig. 3) that both standard and regional algorithms estimate this value with satisfactory accuracy. In the future it is planned to conduct a corresponding test at other wavelengths available from satellite data.

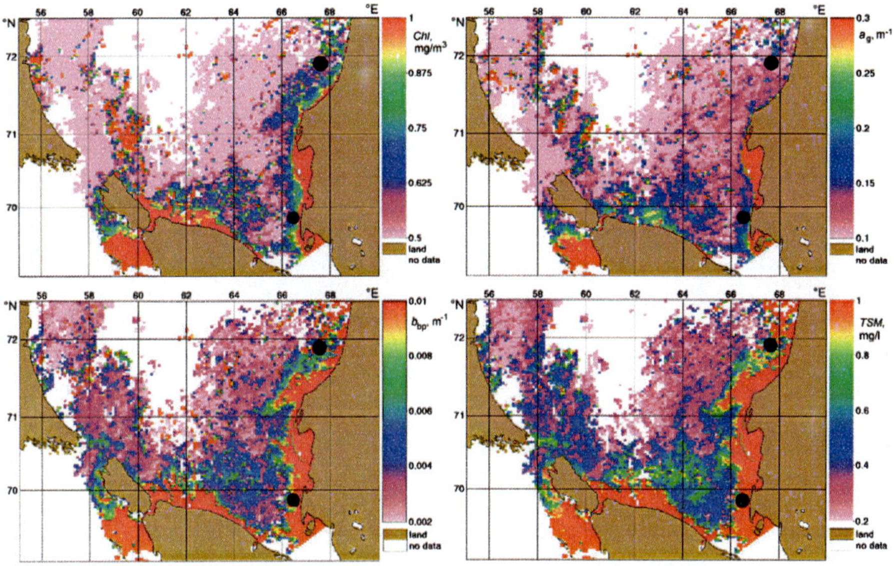

Fig. 2. Spatial distributions of chlorophyll-a concentration (top left), CDOM absorption coefficient $a_g(443)$ (top right), particulate backscattering coefficient $b_{bp}(555)$ (bottom left), and total suspended matter concentration (bottom right). The dots mark the positions of the vessel at the time of the aircraft flight. September 9–10, 2022

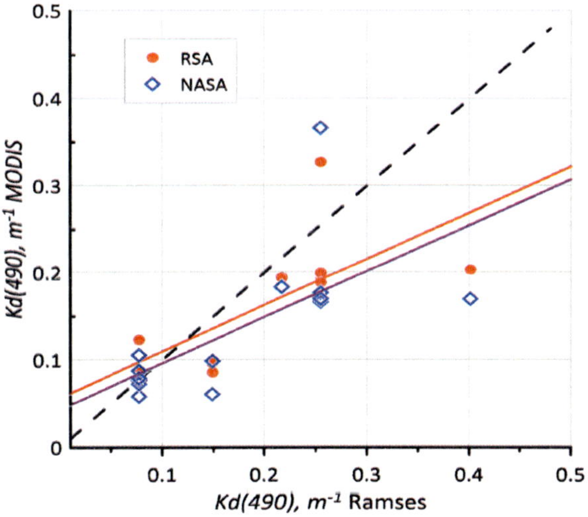

Fig. 3. Correspondence of the $K_d(490)$ calculations results, based on shipboard measurements and MODIS data

4 Conclusion

New current data on the spatial distribution of the seawater bio-optical characteristics and their features of the Barents and Kara seas surface layer have been obtained. These data will be used to validate the results of regional satellite algorithms for assessing the seawater bio-optical characteristics, as well as to evaluate the errors of these algorithms.

Acknowledgements. Shipboard data retrieval was carried out as part of the SIO RAS state assignment (theme No. FMWE-2021–0001), satellite data processing and comparison supported by RSF grant No. 21–77-10059.

References

1. Flint, M.V.: Cruise 54th of the Research vessel akademik mstislav keldysh in the Kara Sea. Oceanology **50**(5), 637–642 (2010)
2. Flint, M.V., Poyarkov, S.G., Rimsky-Korsakov, N.A., Miroshnikov, A.Y.: Ecosystems of siberian arctic seas-2019: spring processes in the kara sea (cruise 76 of the R/V Akademik Mstislav Keldysh). Oceanology **60**(1), 134–137 (2020). https://doi.org/10.1134/S0001437020010105
3. Burenkov, V.I., Vasil'kov, A.P.: The influence of runoff from land on the distribution of hydrologic characteristics of the Kara Sea. Oceanology **34**(5), 591–599 (1995)
4. Burenkov, V.I., Goldin, Y.A., Artemiev, V.A., Sheberstov, S.V.: Optical Characteristics of the kara sea derived from shipborne and satellite data. Oceanology **50**(5), 675–687 (2010)
5. Burenkov, V.I., Goldin, Y.A., Kravchishina, M.D.: The distribution of the suspended matter concentration in the kara sea in september 2007 based on ship and satellite data. Oceanology **50**(5), 798–805 (2010)
6. Glukhovets, D.I., Goldin, Y.A.: Surface desalinated layer distribution in the Kara Sea determined by shipboard and satellite data. Oceanologia **62**(3), 364–373 (2020)
7. Glukhovets, D., et al.: Evaluation of the CDOM absorption coefficient in the arctic seas based on sentinel-3 OLCI data. Remote Sensing **12**(19), 3210 (2020)
8. Kopelevich, O.V., Vazyulya, S.V., Grigoriev, A.V., Khrapko, A.N., Sheberstov, S.V., Sahling, I.V.: Penetration of visible solar radiation in waters of the barents sea depending on cloudiness and coccolithophore blooms. Oceanology **57**(3), 402–409 (2017)
9. Rost, B., Riebesell, U.: Coccolithophores and the biological pump: responses to environmental changes. In: Thierstein, H.R., Young, J.R. (eds.) Coccolithophores, pp. 99–125. Springer, Heidelberg (2004). https://doi.org/10.1007/978-3-662-06278-4_5
10. Goldin, Y.A., Glukhovets, D.I., Gureev, B.A., Grigoriev, A.V., Artemiev, V.A.: Shipboard flow-through complex for measuring bio-optical and hydrological seawater characteristics. Oceanology **60**(5), 713–720 (2020)
11. Kopelevich, O.V., et al.: Bio-optical characteristics of the seas, surrounding the western part of Russia, from data of the satellite ocean color scanners of 1998–2017. In: OOO «VASh FORMAT», Moscow, p. 140 (2018)
12. Sheberstov, S.V., Lukyanova, E.A. A system for acquisition, processing, and storage of satellite and field biooptical data. In: Proceedings of IV International Conference "Current Problems in Optics of Natural Waters". Nizhny Novgorod, pp. 179–183 (2007)
13. Demidov, A.B., Kopelevich, O.V., Mosharov, S.A., Sheberstov, S.V., Vazyulya, S.V.: Modelling Kara Sea phytoplankton primary production: development and skill assessment of regional algorithms. J. Sea Res. **125**, 1–17 (2017). https://doi.org/10.1016/j.seares.2017.05.004

14. Vazyulya, S.V., Kopelevich, O.V., Sheberstov, S.V., Artemiev, V.A.: Satellite estimation of the coefficients of CDOM absorption and diffuse attenuation in the White and Kara seas. Curr. Probl. Remote Sens. Earth Space **11**, 31–41 (2014)
15. Glukhovets, D.I., Ivlev, G.A., Belan, B.D. Calculation of the remote sensing reflectance based on the total upwelling radiance measured from the 'Optik' Tu-134 aircraft laboratory. In: Proceedings of SPIE, 29th International Symposium on Atmospheric and Ocean Optics, Atmospheric Physics (2023)
16. Glukhov, V.A., Goldin, Y.A., Rodionov, M.A.: Method of internal waves registration by lidar sounding in case of waters with two-layer stratification of hydrooptical characteristics. Fundamentalnaya i Prikladnaya Gidrofizika **14**(3), 86–97 (2021)
17. Kravchishina, M.D., et al.: Systems research of sedimentation in the european arctic during the 84th cruise of the research vessel Akademik Mstislav Keldysh. Oceanology **62**(4), 660–663 (2022)

The Impact of Mesoscale Eddies on the Vertical Distribution of Oxygen in the Black Sea

A. A. Kubryakov(✉), S. V. Svishchev, P. N. Lishaev, and E. A. Kubryakova

Marine Hydrophysical Institute of RAS, 299011 Sevastopol, Russia
arskubr@ya.ru

Abstract. The influence of mesoscale eddies of different signs on the seasonal variability and vertical distribution of dissolved oxygen in the Black Sea was investigated based on the method of eddy identification by altimetric measurements and a long-term array of hydrochemical measurements. It has been shown that in z-coordinates, at the depths of the oxycline, the maximum growth is observed in anticyclonic eddies (AE) during the summer season and a decrease in cyclonic eddies (CE). In the upper layers, the situation is opposite, with a decrease in AE and an increase in CE, especially pronounced during the period of spring phytoplankton bloom and associated with the influence of eddies on bioproductivity. At the same time, in isopycnic coordinates, eddies of both signs cause unidirectional changes: a significant increase in oxygen anomaly in the oxycline and a decrease in the sub-oxygen zone, apparently caused by an intensification of shear turbulence mixing in eddy formations.

Keywords: mesoscale eddies · oxygen · Black Sea · Bio-Argo · seasonal variability

1 Introduction

The Black Sea is one of the largest anoxic basins in the world's oceans. The oxygen layer suitable for life occupies the upper 100–200 m, while below the basin is filled with hydrogen sulfide, and oxygen is completely absent. In the upper layers, the variability of oxygen concentration is determined primarily by biological productivity and oxygen solubility. In the lower part of the oxygen zone, it is associated with the redox processes that occur in a number of thin layers with different hydrochemical conditions. The positions of these layers are closely related to water stratification and certain isopycnic layers [1, 2].

Eddy dynamics have a significant influence on the spatial and temporal variability of dissolved oxygen concentration due to several main factors. Firstly, convergent/divergent motions in eddies cause intense vertical movements, significantly affecting the vertical distribution of dissolved oxygen in the basin. The works [3, 4] have shown that anticyclones cause a significant increase in oxygen content in the basin, which is associated with convergence and subsidence of surface water masses in their cores. On the contrary, the ascent of deep water in cyclones contributes to the decrease of O2 in the upper layer.

T. Chaplina (Ed.): CIWO 2023, SPEES, pp. 446–455, 2023.
https://doi.org/10.1007/978-3-031-47851-2_54

Another important effect that eddies have is associated with the intensification of vertical shift of currents in these energetic formations [5, 6]. Such a process causes an intensification of turbulent entrainment at the periphery and the lower boundary of eddies, where orbital velocities are higher, and weakening in the cores of eddies where velocities are minimal. The intensification of turbulent exchange and the associated intensification of diapycnic mixing affect the variability of oxygen not only in vertical but also in isopycnic coordinates. In particular, in the works [7], the authors recorded a sharp penetration of oxygen into the underlying layers. The work [8] based on numerical modeling showed that in AE, the upper boundary of the sub-oxygen zone can drop from isopycnal 1015.5 to isopycnal 1015.9 kg/m^3. Thirdly, eddies modulate the vertical entrainment of biogenic elements, causing changes in oxygen biological production. The influence of eddies on variability of biological characteristics of the Black Sea was discussed in a number of works (see review in [9]). The increase in productivity in cyclones and its decrease in anticyclones can have an important value in the variability of oxygen in the upper photic layer. Finally, the decrease in water temperature in cyclones leads to an increase in oxygen solubility, promoting its more efficient penetration into lower layers. The opposite effect is observed in anticyclones, where the temperature is higher.

At the Marine Hydrophysical Institute of RAS, a large array of measurements of the vertical distribution of oxygen has been accumulated [10]. These data were used to study the long-term variability of oxygen in a number of monographs [11, 12], where it was shown that the dependence between oxygen content and depth exists only for depths not exceeding the thickness of the active layer. Continuous measurements of oxygen concentration in the Black Sea are currently being conducted using Bio-Argo buoys and anchored Aqualog buoys, which allow detailed information on its temporal variability with high vertical resolution [13–15]. In work [7], based on the highest frequency measurements with a discretization of 6 h on an anchored profiling buoy Aqualog, showed significant short-period variability of the vertical position of oxygen.

Since 2010, methods of automatic identification of eddies based on altimeter measurements have been developed for their study. In the Black Sea, such methods have been used in a number of works [16, 17]. Composite analysis of identified eddies, together with hydrological profiles, allows for obtaining new data on the influence of eddies on the thermohaline structure of water [18]. Combining in-situ oxygen measurements with eddy data allows for studying the influence of eddies on the vertical distribution of oxygen. The first such work was carried out in [15], where information on the influence of eddies on average oxygen concentration was provided based on several different arrays of identified eddies and Bio-Argo data.

Intensity of vertical movements, stratification, and productivity have pronounced seasonal variability, which affects the features of the influence of eddies on oxygen. In this work, based on composite analysis of Bio-Argo buoy and satellite altimetry data, the influence of eddies on the variability of oxygen concentration in vertical and isopycnal coordinates is analyzed, as well as the features of the seasonal response of oxygen concentration to the impact of eddies of different signs.

2 Data and Methods

2.1 Bio-Argo Measurements

We use measurements of Bio-Argo floats of oxygen concentration in the upper 200 m water layer of the deep-water part of the Black Sea, limited by the 1500 m isobath for the period from March 2011 to January 2021. The array includes more than 1400 oxygen concentration profiles with high vertical resolution (1 m). The data were downloaded from the IFREMER archive. Dissolved oxygen measurements were made with Aanderaa Oxygen Optode optical oxygen sensors. This method allows for long-term measurements due to the fact that Bio-Argo buoys spend most of their time below the oxic layer, which prevents biofouling.

2.2 Data from the MHI Oceanographic Database

In addition, data on oxygen concentration from the oceanographic database of the Marine Hydrophysical Institute of the Russian Academy of Sciences from 1992 to 2018 were used in this work [10]. This array includes measurements obtained in hydrological studies of Russian, Ukrainian, and Turkish expeditions [19]. The main part of the measurements was obtained using the Winkler method. The archive contains more than 50,000 measurements. A detailed description of the data processing and quality control is given in the study by [20].

2.3 Eddy Identification

Automatic identification method and composite analysis method The automated eddy identification method "winding angle" is used to identify mesoscale eddies from satellite altimetry data in 1992–2015. Details of the algorithm previously used for the study of spatial variability of the Black Sea eddies characteristics, their interannual, and seasonal variability are described in [16, 17]. Overall, there were identified more than 2400 unique anticyclones and 2800 unique cyclones, that corresponded to more than 55,000/50,000 times cyclonic/anticyclonic eddies observed in the daily altimetry dataset in 1992–2021 period. For each identified eddy, we define the following parameters: radius of eddy (R), maximum orbital velocity (Vmax), mean vorticity, eddy lifetime (L), position of the eddy center (xc, yc). Radius R is defined from the eddy area using the formula of the area of a circle.

The oxic structure of the Black Sea eddies was analyzed based on the composite method, which was used in [18] to investigate the thermohaline structure of the Black Sea eddies. Knowing the eddy positions identified from satellite altimetry measurements, we choose all the O2 profiles that were situated on the distance less than 1.2 R away from the eddy center. Totally 197 profiles of O2 were chosen in AEs, 221 profiles of O2 were chosen in CEs. For each measurement in point (x, y), we define its relative coordinates to the eddy center (x_c, y_c)., where $x_r = (x - x_c)/R$; $y_r = (y - y_c)/R$ and distance to the eddy center $r = \sqrt{x_r{}^2 + y_r{}^2}$. . Then the composite structure of the eddy data on temperature, salinity in relative coordinates O2(r, z), was computed on 0.1 R spatial grid at each horizon. Area-averaged profiles of oxygen were computed by averaging O2(r,z) over the eddy radius R.

To determine the anomaly of oxygen concentration from background values, profiles selected outside of eddies were used. Averaged profiles outside of eddies over a certain period were subtracted from profiles within eddies to determine anomalies caused by eddy movements.

3 Results

3.1 Impact of Eddies on the Vertical Distribution of Oxygen in z-coordinates

It is known from previous studies that downwelling in anticyclones lead to accumulation of oxygen-rich surface water in anticyclones [3]. Conversely, upwelling in cyclones leads to the lifting of oxygen-poor water to upper layers. The results of the composite analysis are consistent with these studies. The average profiles of oxygen in anticyclones and cyclones show that in the layer of 30–150 m, oxygen decreases in cyclones and increases in anticyclones, compared with profiles outside of vortices (Fig. 1). This is visually evident in the profiles of oxygen anomalies in Fig. 1b. In cyclones, the maximum decrease in oxygen is observed at depths of 60–70 m, i.e. above the oxycline. Upwelling of oxygen-poor water to these layers leads to a decrease in O2 by an average of 50 μmol/L. In anticyclones, it is located slightly deeper in the layer of 70–80 m, where its anomaly exceeds 100 μmol/L.

In the lower part of the oxygen zone (deeper than 100 m), these anomalies are 10 μmol/L and 20–30 μmol/L for cyclones and anticyclones, respectively. It is noticeable that anticyclones cause a sharper change in oxygen compared to cyclones. In the Black Sea, anticyclones are usually much more intense than cyclones. Apparently, this is why they cause more significant oxygen anomalies compared to cyclones.

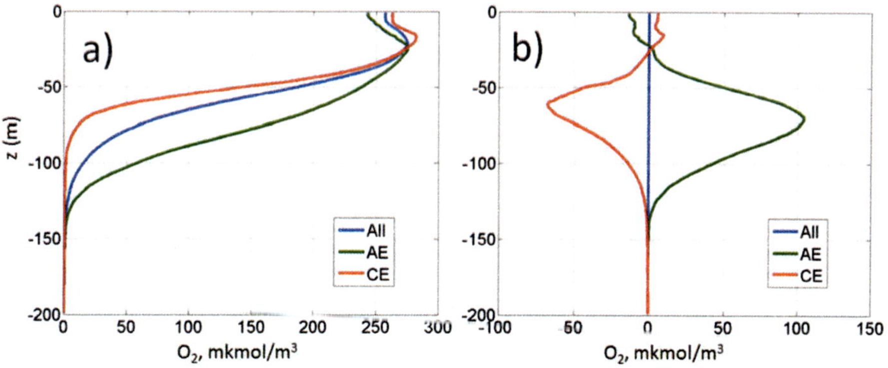

Fig. 1. Composite profile of dissolved oxygen concentration (mkmol/l) (a) and its anomalies (b) in cyclones (red line), anticyclones (green line) compared to waters outside of vortices (blue line) in z-coordinates. (Color figure online)

Seasonal diagrams of oxygen and its anomalies in AE and CE demonstrate significant seasonal fluctuations of O2 response to eddy movements. In all seasons, in the layer of

40–120 m, oxygen concentration in AE is much higher than average. The most significant difference is observed in the summer period when it exceeds 100 μmol/L at depths of 60–80 m. The layer of increased values has a thickness of about 100 m and occupies depths of 120 m. This growth of the anomaly is likely associated with the seasonal intensification of AE in the summer period, which is described in the study.

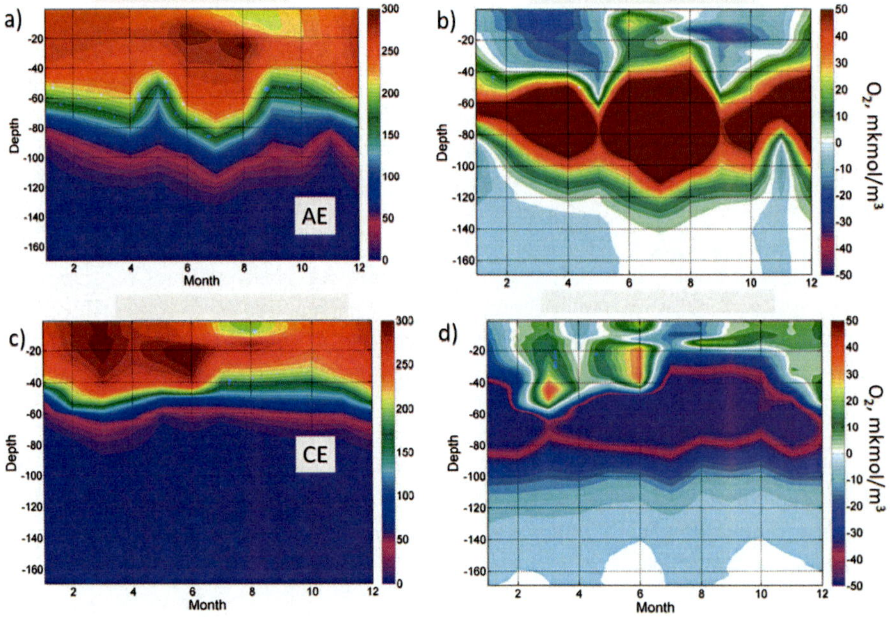

Fig. 2. Seasonal variability of vertical oxygen distribution in anticyclones (a) and cyclones (b) and its anomaly from seasonal variation outside vortexes (c and d, respectively) in z-coordinates.

In cyclones, oxygen anomalies are generally negative. They are less intense and occupy a layer of ~ 20–100 m. The largest anomalies are observed in winter and autumn, when they occupy the most extensive layer (20–100 m), which can also be facilitated by the seasonal strengthening of cyclones in the autumn period. In spring, negative anomalies occupy the smallest layer of 60–100 m.

Interestingly, in the upper euphotic layer (above 30 m), the anomaly of O2 changes sign to the opposite. In cyclones, oxygen growth is generally observed, while in anticyclones, it decreases. The diagrams in Fig. 2d show that positive values are observed in the upper layer of cyclones, especially during the period of spring phytoplankton bloom. The oxygen anomaly in the euphotic layer in the CE region is positive from February to June, reaching 20–30 μmol/L (Fig. 2d). The same increase is noticeable during the autumn phytoplankton bloom in October-December.

At the same time, in AEs, anomalies in the layer of 0–20 m are negative and reach values of −20 to −30 μmol/L (Fig. 2b). The summer surface minimum in AEs covers a much larger layer (0–30 m) and lasts longer from June to October, while in cyclones, it is found in the layer of 0–10 m from July to September.

These features may be related to the influence of eddies on bioproductivity and oxygen formation associated with primary production. CEs cause the uplift of nutrients and increase productivity, leading to more intensive oxygen formation during respiration. This effect should be of the most important value during the phytoplankton bloom period in spring and autumn, corresponding to positive anomalies in the diagram in Fig. 2d. In the spring, phytoplankton bloom usually covers the largest layer (0–50 m), leading to an increase in oxygen in the CEs at these depths of 0–50 m.

In AEs, downward movements cause a weakening of vertical flows of nutrients, which apparently leads to a decrease in oxygen production in the upper layers. An additional effect can be the intensive entrainment of shelf waters with a large amount of dissolved organic matter by AEs, which leads to the consumption of O2 for its oxidation. In addition, the increase in temperature in the AE can cause the decrease in oxygen solubility.

3.2 Impact of Eddies on the Vertical Distribution of Oxygen in Isopycnal Coordinates

Due to the sharp stratification, oxygen concentration in the Black Sea is strongly linked to isopycnal coordinates [2]. The variability of oxygen concentration in isopycnal coordinates is significantly lower than in vertical coordinates. Downward and upward movements cause isopycnal displacement of water, accompanied by the sinking of oxygen-rich water. However, oxygen behaves as a passive tracer that does not penetrate into lower isopycnal layers in most cases. The processes of oxygen incorporation into lower isopycnal layers, associated with diapycnal mixing, have a significant impact on oxygen redistribution, but their causes are poorly understood. These processes can be observed when analyzing changes in oxygen concentration in isopycnal coordinates.

The average profile of oxygen concentration anomaly in AE and CE shows that, unlike z-coordinates, eddies of both signes cause intense growth of oxygen concentration at isopycnals ~ 1014.5–1105 kg/m^3. The results of the study [18] show that there is a maximum velocity shear at pycnocline depths in eddies. Studies [5] show that intense orbital velocities on the periphery of eddies lead to a sharp increase in the coefficient of turbulent mixing. Shear turbulence causes oxygen to penetrate from the upper layers, where the oxygen source is located, to the lower layers. Figure 3 shows that this process leads to an increase in oxygen concentration in both positive and negative eddies. Due to the higher intensity of the AB oxygen anomaly, it is higher in AB and reaches 25 μmol/l in the layer of 1014.5–1014.7 kg/m^3. In cyclones, this maximum is located slightly higher at isopycnals of 1014.4–1014.6 kg/m^3 and is slightly lower in amplitude.

The exception is the deep isopycnal layers in cyclones. In the layers of 1015.5 and below, there is a slight decrease in oxygen concentration. In cyclones, the rising water leads to the lifting of the hydrogen sulfide zone. The intensification of diapycnal mixing leads to the involvement of hydrogen sulfide, the source of which is located below, in the upper layers. The magnitude of this anomaly is small and is less than 5 μmol/l. However, considering that oxygen concentrations at these depths are also very low, such a process can lead to a noticeable lifting of the sub-oxygen and hydrogen sulfide zones upwards.

In anticyclones, this process is less pronounced, but the anomalies in the layer below 1015.7 kg/m^3 are also negative.

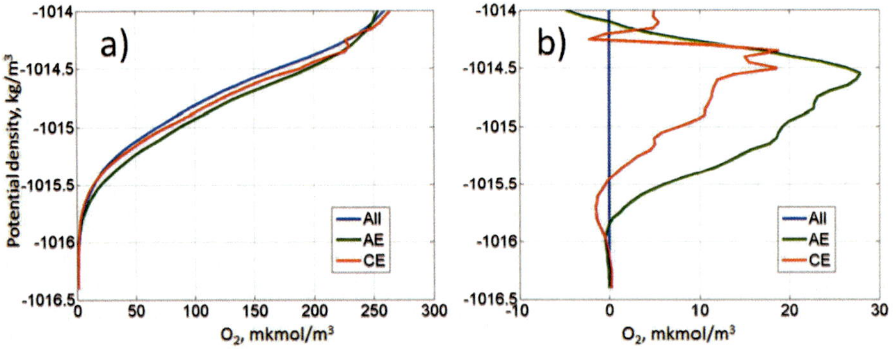

Fig. 3. Composite profile of dissolved oxygen concentration (a) and its anomalies (b) in cyclones (red line), anticyclones (green line) compared to waters outside the eddies (blue line) in σ-coordinates. (Color figure online)

The seasonal variability of oxygen concentration in isopycnal coordinates in eddies (Fig. 4) has significant differences from the mean, which are particularly well observed in the distribution of the oxygen concentration anomaly. The most pronounced increase in oxygen anomaly in the layer of 1014–1015.5 kg/m^3 is observed in the summer in the

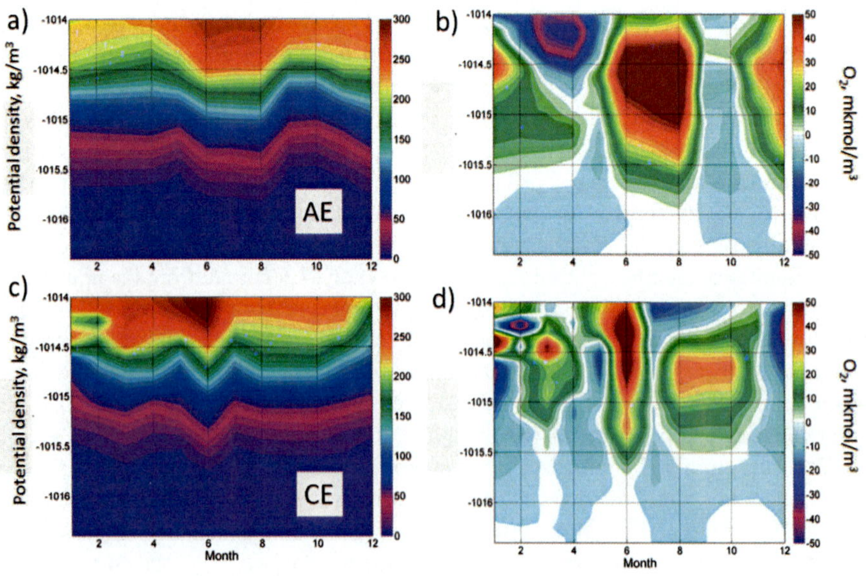

Fig. 4. Seasonal variability of the vertical distribution of oxygen in anticyclones (a) and cyclones (c) and its anomaly from the seasonal cycle outside the eddies (b and d, respectively) in σ-coordinates

Black Sea. At this time, an increase in eddy activity is observed, anticyclonic eddies reach their maximum intensity and size [17]. In addition, the positive oxygen anomaly in these layers increases in winter, during the weakening of stratification, which can also contribute to the strengthening of diapycnal mixing. A characteristic feature in Fig. 4b is the decrease in oxygen in the AEs in isopycnal layers of 1014–1014.5 kg/m^3 during the spring phytoplankton bloom. This decrease apparently reflects a decrease in productivity in the AEs, which in spring occupies the entire photic layer corresponding to these isopycnals. In CEs, the variability is more complex, with some increase in O2 anomaly in the warm period and a decrease in the winter.

4 Conclusions

In this study, based on data on the location of eddies identified from altimeter measurements and a multi-year array of oxygen concentration measurements from oceanographic surveys and Bio-Argo buoys, we investigated the influence of eddies of different sign on the seasonal variability and mean vertical distribution of oxygen in depth and isopycnal coordinates.

The main conclusions are as follows:

- Maximum O2 anomalies in eddies are observed in the intermediate layers at depths of the oxycline (40–120 m). The concentration increase in the AEs is more pronounced than the decrease in the CEs, and the peak anomaly, which reaches 50 and 40 μmol/l, respectively, is located at deeper depths - 75 and 60 m. The largest negative anomalies of O2 are observed in the summer period in the AEs, when the eddy dynamics intensify.
- In the upper layers, the situation is reversed: a decrease in oxygen concentration in the AEs and an increase in the CEs are observed. This feature is most pronounced during the spring and autumn phytoplankton blooms and is likely associated with the influence of eddies on productivity....
- At the same time, in isopycnal coordinates, eddies of both signs cause an increase in oxygen in the layer of 5–1015 kg/m^3, which is apparently associated with an intensification of shear turbulent mixing in intense eddies. The peak anomaly is 15–20 μmol/l and is observed at isopycnals of 1014.5 μmol/l in the CEs and 1014.7 kg/m^3 in the AEs Anomalies are larger in the AEs during their intensification in the summer.
- In the lower isopycnal layers (with a density of more than 1015.5 kg/m^3), a decrease in oxygen is observed, especially sharp in the CE, which may be associated with the turbulent ientrainment of hydrogen sulfide in these layers under the action of mixing.

Acknowledgements. The work was carried out with the support of the Russian Science Foundation grant 21-77-10052.

References

1. Stunzhas, P.A., Yakushev, E.V.: Fine hydrochemical structure of the redox zone in the black sea according to the results of measurements with an open oxygen sensor and with bottle samplers. Oceanology **46**, 629–641 (2006). https://doi.org/10.1134/S0001437006050031

2. Konovalov, S., Tuğrul, S., Baştürk, Ö., Salihoğlu, İ: Spatial isopycnal analysis of the main pycnocline chemistry of the Black Sea: seasonal and interannual variations. In: Özsoy, E., Mikaelyan, A. (eds.) Sensitivity to change: Black Sea, Baltic Sea and North Sea, pp. 197–210. Springer, Dordrecht (1997). https://doi.org/10.1007/978-94-011-5758-2_16
3. Latun, V.S.: Anticyclonic eddies in the Black Sea in the summer of 1984. Soviet J. Phys. Oceanogr. **1**(4), 279–286 (1990). https://doi.org/10.1007/BF02197397
4. Krivosheya, V.G., Titov, V.B., Ovchinnikov, I.M., Kosyan, R.D., Skirta, A.Y.: The influence of circulation and eddies on the depth of the upper boundary of the hydrogen sulfide zone and ventilation of aerobic waters in the Black Sea. Oceanology **40**, 767–776 (2000)
5. Zatsepin, A.G., Golenko, N.N., Korzh, A.O., et al.: Influence of the dynamics of currents on the hydrophysical structure of the waters and the vertical exchange in the active layer of the Black Sea. Oceanology **47**(3), 301–312 (2007). https://doi.org/10.1134/S0001437007030022
6. Podymov, O., Zatsepin, A., Kubryakov, A., Ostrovskii, A.: Seasonal and interannual variability of vertical turbulent exchange coefficient in the Black Sea pycnocline in 2013–2016 and its relation to variability of mean kinetic energy of surface currents. Ocean Dyn. **70**(2), 199–211 (2020). https://doi.org/10.1007/s10236-019-01331-w
7. Ostrovskii, A.G., Zatsepin, A.G., Solovyev, V.A., Soloviev, D.M.: The short timescale variability of the oxygen inventory in the NE Black Sea slope water. Ocean Sci. **14**(6), 1567–1579 (2018). https://doi.org/10.5194/os-14-1567-2018
8. Oguz, T.: Role of physical processes controlling oxycline and suboxic layer structures in the Black Sea: physical control on black sea suboxic layer structure. Glob. Biogeochem. Cycl. **16**(2), 3-1–3-13 (2002). https://doi.org/10.1029/2001GB001465
9. Mikaelyan, A.S., Zatsepin, A.G., Kubryakov, A.A.: Effect of mesoscale eddy dynamics on bioproductivity of the marine ecosystems. Phys. Oceanogr. **27**(6), 590–618 (2020). https://doi.org/10.22449/1573-160X-2020-6-590-618
10. Khaliulin, AKh., Godin, E.A., Ingerov, A.V., Zhuk, E.V., Galkovskaya, L.K., Isaeva, E.A.: Ocean data bank of the Marine Hydrophysical Institute: information resources to support research in the Black Sea coastal zone. Ecol. Saf. Coastal Shelf Zones Sea **1**, 90–96 (2016)
11. Konovalov, S.K., Eremeev, V.N.: Stability and Evolution of Oceanological Characteristics of the Black Sea Ecosystem (EKOSI-Gidrofizika, Sevastopol) (2012)
12. Konovalov, S.K., Vidnichuk, A.V., Orekhova, N.A.: Spatio-temporal characteristics of the hydrochemical structure of the waters of the deep part of the Black Sea In: «The System of the Black Sea», Nauchnyi Mir, Moscow, 2018, pp. 106–119 (2018). https://doi.org/10.29006/978-5-91522-473-4.2018.106
13. Stanev, E.V., He, Y., Staneva, J., Yakushev, E.: Mixing in the Black Sea detected from the temporal and spatial variability of oxygen and sulfide – argo float observations and numerical modelling. Biogeosciences **11**(20), 5707–5732 (2014). https://doi.org/10.5194/bg-11-5707-2014
14. Svishchev, S.V., Kubryakov, A.A.: Impact of winter cooling on the interannual variability in the vertical oxygen distribution in the black sea: evidence from bio-argo floats. Oceanology **62**(2), 143–154 (2022). https://doi.org/10.1134/S0001437022020163
15. Capet, A., Taburet, G., Mason, E., Pujol, M.I., Grégoire, M., Rio, M.H.: Using argo floats to characterize altimetry products: a study of eddy-induced subsurface oxygen anomalies in the black sea. Front. Marine Sci. **9**, 875653 (2022). https://doi.org/10.3389/fmars.2022.875653
16. Kubryakov, A.A., Stanichny, S.V.: Mesoscale eddies in the black sea from satellite altimetry data. Oceanology **55**(1), 56–67 (2015). https://doi.org/10.1134/S0001437015010105
17. Kubryakov, A.A., Stanichny, S.V.: Seasonal and interannual variability of the Black Sea eddies and its dependence on characteristics of the large-scale circulation. Deep Sea Res. Part I: Oceanogr. Res. Papers **97**, 80–91 (2015). https://doi.org/10.1016/j.dsr.2014.12.002

18. Kubryakov, A.A., Bagaev, A.V., Stanichny, S.V., Belokopytov, V.N.: Thermohaline structure, transport and evolution of the Black Sea eddies from hydrological and satellite data. Progr. Oceanogr. **167**, 44–63 (2018). https://doi.org/10.1016/j.pocean.2018.07.007
19. Eremeev, V.N., Konovalov, S.K.: On the formation of the budget and regularities of the oxygen and hydrogen sulfide distribution in Black Sea waters. Morsk. Gidrofiz. Zh. **3**, 5–30 (2006)
20. Svishchev, S.V.: Assessment of the data quality by oxygen in the Black Sea for the period 1923–2007. Sist. Kontrolya Okruzh. Sredy **15**, 208–212 (2011)

Modeling Chlorophyll *a* Concentration for the European Arctic Corridor Based on Environmental Parameters

S. K. Kuzmina[1(✉)], P. V. Lobanova[1], and S. S. Chepikova[2]

[1] Saint Petersburg State University, St. Petersburg 1199034, Russia
so.k.kuzmina@gmail.com

[2] State Hydrological Institute, St. Petersburg 2199004, Russia

Abstract. This study analyzes remotely sensed chlorophyll *a* (Chl *a*) concentration as an indicator of ocean productivity in the European Arctic Corridor (the Norwegian, Greenland, and Barents Seas), and its connection to physical environmental parameters: Photosynthetically Active Radiation (PAR), Sea Surface Temperature (SST), Mixed Layer Depth (MLD) and Sea Surface Salinity (SSS). Using the Random Forest Machine Learning algorithm in the Classifier modification we created models describing a correlation between Chl *a* and environmental parameters, and retrieved a total area of high-productivity zones (Chl *a* is more than 1 mg m^{-3}) based on these correlations.

Keywords: Ocean productivity · European Arctic Corridor · Modelling · Ocean colour · Remote sensing · Barents Sea · Norwegian Sea · Greenland Sea

1 Introduction

The European Arctic Corridor accounts for almost 50% of the total primary production (PP) of the Arctic region [1]. In recent years, PP of polar seas has been changing due to climate instability which is the most dramatic in the Arctic [1, 2]. The main problem in studying PP in the ocean as a whole, and in the Arctic especially is that in situ data needed for statistically significant analysis of its spatial and temporal variability is lacking: one of the solutions would be to use satellite-derived data to study the productivity of the Arctic [3]. The study of ocean productivity based on satellite-derived data provides the basis for its modeling for the Arctic as a whole: having determined the contribution of various environmental factors to chlorophyll *a* (Chl *a*) variability (as it can indicate the water area's productivity), it is possible to develop algorithms for estimating the productivity of Arctic waters [3, 4].

Since productivity of autotrophic organisms is determined by photosynthesis, for which the pigment Chl *a* is needed, it is influenced by those environmental factors that contribute to the maximum rate of photosynthesis, and phytoplankton growth: light [5], nutrients (entering the upper ocean layers with river runoff, as a result of upwelling and horizontal advection, etc. [6]), and sea surface temperature (SST) [7].

T. Chaplina (Ed.): CIWO 2023, SPEES, pp. 456–462, 2023.
https://doi.org/10.1007/978-3-031-47851-2_55

The study region is characterized by two phytoplankton blooms: spring and autumn. In the Greenland and Norwegian Seas, the onset of the bloom happens later than in the Barents Sea (June and May, respectively); the autumnal bloom is usually more intense than the spring one, and this intensity is ever-increasing due to climate changes in the Arctic [2]. Chl *a* concentration is often greater in the Barents and Norwegian Seas than in the Greenland Sea [8]; however, the spring bloom in the Norwegian Sea is shorter [9].

This work aims to identify environmental factors determining the European Arctic Corridor's productivity (Norwegian, Greenland and Barents Seas), and to create models describing the position of productive waters using the Random Forest machine learning algorithm in the Classifier modification on a base of satellite derived data.

2 Materials and Methods

The following satellite and reanalysis data were used in the study: 1) ocean colour data from the OC-CCI database, version 6 (https://www.oceancolour.org) – chlorophyll *a* concentration (Chl *a*, [mg m^{-3}]); 2) mixed layer depth (MLD, [m]) calculated using a method of Dukhovsky D.S. presented in [10] on a base of vertical density profiles calculated using salinity and temperature data from the EN4 Hadley Center and ARMOR data bases; 3) sea surface salinity (SSS, [psu]) from the MIRAS satellite SMOS database from the ESA (https://earth.esa.int); 5) sea surface temperature (SST, [°C]) from NASA's PODAAC MUR SST data-base (https://podaac.jpl.nasa.gov/MEaSUREs-MUR); 6) photosynthetically active radiation (PAR, [E m^{-2} day^{-1}]) from the MODIS-Aqua spectroradiometer from NASA's OBPG database (https://oceancolor.gsfc.nasa.gov).

All data were interpolated to a grid with a 4x4 km spatial resolution, and averaged monthly for the 2010 to 2019 time period. This was done to eliminate gaps in data caused by atmospheric interference such as, for example, cloud cover for ocean colour data.

In this study, we consider modeling the productivity of the ocean as modeling high-productivity zones (HPZ). We determine HPZ using a threshold value of Chl *a* concentration in a grid cell (4x4 km) greater than 1 mg m^{-3} [11]. Therefore, in this study, we suggest viewing productivity as "fertility".

As the Machine Learning we used Random Forest (RF) algorithm in the Classifier modification. Based on the results, quantitative estimates (f1-score, recall (R), and precision (P)) of the quality of the model were obtained:

$$\begin{aligned} \mathrm{f1-score} &= (2 \cdot \mathrm{P} \cdot \mathrm{R})/(\mathrm{P}+\mathrm{R}) \cdot 100\% \\ \mathrm{Recall\ (R)} &= \mathrm{TrP}/(\mathrm{TrP}+\mathrm{FalN}) \cdot 100\%, \\ \mathrm{Precision\ (P)} &= \mathrm{TrP}/(\mathrm{TrP}+\mathrm{FalP}) \cdot 100\%, \end{aligned}$$

where true positive is TrP, false positive is FalP, and false negative values are FalN, f1-score – the weighted average between recall (R) and precision (P). Resulting metrics describe the pro-portion of correctly predicted cases: thus, 100% is the best indicator for each metric, 0% – the worst.

The resulting data sets were divided into training and test data sets using a ratio conventional for Machine Learning of 2:1 respectively for each sea. For the second validation set, we used all available data except year 2019 as a training set to maximize the accuracy of our algorithms' performance by maximizing the amount of training data available to the algorithm.

3 Results and Discussion

Using the Random Forest Machine Learning algorithm, we developed models to retrieve temporal variability of high- and low-productivity zones in the study area. Model Quality Metrics (MQMs) exceed 90% for each sea for the whole test sample (Table 1). Concurrently, the models for the Greenland and Barents Seas show slightly higher MQMs (by 3–6%) than the model for the Norwegian Sea.

Based on the results of the models, the environmental parameters' input to the variableness of the productivity of the area is estimated (Table 1). Here, PAR, SSS and SST make the greatest contribution to the temporal and spatial variability of Chl *a*: for PAR, this contribution was 28–32%, depending on the sea, for SSS it was 26–31%, and for SST it was 25–29%, however, MLD determined only 12–17% of the variability.

To further evaluate the performance of the models, we validated them for year 2019, which is the last year available for the analysis. For this, a training set was all available data except for year 2019, and a test one was the 11 available months of year 2019 (January-November). MQMs for this data set retained high values, although decreased slightly: precision: 67–71%, recall: 66–68%, f1-score: 66–68%.

According to the analysis of the models, all three models quite accurately represent the general nature of temporal variability (Table 2) of high-productivity zones. This is achieved best by the model for the Greenland Sea, which almost always correctly shows total extension of HPZs during the year, however, it somewhat underestimates their size during spring and autumn blooms. In second place is the model for the Norwegian Sea, which underestimates total extension of HPZs during spring-autumn period. The model for the Barents Sea estimates the total area of HPZs correctly only during a spring bloom (March-June).

To improve and illustrate our understanding of the conditions influencing Chl *a* in each sea, we produced hexbin plots (Fig. 1) showing two of the most prevalent factors influencing productivity for each sea and their relationship to Chl *a* concentration: for the Barents and Greenland Seas SSS and PAR are the more important factors, while for the Norwegian Sea – SST and PAR take the lead.

Table 1. Contribution (in %) of environmental parameters to the variability of the productivity. Here SST stands for Sea Surface Temperature, PAR stands for Photosynthetically Active Radiation, MLD stands for Mixed Layer Depth. Metrics in brackets refer to the metrics for two classes – when Chl *a* is less than 1 mg m^{-3}, and when Chl *a* is more than 1 mg m^{-3}. Metrics in bold refer to a mean metric between the two classes.

Parameter	Barents Sea	Norwegian Sea	Greenland Sea
PAR	29%	28%	32%
SST	28%	29%	25%
SSS	31%	26%	28%
MLD	12%	17%	14%
Model Quality Metrics			
Precision	**0.96** (0.95; 0.97)	**0.93** (0.95; 0.90)	**0.96** (0.97; 0.95)
Recall	**0.96** (0.93; 0.98)	**0.91** (0.97; 0.86)	**0.95** (0.97; 0.93)
f1-score	**0.95** (0.94; 0.97)	**0.92** (0.96; 0.88)	**0.95** (0.97; 0.94)
Data set size in 4x4 km pixels, training/test data	1774134/887067	613026/306513	862802/431401

In the Norwegian and Greenland Seas just before the spring bloom begins, when phytoplankton are still light limited, a thin layer above the shallow pycnocline can be formed due to the relatively fresh waters coming from melting glaciers on land as well as from melting sea ice [12]. This layer is relatively fresh and warm; it lies very close to the surface, so phytoplankton can comfortably photosynthesize in it even during limited light conditions. In addition, mainland meltwater contains a lot of nutrients phytoplankton uses to grow. For instance, silica compounds, which are the base for the skeleton of diatoms. Oliver et al., 2018 [13] modeled the same conditions for the Labrador Sea, confirming the increase in PP when glacial water of Greenland flowed into the sea.

In addition, meltponds formed on sea ice provide phytoplankton with comfortable conditions for productivity due to the presence of nutrients suspended in the ice meltwater and abundant light penetrating the shallow ponds [14]. These conditions are clearly seen in Fig. 1 in areas where PAR is between 10–20 E m^{-2} day^{-1}, salinity is relatively low (below 33 psu), and SST is relatively high (above 6°C). During summer, photosynthesis is limited by nutrients. Thusly, despite high illumination, there are low Chl *a* values corresponding to waters with relatively high PAR, SST and salinity (Fig. 1).

Table 2. Validation results for the proposed models. Comparison of the estimated and remotely sensed total area of high-productivity zones (HPZ) based on chlorophyll *a* concentration in 2019. Data size is in number ($\times 10^3$) of HPZ 4 × 4 km cells. "–" means the absence of data for autumn-winter months.

Month\Sea	Barents Sea		Norwegian Sea		Greenland Sea	
	Satellite data	Estimated data	Satellite data	Estimated data	Satellite data	Estimated data
1	-	-	-	-	-	-
2	-	-	0,2	0,2	0,2	0,2
3	7,4	7,3	0,9	0,8	0,4	0,3
4	43,0	37,2	32,292	30,9	12,5	9,2
5	147,3	149,3	27,325	21,7	68,8	54,5
6	92,6	85,4	106,583	92,9	105,0	82,3
7	55,8	44,0	54,090	27,6	46,3	50,2
8	49,4	56,2	44,141	23,8	28,5	25,9
9	99,7	70,7	98,500	64,7	70,0	37,8
10	5,2	5,2	57,401	57,4	31,6	31,6
11	-	-	-	-	-	-

Local phytoplankton blooms are documented during the intrusion of dynamically active, salty and nutrient-rich waters of the Atlantic Ocean into the area of study [15]. Similar conditions occur after vertical water mixing in deep convection regions and the Greenland and Norwegian Seas' frontal zones (high illumination, high salinity and low temperature zones in Fig. 1). For example, Hansen et al., 2010 [16] described an increase in PP in anticyclonic eddies near the Norwegian Sea fronts. Vertical mixing causes upwelling of nutrients into the surface layer for consumption by phytoplankton.

In autumn, when illumination decreases again and salinity is relatively high, second bloom can take place due to the beginning of winter convection. This is especially prevalent for the Barents Sea (according to Fig. 1): high Chl *a* is characteristic for the light-limited areas (PAR is about 20–30 E m^{-2} day^{-1}) with high salinity (above 30 psu).

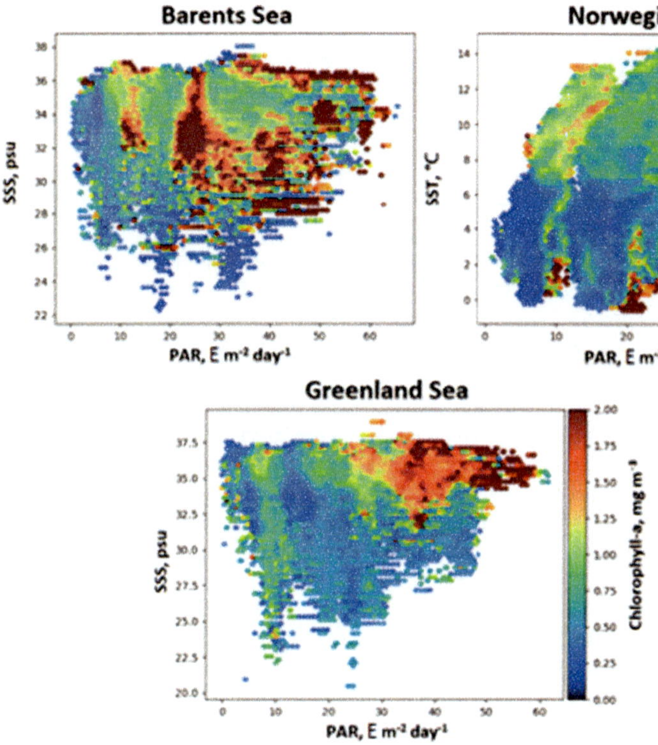

Fig. 1. Hexagonal scatter-plots for the Greenland, Barents and Norwegian Seas: chlorophyll *a* concentration (coloured scale) [mg m^{-3}] and environmental parameters: Sea Surface Salinity (SSS) [psu], Sea Surface Temperature (SST) [°C] and Photosynthetically Active Radiation (PAR) [E m^{-2} day^{-1}].

4 Conclusion

According to the results of the model, the greatest contribution to the variability of Chl *a* is made by the variability of PAR (contribution – 28–32%), SSS (26–31%) and SST (25–29%), the contribution of MLD is almost twice as low (12–17%). Together, these parameters describe the conditions that most favorably affect the productivity of the area: the availability of light in the spring and autumn periods and the availability of nutrients with vertical mixing in the summer and autumn periods. All three models predict the overall temporal variability of HPZs with an accuracy comparable to observed values (depending on the sample, from 66 to 96% of the results were correct).

Acknowledgements. This research was funded by the Saint Petersburg State University, project N 94033410.

References

1. Oziel, L., Massicotte, P., Babin, M., Devred, E.: Decadal changes in arctic ocean chlorophyll a: bridging ocean color observations from the 1980s to present time. Rem. Sens. Environ. **275**, 113020 (2022). https://doi.org/10.1016/j.rse.2022.113020
2. Lewis, K.M., van Dijken, G.L., Arrigo, K.R.: Changes in phytoplankton concentration now drive increased arctic ocean primary production. Science **369**(6500), 198–202 (2020). https://doi.org/10.1126/science.aay8380
3. Lee, Y.J., et al.: An assessment of phytoplankton primary productivity in the Arctic Ocean from satellite ocean color/in situ chlorophyll- a based models. J. Geophys. Res. Oceans **120**(9), 6508–6541 (2015). https://doi.org/10.1002/2015JC011018
4. Kuzmina, S., Lobanova, P., Bashmachnikov, I.: Spatial and temporal variability of chlorophyll-a and its relation to physical and biological parameters: a case study for the European Arctic Corridor. Pices-2022: Book of Abstracts, Busan, Korea. Busan, Korea, p. 127 (2022)
5. Morel, A.: Available, usable, and stored radiant energy in relation to marine photosynthesis. Deep Sea Res. **25**(8), 673–688 (1978). https://doi.org/10.1016/0146-6291(78)90623-9
6. Sverdrup, H.U.: On conditions for the vernal blooming of phytoplankton. ICES J. Marine Sci. **18**(3), 287–295 (1953). https://doi.org/10.1093/icesjms/18.3.287
7. Talling, J.F.: Photosynthetic characteristics of some freshwater plankton diatoms in relation to underwater radiation. New Phytol. **56**(1), 29–50 (1957). https://doi.org/10.1111/j.1469-8137.1957.tb07447.x
8. Bo, Q., Gabric, A.J.: The multi-year comparisons of chlorophyll and sea ice in Greenland sea and Barents sea and their relationships with the North Atlantic Oscillation. J. Marine Syst. **231**, 103749 (2022). https://doi.org/10.1016/j.jmarsys.2022.103749
9. Børsheim, K.Y., Milutinović, S., Drinkwater, K.F.: TOC and satellite-sensed chlorophyll and primary production at the Arctic front in the Nordic Seas. J. Marine Syst. **139**, 373–382 (2014). https://doi.org/10.1016/j.jmarsys.2014.07.012
10. Bashmachnikov, I.L., Fedorov, A.M., Vesman, A.V., Belonenko, T.V., Koldunov, A.V., Dukhovskoy, D.S.: Thermohaline convection in the subpolar seas of the North Atlantic from satellite and in situ observations. Part 1: localization of the deep convection sites. Sovremennye problemy distantsionnogo zondirovaniya Zemli iz kosmosa **15**(7), 184–194 (2018). https://doi.org/10.21046/2070-7401-2018-15-7-184-194
11. Balch, W., Evans, R., Brown, J., Feldman, G., McClain, C., Esaias, W.: The remote sensing of ocean primary productivity: use of a new data compilation to test satellite algorithms. J. Geophys. Res. Oceans **97**(C2), 2279–2293 (1992). https://doi.org/10.1029/91JC02843
12. Levasseur, M.: Impact of Arctic meltdown on the microbial cycling of Sulphur. Nat. Geosci. **6**(9), 691–700 (2013). https://doi.org/10.1038/ngeo1910
13. Oliver, H., et al.: Exploring the potential impact of Greenland meltwater on stratification, photosynthetically active radiation, and primary production in the Labrador sea. J. Geophys. Res. Oceans **123**(4), 2570–2591 (2018). https://doi.org/10.1002/2018JC013802
14. Arrigo, K.R., et al.: Massive phytoplankton blooms under Arctic Sea Ice. Science **336**(6087), 1408–1408 (2012). https://doi.org/10.1126/science.1215065
15. Haug, T., et al.: Future harvest of living resources in the arctic ocean north of the Nordic and Barents seas: a review of possibilities and constraints. Fish. Res. **188**, 38–57 (2017). https://doi.org/10.1016/j.fishres.2016.12.002
16. Hansen, C., Kvaleberg, E., Samuelsen, A.: Anticyclonic eddies in the Norwegian sea; their generation, evolution and impact on primary production. Deep Sea Res. Part I: Oceanogr. Res. Papers **57**(9), 1079–1091 (2010). https://doi.org/10.1016/j.dsr.2010.05.013

Optical and Microphysical Characteristics of Tropospheric Aerosol over Kara Sea using Measurements Onboard TU-134 "Optic" Aircraft Laboratory in 2014, 2020, and 2022

P. N. Zenkova(✉), B. D. Belan, D. G. Chernov, D. M. Kabanov, and I. A. Kruglinsky

V. E. Zuev Institute of Atmospheric Optics SB RAS, 634055 Tomsk, Russia
zpn@iao.ru

Abstract. We present the results from the study of the joint measurements of the aerosol optical and microphysical characteristics onboard TU-134 "Optik" aircraft laboratory and research vessel "Akademik Mstislav Keldysh", carried out in the Kara Sea in September 2022. The mass concentrations of aerosol M_A and absorbing substance M_{eBC} in the near-surface and 9-km layers of the atmosphere were obtained. The results from aircraft sensing are compared with analogous data, obtained during aircraft expeditions in the Kara Sea in 2014 and 2020. Also, our estimates are compared with MERRA-2 reanalysis data.

Keywords: atmospheric aerosol · the Arctic · Kara Sea · Black Carbon · RV "Akademik Mstislav Keldysh" · aircraft laboratory

1 Introduction

The Arctic is the most vulnerable region on the globe, facing global climate change. Studies in 2020 [1] indicate that in the last 43 years the atmosphere of the Arctic warned almost a factor of four faster than, on the average, the whole planet. This accelerating warming is called the Arctic amplification. These processes occur under the influence of many climatic factors, and atmospheric aerosol is one them [1, 2].

In recent time, a lot of papers are devoted to the study of Arctic aerosol [1–9]. However, the majority of publications consider only the near-ground or columnar values of the optical parameters. Also, the gas and aerosol exchange between the atmosphere and sea water remain unclear in these studies. The lacking or episodic observation time series of the vertical distributions of aerosol constituents of the atmosphere complicate the simulation of the atmospheric aerosol processes and make it very difficult to predict the environmental changes [1, 2, 10]. These data are replenished using observations, carried out onboard aircraft laboratory [9].

T. Chaplina (Ed.): CIWO 2023, SPEES, pp. 463–469, 2023.
https://doi.org/10.1007/978-3-031-47851-2_56

Complex studies of the gas and aerosol composition of the troposphere over the Kara Sea were carried out in October 2014 and September 2020 and 2022 onboard the TU-134 "Optik" aircraft laboratory. At the same time, synchronous measurements of the aerosol microphysical and optical properties have been successfully conducted in 2022 onboard RV "Akademik Mstislav Keldysh" and onboard TU-134 "Optik" aircraft laboratory.

2 Material Experiments and Data Representation

Aircraft Laboratory. The routes of the TU-134 "Optik" aircraft laboratory are shown in Fig. 1. In the framework of the YAK-AEROSIB project, the atmospheric sensing in the eastern sector of the Kara Sea was carried out on October 15, 2014 and in the western sector on October 16, 2014 [11]. During the comprehensive experiment on the study of aerosol characteristics un the Russian sector of Arctic, in September 2022 [9], simultaneous measurements were performed for the first time from onboard the research vessel "Akademik Mstislav Keldysh" and the TU-134 "Optik" airborne laboratory. On September 9 and 10, 2022, the aircraft twice flew directly above flew directly above the ship at the altitude of 200 m.

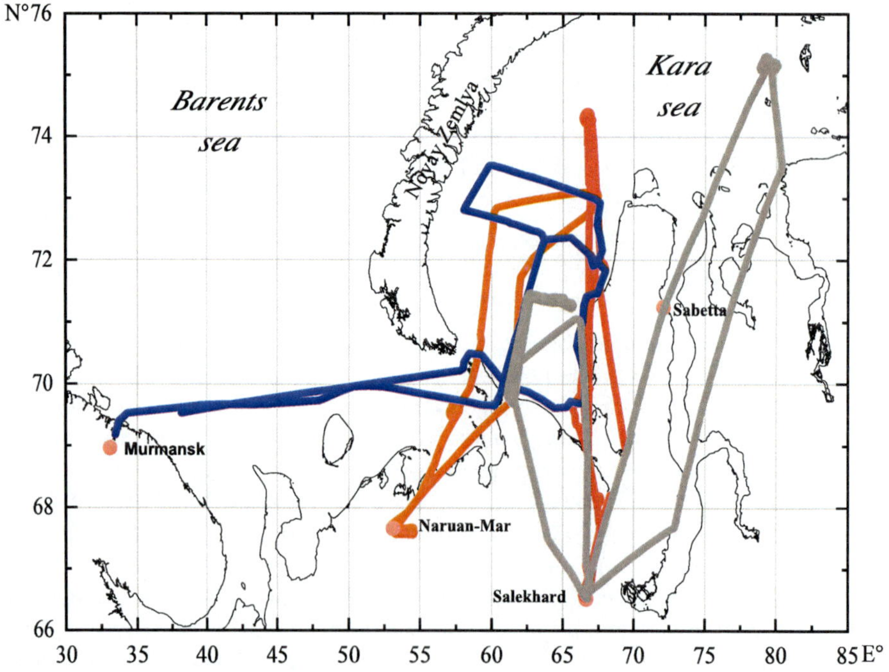

Fig. 1. Map of the routes of the TU-134 "Optik" aircraft laboratory in 2014 (gray), in 2020 (orange), in 2022 (red), and of the route 89.1 of cruise of RV "Akademik Mstislav Keldysh" (blue) in September 2022. (Color figure online)

The aircraft-laboratory flights toward the Kara Sea were carried out from Salekhard in 2014 and 2022 and from Naryan-Mar in 2020. The minimum altitude of the aircraft-laboratory was 200 m above sea level, and the maximum was 9–10 km. Between the extreme points, several 10-min long horizontal flights were performed at the altitudes of 200 m, 5 km, and 9 km.

The aerosol complex onboard the aircraft laboratory comprised aethalometer MDA-02, developed in Institute of Atmospheric Optics, Siberian Branch, Russian Academy of Science [9], for the measurements of the mass concentration of absorbing substance M_{eBC} (equivalent of black carbon, eBC) [12]; and nephelometer PhAN-M for recording the coefficient of the angular scattering by "dry matter" of submicron particles μ_d ($\lambda = 0.53$ μm, $\varphi = 45°$) at the wavelength of $\lambda = 0.53$ μm [9]. The mass concentration of submicron aerosol M_A was estimated empirically from the formula $M_A = 2.4^* \mu_d$ ($\lambda = 0.53$ μm, $\varphi = 45°$), for the particle density of 1.5 g/cm^3. The TU-134 "Optik" aircraft laboratory, instrumentation, and the methods for calibration and measurements were described in [9].

Research Vessel. From 6 to 19 September, 2022, within the program of the 89.1st cruise of RV "Akademik Mstislav Keldysh" (Fig. 1), we carried out measurements of aerosol physical and chemical characteristics in the Kara Sea. We used aethalometer MDA-02 and Grimm-11D aerosol laser spectrometer [8] which were placed at an altitude of 10–12m above sea level. The aethalometer was used to measure the mass concentration of absorbing substance in aerosol in terms of elemental black carbon eBC. The Grimm-11D spectrometer was used to measure and calculate: volumes of fine and coarse fractions of aerosol (with radii smaller and larger than 0.5 μm) V_f and V_c; the volume particle distribution functions dV/dr; and the total number concentration of particles with radii of 0.15–18 μm N_A. Aerosol and eBC concentrations in expedition were measured automatically every hour. Each measurement cycle lasted for 10–20 min. The measurement conditions were described in more detail in [8].

3 Results and Discussion

Aircraft Laboratory. The vertical distributions of the mass concentrations of aerosol and eBC, obtained in the frameworks of three Arctic aircraft expeditions, are presented in Fig. 2.

The vertical distributions of M_A and M_{eBC} over the eastern sector of the Kara Sea on October 15, 2014 are characterized by the values, decreasing with the altitude, and a well-defined mixing layer up to 4.5 km. Over the western sector, the vertical distribution of M_A shows a pronounced mixing layer up to 4.5 km, with the concentrations a factor of two larger than in the eastern sector of the Kara Sea. The altitude distribution of M_{eBC} is characterized by low concentrations ~ 80 ng/m^3 near the surface (0.2–1 km); while at the altitudes from 2 to 5 km, the mass concentration of absorbing substance increases to 100 ng/m^3. The correlation coefficient between M_A and M_{eBC} for the total dataset is 0.98 (0.04) for the eastern (western) sector of the Kara Sea. The high correlation coefficient seems to be indicative of the fact that, in this case, the measurements were carried in a single air mass, formed earlier and transported to the study region.

The vertical distributions of M_A and M_{eBC} over the Kara Sea in 2020 are characterized by the values, decreasing with the altitude, and a well-defined mixing layer up to 5 km. The correlation coefficient between M_A and M_{eBC} over the total dataset for the Kara Sea is 0.67. The large mass concentrations in the lower atmospheric layers can be explained by the existence of the sources of aerosol and absorbing substance on the Earth's surface. The analysis of the back trajectories, obtained using HYSPLIT model (https://ready.arl.noaa.gov/HYSPLIT.php), showed that the air masses during flights came to this region from the European part of the continent.

The measurements in the aircraft expedition were carried out in the western part of the Kara Sea near the Yamal Peninsula. In the vertical distribution of aerosol on September 9, 2022, there was a poorly defined mixing layer with the low content of submicron aerosol in the surface layer (M_A of about 8 μg/m^3) and a gradual decrease to 1 μg/m^3 at the altitude of 3.5 km. The concentrations of submicron aerosol were found to increase at the altitudes of 5.5–7 km. The concentrations M_{eBC} at the altitudes up to 2 km were about 450 ng/m^3 and decreased further upward to 25 ng/m^3 at the altitude of 7.5 km. In the layer of 7.5–8.5 km, the concentrations M_{eBC} rapidly increased to 300 ng/m^3.

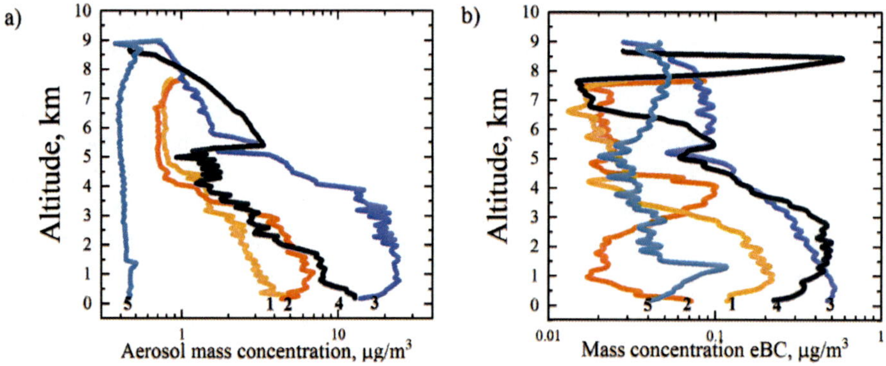

Fig. 2. Vertical profiles of the mass concentrations of (a) submicron aerosol and (b) eBC over the Kara Sea ~~basin~~ on October 15, 2014 (1), October 16, 2014 (2), September 6, 2020 (3), September 9, 2022 (4), and September 10, 2022 (5).

The vertical profiles of the concentrations of submicron aerosol and absorbing substance obtained on September 10 were also characterized by weakly pronounced mixing layer: M_A up to 3.5 km, and M_{eBC} up to 3 km. The parameter M_A gradually decreased from 3 μg/m^3 in the near-ground layer to 0.4 μg/m^3 at the altitude of 3.5 km. The concentrations of absorbing substance were at the altitudes 2 to 5.5 km increase to 280 ng/m^3. Considering back trajectories, calculated be means of the HYSPLIT model (https://ready.arl.noaa.gov/HYSPLIT.php) (Fig. 3), one can see that the air masses came to the region of measurements from the area of the North Pole, North Atlantic, and the north of Canada.

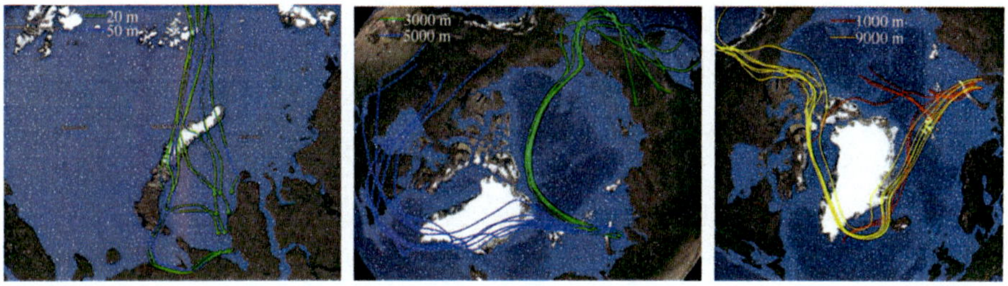

Fig. 3. The 7-day back trajectories of the air mass motion, carried to the altitudes of 20, 50, 1000, 3000, 5000, and 9000 m over the Kara Sea basin on September 9, 2022.

The spatial distribution of the columnar content of absorbing substance $M_{eBC}(col)$ up to 9 km is presented in Fig. 4a. The columnar content of absorbing substance over the water area in the Russian Arctic was determined using the data, recorded during ascents and descents of the aircraft in the altitude range from 200 m to 9 km and back. Over the Kara Sea, the $M_{eBC}(col)$ values varied from 0.05 mg/m^2 to 1.9 mg/m^2 in 2014, from 1.12 mg/m^2 to 2.54 mg/m^2 in 2020, and from 0.41 mg/m^2 to 1.5 mg/m^2 in 2022.

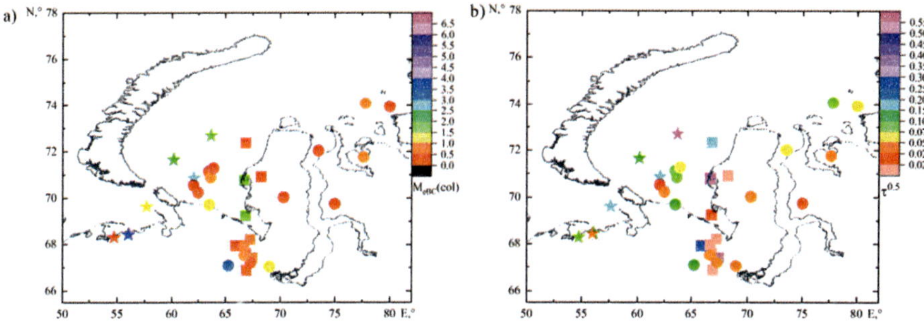

Fig. 4. Spatial distribution of the columnar mass concentration of absorbing substance (a) and aerosol optical depth at the wavelength of 0.5 μ m (b) up to 9 km from the data obtained onboard aircraft laboratory in 2014 (circle), 2020 (asterisk), and 2022 (square).

Synchronous Measurements Onboard Aircraft Laboratory and Research Vessel "Akademik Mstislav Keldysh". Studies of atmospheric aerosol onboard RV "Akademik Mstislav Keldysh" were carried out from September 6 to 19, 2022. In this period, the average aerosol characteristics were: the mass concentration of absorbing substance $M_{eBC} = 49$ ng/m^3, the volume concentration of fine (submicron) particles $V_f = 0.29$ μm^3/cm^3, and the volume concentration of coarse particles $V_c = 4.18$ μm^3/cm^3 (Fig. 5). The maximum aerosol characteristics were observed ($M_{eBC} = 225$ ng/m^3, $V_f = 0.87$ μm^3/cm^3) in the region of the Kara Gates and near Kolguyev Island. This increase was due to the anthropogenic effect. The aerosol characteristics in the period of the expedition somewhat differed from long-term average values in the Kara Sea [8]; for instance, the concentrations M_{eBC} were a factor of 1.4 larger.

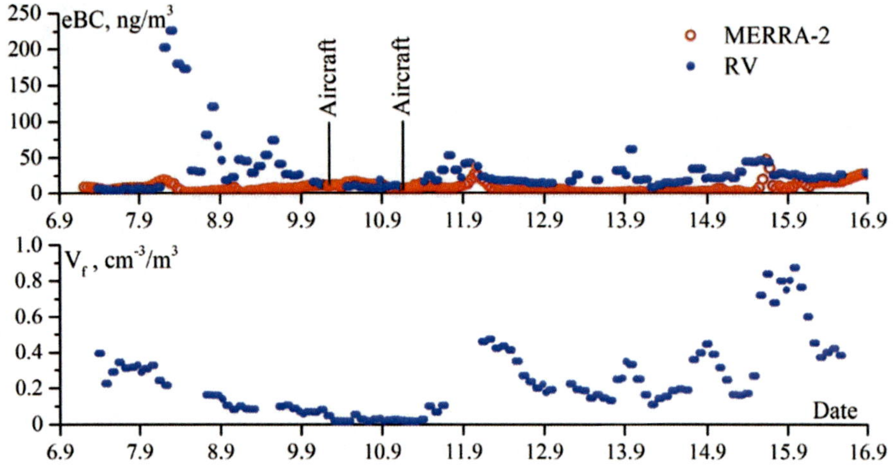

Fig. 5. Time behavior of the aerosol characteristics (eBC, Vf) in 89th cruise of RV "Akademik Mstislav Keldysh" and from MERRA-2 reanalysis data (https://giovanni.gsfc.nasa.gov/giovanni).

We will consider in more detail the aerosol characteristics, obtained on September 9 and 10, i.e., in the periods of simultaneous RV and aircraft measurements. The salient feature of measurements in the near-water layer is low values of the aerosol characteristics: *MeBC* varied from 15 to 52 ng/m^3, the average was 15 ng/m^3; *Vf* varied from 0.02 to 0.1 μm^3/cm^3, the average was 0.04 μm^3/cm^3. If extrapolate the aircraft data to the sea surface, *MeBC* was 168 ng/m^3 on September 9 and 36 ng/m^3 on September 10.

The concentrations M_{eBC} calculated from the MERRA-2 reanalysis data (https://giovanni.gsfc.nasa.gov/Giovanni), quite well agree with the RV data (Fig. 2); the concentrations varied from 5 to 15 ng/m^3 (the average being 9 ng/m^3) in the surface layer. The situation was strongly different on September 8–9, when the vessel moved from Kolguev Island to Kara Gates. The columnar content *eBC* was 0.26 mg/m^2. Based on the aircraft sensing data, the columnar content *eBC* up to 9 km, was 0.43 mg/m^2.

4 Conclusions

The vertical distributions of the mass concentrations of submicron aerosol and eBC, recorded in three expeditions, strongly differ because they are carried out under different synoptic and seasonal conditions. In addition, it is possible that at different altitudes they belong to different air masses, so their distribution is inhomogeneous. This is quite expected because, as air mass moves over territories where there are strong, permanently acting eBC sources, the total composition of submicron aerosol is formed due to natural processes. In this case the principal quantity of absorbing particles in transferred to the region of observations by remote transport from far industrial regions. This process is imprinted also on the chemical composition of aerosol, sampled at different altitudes [9, 11].

The aircraft data, obtained on September 9 and 10, 2022, and extrapolated to the sea level, exceed several-fold the ship-based measurements. Seemingly, this is because of the effect of local pollution source, i.e., emissions from ship chimneys and ventilation shafts.

Acknowledgement. Studies in marine expedition were carried out in the framework of the Russian Science Foundation project no. 21-77-20025; and aircraft observations were supported by the project from the Ministry of Science and Higher Education of the Russian Federation (agreement no. 075-15-2021-934) "Studying the anthropogenic and natural factors of variations in air composition and environmental entities in Siberia and Russian sector of the Arctic under the conditions of rapid climate changes, using unique scientific installation Aircraft laboratory Optik Tu-134".

References

1. Rantanen, M., et al.: The Arctic has warmed nearly four times faster than the globe since 1979. Nat. Commun. Earth Environ. **3**(1), 168–178 (2022)
2. Schmale, J., Zieger, P., Ekman, A.M.L.: Aerosols in current and future Arctic climate. Nat. Climate Change **11**(2), 95–105 (2021). https://doi.org/10.1038/s41558-020-00969-5
3. Vinogradova, A.A., Vasileva, A.V., Ivanova, Y.A.: Air pollution by Black Carbon in the region of Wrangel Island: comparison of Eurasian and American sources and their contributions. Atmos. Ocean. Opt. **34**(2), 97–103 (2021). https://doi.org/10.1134/S1024856021020111
4. Zhuravleva, T.B., Artyushina, A.V., Vinogradova, A.A., Voronina, Yu.V.: Black Carbon in the near-surface atmosphere far away from emission sources: comparison of measurements and MERRA-2 reanalysis data. Atmos. Ocean. Opt. **33**(6), 591–601 (2020). https://doi.org/10.1134/S1024856020060251
5. Vinogradova, A.A., Titkova, T.B.: Atmospheric Black Carbon and surface Albedo in the Russian Arctic during spring. Atmos. Ocean. Opt. **33**(3), 260–266 (2020). https://doi.org/10.1134/S1024856020030136
6. Vinogradova, A.A., Titkova, T.B., Ivanova, Y.A.: Episodes with anomalously high black carbon concentration in surface air in the region of Tiksi station, Yakutiya. Atmos. Ocean. Opt. **32**(1), 94–102 (2019). https://doi.org/10.1134/S1024856019010184
7. Schulz, H., et al.: High Arctic aircraft measurements characterising black carbon vertical variability in spring and summer. Atmos. Chem. Phys. **19**(4), 2361–2384 (2019). https://doi.org/10.5194/acp-19-2361-2019
8. Sakerin, S.M., et al.: Spatiotemporal variations in atmospheric aerosol characteristics over the Kara, Barents, Norwegian, and Greenland seas (2018–2021 expeditions). Atmos. Ocean. Opt. **35**(6), 651–660 (2022). https://doi.org/10.1134/S1024856022060203
9. Belan, B.D., et al.: Integrated airborne investigation of the air composition over the Russian sector of the Arctic. Atmospheric Measurement Techniques **15**(13), 3941–3967 (2022). https://doi.org/10.5194/amt-15-3941-2022
10. Yang, Q., Bitz, C.M., Doherty, S.J.: Offsetting effects of aerosols on Arctic and global climate in the late 20th century. Atmos. Chem. Physics **14**(8), 3969–3975 (2014). https://doi.org/10.5194/acp-14-3969-2014
11. Antokhina, O.Y., et al.: Vertical distributions of gaseous and aerosol admixtures in air over the Russian Arctic. Atmos. Ocean. Opt. **31**(3), 300–310 (2018). https://doi.org/10.1134/S102485601803003X
12. Petzold, A., et al.: Recommendations for reporting "black carbon" measurements. Atmos. Chem. Phys. **13**(16), 8365–8379 (2013). https://doi.org/10.5194/acp-13-8365-2013

Oceanological Technology and Instrumentation

Application of Microwave Resonant Measuring Transducer for Express Assessment of Water Salinity

D. A. Poletaev[1] (✉), B. V. Sokolenko[1], and V. I. Voytitsky[2]

[1] V.I. Vernadsky Crimean Federal University, 295007 Simferopol, Russia
poletaevda@cfuv.ru

[2] Moscow Timiryazev Agricultural Academy, 127434 Moscow, Russia

Abstract. In this paper a precision water salinity sensor based on microwave resonator is proposed. The geometrical dimensions of the sensor aperture are optimised. Experimental verification of sensitivity is carried out and operational characteristics of the device are determined. The methodology of work with the development is proposed.

Keywords: Sensor · Microwave · Resonator · Water Salinity

1 Introduction

Salinity is a relevant indicator that requires precise measurement for various aquatic environments. Analysis of such solutions is relevant for seas, oceans, water bodies for economic purposes [1, 2]. Gravimetric, conductometric methods of salt content analysis are widely used. However, gravimetric methods are very time-consuming and approximate in content. Conductometric methods are express, but require additional maintenance in terms of assessing the degree of oxidation of sensor elements (corrosion). Therefore, modern rapid methods for assessing the qualitative and quantitative composition of water are required.

The main functional unit of the equipment for microwave diagnostics of materials is a sensor, which includes a source of electromagnetic waves, a measuring transducer and information signal extraction devices. The sensors based on resonators are the most widespread [3, 4]. In this case, to ensure non-contact rapid measurements, the electromagnetic field of the resonator probes the sample through a hole in one of the walls. Such a resonant measuring transducer belongs to the aperture type [5–7].

Taking into account the demand for microwave diagnostic methods and their possible wide use, numerical and analytical modelling of a resonant measuring transducer, which, in particular, allows to calculate its conversion characteristics - the source material for further computer processing and visualisation of measurement results, is relevant. Multiparametric optimisation, quantitative study of metrological parameters, justification of application of new types of resonant measuring transducers and their possibilities, also need creation of adequate numerical model.

T. Chaplina (Ed.): CIWO 2023, SPEES, pp. 473–482, 2023.
https://doi.org/10.1007/978-3-031-47851-2_57

Microwave diagnostic methods are widely used in science, technology and national economy. The widespread use of equipment based on these methods is due to the indestructibility of the sample, the absence of physical contact, the possibility of obtaining express data during research [3]. To the greatest extent such equipment is in demand in agro-industrial farms (moisture meters), in various fields of science and production (microwave microscopy), and in biology (to study the characteristics of objects) [5–9]. An integral component of the equipment when using near-field microwave methods are resonant measuring transducers, the design and geometry of which are determined by the field of application. Taking into account the variety of sensors used, we consider the design of a resonant measuring transducer with a coaxial measuring aperture, which has found application for studies of distributed structures, averaging the measured parameters over the volume and area of the object under study, which are liquid solutions.

Measured characteristics of liquid media, as well as other dielectric objects, are electrophysical parameters: relative permittivity and dissipation factor. These quantities are uniquely related to information signals of the resonant measuring transducer: goodness of fit, resonant frequency and their relative changes.

Direct determination of electrophysical parameters of dielectric objects is very labour-intensive. This is due to the necessity of preliminary calibration of the sensor with specific geometrical dimensions, as well as to the peculiarities of the experiments: the measured parameters are strongly influenced by the tightness of the object of study to the aperture, external factors. In addition, it is required to optimise the parameters of the measuring transducer to ensure simultaneously high sensitivity while maintaining the averaged nature of measurements.

Averaging of the obtained experimental values is easily achievable in traditional closed resonators of dielcometric methods [7]. However, the necessity to place an object inside the resonator each time requires considerable time expenditure and complicates its application for series of samples.

Rapid monitoring of electrophysical parameters actually requires an external, relative to the sensor, location of the object and real-time measurements. Conceptually, this could be a portable instrument easily applicable to water salinity measurement.

There is a large number of experimental and theoretical works related to the study of the application of resonator transducers for measuring electrophysical parameters [3–7]. However, these works do not pay due attention to the optimisation of transducer sensitivity for different media (especially liquid media). The theoretical justification of the choice of parameters of the non-dissipative measuring transducer is based on simplified models, allowing to take into account only the vibrational losses in the resonator system. Modern direct numerical methods make it possible to substantially refine the model of the non-dissonant measuring transducer, i.e., to take into account the geometry of the aperture, location, dimensions, and other characteristics of the object under study and to construct the estimated theoretical characteristics of the transformation, linking the information signals with the electrophysical parameters of the object under study [4].

The aim of the work is to optimise the parameters of the dissonant measuring transducer used for measuring the salinity of water medium, as well as to propose a method of working with this device.

2 Material and methods

Figure 1 presents the design of the considered dissonant measuring transducer with an adjacent object. The model includes a section of coaxial waveguide, a sample of thickness h_2 with electrophysical parameters ε_1, tg δ_1.

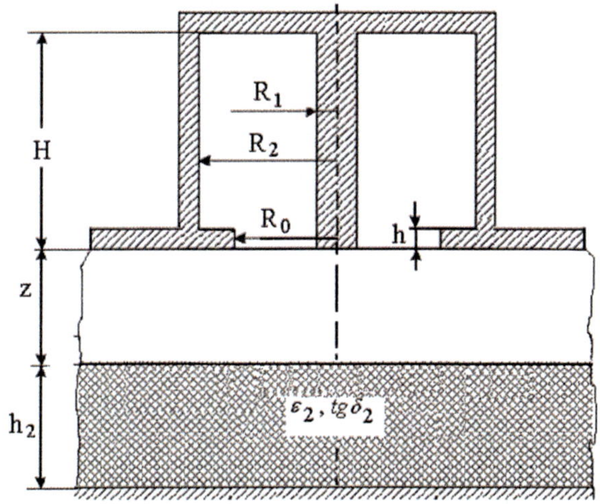

Fig. 1. Model of resonant measuring transducer

For practical measurements, it is important that the electrophysical parameters of the sample have as large an influence as possible on the loaded goodness of the nonlinear measuring transducer. In this case, the loaded goodness of fit should not be less than 100 [4]. From the theory of coaxial transmission lines [5] it is known that the minimum attenuation coefficient in a coaxial line is achieved at the ratio of radii: $R_1 / R_2 = 0.28$. Obviously, the maximum goodness of the quarter-wave resonator will also be achieved at this ratio. When the minimum values are reached, the measurement accuracy drops sharply.

The dispersion maximum for water occurs at a frequency of about 10 GHz [4]. Preliminary analysis allows us to choose the geometrical dimensions of the model: $H / \lambda = 1.25$; $R_2 / \lambda = 0.17$; $h / H = 5.6 \cdot 10^{-3}$; $z = \infty$ (the quarter-wave resonator is loaded on free space); the conductivity of the resonator walls is taken equal to $\sigma = 5.8 \cdot 10^6$ Sm/m, where λ is the wavelength of microwave oscillation. The conductivity value was chosen to be less than the conductivity of pure copper for modelling consideration of wall roughness. The influence of other geometrical dimensions and parameters of the resonator, aperture and object of study was investigated using a numerical model based on the finite element method. It provides for the solution of wave equations with specified boundary conditions [10] when the three-dimensional space is divided into tertrahedrons. It should be noted that the peculiarities of modelling and finite element mesh construction can be presented in a separate large work.

It follows from general physical concepts that the choice of the radius R_0 of the aperture can significantly change the goodness of the resonator measuring transducer. Figure 2 presents the dependences of the goodness and resonant frequency of the resonator on the value of R_0, calculated in the absence of the object of study.

As can be seen from the graphs in Fig. 2, decreasing R_0 of the aperture reduces the radiation losses to free space. The necessity of taking radiation losses into account is most clearly demonstrated by the dependence of the goodness of fit on R_0 of the aperture (Fig. 2a). Even at small size R_0 the goodness of fit when taking radiation into account and without taking it into account differ almost by a factor of 2. It should be noted that the goodness of the resonator decreases by a factor of 1.5 in comparison with the one in which the optimum ratio of radii is maintained. This decrease in the goodness of fit has to be considered forced, when designing a measuring transducer for the study of objects with a high tangent of the dielectric loss angle. From the graph in Fig. 2b we can see that taking into account the radiation practically does not affect the resonance frequency.

Figure 3 presents the dependences of the quality factor and resonant frequency of the measuring transducer on the longitudinal size of the aperture-forming part of the coaxial waveguide h, at different values of R_0.

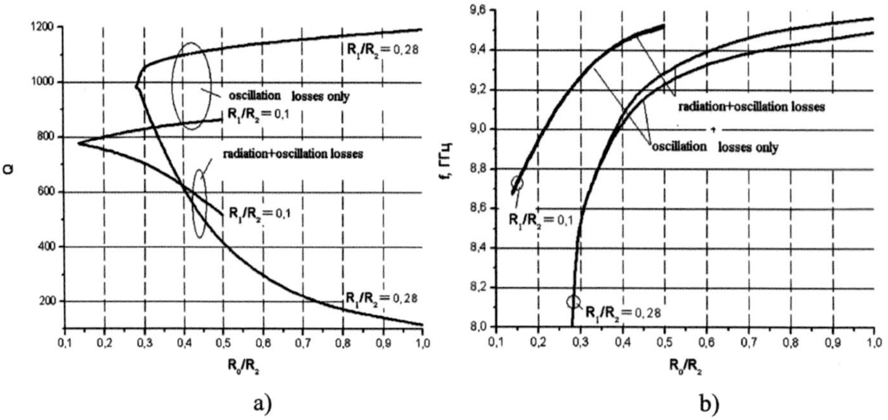

Fig. 2. Dependence of a) goodness and b) frequency from the value of R0

The character of the goodness of fit change (Fig. 3a) is resonant. A sharper resonance peak corresponds to a smaller R_0, which is consistent with general physical concepts. The decrease of the resonance frequency (Fig. 3b) is caused by the growth of the aperture capacitance, while increasing its longitudinal size of the aperture-forming part of the coaxial waveguide.

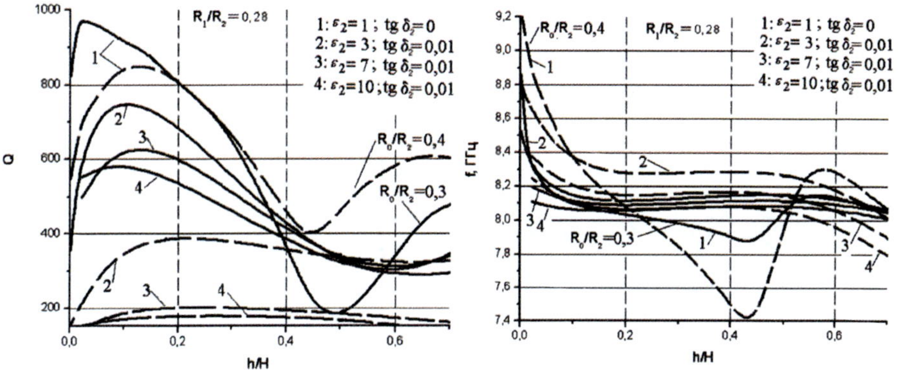

a) b)

Fig. 3. Dependence of a) goodness, b) resonant frequency from the aperture dimensions

As can be seen from the graph (Fig. 3a), the goodness of the resonator in the study object with the parameters: $\varepsilon_1 = 3$; tg $\delta_1 = 0.01$ remains quite high even at aperture openings $R_0 / R_2 > 0.3$. A significant change of the goodness of fit with the object of study, at $h / H \approx 0.4$ (which corresponds to $H / \lambda = 0.5$), relative to the goodness of fit calculated with free space loading is clearly visible (Fig. 3a). This fact is consistent with the physical representations and indicates the adequacy of the numerical model. At the same time, even at $h / H \approx 0.4$, the change in the goodness of the resonator is 2 times smaller than at $h \approx 0$. This is due to the significant field sag from the aperture. From the graph in Fig. 3, a, a very important conclusion follows: the largest change in goodness of fit is provided at $h / H < 0.05$.

It is reasonable to provide for the possibility of investigating a wide range of water salinity (from 0 to 40 ppm) [1]. Quantitatively it can be characterised by the change of relative dielectric permittivity from 3 to 10 and the tangent of dielectric loss angle from 0.01 to 0.4, at $\lambda = 3$ cm. Thus, numerical calculation is reasonable to perform in this range of variation of electrophysical parameters. These data are applied to optimise the geometrical dimensions of the resonant transducer.

Calculations were carried out for the model with parameters: $H / \lambda = 1.25$; $R_2 / \lambda = 0.17$; $z = 0$; $h = 0$; $h_{212} / \lambda = 0.14$; $h / H = 0.01$; conductivity of the resonator walls $\sigma = 5.8 \cdot 10^6$ Sm/m; $\lambda = 3$ cm, when changing the aperture value R_0 / R_2 from 0.29 to 1, for objects whose parameters vary within the above mentioned limits. Graphs of dependence of goodness of fit and sensitivity of the measuring transducer on the value R_0 of the aperture are shown in Fig. 4. The sensitivity by goodness of fit $\delta Q/Q$ was calculated from the change of the tangent of the angle of dielectric losses dtg $\delta_2 = 0.01$, with the value of tg $\delta_2 = 0.01$ and constant relative dielectric permittivity.

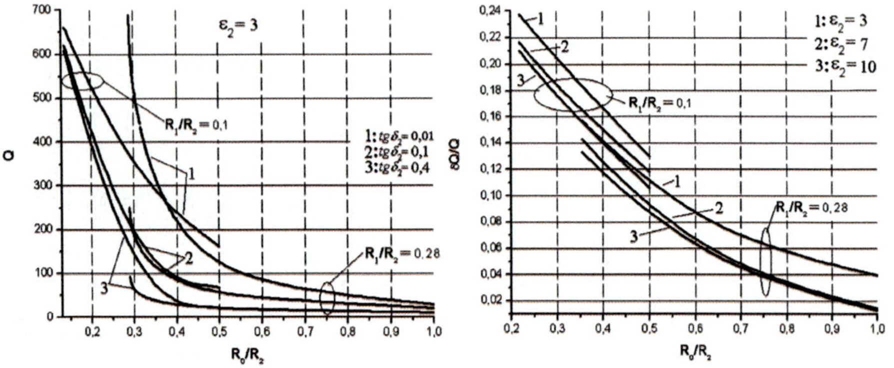

a) b)

Fig. 4. Dependence of a)Q and b)δQ/Q from aperture size and object parameters

The goodness of the measuring transducer (Fig. 4a) depends to a greater extent on the aperture value, which determines the magnitude of radiative losses. Sensitivity of the transducer (Fig. 4b) increases with decreasing R_0 / R_2.

Figure 5 presents the dependences of the resonant frequency and the change of the resonant frequency at variation of the aperture value R_0 / R_2.

As can be seen from the graphs (Fig. 4 and Fig. 5), the goodness of fit in the presence of an object, convenient in metrological terms [4], is achieved in a narrow range of values R_0. At the same time, the sensitivity to samples with high dielectric permittivity turns out to be insufficient. Therefore, such a resonator measuring transducer should be optimised separately for objects with low and high dielectric constant values.

From the Fig. 6 we can see that a larger value of the relative permittivity of the sample increases the value of the system goodness of fit, due to the "locking" of energy inside the resonator.

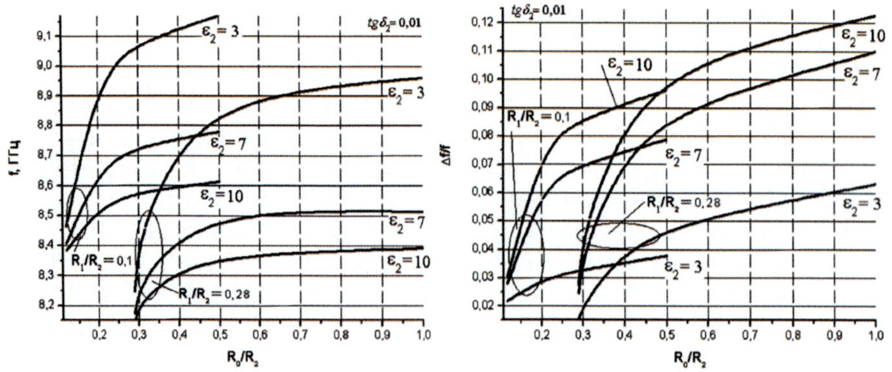

a) b)

Fig. 5. Dependence of a)f and b)Δf/f from aperture size and object parameters

Figure 6 and Fig. 7 present the calculated conversion characteristics with different geometrical dimensions of the aperture.

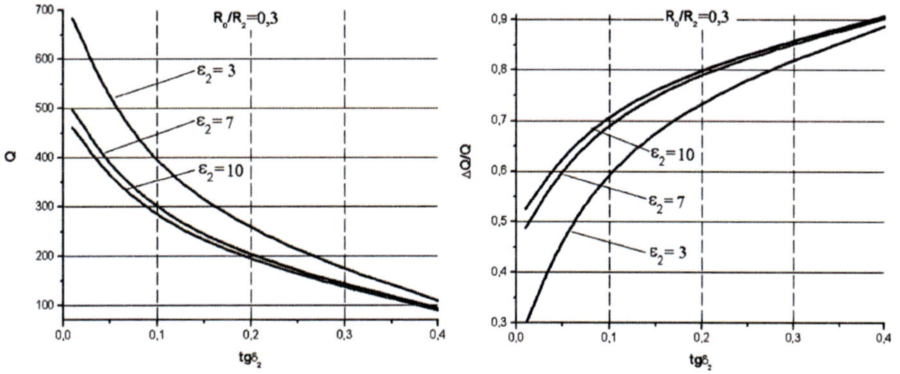

a) b)

Fig. 6. Dependence of a) Q and b) ΔQ/Q from object parameters

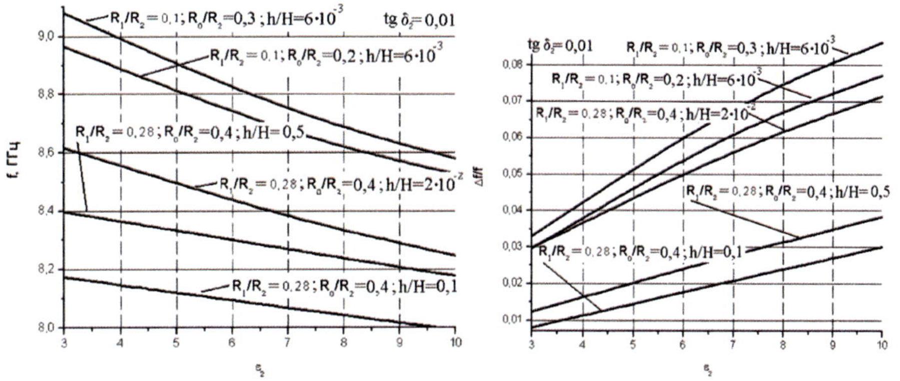

a) b)

Fig. 7. Dependence of a) f and b) Δf/f from object parameters

3 Results and Discussion

On the basis of the numerical model the size optimisation of the resonant measuring transducer is carried out. A prototype of the device was manufactured according to the optimised geometrical dimensions: the ratio of the coaxial aperture radii – 0.4; the ratio of the inner and outer radii of the coaxial resonator - 0.28; the ratio of the coaxial resonator height to the wavelength – 1.25. The material of the coaxial resonator and walls is steel with a copper coating, the thickness of which is greater than the skin layer at the operating frequency. A vector circuit analyser P4226 with a standard connection scheme of the analysed device - resonator - was used as a measuring complex. Prepared water samples with the following salinity values were used as objects: 0 ppm; 10 ppm; 20 ppm; 40 ppm. Such values are typical for the majority of gravimetric meters. Preparation of the study objects was carried out by dissolving the weighed table salt in a measured volume of distilled water. The masses of the component were measured with scales with an accuracy of 0.1 g. Then the liquid was poured into polyethylene bags (thickness not more than 300 μm), which were placed near the aperture of the measuring transducer. It was preliminarily diagnosed that there was no influence of polyethylene of such thickness on the readings of the instrument. Each measurement was carried out several times with subsequent averaging of the results. It should be noted that the scatter of numerical values of the experimental results was not more than 2%. This emphasises the high performance characteristics of the device. During the experiment it was revealed that the investigated object has a significant influence on the characteristics of the sensor.

Figure 8 presents the interpolated values of the relative changes in the resonance frequencies of the sensor with the breakdown, relative to the frequency at open aperture. The dots in Fig. 8 mark the values of information signals obtained during the experiment. Interpolation was performed on these values.

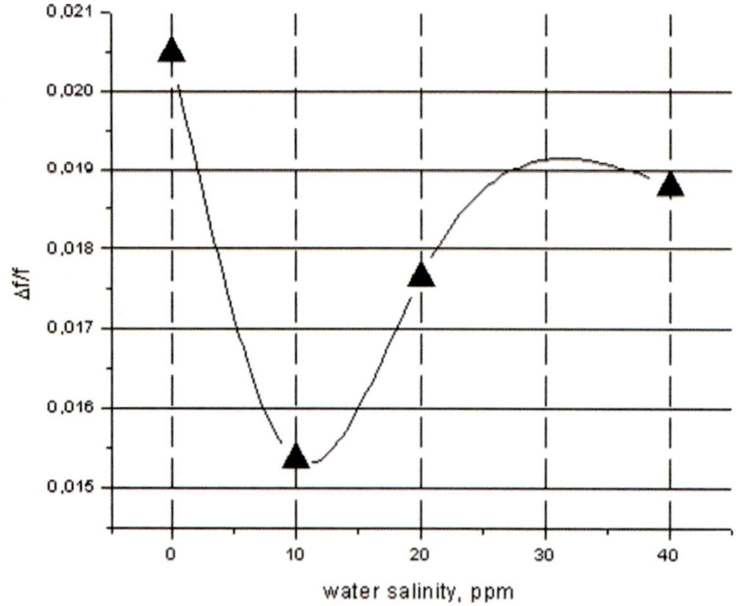

Fig. 8. Values of sensor frequency variation for different water samples

As can be seen from Fig. 8, the obtained values cannot unambiguously diagnose water salinity due to the non-monotonic character of the curve. The type of dependence is most likely caused by the peculiarities of the aperture sensor design. Small values of the dielectric loss angle tangent (at low salt concentrations) practically do not limit the electromagnetic energy output from the aperture. At the same time, the high value of ε of water at microwave affects the reduction of the resonant frequency of the device. As tg δ increases, shielding of the aperture is observed and the electromagnetic energy loss decreases. At the same time, a high value of the dielectric permittivity of water at (ultra high frequency) UHF has less influence on the change of the resonant frequency. At the same time, the revealed dependence (in Fig. 8) can be very informative in the range from 10 ppm to 30 ppm.

Figure 9 presents the interpolated values of relative changes in the quality factor of the resonant measuring transducer, relative to the quality factor at open aperture. The dots in Fig. 9 mark the values of information signals obtained during the experiment. The interpolation was performed on these values.

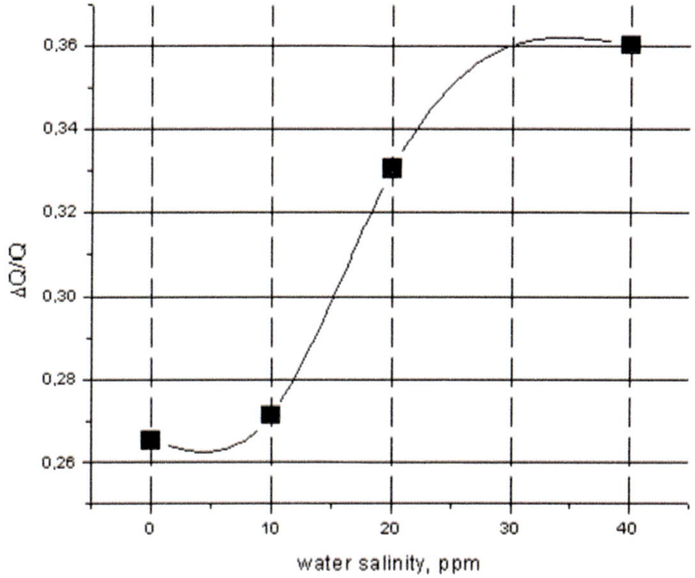

Fig. 9. Values of the variation of the sensor goodness for different water samples

As can be seen from Fig. 9, the obtained characteristic is monotonic, which simplifies the application of the described sensor for practical measurements. The growth of the quality factor variation with the growth of water salinity fits into the existing physical concepts. Thus, the growth of water salinity leads to an increase in the dissipation factor.

The technological process of water salinity measurement on the proposed device can be carried out according to the following methodology:

1. Preliminary calibration of the sensor using solutions with known salt content. This process compares the change in goodness of fit of the sensor to the salinity of the water.
2. Determination of material parameters. The resonant measuring transducer controls the current value of water salinity according to the calibration data of the change of goodness.

4 Conclusion

A precision water salinity sensor is proposed and modelled, geometrical dimensions are optimised. A prototype of the measuring transducer was manufactured. A set of practical measurements of prepared water samples was carried out. The measurement time is estimated to be not more than 100 μs. It assumes application of the developed transducer in a complex of flowing automatic meters, carrying out control of water salinity in real time. As a result of modelling and experimental verification it is established that this device is characterised by high sensitivity, wide range of measurements, expressiveness. In addition, the methodology of technological measurements of water salinity using the developed device is proposed.

Acknowledgements. This work has been supported by the grants the Russian Science Foundation, RSF 22-22-20126 and Crimea region.

References

1. Zaburdaev, V.: To the question of methodology for determining the absolute salinity of seawater with salt composition different from standard seawater. Environ. Control Syst. **20**, 27–32 (2014)
2. Derraik, J.G.B.: The pollution of the marine environment by plastic debris: a review. Marine Pollut. Bull. **44**(9), 842–852 (2002). https://doi.org/10.1016/S0025-326X(02)00220-5
3. Egorov, V.: Resonance methods of the dielectric research on the microwave. Instrum. Techn. Exp. **2**, 5–38 (2007)
4. Chen, L.: Microwave Electronics Measurement and Materials Characterisation, p. 537. John Wiley & Sons, New Yor (2004)
5. Detinko, M.: Physical Bases of Non-destructive Microwave Resonator Method of Local Control of Electrophysical Parameters of Semiconductors, p. 30. Tomsk University, Tomsk (1988)
6. Zavyalov, A.: Measurement of Material Parameters at Ultrahigh Frequencies, p. 213. Tomsk, TSU (1985)
7. Brandt, A.: Investigation of Dielectrics at Ultrahigh Frequencies, p. 404. Fizmatizdat, Moscow (1963)
8. Hong, S.: Tunable resonance cavity control in a near-field scanning microwave microscope. J. Korean Phys. Soc. **40**(5), 861–865 (2002)
9. Kim, J., Kim, M.S., Lee, K., Lee, J., Cha, D., Friedman, B.: Development of a near-field scanning microwave microscope using a tunable resonance cavity for high resolution. Measure. Sci. Technol. **14**(1), 7–12 (2003). https://doi.org/10.1088/0957-0233/14/1/302
10. Grigoriev, A.: Electrodynamics and Microwave Technique, p. 335. M. High School (1990)

Investigation of Winter Plankton Ecology of Lake Baikal Using Complex Instrumental Methods

T. P. Rzhepka[1] (✉) and E. Y. Naumova[2]

[1] Irkutsk National Research Technical University, 664033 Irkutsk, Russia
tasyarjpk@gmail.com
[2] Limnological Institute of the Siberian Branch of RAS, 664033 Irkutsk, Russia

Abstract. Continuous observations of changes in ecosystems are necessary due to global climate change, which affects the state of water bodies and has affected even the deepest lake of the planet - Lake Baikal. For the tasks of establishing continuous automated monitoring of Lake Baikal, daily observations of plankton behavior were carried out. Within the framework of the publication we give a brief description and the first results of the complex work on the evaluation of the efficiency of using the immersed holographic camera and hydroacoustic instruments for monitoring the amount of zooplankton in Lake Baikal.

Keywords: Monitoring · Submersible holographic camera · Jedi network · Lake Baikal · Winter plankton

1 Introduction

The work was conducted on March 14 and 15, 2019 in the Southern Basin of Lake Baikal, 3.5 km from Cape Ivanovsky. Observations were made every 4 h throughout the day. Samples of mesozooplankton were collected with a closing Jedi net with a mesh size of 88 μm (open diameter 35.7 cm) along horizons 0–10, 10–25, 25–50, 50–100 m, fixed in 4% formalin, concentrated by sedimentation and then counted under a light microscope [1].

2 Material and Methods

Hydroacoustic survey was carried out by modernized echo sounder FCV-1100, It is modernized and has the following parameters: pulse duration - 0.3 ms, pulse repetition rate - 5 Hz, signal frequency - 28 kHz. Single-beam hydroacoustic transducer 3 dB has a beam width of 12°. The sonar configuration was as follows: signal frequency - 28 kHz, pulse repetition rate -5 Hz, pulse duration -0.3 ms. The beamwidth of the 3 dB single-beam hydroacoustic transducer was 12°. Water temperature, chlorophyll, photosynthetically active radiation and other parameters were measured by AAQ-Rinko

T. Chaplina (Ed.): CIWO 2023, SPEES, pp. 483–490, 2023.
https://doi.org/10.1007/978-3-031-47851-2_58

probe. RBR-Company RBR TR-1000 (resolution 0.002 °C, accuracy 0.05 °C) and TR-1060 (resolution 0.0001 °C, accuracy 0.002 °C) loggers were used to measure water temperature. The polygon consisted of four points where measurements were taken at depths of 45, 50, and 55 m. The measurement discreteness was 5 min. Another RBRduo logger (resolution 0.00005 °C, accuracy 0.002 °C, sampling rate 5 counts per second) was attached to the Jedi network. An immersion holographic camera was used simultaneously [2].

3 Results and Discussion

We present data from a single vertical 12.00 h. March 14, 2019. The temperature profile at the time of the survey was normal for the subglacial period (Fig. 1 (1)).

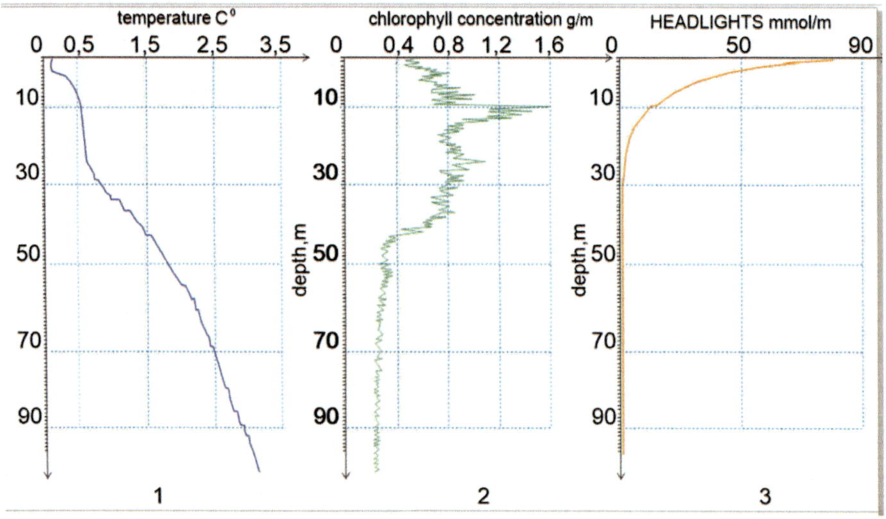

Fig. 1. Variation of temperature, °C (1), chlorophyll a concentration, g/L (2), and photosynthetically active radiation (PAR), μmol/m^2 s (3), by depth on March 14, 2019 at 13.00, m. Ivanovsky. Here and further vertically - depth (m)

Immediately below the ice layer, the temperature was 0.2 °C. Descending to a depth of 4–5 m, a slight warming to 0.5 °C was observed. Up to a depth of 25 m no changes were observed and the temperature was the same 0.5 °C in the epilimnion layer.

Deeper than 25 m, the so-called seasonal layer of increased temperature (3.2 °C per 100 m) occurred [3]. The index of photosynthetic active radiation gradually decreased, the maximum depth of radiation penetration was 71 m. The amount of chlorophyll a increased until a depth of 23 m and then decreased until a depth of 98 m (Figs. 1(2),1(3)). The immersion depths of the Jedi net and holographic camera were monitored by echosounder readings (Fig. 2).

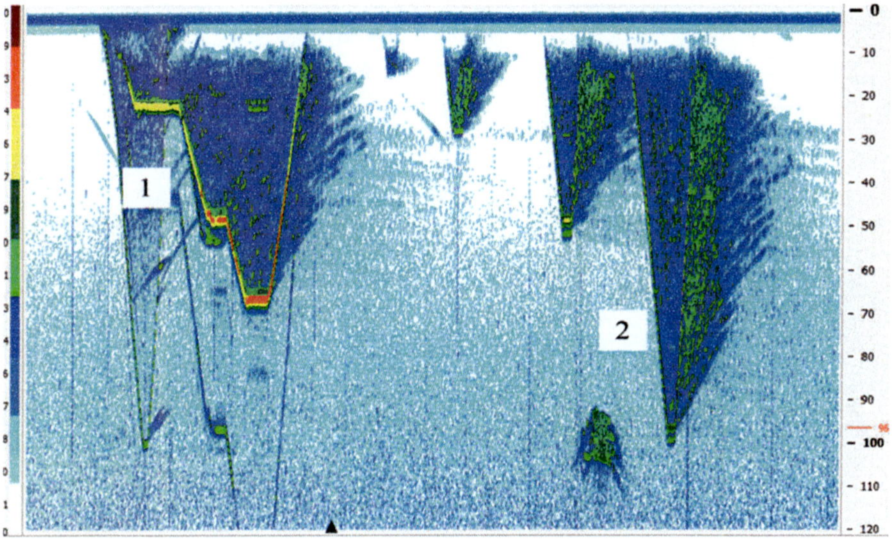

Fig. 2. Echogram obtained on March 14, 2019 at 13.00, m. Ivanovsky. 1 - signal from the submersible holographic unit, 2 - four horizons of the Jedi net dive.

It was assumed that water density affects the vertical distribution of mesoplankton. To test this hypothesis, 6 break points (Fig. 3) at depths of 2, 5, 25, 25, 30, 39, 58 m were marked on the graph [2].

Due to free convection temperature, the lower limit of daily mixing is noted at 5 m depth and is clearly distinguished in all indicators. From 23 m depth the seasonal thermocline begins. This layer is up to 30 m thick (i.e. extends for 7 m).

Considering the places of zooplankton distribution in March and comparing the vertical distributions, we can see that rotifers prefer the epilimnion, where water is well mixed (diurnally and seasonally) and there is enough light. They are much less abundant in the thermocline.

In March, during the subglacial period of the study, the main component of mesoplankton was *E. baikalensis* of different age stages and rotifers of several species. All age stages of Epishura were present in all horizons during the whole observation period.

At the time of the beginning of the observation (13.00 h on March 14), the number of zooplankton in the 0–10 m layer was 459 eq./m^3 (Fig. 4). In the 10–25 m layer, the numbers decrease to 218 eq./m^3, in the 25–50 m layer - 98 eq./m^3 and in the 50–100 m layer - 169 eq./m^3. *E. baikalensis* naupliuses dominate in zooplankton abundance (0–10 m - 413 eq./m^3, deeper than 50 m the values drop to 164 eq./m^3. The results of net collections show that the surface water layer contained about 40% of the total number of rotifers from 0 to 100 m depth and almost a quarter of all naupliuses. Copepodites of *Epischura baikalensis* are distributed in a similar manner.

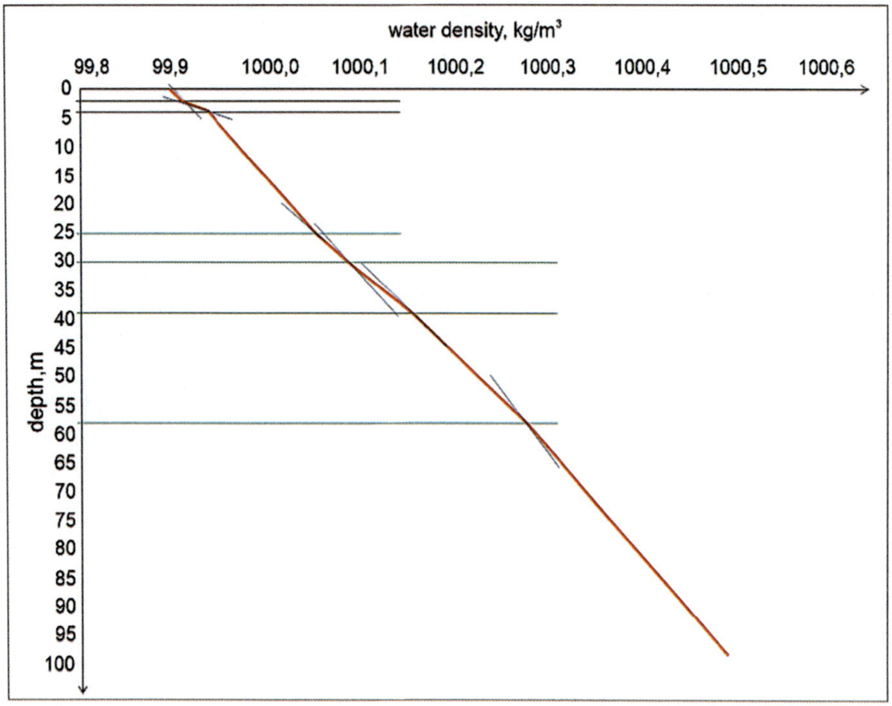

Fig. 3. Vertical distribution of water density (orange curve, kg/m^3), its break point (marked by intersection of blue lines). Green lines are drawn to determine the depth of the fracture point location

The daily dynamics of winter-spring rotifers was poorly expressed (Fig. 5).

In contrast to winter-spring rotifers, rotifers of the year-round complex moved with water masses during the day [4]. At 17.00, the maximum number of year-round rotifers was observed in layer 10–25 and amounted to 20 specimens/m^2. The minimum number in layer: 0–10, 25–50, 50–100 m. At 21.00 the maximum number of year-round rotifers is observed in the layer 50–100 m, amounting to 70 specimens/m^2. The minimum number in the layer 10–25 is 20 exv./m^2. At night time (01.00), the maximum number of year-round rotifers is observed in the layer 10–25 m, amounting to 50 specimens/m^2. The minimum in the 25–50 layer is 20 specimens/m^2. At the station at 05.00, the maximum number of year-round rotifers is observed in the 0–10 m and 50–100 m layer, amounting to 40 specimens/m^2. The minimum number in the 25–50 layer is 10 specimens/m^2. In the morning at 09.00, the maximum number of year-round rotifers is observed in the 0–10 m layer, amounting to 70 specimens/m^2. The minimum number in the layer 25–50 is 10 specimens/m^2. At 13.00 on March 15, the maximum number of year-round rotifers is observed in the layer 0–10 m, amounting to 40 specimens/m^2. The minimum number in the layer 25–50 is 10 specimens/m^2. Further, at 17.00, the maximum number of round rotifers is observed in layers 0–10 and 25–50 m, amounting to 30 specimens/m^2. The minimum number in layer 10–25 is 10 specimens/m^2.

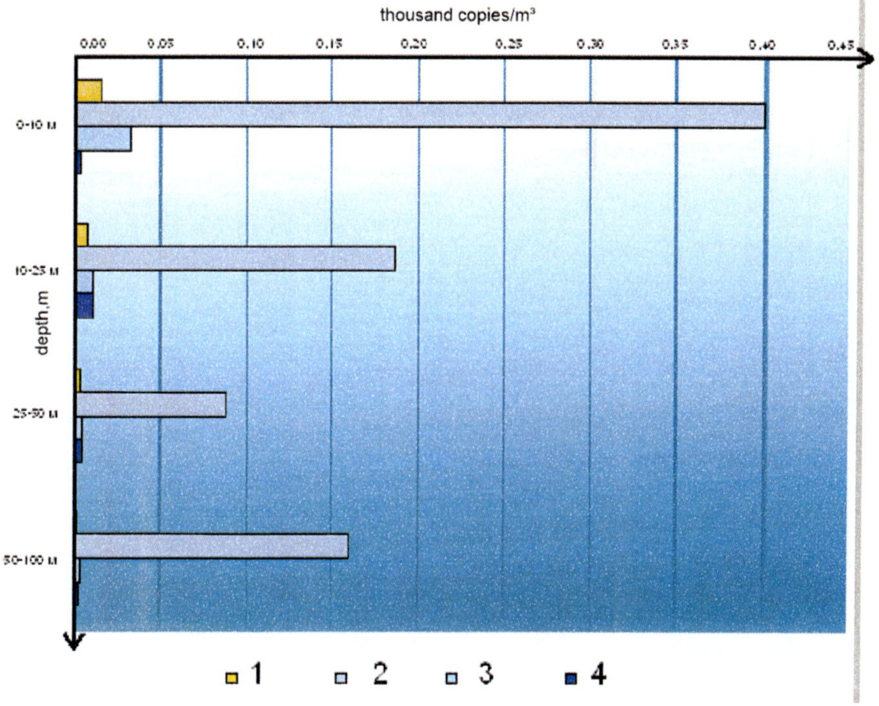

Fig. 4. Distribution of concentrations of zooplankton groups thousand eq./m^3 by depth on March 14, 2019 at 13.00, m. Ivanovsky. 1 - total concentration of rotifers, 2 - nauplial stages of *Epischura baikalensis*, 3 - copepodites of *E. baikalensis*, 4 - adult crustaceans of *E. baikalensis*

The data obtained by the holographic camera were computed semi-automatically from the focused images [5]. At a depth of 5 m, at the station 13.00 on March 14, up to 15 thousand particles/m^3 were observed, this indicator coincided with the maximum of acoustic density and simultaneously with the maximum of mesoplankton concentration (layer 0–10 m). At a depth of 10 m, the presence of diatom algae colonies displayed an increase in the number of particles [6]. At a depth of 23 m the number of particles reaches 20 thousand /m^3, then with increasing depth the number of particles gradually decreased (up to 7 thousand /m^3), but by 70 m depth there was again an increase (15 thousand /m^3).

E.baikalensis nauplii mainly inhabit a narrow 3–5 m layer in the upper part of the epilimnion, where daily mixing takes place. In the lower part of the thermocline (deeper than 50 m) the high concentration of nauplii practically coincides with the same increased concentration of adult stages of *E.baikalensis*. It should be assumed that this is what was registered by the holographic method (peaks deeper than 50 m) [7].

The results show that the adult stages of *E. baikalensis* crustaceans are located in the transition horizon from the epilimnion to the thermocline (23–25(30) m). This is most likely explained by the settling of diatom algae colonies in the layer of the largest vertical gradients of water temperature and density [8].

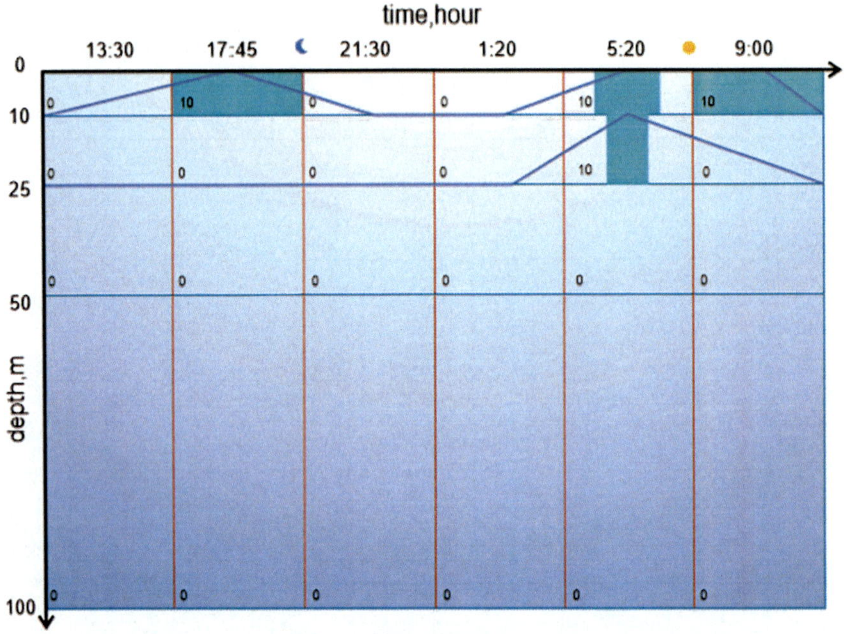

Fig. 5. Daily vertical migrations of winter-spring rotifers (in %, eq./m^2 in layers 0-100m) on March 14–15, 2019.

At 12 a.m. the maximum number of particles was up to 19000 pcs./m^3 in the thermocline zone. This may also be related to the presence of epishura and year-round rotifers, algae at the horizon from 10–25 m. The minimum particle number was 6000 pcs./m^3. The minimum number of particles is 6000 pcs/m^3 at a depth of 0 m, directly under the ice.

At 4 a.m. two peaks are observed; at a depth of 10 m–18000 pcs./m^3, deeper, the same as before, at a depth of 22 m. The particle counts then drop with depth to 9000 eq/m^3. The upper peak is most likely related to the morning rise of rotifers and older age stages of epishura.

At 8 a.m. the maximum particle count is at 28 m depth at 19,000 eq/m3 and the minimum at 62 m depth at 6,000 eq/m3.

At 12 noon on March 15, it was possible to measure particle counts to a depth of 180 m. The highest was at a depth of 20 m–21000 eq./m^3. The minimum index of particles is at a depth of about 100 m and was 5000 eq./m^3. We would like to note an increase in the number of particles at the horizon 140–150 m, which needs further investigation.

4 Conclusion

Only an integrated approach to the use of various instrumental methods in the monitoring study of winter plankton ecology will make it possible to draw the most objective conclusions and develop measures to prevent negative phenomena associated with both global climate change and anthropogenic impact on Lake Baikal.

As a result of the study, the following conclusions can be drawn:

1) The composition and abundance of zooplankton was comparable to previously obtained data. *Epischura baikalensis* naupliuses dominated at all depths. Nauplius abundance at 0–10 m depth varies greatly depending on the time of day. In the daytime, their number at the surface is almost twice as high as at night. At a depth of 10–25 m, the number of naupliuses increases at night. Deeper than these marks, the diurnal dynamics is not so obvious.
2) The maximum of mesozooplankton was also recorded in both acoustic and holographic data at a depth of about three meters. A maximum of photosynthetic organisms was observed above the thermocline and is reflected in the holographic camera data at a depth of 23 m. Experimental results, which were obtained using complex collection methods show that the distribution in the water column of mesoplankton is very closely related to hydrographic parameters. First of all, these are water temperature and density, then - illumination and various dynamic processes.
3) The holographic camera allows us to identify micro-layers with increased plankton concentration, which cannot be detected by other methods (plankton nets, bathometers). The use of these data can expand our knowledge of Lake Baikal plankton ecology.

In connection with the above conclusions, techniques for use in remote sensing studies of zooplankton need to be developed for further effective use.

Nowadays, remote sensing methods are coming to the forefront and it will automate the whole process of monitoring not only Lake Baikal, but also other water bodies, as well as make observations more widespread [9].

By what kinds of plankton species are in the water, how they behave, we can understand how dynamic this ecosystem is and whether it is moving towards blossoming or degrading [10]. This is interesting from both a scientific and an ecological point of view. And the holographic camera is one way to make this event a success.

The holographic camera is one way to make this event a success.

Acknowledgements. The work was supported by the project of the Ministry of Science and Higher Education of Russia 0279-2021-0005 "Study of transformations in the state of water bodies and watercourses of Eastern Siberia in seasonal and long-term aspects in the context of climate change, geological environment and anthropogenic pressures".

References

1. Naumova, E.Y., et al.: Investigation of the ecology of winter plankton of Lake Baikal using complex instrumental methods. South Russia Ecol. Dev. **16**(3), 59–67 (2021). https://doi.org/10.18470/1992-1098-2021-3-59-67

2. Rzhepka, T., Troitskaya, E., Makarov, M., et al.: Preliminary results of a comprehensive study of the vertical distribution of zooplankton using a submersible holographic camera. In: Integrated Research of the World Ocean: Proceedings of the VI All-Russian Scientific Conference of Young Scientists, Moscow, April 18–24, 2021, pp. 530–531. Shirshov Institute of Oceanology. P.P. Shirshov Institute of Oceanology of the Russian Academy of Sciences, Moscow (2021)
3. Rusanovskaya, OO, et al.: Comparison of pelagic rotifers in the 0–50 m and 50–250 m layers in the water area of southern Baikal Lake in 2017. XXI Century. Technos. Saf. **4**(4), 422–429 (2019). https://doi.org/10.21285/2500-1582-2019-4-422-429
4. Arov, I.V., Misharina, E.A.: Modern state of knowledge and ways of genesis of the fauna of rotifers (Rotifera) of Lake Baikal. Bull. Irkutsk State Univ. Ser. Biol. Ecol. **25**, 70–90 (2018). https://doi.org/10.26516/2073-3372.2018.25.70
5. Dyomin, V.V., Kamenev, D.V.: Methods of processing and retrieval of information from digital particle holograms and their application. Radiophys. Quan. Electron. **57**(8–9), 533–542 (2015). https://doi.org/10.1007/s11141-015-9536-y
6. Timoshkin, O., et al.: Atlas and identifier of Baikal pelagobionts (with brief essays on their ecology) – Novosibirsk, p. 694. Siberian Publishing Company of the Russian Academy of Sciences, Nauka (1995)
7. Arov, I., et al.: Annotated list of the fauna of Lake Baikal and its catchment area. Nauka, Novosibirsk, Lake Baikal. Book, vol. 1, pp. 329–376 (2001)
8. Eggert, M.: Daily vertical distribution of winter zooplankton in the Pelagial of Lake Baikal. Hydrobiol. Zhurn. **9**(1), 36–46 (1973)
9. Brown, K.P., et al.: Human impact and ecosystemic health at Lake Baikal. WIREs Water **8**(4), 1–36 (2021). https://doi.org/10.1002/wat2.1528
10. Lombard, F., et al.: Globally consistent quantitative observations of planktonic ecosystems. Front. Marine Sci. **6**, 1–21 (2019). https://doi.org/10.3389/fmars.2019.00196

Author Index

A
A. N. Demidov 35
Aglova, E. A. 439
Alekseeva, N. K. 233
Artemev, G. M. 207
Artemiev, V. A. 439
Atadzhanova, O. A. 191

B
Bashmachnikov, I. L. 42, 104
Basova, E. D. 332
Batalin, G. A. 207
Belan, B. D. 463
Belanov, M. A. 363
Belogurova, R. E. 370
Belonenko, T. V. 111, 150
Belyaev, N. A. 287
Blagodatskikh, E. A. 191
Bobylev, L. P. 104
Bocherikova, I. Yu. 376
Bochkarev, A. V. 326
Borisenko, G. V. 239
Borovkova, A. D. 381
Bubnova, E. S. 280
Budarova, V. Yu. 387
Budyansky, M. V. 59, 111
Bulgakov, K. Yu. 118
Bulokhov, A. V. 308
Burdiyan, N. V. 387

C
Chepikova, S. S. 456
Chernov, D. G. 463
Chubarenko, I. P. 376, 412
Chukharev, A. M. 48
Chultsova, A. L. 295

D
Demyshev, S. G. 29
Deryagin, D. N. 439
Deshova, D. V. 3
Didenkulova, E. G. 160
Donets, M. M. 363
Doroshenko, Yu. V. 387

E
Egorova, E. S. 11
Elkin, D. N. 19
Ershova, A. A. 404
Evstigneeva, N. A. 29

F
F. N. Gippius 35
Fedorova, A. A. 233
Fedulov, V. Y. 287
Filippova, T. A. 396
Frolova, N. S. 141

G
Gareev, B. I. 207
Glitko, O. V. 439
Glukhov, V. A. 439
Glukhovets, D. I. 439
Golikov, A. V. 207
Golubeva, E. A. 404
Gordeeva, S. M. 3
Gurov, K. I. 245
Gurova, Y. S. 245

I
Ivanov, B. V. 126
Ivanov, V. V. 135

K
K. V. Artamonova 35
Kabanov, D. M. 463
Kaledina, A. S. 42
Karpova, E. P. 370
Kazakov, D. A. 48
Kazakova, U. A. 254
Khudyakova, S. P. 150

T. Chaplina (Ed.): CIWO 2023, SPEES, pp. 491–493, 2023.
https://doi.org/10.1007/978-3-031-47851-2

Kil, A. O. 332
Kivva, K. K. 59, 262
Klimenko, S. K. 439
Klyuvitkin, A. A. 308
Kodryan, K. V. 262
Kokorina, A. V. 160
Koltovskaya, E. V. 303
Kornilova, R. V. 233
Korzhuev, V. A. 69
Kostyleva, A. V. 287
Kravtsova, K. V. 80
Krivoshlyk, P. N. 412
Kruglinsky, I. A. 463
Kruglova, E. E. 95
Kryzhova, K. A. 268
Kubryakov, A. A. 446
Kubryakova, E. A. 446
Kudinov, A. A. 341
Kulik, K. V. 141
Kuzmina, S. K. 456

L
Latonin, M. M. 104
Lishaev, P. N. 446
Lobanova, P. V. 456
Lobchuk, O. I. 376

M
Malysheva, A. A. 111
Matveeva, T. V. 326
Migdisova, I. A. 308
Mingazov, G. Z. 207
Mironov, Ye. U. 11
Miroshnichenko, O. N. 418
Mukhanov, V. S. 217
Muratova, A. A. 280
Murzakova, Y. V. 287
Myslenkov, S. A. 95

N
Naumova, E. Y. 223, 483
Nemirovskaya, I. A. 303
Nikulina, A. L. 233
Novigatsky, A. N. 308
Novikhin, A. E. 233

O
Orekhova, N. A. 245

P
Paltsev, I. O. 341
Panin, A. V. 319
Pankratov, D. V. 363
Paraskiv, A. A. 418
Pavlov, M. I. 48
Pavlova, M. A. 439
Podrezova, N. A. 80
Poletaev, D. A. 473
Polovkov, V. V. 341, 350
Poludetkina, E. N. 332
Polukhin, A. A. 239, 254, 280, 287
Prants, S. V. 59

R
Romashchenko, D. D. 118
Rudenko, N. O. 126
Rybalko, A. E. 341
Rybalko, A. Y. 332
Ryzhov, I. V. 233
Rzhepka, T. P. 483

S
S. B. Krasheninnikova 35
Sabirov, R. M. 207
Sahling, I. V. 439
Seliverstova, A. M. 198, 295
Semikolennykh, D. V. 319
Semkin, P. Yu. 268
Sharatunova, M. V. 135
Sharmar, D. V. 239
Shchelkanov, M. Yu. 363
Shchur, A. A. 326
Shchur, N. A. 326
Sidorov, I. G. 418
Smirnov, N. A. 233
Smirnov, Yu.Yu. 326
Sokolenko, B. V. 473
Starodymova, D. P. 308
Sukhanova, L. V. 223
Svergun, E. I. 141
Svishchev, S. V. 446

T
Tabyrtsa, S. N. 350
Tikhonova, E. A. 387, 425
Tkachenko, Y. S. 425
Tokarev, M. Y. 332
Travkin, V. S. 150

Tregubov, A. S. 160
Tsedrik, S. V. 175
Tsygankov, V. Yu. 363, 381

U
Ufimtseva, M. A. 217
Uleysky, M. Yu. 59
Ustinova, T. S. 181

V
Vasechkina, E. F. 396
Voytitsky, V. I. 473

Y
Yumasheva, A. K. 332

Z
Zaidykov, I. Y. 223
Zakharov, D. V. 207
Zatsepin, A. G. 19
Zenkova, P. N. 463
Zholondz, A. S. 350
Zimin, A. V. 141, 191
Zuev, O. A. 198, 295